RAS Family GTPases

PROTEINS AND CELL REGULATION

Volume 4

Series Editors: Professor Anne Ridley
Ludwig Institute for Cancer Research
and Department of Biochemistry and Molecular Biology
University College London
London
United Kingdom

Professor Jon Frampton
Professor of Stem Cell Biology
Institute for Biomedical Research,
Birmingham University Medical School,
Division of Immunity and Infection
Birmingham
United Kingdom

Aims and Scope

Our knowledge of the ways in which a cell communicates with its environment and how it responds to information received has reached a level of almost bewildering complexity. The large diagrams of cells to be found on the walls of many a biologist's office are usually adorned with parallel and interconnecting pathways linking the multitude of components and suggest a clear logic and understanding of the role played by each protein. Of course this two-dimensional, albeit often colourful representation takes no account of the three-dimensional structure of a cell, the nature of the external and internal milieu, the dynamics of changes in protein levels and interactions, or the variations between cells in different tissues.

Each book in this series, entitled "*Proteins and Cell Regulation*", will seek to explore specific protein families or categories of proteins from the viewpoint of the general and specific functions they provide and their involvement in the dynamic behaviour of a cell. Content will range from basic protein structure and function to consideration of cell type-specific features and the consequences of disease-associated changes and potential therapeutic intervention. So that the books represent the most up-to-date understanding, contributors will be prominent researchers in each particular area. Although aimed at graduate, postgraduate and principle investigators, the books will also be of use to science and medical undergraduates and to those wishing to understand basic cellular processes and develop novel therapeutic interventions for specific diseases.

RAS Family GTPases

Edited by

Channing Der

University of North Carolina at Chapel Hill, U.S.A.

Springer

A C.I.P. Catalogue record for this book is available from the Library of Congress.

ISBN-10 1-4020-4328-7 (HB)
ISBN-13 978-1-4020-4328-4 (HB)
ISBN-10 1-4020-4708-8 (e-book)
ISBN-13 978-1-4020-4708-4 (e-book)

Published by Springer,
P.O. Box 17, 3300 AA Dordrecht, The Netherlands.

www.springer.com

Printed on acid-free paper

CONTENTS

PREFACE

Ras proteins are arguably some of the most intensely studied proteins in biology. First identified as the gene products of the cancer-causing genes of highly oncogenic retroviruses, Ras proteins have been the subjects of this intensive research effort since 1983, when their involvement in human cancers was first revealed (Chapter 1). The lofty position of Ras in research is merited by (1) their status as the proteins encoded by the oncogenes most frequently mutated in human cancers, (2) their key role as signaling nodes that relay the actions of diverse extracellular signals to equally diverse cytoplasmic signaling networks, and (3) their significance as the founding members of a large superfamily of small GTPases that now number over 150 human members, with orthologs conserved in worms, flies, yeast and plants. Much is now known about the regulation and biology of Ras proteins in normal and neoplastic cells, yet new revelations about Ras continue to emerge at a remarkable pace. This volume reports on the state-of-the art of many facets of the Ras branch of the Ras superfamily. Many essential features of regulation of the Ras GTPase as a GDP/GTP-regulated binary switch, and of the downstream effector signaling of Ras, were initially established in 1993. More regulators and effectors continue to be identified, and an update on the ever expanding roster of Ras regulators is provided (Chapter 2). Structural studies have aided our understanding of how these proteins regulate and facilitate Ras activity (Chapter 3). The Raf-MEK-ERK mitogen-activated protein kinase cascade is the first identified and perhaps the most important Ras effector pathway. Ras regulation of this pathway is complex, with new components added to this once simple linear cascade (Chapter 4). Furthermore, other functionally distinct Ras effector pathways have emerged (Chapter 5), several of which serve important roles in promoting Ras-mediated oncogenesis (Chapter 6). A well-recognized theme is that Ras function is linked with Ras-related proteins, most notably the Rho family of GTPases (Chapter 7). One area still in its relative infancy involves the identity of the gene targets of Ras signaling (Chapter 8). The genetic study of Ras in invertebrate species has contributed immensely to our delineation of Ras regulation and signaling, and two chapters summarize the power of these systems for Ras studies (Chapter 9 and 10). Recent technologic breakthroughs have fostered the development of new cell culture and mouse model systems to study Ras in mammalian cells (Chapters 11 and 12). Finally, when Ras-related proteins were first identified, an obvious question was whether these proteins also function as oncogene proteins. For some members of the Ras branch, involvement in oncogenesis has indeed been established (Chapter 13). However, despite the strong biochemical similarity and structure of Ras family proteins, other members have strikingly divergent functions (Chapter 14). Finally, an important long-term goal in the study of Ras has been the development of anti-Ras drugs for cancer treatment, and a status report is provided on this topic (Chapter 15). In summary, this volume is merely a current progress report on our knowledge of a family of fascinating proteins that will continue to evolve and to reveal new roles and mechanisms in cell physiology.

CONTRIBUTORS

Mohammad Reza Ahmadian Max-Planck-Institute of Molecular Physiology, Department Structural Biology, Otto-Hahn-Strasse 11, 44227 Dortmund, Germany

Benjamin Boettner Cold Spring Harbor Laboratory, 1 Bungtown Road, Cold Spring Harbor, NY 11724

Emily J. Chenette University of North Carolina at Chapel Hill, Curriculum in Genetics and Molecular Biology, Department of Pharmacology and Lineberger Comprehensive Cancer Center, Department of Pharmacology, Chapel Hill, NC 27599-7295

Melissa J. Comiso Department of Microbiology, Immunology, and Molecular Genetics, 1602 Molecular Sciences Building, University of California, Los Angeles, California 90095-1489

Christopher M. Counter Duke University Medical Center, C225 LSRC, Box 3813, Durham, NC 27710

Adrienne D. Cox University of North Carolina at Chapel Hill, Departments of Radiation and Oncology and Pharmacology, Chapel Hill, NC 27599.

Ira O. Dar National Cancer Institute, Laboratory of Protein Dynamics and Signaling, Frederick, MD 21701

Jean de Gunzburg Inserm U528, Institut Curie-Section de Recherche, 26 rue d'Ulm, 75005, Paris, France

Channing J. Der University of North Carolina at Chapel Hill, Curriculum in Genetics and Molecular Biology, Department of Pharmacology and Lineberger Comprehensive Cancer Center, Department of Pharmacology, Chapel Hill, NC 27599-7295

Radovan Dvorsky Max-Planck-Institute of Molecular Physiology, Department Structural Biology, Otto-Hahn-Strasse 11, 44227 Dortmund, Germany

Dennis J. Eastburn Howard Hughes Medical Institute, Department of Molecular Cellular and Developmental Biology, University of Colorado at Boulder, USA

Dennis Fiegen Max-Planck-Institute of Molecular Physiology, Department Structural Biology, Otto-Hahn-Strasse 11, 44227 Dortmund, Germany

James J. Fiordalisi University of North Carolina at Chapel Hill, Department of Radiation and Oncology, Chapel Hill, NC 27599.

Gourisankar Ghosh Department of Chemistry & Biochemistry, University of California, San Diego, Mail Code 0359, Urey Hall 5230, 9500 Gilman Drive, La Jolla, CA 92093-0359

Min Han Howard Hughes Medical Institute, Department of Molecular Cellular and Developmental Biology, University of Colorado at Boulder, USA

Tom Huxford Department of Chemistry & Biochemistry, University of California, San Diego, Mail Code 0359, Urey Hall 5230, 9500 Gilman Drive, La Jolla, CA 92093-0359

Antoine E. Karnoub Whitehead Institute for Biomedical Research Nine Cambridge Center Cambridge MA 02142

Michael A. Jocobetz, Jr. Abramson Cancer Research Institute and Avramson Cancer Center, Department of Medicine and Cancer Biology, University of Pennsylvania, Philadelphia, PA 19104

Kian-Huat Lim Duke University Medical Center, C225 LSRC, Box 3813, Durham, NC 27710

Anu K. Moorthy Department of Chemistry & Biochemistry, University of California, San Diego, Mail Code 0359, Urey Hall 5230, 9500 Gilman Drive, La Jolla, CA 92093-0359

Deborah K. Morrison National Cancer Institute, Laboratory of Protein Dynamics and Signaling, Frederick, MD 21701

Gretchen A. Repasky Colorado College, Department of Chemistry, Colorado Springs, CO 80903

Patrick J. Roberts University of North Carolina at Chapel Hill, Lineberger Comprehensive Cancer Center Division of Pharmacotherapy, Department of Pharmacology, Chapel Hill, NC 27599-7295

Jose M. Rojas Centro de Investigación del Cáncer, IBMCC (USAL-CSIC), Campus Unamuno, Universidad de Salamanca, 37007 Salamanca, Spain

Eugenio Santos Unidad de Biología Celular, Centro Nacional de Microbiología, Instituto de Salud Carlos III, 28220 Majadahonda, Madrid, Spain

Reinhold Schäfer Laboratories of Molecular Tumor Pathology and of Functional Genomics, Charité, Universitätsmedizin Berlin, Schumannstr. 20/21, D-10117 Berlin, Germany

Christine Sers Laboratories of Molecular Tumor Pathology and of Functional Genomics, Charité, Universitätsmedizin Berlin, Schumannstr. 20/21, D-10117 Berlin, Germany

Angel P. Tabancay, Jr. Department of Microbiology, Immunology, and Molecular Genetics, 1602 Molecular Sciences Building, University of California, Los Angeles, California 90095-1489

Fuyuhiko Tamanoi Department of Microbiology, Immunology, and Molecular Genetics, 1602 Molecular Sciences Building, University of California, Los Angeles, California 90095-1489

Oleg I. Tchernitsa Laboratories of Molecular Tumor Pathology and of Functional Genomics, Charité, Universitätsmedizin Berlin, Schumannstr. 20/21, D-10117 Berlin, Germany

David A. Tuveson Abramson Cancer Research Institute and Avramson Cancer Center, Department of Medicine and Cancer Biology, University of Pennsylvania, Philadelphia, PA 19104

Linda Van Aelst Cold Spring Harbor Laboratory, 1 Bungtown Road, Cold Spring Harbor, NY 11724

Tal Z. Zaks Abramson Cancer Research Institute and Avramson Cancer Center, Department of Medicine and Cancer Biology, University of Pennsylvania, Philadelphia, PA 19104

CHAPTER 1

RAS STORIES: THE STATE OF THE ART

PATRICK J. ROBERTS AND CHANNING J. DER

Lineberger Comprehensive Cancer Center, Division of Pharmacotherapy and Department of Pharmacology, Chapel Hill, NC 27599-7295

Abstract: It has been nearly four decades since the potent tumor-inducing Harvey and Kirsten murine sarcoma viruses were first discovered. Subsequently, it was the pioneering work by Scolnick and colleagues during the 1970's that first established the identity of cellular Ras genes as oncogenes and their gene products as guanine nucleotide binding proteins. However, it was not until 1982, when mutationally activated forms of human Ras genes were found in human tumor cell lines that Ras became the subject of intense research to elucidate the genetic basis of cancer and to develop novel target-based cancer therapeutics. Since that time great advances have been made in our understanding of Ras function and biology. These advances include landmark discoveries implicating Ras proteins as key components in signal transduction and as the founding members of a large superfamily of small GTPases that number over 150 human proteins. This chapter will provide a historical prospective of the various aspects of Ras and Ras signaling that are the topics of this book

Keywords: Prenylation, oncogenes, GTPases, signal transduction, cancer

1. INTRODUCTION

The storied history of the Ras family dates back over forty years to the discoveries by Harvey, and later Kirsten, that the highly oncogenic HaMSV and KiMSV sarcoma retroviruses caused rapid tumor formation in rats (Harvey, 1964; Kirsten, 1967) (Figure 1). The viral oncogenes responsible for their oncogenic properties, named Harvey and Kirsten r*as* (H-*ras* and K-*ras*), for rat sarcoma, were actually transduced and altered versions of rat genes that encode enzymes with intrinsic guanine nucleotide binding and GTPase activity (Figure 2A). Together with the most famous of these viral oncogenes, the Src oncogene, the Ras oncogenes comprised a large, diverse roster of normal cellular genes that were converted by retroviruses into potent oncogenes. This pivotal discovery, for which Bishop and Varmus received

C. Der (ed.), RAS Family GTPases, 1–14.

the Nobel Prize in Physiology or Medicine in 1989, prompted the concept that cancer arises from "good genes gone bad". However, it was not until 1982, when activated forms of human H-*ras* and K-*ras* genes (and a third ras gene member, N-*ras*) were found in human tumor cell lines that Ras became the subject of intense

1964	Harvey murine sarcoma virus (HaSV) is isolated from a rat tumor
1967	Kirsten murine sarcoma virus (KiSV) is isolated from a rat tumor
1979	Transforming genes of HaSV and KiSV identified as transduced normal genes
	Viral and cellular ras genes encode 21 kDa p21 proteins
1980	Viral H-Ras is a GTP-binding protein
1982	Mutated and transforming H-Ras and K-Ras genes found in human cancers
	Ras is activated by single amino acid substitutions
1983	Chemically-induced rodent tumors contain ras mutations
	Ras requires cooperation with other oncogenes to transform normal rodent cells
1984	Ras requires membrane association for transformation
	Mutant Ras has impaired intrinsic GTPase activity
	Epidermal growth factor (EGF) stimulation causes Ras activation
1985	RhoA identified fortuitously as the first Ras-related protein
1987	Ras GAP activity is found and mutated Ras is GAP-insensitive
	Ras transgene causes tumors in mice
1988	Crystal structure of Ras and switch I & II conformation regions identified
1989	Ras is modified by farnesyl isoprenoid
1990	Neurofibromatosis type I tumor suppressor is a Ras GAP
	FTase that catalyzes Ras isoprenoid modification is identified
1992	Mammalian Ras GEFs (RasGRF and Sos) identified
1993	Raf is identified as a Ras effector
	Grb2:Sos complex promotes EGF receptor activation of Ras
1994	PI3K identified as a critical effector of Ras transformation
	Ral GEFs identified as Ras effectors
	FTase inhibitors (FTIs) block Ras transformation
1995	Ras requires Rho GTPases for transformation
	FTIs cause regression of Ras transgene-induced mouse tumors
	K-ras, but not H-ras or N-ras, is essential for mouse development
1997	FTIs in phase I clinical trials
	K-Ras, but not H-Ras or N-Ras, is essential for mouse development
1998	RhoA inhibition of p21CIP1 allows Ras transformation
1999	Telomerase and SV40T/t antigen expression allows Ras to transform human cells
	Ras is required for tumor maintenance
2001	Ras effector Rin1 is a GEF for Rab5 small GTPase
	Somatic K-Ras activation causes lung cancer in mice
2002	B-Raf mutations found in human cancers
	Ral GEF effector is sufficient for Ras transformation of human cells
	Tiam1 identified as a Ras effector required for transformation
	Phospholipase C epsilon identified as Ras effector
2004	Activating mutations found in the p110 catalytic subunit of PI3K
	PLCε deficiency impairs Ras-induced mouse tumor formation
2005	RalGEF deficiency impairs Ras-induced mouse tumor formation
	A gene expression profile defines K-Ras activation in mouse and human lung cancer

Figure 1. Timeline of major discoveries in Ras research. An emphasis has been placed on the discoveries related to the roles of Ras in cancer and in signal transduction, and to the topics covered by chapters in this book

research interest (Malumbres and Barbacid, 2003) (Figure 2B). This interest has been fueled by the critical contribution of Ras in human oncogenesis, its central involvement in signal transduction (Figure 3), and because Ras is the founding member and prototype of superfamily of small GTPases (Figure 4).

The first glimpses of the function of Ras came from initial biochemical studies of the viral Ras proteins by Scolnick and colleagues in the 1970s, which demonstrated that Ras bound guanine nucleotides. Ras proteins function as biological switches that serve as critical regulators of cytoplasmic signaling networks (Vetter and Wittinghofer, 2001). Similar to the heterotrimeric G protein alpha subunits, the Ras family of proteins function as binary molecular switches where binding to guanine diphosphate (GDP) or guanine triphosphate (GTP) control whether they are "off" or "on", respectively (Figure 5). These proteins possess high-affinity for binding GDP and GTP, and GTP hydrolysis activity. However their intrinsic GTP hydrolysis and GDP/GTP exchange activities are too low to

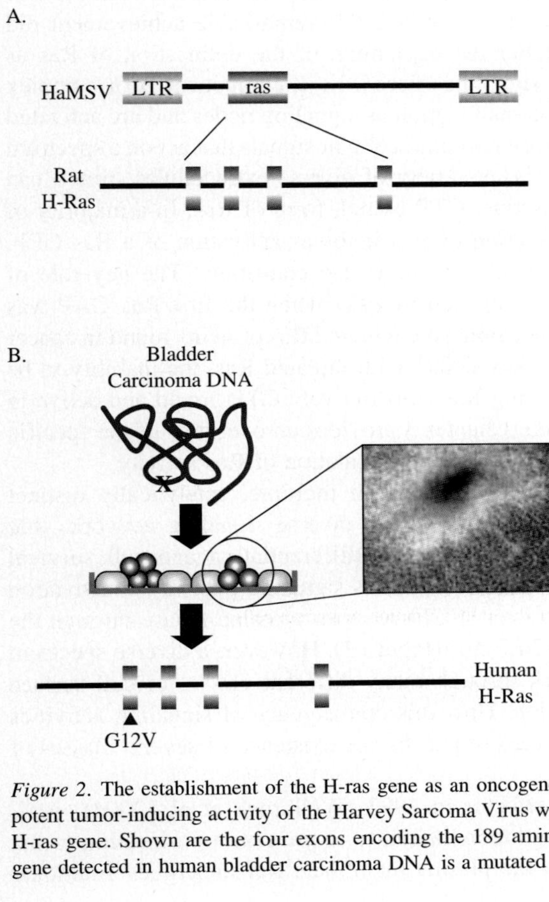

Figure 2. The establishment of the H-ras gene as an oncogene. (A) The viral gene responsible for the potent tumor-inducing activity of the Harvey Sarcoma Virus was derived originally from the rat cellular H-ras gene. Shown are the four exons encoding the 189 amino acid H-Ras protein. (B) Transforming gene detected in human bladder carcinoma DNA is a mutated H-ras gene

account for the rapid GDP/GTP cycling that occurs following extracellular stimulation. Thus, GDP/GTP cycling is controlled by two main classes of regulatory proteins that accelerate these intrinsic activities, guanine nucleotide exchange factors (GEFs) and GTPase activating proteins (GAPs). GEFs function as positive regulators of Ras proteins by promoting the displacement of GDP and the binding of GTP (Mitin et al., 2005). GAPs, the negative regulators, accelerate the intrinsic GTPase activity to promote the formation of the inactive GDP-bound form (Bernards and Settleman, 2004). When bound to GTP these proteins undergo a conformational change that allows them to interact with target downstream effectors (Herrmann, 2003). Chapter 2 will discuss the state-of-the-art concerning the structural aspects of Ras, including fundamental principles of its intrinsic functions, the regulation provided by GEFs and GAPS, and the interaction with downstream effectors.

In 1993, the culmination of discoveries from biochemical and biological studies in mammalian cells and genetic studies in invertebrates established that Ras is a central component in a signaling pathway that connected the cell surface with the nucleus (Egan, 1993) (Figure 1). However, this remarkable achievement did not mark the completion, but rather the beginning, of the delineation of Ras as a signaling protein. This linear signaling cascade is a component of a complex signaling network, where Ras proteins function as signaling nodes and are activated by upstream signaling initiated by diverse extracellular signals that act on a spectrum of cell surface receptors (Figure 5). A variety of diverse extracellular stimuli can cause transient formation of the active, GTP-bound, form of Ras. In a majority of situations, receptor-mediated activation of Ras involves activation of a Ras GEF, although inactivation of Ras GAP activity may also contribute. The key role of Ras GAPs in Ras function was established in 1987 when the first Ras GAP was identified and shown to be inactive against the mutated Ras proteins found in cancer cells. This finding established the key defect with mutated Ras, the inability to be inactivated by Ras GAP, thus leaving Ras constitutively GTP-bound and active in the absence of extracellular stimuli. Chapter 3 provides an overview of the specific roles and functions of Ras GEFs and GAPs in regulation of Ras activity.

Once activated, the Ras proteins interact with multiple, catalytically distinct downstream effectors (Figure 5), which regulate diverse signaling networks that control gene expression, cell proliferation, cell differentiation and cell survival (Repasky et al., 2004). The best characterized Ras signaling pathway is activation of Ras by the epidermal growth factor (EGF) receptor tyrosine kinase through the Grb2 adaptor protein and the RasGEF Sos (Figure 3). However, a diverse spectrum of extracellular stimuli, that work through many different classes of cell surface receptors, also cause Ras activation. How this convergence of signaling activities can cause Ras activation is explained in part by the existence of several classes of Ras GEFs.

Raf was identified as a Ras effector in 1993 (Wellbrock et al., 2004). Activated Ras binds to the Raf serine/threonine kinase to promote translocation to the plasma membrane (Figure 3). At the plasma membrane Raf undergoes additional

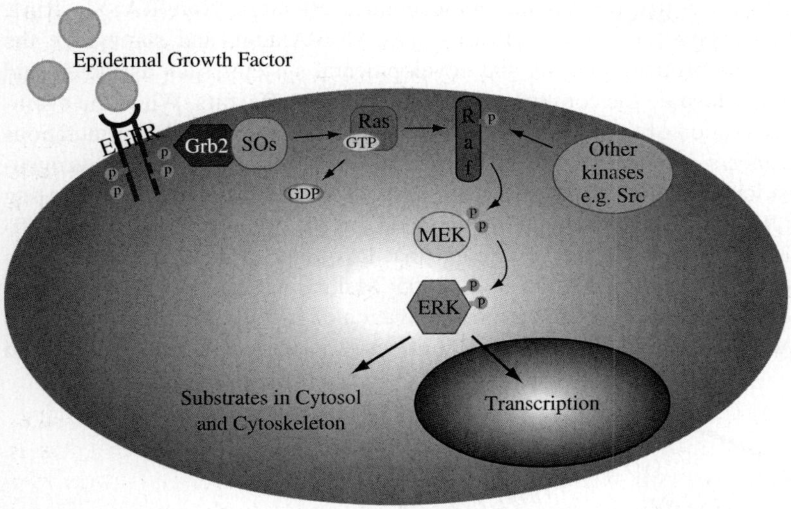

Figure 3. Receptor tyrosine kinase activation of the Ras–Raf–MEK–ERK signaling pathway. EGF induces receptor dimerization and autophosphorylation (P) of the epidermal growth factor receptor (EGFR). EGFR utilizes the RasGef Sos to catalyze the exchange of Ras-GDP for GTP resulting in a conformational change in Ras which allows Raf binding. Ras then binds Raf and recruits it to the cell membrane, where Raf activation takes place. Activated Raf phosphorylates and activates MEK (MAPK/ERK kinase), which in turn phosphorylates and activates extracellular-signal-regulated kinase (ERK). Activated ERK has many substrates in the cytosol and nucleus

phosphorylation and other events that result in full Raf activation. Raf phosphorylates and activates the MEK1 and MEK2 dual specificity protein kinase, which then phosphorylates and activates the ERK1 and ERK2 mitogen-activated protein kinases (MAPKs). Activated ERK translocates to the nucleus, where it phosphorylates Ets-family transcription factors, ultimately leading to changes in gene expression. The identification of mutationally activated Raf in human cancers in 2002 provided further validation of the importance of the MAPK pathway in Ras-mediated oncogenesis (Davies et al., 2002). This pathway is often depicted as a simple linear pathway that connects the cell surface with the nucleus. However, this is a gross simplification, since signal transmission through this signaling pathway is highly regulated both at the level of the individual components as well as by scaffolding proteins and signaling modulators (e.g., KSR). All of these components work in concert to control the efficiency, spatiotemporal dynamics, and levels of signaling through this signaling pathway. Chapter 4 will provide an in depth review of Raf-MEK-ERK signaling regulation, with an emphasis on the complex nature of Raf regulation.

Additional complexity of Ras signaling is introduced by the fact that Raf is but one of a multitude of effector classes utilized by Ras (Repasky et al., 2004). In particular, phosphatidylinositol 3-kinase (PI3K) was established to be an important effector of Ras in 1994 and mutationally activated PI3K found in human cancers in 2004. Chapter 5 will discuss the non-canonical effectors, Nore/RASSF, Rin1, Tiam1, phospholipase C epsilon (PLCε), and AF-6/Afadin, and summarize the results from biochemical analysis and developmental systems such as mouse and *Drosophila* to illustrate the conserved functions of these effectors. While the evolutionary conservation of the Raf-MEK-ERK cascade, together with B-Raf mutations in human cancers, implicate the MAPK cascade as a key mediator of Ras oncogenesis, at least four other Ras effector pathways have also been implicated in promoting oncogenic Ras function (Figure 5). For example, recent studies utilizing mice deficient in Tiam1 or PLCε expression found that Ras-induced tumor formation was greatly impaired (Gonzalez-Garcia et al., 2005; Malliri, A., et al., 2002). Chapter 6 summarizes recent findings of mutational activation of B-Raf in human cancers and examines the importance of the non-B-Raf effectors in Ras-mediated signaling and

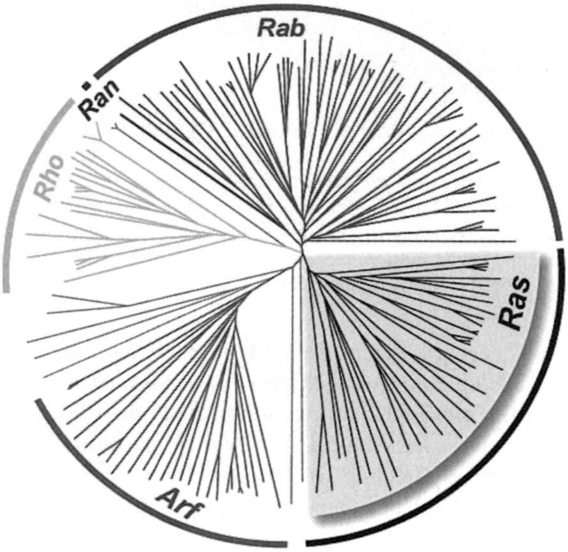

K-Ras4A	Rit1	Rap2C	Rem2	RASL11B
K-Ras4B	Rit2	Noey2	Rem1	Rheb
H-Ras	RalA	Di-Ras1	Gem	Rheb2
N-Ras	RalB	Di-Ras2	Rad	κB-Ras1
E-Ras	Rap1A	RasD1	Rerg	κB-Ras2
M-Ras	Rap1B	RasD2	Ris	
R-Ras	Rap2A	RRP22	FLJ22655	
TC21	Rap2B	RASL10B	RASL11A	

Figure 4. The Ras Superfamily and the Ras subfamily. The human Ras superfamily is composed of over 150 members which share common sequence, structural and biochemical characteristics. Based on sequence similarities and functional relationships, the family can be divided into six sub-families: Rab, Ras, Arf, Rho, Ran and others. The Ras family is made up of 36 genes which encode 39 different protein isoforms

malignant transformation. In particular, a growing body of evidence supporting a role of the RalGEF-Ral pathway has begun to accumulate (Gonzalez-Garcia et al., 2005; Hamad et al., 2002; Lim et al., 2005; Rangarajan et al., 2004).

The cellular consequences of Ras-mediated oncogenesis include alterations in actin cytoskeletal organization, cell morphology and cell motility, aberrant cell cycle control, increased proliferation, reduced apoptotic response, and increased invasive properties. These changes are mediated in part by deregulation of Rho GTPase activation. Rho (Ras homologous) proteins, identified fortuitously in 1985 as the first mammalian Ras-like proteins (Madaule and Axel, 1985), are a distinct subfamily of the Ras superfamily which regulates actin organization, cell cycle progression, and gene expression (Etienne-Manneville and Hall, 2002). Unlike Ras proteins, direct evidence for the involvement of Rho proteins in human cancer is limited, as mutationally activated Rho GTPases have not been identified in human cancers (Sahai and Marshall, 2002). Despite this, there is substantial evidence to support a role for Rho proteins and their regulators in oncogenesis. In particular, studies in 1995 found evidence that activated Ras utilizes Rho family GTPases to induce oncogenesis (Khosravi-Far et al., 1995; Qiu et al., 1995a; Qiu et al., 1995b). Chapter 7 is focused on the involvement and contributions of Rho family GTPases in Ras-mediated oncogenesis.

Thus far a brief description of the regulation of Ras signaling and the various Ras effectors has been provided. Activation of many of these effectors ultimately controls the regulation of transcription factor function and gene expression. Historically, changes in gene expression resulting from oncogenic Ras have been determined by evaluation of specific candidate genes whose functions may contribute to specific facets of Ras transformation (Ulku and Der, 2003). Recently, technological advances have allowed transcriptome-wide screening approaches using gene microarray analyses and differential gene expression procedures. For example, Schafer and colleagues found that the upregulation and downregulation of hundreds of genes are associated with Ras transformation (Zuber et al., 2000).

The genetic analyses of invertebrate model systems have contributed immensely to our understanding of Ras regulation and signaling (Figure 6). *Caenorhabditis elegans* has been an integral genetic system to dissect components of Ras signal transduction. Features that define the utility of this system include the invariant cell lineage, ability to study individual cells, and ease of application of genetic and molecular analyses. The identification of a worm ortholog of Ras, designated Let-60, and its involvement in vulval development in 1990 was an important finding in *C. elegans* (Beitel et al., 1990; Han and Sternberg, 1990). The ability of hyperactivation of Let-60 to cause a multivulva morphologic phenotype facilitated the development of genetic screens to identify upstream regulators, and downstream effectors, of Ras signaling. Chapter 9 provides a comprehensive account of our current understanding of Ras signaling in *C. elegans*.

Another model system that has been instrumental in the understanding and dissection of the Ras superfamily of proteins has been the unicellular eukaryotic yeast organism. Chapter 10 details the Ras family GTPases of budding and fission yeast

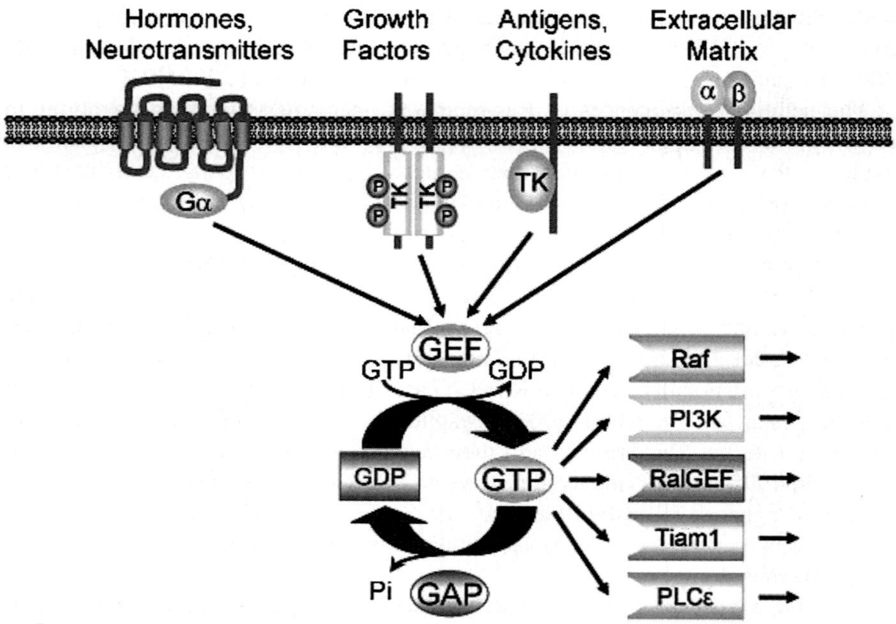

Figure 5. Ras is a signaling node. Ras proteins function as GDP/GTP-regulated binary switches that cycle between the inactive, GDP-bound and active, GTP-bound states. Diverse extracellular stimuli stimulate cell surface receptors, including G protein-coupled receptors, receptor tyrosine kinases, tyrosine kinase-linked cytokine receptors and integrins. In a majority of cases, receptor-stimulated activation of Ras involves stimulation of a Ras GEF and GDP/GTP exchange to promote Ras-GTP formation. Ras GAPs stimulate GTP hydrolysis to promote Ras-GDP formation. Ras-GDP and Ras-GTP differ in protein conformation, with the GTP-bound protein binding preferentially to downstream effectors (E). Shown are the effectors with demonstrated roles in Ras-mediated oncogenesis

species, *Saccharomyces cerevisiae* and *Schizosaccharomyces pombe*, respectively. Much of our knowledge of the Ras superfamily of proteins originated from observations in yeast and have subsequently been carried forward and investigated in higher eukaryotic systems. For example, the first Ras GEF, the yeast CDC25 protein, was identified in *S. cerevisiae* (Broek et al., 1987). This information, together with the discovery of a Ras GEF in *Drosophila* (Sos) (Bonfini et al., 1992), provided the key clues that allowed the discovery of mammalian Ras GEFs in 1992. This chapter describes studies in *S. cerevisiae* and *S. pombe*, in which the Ras family proteins Ras, Rsr, and Rheb regulate growth and differentiation, Rsr is involved in budding, and Rheb plays an important role in nutrient uptake and cell cycle regulation.

While *C. elegans*, yeast and other invertebrate genetic systems have provided invaluable information regarding regulators and effectors of Ras signaling, their use in studying Ras-mediated oncogenesis is limited. Instead, studies of mechanisms of Ras-mediated oncogenesis have utilized cell culture model systems and mouse models. Each approach has strengths and weaknesses. In light of possible species

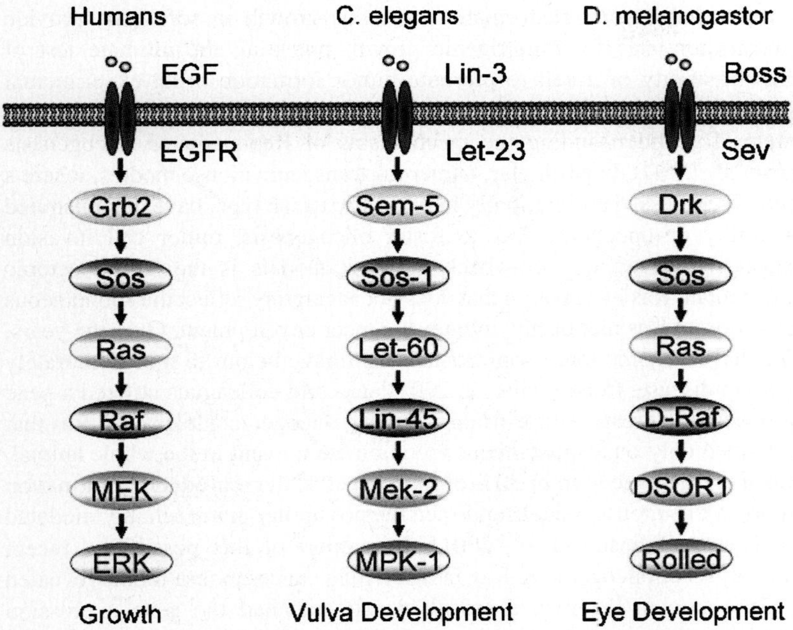

Figure 6. Conservation of Ras signaling. In human cells, EGF receptor activation of Ras is mediated by a Grb2-Sos complex, which facilitates Sos, a Ras GEF, activation of Ras. Activated Ras then binds to and promotes the activation the Raf serine/threonine. Raf then phosphorylates and activates the MEK1 and MEK2 dual specificity kinases, which then phosphorylate and activate the ERK MAPKs. Activated ERks then translocate from the cytoplasm into the nucleus, where they phosphorylate and activate Ets family transcription factors and promotes changes in gene expression. This pathway is conserved in both *C. elegans* and *Drosophila*, with homologous components functioning both upstream and downstream of Ras

differences in mechanisms of oncogenesis and in Ras signaling and function, the use of human epithelial cell culture has the obvious advantage of studying the "right" cell type. Thus, while many of the key findings in Ras function has come from the use of the legendary NIH 3T3 mouse fibroblast cell line, recent studies have emphasized the use of more physiologically relevant cell culture models. An important advance in cell culture approaches to study Ras was the study by Weinberg and colleagues in 1999, where primary normal human epithelial cells, the cell type from which the majority of human cancers arise, were immortalized and rendered sensitive to Ras transformation. These and other immortalized human cell types have provided powerful cell culture models to study Ras signaling and oncogenesis (Hahn and Weinberg, 2002). In Chapter 11, the use of such cell culture models to dissect the role of specific effectors in Ras transformation is described.

An obvious deficiency in cell culture systems is that they are homogenous, two-dimensional cell populations that do not recapitulate the in vivo tumor-stromal interactions known to greatly influence tumor cell growth. Furthermore, while the various in vitro assays for transformation, such as growth in soft agar, provide strong in vitro correlates for tumorigenic growth potential, the ultimate test of this remains the ability of a cell to promote tumor formation in an experimental animal (Rangarajan et al., 2004). Mouse model systems have provided powerful model systems for understanding the mechanisms of Ras-mediated oncogenesis (Rangarajan et al., 2004). In particular, numerous transgenic mouse models, where a Ras transgene is expressed preferentially in a specific tissue type, have demonstrated the potent ability of oncogenic Ras to foster oncogenesis, tumor cell invasion and metastasis. However, one drawback of these models is the forced ectopic expression of mutant Ras in a fashion that does not accurately reflect the spontaneous somatic activation of Ras that occurs in human cancer development. Over the years, these models have become more sophisticated and have begun to more accurately resemble human disease. In particular, in 2001 Jacks and colleagues utilized a gene targeting procedure to create mouse strains carrying oncogenic alleles of K-Ras that could be activated only on a spontaneous recombination event in the whole animal. Since K-Ras mutations are seen in 30% of lung cancers, the reproducible formation of lung tumors in this mouse model supported the possibility that it reliably modeled the human disease (Johnson et al., 2001). In support of this possibility, recent microarray analysis of oncogenic K-Ras-induced lung cancer mouse model revealed that gene expression in the mouse model closely matched the gene expression pattern seen in patients with K-Ras mutations (Sweet-Cordero et al., 2005). The authors concluded that this gene expression profile represented an oncogenic K-Ras "signature". This also demonstrated a new level of accuracy in the development of lung cancer mouse models. Chapter 12 discusses the use of mouse models with endogenous K-Ras activation to model human cancers.

Ras proteins are the founding members of a large superfamily of small GTPases that includes over 150 human small GTPases (Colicelli, 2004; Wennerberg et al., 2005) (Figure 4). Based on sequence and functional similarities, this large group of proteins can be subdivided into six main subfamilies (Ras, Rho, Ran, Rab, ARF, and others). The Ras subfamily of proteins (36 human members) share a high degree of sequence similarity and common conserved structural features. Particularly, they share a group of conserved G box GDP/GTP-binding motif elements beginning at the N-terminus: G1 (GXXXXGKS/T), G2 (T), G3 (DXXGQ/H/T), G4 (T/NKXD), and G5 (C/SAK/L/T) (Bourne et al., 1991) (Figure 7). These elements make up the G domain and explain the conserved biochemical properties (GDP/GTP-binding and GTP hydrolysis) of this family of proteins. In particular, the strong sequence conservation of the G3 box provided the basis for the discovery of many key members of the Ras superfamily. Degenerate oligonucleotide probes corresponding to this sequence identified a Ras-like (Ral) gene in 1986 (Chardin and Tavitian, 1986), *ras*-related genes in rat brain (Rab) in 1987 (Touchot et al., 1987) homologous to a previously identified ras-like gene in yeast (YPT) in 1983 (Gallwitz et al., 1983), and several in from a human teratocarcinoma expression library (TC21, Ran/TC4, and TC10) that were discovered in 1990 (Drivas et al., 1990). A variety of other library screening approaches, biochemical analyses and more recently, database searches, have added to the current roster of 157 human Ras family small GTPases, many with orthologs found in invertebrate species and plants, but not bacteria.

The high degree of structural and biochemical conservation amongst the Ras subfamily of proteins has challenged scientists to understand the features that distinguish these proteins from one another and allow each member to carry out unique functions within the cell. Another challenge has been the determination of how specific family members differentially recognize multiple downstream effectors. Data from mutational analysis and crystallography have led the way in our understanding of the structure-function relationships that dictate these issues of specificity.

This book will focus on Ras as well as key members of the Ras subfamily of proteins which is comprised of the prototypical Ras proteins, as well as the Ras-related Ral, Rap, κB-Ras and R-Ras protein (Wennerberg et al., 2005). In addition to their structural and biochemical relationships with Ras proteins, many Ras superfamily proteins share functional relationships with Ras. For example, Ras GEFs and GAPs also serve as regulators of other Ras family proteins (Chapter 2). The Ral Ras family and the Rac Rho family small GTPases are connected to Ras via the Ras effector Tiam1 (Chapters 6, 7 and 13). Chapter 13 will focus on the Ras branch of the superfamily, including the Rap, Ral, R-Ras, Rin and Di-Ras proteins, and discuss their biochemical properties, structural features, interacting partners, and cellular functions. In addition to the conserved G box domain described above, a majority of Ras family proteins also terminate with a C-terminal CAAX motif (where C = cysteine, A = aliphatic amino acid, and X = terminal amino acid) that specifies post-translational isoprenoid modifications. This tetrapeptide motif is found in a majority of Ras family proteins and is a crucial component of a larger C-terminal sequence that is essential for proper membrane interaction and subcellular localization of Ras and Ras family proteins (Figure 7). The CAAX-signal modification of Ras proteins, the covalent addition of a farnesyl isoprenoid lipid to the cysteine residue, is critical for Ras membrane association and function.

Recently, two additional members of the Ras superfamily, κB-Ras1 and κB-Ras2, have been identified. These are most closely related to the Ras subfamily of proteins, displaying nearly 45% sequence similarity (Fenwick et al., 2000). κB-Ras1 and κB-Ras2 are known to inhibit the transcription factor NF-κB, but the mechanism of this inhibition has not been fully elucidated. These proteins are thought to have intrinsic GTPase activity similar to the other Ras proteins, while lacking many of the additional structural motifs like the C-terminal CAAX sequence responsible for post-translational lipid modifications. Chapter 14 reviews the limited information regarding these proteins, including their structural characteristics, proposed regulatory mechanisms, and their role in controlling NF-κB.

Overall, this book has detailed the structural and biochemical properties of the Ras family of small GTPase proteins, has provided an in depth review of their mechanisms of regulation and has summarized our current understanding of how they utilize their different effectors. This detailed summary has included extensive evidence highlighting their role in cancer and explaining the mechanisms of

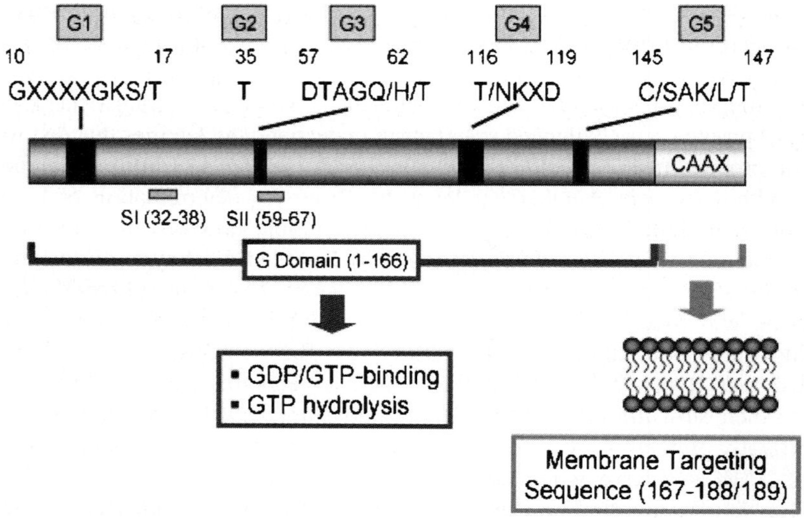

Figure 7. Functional domains of Ras. Ras proteins are 188/189 amino acid proteins. The N-terminal 166 residues comprise the G domain, which contains the consensus GDP/GTP-binding motifs, and is responsible for GDP/GTP-binding and GTP hydrolysis. Also within the G domain are the two regions of Ras that "switch" in conformation during the GDP/GTP cycle, and hence, designated switch I and II (SI and SII). The residue numbers for the GDP/GTP binding motifs and the switch regions correspond to Ras residue numbers. The remaining approximately 20 amino acids comprise the C-terminal plasma membrane targeting sequence. Within this sequence is the C-terminal CAAX tetrapeptide sequence, which signals for series of three posttranslational processing steps, including modification by a farnesyl isoprenoid

Ras-mediated oncogenesis. An obvious important long-term goal for Ras studies is to identify approaches that may be exploited for the development of anti-Ras therapeutic strategies for cancer treatment (Cox and Der, 2002). The last chapter of this book, Chapter 15, concludes with a summary of the approaches that have been considered and are ongoing to block Ras function for cancer treatment. It also discusses the difficulties that have slowed the development of these treatments as well as the current status of efforts to block oncogenic Ras. In particular, the discussion of the development of farnesyltransferase inhibitors shows the promises and problems commonly associated with target-based drug discovery (Sebti and Der, 2003). First identified by Goldstein, Brown and colleagues in 1999, the farnesyltransferase enzyme that catalyzes this modification has been the target of intense drug discovery efforts to develop anti-Ras drugs for cancer treatment. However, while potent inhibitors of farnesyltransferase have been developed and analyzed extensively in preclinical and clinical studies, and found to display anti-tumor activity, their development as anti-Ras drugs has met with unexpected complexities and disappointments. While there are currently no anti-Ras drugs in clinical use, the continued elucidation of Ras function from the various experimental strategies highlighted in this book is certain to continue to foster and fuel these efforts.

In summary, as Ras research progresses through its third decade, what has been clear is the significant contributions that these studies have made in our understanding of normal and neoplastic signaling mechanisms and cell biology (Figure 1). What is anticipated in the future are more exciting and unexpected discoveries.

ACKNOWLEDGEMENTS

We thank Misha Rand for assistance in manuscript and figure preparation. We apologize for not being able to cite original work of many colleagues due to space constraints. Our research studies were supported by grants from the NIH to CJD (CA42978, CA67771 and GM65533).

REFERENCES

Bernards, A. and Settleman, J. (2004) GAP control: regulating the regulators of small GTPases. Trends Cell Biol. *14*, 377-385.

Bonfini, L., Karlovich, C.A., Dasgupta, C. and Banerjee, U. (1992) The Son of sevenless gene product: a putative activator of Ras. Science *255*, 603-606.

Broek, D., Toda, T., Michaeli, T., Levin, L., Birchmeier, C., Zoller, M., Powers, S. and Wigler, M. (1987) The S. cerevisiae CDC25 gene product regulates the RAS/adenylate cyclase pathway. Cell *48*, 789-799.

Chardin, P. and Tavitian, A. (1986) The ral gene: a new ras related gene isolated by the use of a synthetic probe. EMBO J. *5*, 2203-2208.

Colicelli, J. (2004) Human RAS superfamily proteins and related GTPases. Sci. STKE *2004*, RE13.

Cox, A.D. and Der, C.J. (2002) Ras family signaling: therapeutic targeting. Cancer Biol. Ther. *1*, 599-606.

Davies, H., Bignell, G.R., Cox, C., Stephens, P., Edkins, S., Clegg, S., Teague, J., Woffendin, H., Garnett, M.J., Bottomley, W., Davis, N., Dicks, E., Ewing, R., Floyd, Y., Gray, K., Hall, S., Hawes, R., Hughes, J., Kosmidou, V., Menzies, A., Mould, C., Parker, A., Stevens, C., Watt, S., Hooper, S., Wilson, R., Jayatilake, H., Gusterson, B.A., Cooper, C., Shipley, J., Hargrave, D., Pritchard-Jones, K., Maitland, N., Chenevix-Trench, G., Riggins, G.J., Bigner, D.D., Palmieri, G., Cossu, A., Flanagan, A., Nicholson, A., Ho, J.W., Leung, S.Y., Yuen, S.T., Weber, B.L., Seigler, H.F., Darrow, T.L., Paterson, H., Marais, R., Marshall, C.J., Wooster, R., Stratton, M.R. and Futreal, P.A. (2002) Mutations of the BRAF gene in human cancer. Nature *417*, 949-954.

Drivas, G.T., Shih, A., Coutavas, E., Rush, M.G. and D'Eustachio, P. (1990) Characterization of four novel ras-like genes expressed in a human teratocarcinoma cell line. Mol. Cell. Biol. *10*, 1793-1798.

Etienne-Manneville, S. and Hall, A. (2002) Rho GTPases in cell biology. Nature *420*, 629-635.

Fenwick, C., Na, S.Y., Voll, R.E., Zhong, H., Im, S.Y., Lee, J.W. and Ghosh, S. (2000) A subclass of Ras proteins that regulate the degradation of IkappaB. Science *287*, 869-873.

Gallwitz, D., Donath, C. and Sander, C. (1983) A yeast gene encoding a protein homologous to the human c-has/bas proto-oncogene product. Nature *306*, 704-707.

Gonzalez-Garcia, A., Pritchard, C.A., Paterson, H.F., Mavria, G., Stamp, G. and Marshall, C.J. (2005) RalGDS is required for tumor formation in a model of skin carcinogenesis. Cancer Cell *7*, 219-226.

Hahn, W.C. and Weinberg, R.A. (2002) Rules for making human tumor cells. N. Engl. J. Med. *347*, 1593-1603.

Hamad, N.M., Elconin, J.H., Karnoub, A.E., Bai, W., Rich, J.N., Abraham, R.T., Der, C.J. and Counter, C.M. (2002) Distinct requirements for Ras oncogenesis in human versus mouse cells. Genes Dev. *16*, 2045-2057.

Harvey, J.J. (1964) An unidentified virus which causes the rapid production of tumors in mice. Nature *204*, 1104-1105.

Herrmann, C. (2003) Ras–effector interactions: after one decade. Curr. Opin. Struct. Biol. *13*, 122-129.

Johnson, L., Mercer, K., Greenbaum, D., Bronson, R.T., Crowley, D., Tuveson, D.A. and Jacks, T. (2001) Somatic activation of the K-ras oncogene causes early onset lung cancer in mice. Nature *410*, 1111-1116.

Khosravi-Far, R., Solski, P.A., Clark, G.J., Kinch, M.S. and Der, C.J. (1995) Activation of Rac1, RhoA, and mitogen-activated protein kinases is required for Ras transformation. Mol. Cell. Biol. *15*, 6443-6453.

Kirsten, W.H. and Mayer, L.A. (1967) Morphological responses to a murine erythroblastastosis virus. J. Natl. Cancer Inst. *39*, 311.

Lim, K.H., Baines, A.T., Fiordalisi, J.J., Shipitsin, M., Feig, L.A., Cox, A.D., Der, C.J. and Counter, C.M. (2005) Activation of RalA is critical for Ras-induced tumorigenesis of human cells. Cancer Cell *7*, 533-545.

Madaule, P. and Axel, R. (1985) A novel ras-related family. Cell *41*, 31-40.

Malumbres, M. and Barbacid, M. (2003) RAS oncogenes: the first 30 years. Nat. Rev. Cancer *3*, 459-465.

Mitin, N., Rossman, K.L. and Der, C.J. (2005) Signaling interplay in Ras superfamily function. Curr. Biol. *15*, R563-R574.

Qiu, R.G., Chen, J., Kirn, D., McCormick, F. and Symons, M. (1995a) An essential role for Rac in Ras transformation. Nature *374*, 457-459.

Qiu, R.G., Chen, J., McCormick, F. and Symons, M. (1995b) A role for Rho in Ras transformation. Proc. Natl. Acad. Sci. U. S. A. *92*, 11781-11785.

Rangarajan, A., Hong, S.J., Gifford, A. and Weinberg, R.A. (2004) Species- and cell type-specific requirements for cellular transformation. Cancer Cell *6*, 171-183.

Repasky, G.A., Chenette, E.J. and Der, C.J. (2004) Renewing the conspiracy theory debate: does Raf function alone to mediate Ras oncogenesis? Trends Cell Biol. *14*, 639-647.

Sahai, E. and Marshall, C.J. (2002) RHO-GTPases and cancer. Nat. Rev. Cancer *2*, 133-142.

Sebti, S.M. and Der, C.J. (2003) Opinion: Searching for the elusive targets of farnesyltransferase inhibitors. Nat. Rev. Cancer *3*, 945-951.

Sweet-Cordero, A., Mukherjee, S., Subramanian, A., You, H., Roix, J.J., Ladd-Acosta, C., Mesirov, J., Golub, T.R. and Jacks, T. (2005) An oncogenic KRAS2 expression signature identified by cross-species gene-expression analysis. Nat. Genet. *37*, 48-55.

Touchot, N., Chardin, P. and Tavitian, A. (1987) Four additional members of the ras gene superfamily isolated by an oligonucleotide strategy: molecular cloning of YPT-related cDNAs from a rat brain library. Proc. Natl. Acad. Sci. U. S. A. *84*, 8210-8214.

Ulku, A.S. and Der, C.J. (2003) Ras signaling, deregulation of gene expression and oncogenesis. Cancer Treat Res. *115*, 189-208.

Vetter, I.R. and Wittinghofer, A. (2001) The guanine nucleotide-binding switch in three dimensions. Science *294*, 1299-1304.

Wellbrock, C., Karasarides, M. and Marais, R. (2004) The RAF proteins take centre stage. Nat. Rev. Mol. Cell Biol. *5*, 875-885.

Wennerberg, K., Rossman, K.L. and Der, C.J. (2005) The Ras superfamily at a glance. J. Cell Sci. *118*, 843-846.

Zuber, J., Tchernitsa, O.I., Hinzmann, B., Schmitz, A.C., Grips, M., Hellriegel, M., Sers, C., Rosenthal, A. and Schafer, R. (2000) A genome-wide survey of RAS transformation targets. Nat. Genet. *24*, 144-152.

CHAPTER 2

RAS-GEFs AND RAS GAPs

JOSE M. ROJAS[1] AND EUGENIO SANTOS[2]

[1] *Centro de Investigación del Cáncer. IBMCC (USAL-CSIC). Campus Unamuno. Universidad de Salamanca. 37007 Salamanca. Spain.*

[2] *Unidad de Biología Celular. Centro Nacional de Microbiología. Instituto de Salud Carlos III. 28220 Majadahonda, Madrid, Spain*

Abstract: Ras proteins (H-, N-, and K-Ras) are critical components of signal transduction pathways leading from cell-surface receptors to the control of cell proliferation, differentiation or death. Normal Ras proteins exist in equilibrium between an active (Ras-GTP) and an inactive (Ras-GDP) state. Once activated, Ras stimulates a multitude of downstream signaling pathways, but different data about Ras location to plasma membrane subdomains and new roles for some docking/scaffold proteins, point to signaling specificities of the different Ras proteins. Studies with knockout mice strains have revealed that *Kras* (but not *Nras* or *Hras*) is necessary and sufficient for development of the animals to the adult stage. Although Ras proteins possess intrinsic GTPase and GDP/GTP exchange activities, they are too low to account for the rapid and transient GDP/GTP cycling that occurs during mitogenic stimulation. Then, Ras function requires regulatory proteins that control the GDP/GTP cycling rate. These regulatory proteins include GTPase activating proteins (Ras-GAPs), which stimulate hydrolysis of bound GTP to GDP, and guanine nucleotide exchange factor proteins (Ras-GEFs), which promote the replacement of bound GDP with GTP. We undertook this review to analyze the current understanding of the mammalian Ras-GAPs and Ras-GEFs functions, focusing on the possible physiological specificities of each Ras-GAP/Ras-GEF family member. Furthermore, we analyzed new mechanisms of Rac activation due to covalent-interaction with hydrophobic molecules as NO and cyclopentenone prostaglandins

Keywords: Ras, p120-GAP, NF1, Ras-GRF, Sos, Ras-GRP, NO, CyPGs

1. INTRODUCTION

The normal Ras cycle involves an actively regulated equilibrium between the ON (Ras-GTP) and the OFF (Ras-GDP) conformations of Ras proteins (Figure 1). A variety of extracellular signals are able to trigger increases of the intracellular concentration of the active Ras-GTP complex, where the conformational change

C. Der (ed.), RAS Family GTPases, 15–43.

induced by GTP binding enabling Ras interaction with target downstream effectors (Bivona and Philips, 2003; Downward, 2003; Hancock, 2003; Rojas and Santos, 2002).

The intrinsic GTPase and GDP/GTP exchange activities of Ras proteins are too low to account for the rapid and transient GDP/GTP cycling occurring during mitogenic stimulation *in vivo* (Bivona and Philips, 2003; Downward, 2003; Hancock, 2003; Rojas and Santos, 2002). Therefore, normal Ras function requires the participation of distinct regulatory proteins in order to effectively control the GDP/GTP cycling rate. Indeed, the extent and duration of Ras activation in cells depends on the interplay between a variety of negative and positive regulators of the Ras cycle. The negative regulators are GTPase activating proteins (Ras-GAPs), capable of multiplying the intrinsic GTPase activity of Ras proteins, thus leading to rapid hydrolysis of bound GTP (Wittinghofer et al., 1997). The positive regulators are guanine nucleotide exchange factors (Ras-GEFs), able to promote displacement of bound GDP, thus allowing association of Ras proteins with the more abundant intracellular GTP (Downward, 1996) and resulting in accumulation of the active Ras-GTP complex (Figure 1).

This article will concentrate on reviewing functional aspects of the various families of cellular proteins possessing GAP or GEF activity on any of the classical members of the Ras subfamily of proteins able to induce malignant transformation of mammalian cells (mainly H, N and K-Ras). GAPs and GEFs for other members of the different Ras families will be the subject of separate reviews in this book series.

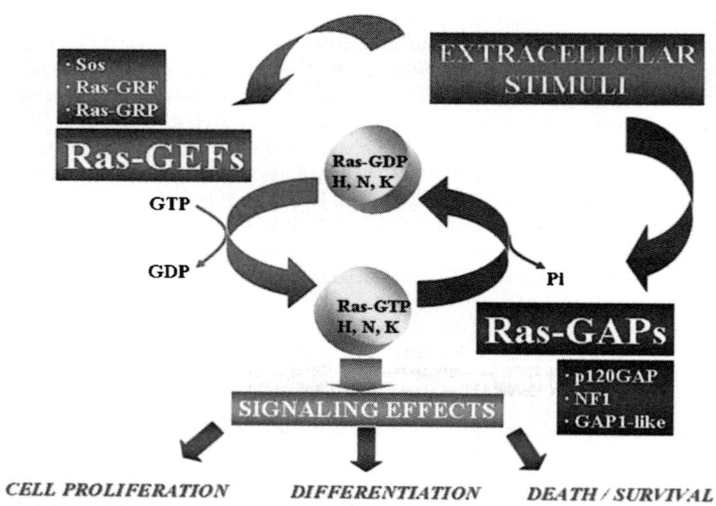

Figure 1. The Ras GTPase Cycle. OFF/ON cycle of Ras proteins between an "inactive state" (Ras-GDP) and an "active state" (Ras-GTP). Guanine nucleotide exchange factors (GEFs) and GTPase-activating proteins (GAPs) act, respectively, as positive and negative regulators of the Ras cycle, thus controlling internalization of a variety of extracellular signals which themselves may trigger different cellular responses including proliferation, differentiation or cell death

2. RAS GAPs

2.1 Structural/Functional Aspects of Ras GAPs

p120GAP, NF1 (neurofibromin) and GAP1 are the main, distinct, GAPs currently identified for Ras family proteins (Bernards, 2003) (Figure 2). All Ras-GAPs encompass three prominent domains within the conserved area of their catalytic domains (RasGAP domain): the arginine-finger loop, the invariant phenylalanine-leucine-arginine (FLR) motif and the alpha7/variable loop, containing structural fingerprints governing the GAP function (Figure 3). The finger loop is crucial for stabilization of the transition state of the GTPase reaction (Ahmadian et al., 2003). This function is controlled by amino acids proximal to the arginine residue that are different for p120GAP and NF1 (Ahmadian et al., 2003). These residues control the specific orientation of the arginine finger within the Ras active site (Ahmadian et al., 2003). The oxygen of the carbonyl group in the backbone of the arginine-finger is able to interact with a water molecule in the active site thus forming a bridge between the NH_2 group of Ras glutamine 61 and the gamma-phosphate of GTP (Resat et al., 2001) (Figure 3B). Thus, the arginine-finger may play a dual role in generating the nucleophile as well as stabilizing the transition state (Resat et al., 2001). The FLR motif indirectly contributes to GTPase stimulation by forming a scaffold stabilizing Ras switch regions (Ahmadian et al., 2003) (Figure 3C). Therefore, mutations of this FLR motif can result in GAP proteins defective in catalysis but not in binding to Ras (Brownbridge et al., 1993). The alpha7/variable loop determines the specificity

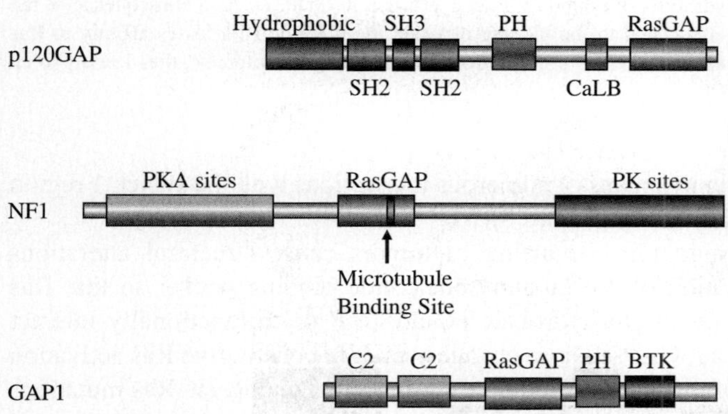

Figure 2. Domain structure of mammalian Ras GAP proteins. Structural schematics of the main families of GAP (GTPase Activating Proteins) –p120GAP, NF1 and GAP1- of Ras in mammalian cells. "Hydrophobic" designates the amino-terminal region of p120GAP, rich in hydrophobic amino acids. SH2 and SH3 are domains with homology to the regions 2 and 3 of Src, respectively. CaLB denotes a domain with homology to the CaLB region of PLA_2, C_2 is the high-affinity Ca^{+2}-dependent phospholipid-binding domain, PH corresponds to the pleckstrin homology domain and BTK is a Bruton's tyrosine kinase motif. The region termed RasGAP depicts the GAP catalytic domain. The amino-terminal and carboxyl-terminal regions of NF1 contain several PKA and PK (protein-kinases) phosphorylation sites, respectively

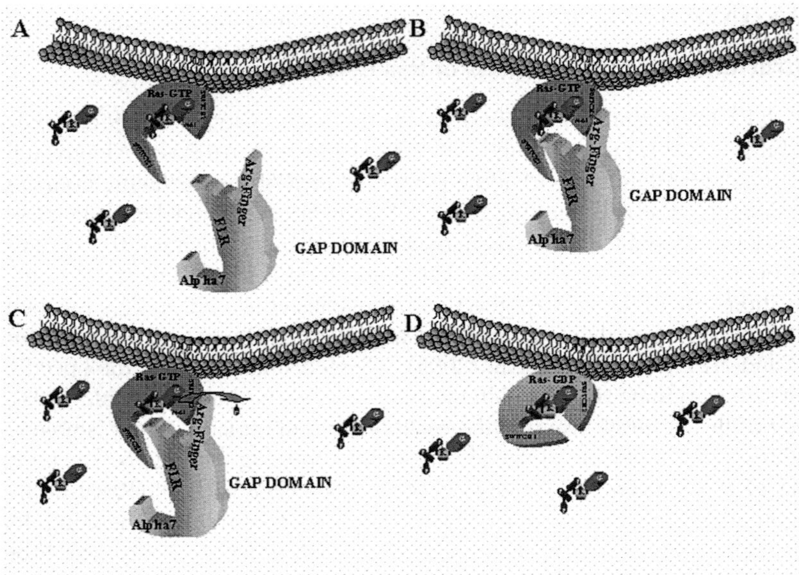

Figure 3. Model of GAP-dependent action on Ras. *A*, Ras bound to GTP is in the active conformation state due to the accessibility of Ras switch 1 to the downstream Ras effector proteins. *B*, The GAP domain of the Ras-GAPs interacts with switch 1 of Ras-GTP through the FLR motif and the alpha7/variable loop. In addition, the arginine-finger loop interacts with a water molecule in the active site, thus bridging the NH_2 group of the Ras residue glutamine 61 (N-61) to the gamma-phosphate of GTP. *C*, The arginine-finger loop stimulates the hydrolysis of bound GTP, converting it to GDP. *D*, As a consequence of the Ras structural modification associated to the binding of GDP, the GAP domain losses affinity to Ras and the Ras switch 1 becomes inaccessible to interaction with Ras effectors proteins, thus leading to an inactive conformation state

of Ras-RasGAP binding by means of numerous interactions with the switch 1 region of Ras (Ahmadian et al., 2003) (Figure 3).

The common *ras* mutations occurring in tumors cause structural alterations affecting the conformation of the guanine nucleotide binding pocket so that Ras proteins are rendered unable to hydrolyze bound GTP or to functionally interact with, and be activated by, GAPs. Those changes result in constitutive Ras activation through stabilization of the Ras-GTP complex. All known oncogenic Ras mutations (found in 25-30% of human tumors) map to a small subset of amino acids located around positions 12 and 61 of the primary sequence of these proteins.

2.2 p120GAP Proteins

The protein product of the p120GAP gene is expressed ubiquitously as a 120 Kda protein. Of the two existing p120GAP isoforms, most cells express type I, whereas type II is an alternatively spliced form detected in placental trophoblasts

(Trahey et al., 1988). In addition to their role as negative regulators for conventional Ras and R-Ras proteins (Ohba et al., 2000), it is believed that p120GAPs may also operate as downstream effectors of Ras (McCormick, 1998). The first evidence in this regard came from the observation that oncogenic Ras mutants still require GAP interaction to display their transforming activity. Such an effector function of p120GAP is dependent on the amino-terminal regulatory domain, which is able to interact with receptor and nonreceptor tyrosine kinases, as well as with phosphorylated proteins (Ellis et al., 1995; Tocque et al., 1997). These interactions are mediated through an adaptor region close to the hydrophobic amino-terminal stretch of p120GAP (Figure 2) constituted by one Src homology (SH) 3 domain flanked by two Src homology (SH) 2 domains. The SH3 domain binds to specific proline-rich sequences (Duchesne et al., 1993), whereas the SH2 domains recognize specific protein motifs containing phospho-tyrosine residues such as those located in the cytoplasmic region of activated RTKs or non-RTK phospho-tyrosine proteins such as Shc, IRS-1, Syp, and Sam68 (Rojas and Santos, 2002; Sanchez-Margalet and Najib, 2001). For example, tyrosine-phosphorylated Sam68 has been reported to link p120GAP (through its C-terminal SH2 domain) to the PI3K signaling pathway (Rojas and Santos, 2002; Sanchez-Margalet and Najib, 2001). Missense mutations within the carboxyl-terminal SH2 domain of p120GAP have been reported to occur in human basal cell carcinoma (Friedman et al., 1993).

The murine p120GAP KO results in embryonic lethality, abnormal vascularization, and enhanced neuronal apoptosis (Henkemeyer et al., 1995). Ectopic expression of p120GAP in wing imaginal disc of *Drosophila melanogaster* resulted in smaller wings with fewer cells (Feldmann et al., 1999). Likewise, *Drosophila* Sprouty (dSpry), a proline-rich protein, has been reported to be able to inhibit Ras-MAPK signalling (Casci et al., 1999) by binding to the SH3 domains of Drk (Drosophila homologue of Grb2) and Gap-1 (homologue of p120GAP). However, a similar *in vivo* interaction between Spry and p120GAP in mammalian cells is yet to be demonstrated.

Besides the SH3 and SH2 domains, the p120GAP amino-terminal region contains also a pleckstrin homology (PH) domain possessing phospholipid binding affinity, and a Ca^{+2}-dependent lipid binding domain (CaLB) (Figure 2) (homologous to the CaLB region of PLA_2) which mediates protein-protein interaction between p120GAP and annexin VI and also binds to negatively charged phospholipids (Chow et al., 1999; Chow and Gawler, 1999). Fibroblasts derived from p120GAP null mouse embryos (p120GAP -/-) (Henkemeyer et al., 1995; Koehler and Moran, 2001b; van der Geer et al., 1997) exhibit prolonged MAPK activation after epidermal growth factor (EGF)-, but not lysophosphatidic acid-, stimulation, and p120GAP needs its PH and CaLB domains to down-regulate the Ras/MAPK pathway (Koehler and Moran, 2001b). Similarly, p120GAP and RACK1 are known to undergo *in vivo* interaction dependent upon both the PH- and CaLB- domain of GAP (Koehler and Moran, 2001a).

The carboxyl-terminal region of p120GAP contains the catalytic domain essential for GAP function (Figure 2). The GAP activity of p120GAP (and also of NF1) increases the GTPase rate of Ras proteins by a factor of up to 10^5 (Gutmann and Collins, 1993). This GAP rate is inhibited by arachidonic and phosphatidic acids, probably by competition in the Ras binding process (Sermon et al., 1996).

2.3 NF1 Proteins

Neurofibromin (human chromosomal localization 17q11.2) shares both sequence identity and substrate specificity with the p120GAP carboxyl-terminal catalytic domain (Figure 2). The protein product of the *NF1* gene is expressed as a 250 Kda protein in brain, spleen, kidney, testis and thymus. Subcellular localization experiments have demonstrated that neurofibromin is associated with cytoplasmic microtubules (Gutmann and Collins, 1992) (Figure 2). The central portion (400 residues) corresponds to the catalytic RasGAP domain, showing high homology with the GAP activity domain of p120 (Ballester et al., 1990; Martin et al., 1990; Scheffzek et al., 1998) (Figure 2). Less is known about the functions of NF1 and it can be assumed that each protein mediates distinct pathways. While growth factors stimulate tyrosine phosphorylation of p120GAP, serine and threonine phosphorylation has been reported for NF1. In addition to its role as a negative regulator of Ras activity, NF1 regulates proliferation and survival of precursors and lineage-restricted myeloid progenitors in response to multiple cytokines by modulating Ras output (Zhang et al., 1998).

NF1 is related to a familial cancer syndrome termed Von Recklinghausen neurofibromatosis, or neurofibromatosis type 1, which is a common autosomal dominant disorder affecting 1 in 3000 individuals (Gutmann, 1998). Missense mutations found in neurofibromatosis type 1 patients map to the RasGAP domain (Scheffzek et al., 1998). The syndrome is diagnosed clinically by finding two or more of the following: café au lait spots, neurofibromas, freckling in non-sun-exposed areas, optic glioma, Lisch nodules, distinctive bony lesions, or a first-degree relative with NF1. Less common manifestations can also include short stature and macrocephaly. NF1 patients can also have learning disabilities, seizures, scoliosis, hypertension, plexiform neurofibromas, or pheochromocytomas. Somatic mutations in the *NF1* gene that result in an absence of neurofibromin expression have been described for a variety of tumor types as malignant melanomas, neuroblastomas, pheochromocytomas, and neurofibrosarcomas (Glover et al., 1991; Seizinger, 1993). Loss of NF1 gene is found in some patients with juvenile chronic myelogenous leukemia. Deficiency in NF1 also induces myeloproliferative disease through Ras-mediated hypersensitivity to granulocyte-macrophage colony-stimulating factor (GM-CSF) (Largaespada et al., 1996). Likewise, NF1 -/- mouse embryos die between embryonic days 12.5 and 13.5 and show aberrant growth of hematopoietic cells, suggesting that NF1 is required to downregulate Ras activation in myeloid cells exposed to GM-CSF, interleukin-3 (IL-3), or stem cell factor (Bollag et al., 1996). Furthermore, NF1 inactivation cooperates with N-Ras in lymphogenesis activating Erk by a mechanism independent of its GTPase activity (Mangues et al., 1998). NF1

+/- mice exhibit increased numbers of brain astrocytes with abnormal attachment, spreading and motility properties (Gutmann et al., 1999; Gutmann et al., 2001), and deficits in learning and spatial memory (Costa et al., 2001). The NF1 +/- cells showed increase Ras pathway activation (Bajenaru et al., 2001) that was reversed by the introduction of a wild-type NF1 RasGAP domain, but no such effect was obtained with p120GAP (Hiatt et al., 2001), suggesting the critical and non-redundant role of neurofibromin in ras regulation (Dasgupta and Gutmann, 2003).

Unlike NF1-deficient mice, *Drosophila NF1* null mutants are viable, but reduced in size. Studies in *Drosophila* have shown that neurofibromin regulates G-protein-stimulated adenylate cyclase activity (The et al., 1997). Therefore, neurofibromin may have both Ras-dependent (i.e., the regulation of circadian rhythm) (Williams et al., 2001) and Ras-independent (i.e., the cellular response to neuropeptides) (Guo et al., 1997) functions in *Drosophila*. Likewise, different results from NF1-deficient murine neurons suggest that some of the neurofibromatosis type 1 clinical symptoms (see above), such as short stature and learning disabilities, may result from other non-Ras neurofibromin functions whereas other features, like tumor formation, involve hyperactivation of Ras proteins (Dasgupta and Gutmann, 2003; Tong et al., 2002).

Several neurofibromin isoforms have been identified that arise from alternative splicing (Cawthon et al., 1990) and display different expression patterns depending on tissue and developmental state. One of the most common alternative splicing events involves exon 23a, which inserts an additional 21 amino acids into the RasGAP domain (Dasgupta and Gutmann, 2003). Targeted disruption of exon 23a resulted in learning impairments in mice, similar to NF1 +/- mice (Costa et al., 2001), although probably not as a result of affected RasGAP activity but rather because of reduced neurofibromin activity attributable to exon 23a function, such as cAMP regulation (Dasgupta and Gutmann, 2003).

Several studies have suggested that neurofibromin may suppress cell growth through mechanisms unrelated to Ras regulation. Indeed, in NIH 3T3 cells, overexpression of NF1 resulted in a threefold reduction in cell growth without any changes in Ras activity (Johnson et al., 1994). Similarly, overexpression of full-length neurofibromin in a human colon carcinoma cell line resulted in reduced tumor growth in nude mice. In these experiments, the observed cooperation emphasizes the importance of searching for additional functions of NF1.

2.4 GAP1-like Proteins

GAP1 was identified as a negative regulator in *Drosophila* eye development (Gaul et al., 1992). The mammalian members of the GAP1 family ($GAP1^{m}$, $GAP1^{IP4BP}$, CAPRI and RASAL) share a common molecular design of amino-terminal, tandem C_2 domains (with homology to the high-affinity Ca^{+2}-dependent phospholipid-binding C_2 domains from synaptotagmin III and protein kinase C βII), a carboxyl-terminal PH domain adjacent to a Bruton's tyrosine kinase motif, and a central catalytic RasGAP domain (Allen et al., 1998; Bernards, 2003; Cullen et al., 1995;

Cullen and Lockyer, 2002; Lockyer et al., 1997; Lockyer et al., 1999; Maekawa et al., 1994; Maekawa et al., 1993; Walker et al., 2002; Yamamoto et al., 1995) (Figure 2). However the GAP1-like proteins differ in several aspects. Thus, the C_2 domains of $GAP1^{m}$ and $GAP1^{IP4BP}$ lack residues required for Ca^{+2} coordination, whereas the corresponding domains of CAPRI and RASAL display high Ca^{+2}-binding affinity (Bernards, 2003; Lockyer et al., 2001; Walker et al., 2004; Walker et al., 2002). This difference may explain the Ca^{+2}-mobilization induced membrane translocation of CAPRI and RASAL, but not of $GAP1^{m}$ or $GAP1^{IP4BP}$ (Bernards, 2003; Lockyer et al., 2001; Walker et al., 2004). Furthermore, the constitutive plasma membrane association of $GAP1^{IP4BP}$ has been attributed to its PH domain serving as a high affinity PIP2 binding site (Bernards, 2003; Cozier et al., 2000). In sharp contrast, the PH domain of $GAP1^{m}$ binds PIP3 and has been related to the PI3K-dependent membrane translocation of $GAP1^{m}$ (Bernards, 2003; Lockyer et al., 1999). Moreover, it has been described that $GAP1^{m}$ also interacts (through its PH domain) with the heterotrimeric G protein subunit Gα12, and its RasGAP activity was stimulated by this interaction (Bernards, 2003; Jiang et al., 1998).

Although it has been reported that p120GAP may be regulated by Ca^{+2}-binding (Filvaroff et al., 1992; Gawler et al., 1995a; Gawler et al., 1995b), others have questioned this possibility (Clark et al., 1995). Nevertheless, the prototypical Ca^{+2}-triggered Ras-GAP are CAPRI and RASAL (Lockyer et al., 2001; Walker et al., 2004). In unstimulated cells, CAPRI and RASAL are cytosolic and inactive Ras-GAPs, and upon agonist-evoked increase in the concentration of intracellular free Ca^{+2} both proteins undergo a rapid translocation to the plasma membrane mediated by their C_2 domains (Lockyer et al., 2001; Walker et al., 2004). This plasma membrane association is transient in the case of CAPRI and does not sense oscillations in the concentration of intracellular free Ca^{+2}, whereas RASAL oscillates between the plasma membrane and the cytosol in synchrony with simultaneously measured repetitive Ca^{+2} spikes (Lockyer et al., 2001; Walker et al., 2004).

3. RAS GEFs

3.1 Structural/Functional Aspects of Ras GEFs

The guanine nucleotide exchange factors (GEFs) promote formation of the active Ras-GTP complex by inducing dissociation of bound GDP to allow association of the more abundant GTP, thus increasing the rate of intracellular exchange of GDP for GTP. At least three different protein families exhibiting GEF activity toward Ras, such as Sos, Ras-GRF and Ras-GRP (formerly denominated Cal-DAG GEF) (Figure 4), have been identified in mammalian cells (Boguski and McCormick, 1993; Ebinu et al., 1998; Rojas and Santos, 2002). All these involve modular proteins containing several conserved domains. Two domains are essential for specifically activating Ras subfamily members and are therefore present in all of them: the Ras exchange motif (REM), involved in the stabilization of binding to Ras, and the CDC25 homology domain (CDC25-H), containing the catalytic domain, which was originally identified as a region in the *Saccharomyces*

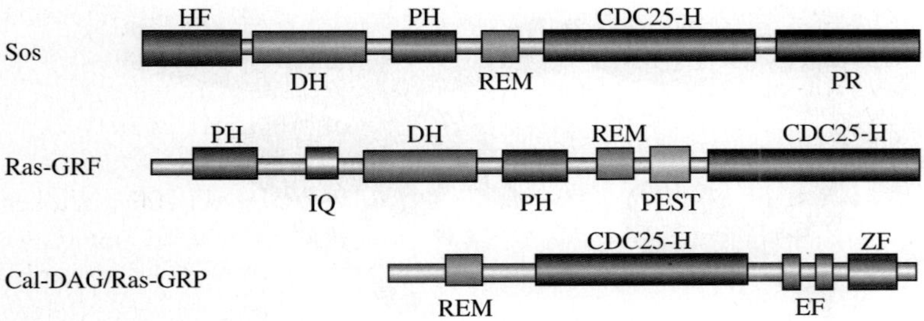

Figure 4. Domain structure of mammalian of Ras GEF proteins. Schematics of the main Ras-GEF (Guanine Nucleotide Exchange Factors) protein families (Sos, Ras-GRF, and Ras-GRP/CalDAG GEF) in mammalian cells. HF, Histone Folding motif. DH, Dbl-homology. PH, Pleckstrin homology. REM, Ras exchanger motif. CDC25-H, CDC25 homology (catalytic GEF domain). PR, Proline-rich domain. IQ, Calmodulin binding domain. PEST, PEST-like region. EF, Ca^{2+} binding domain. ZF, Zinc Finger domain (diacylglycerol binding domain)

cerevisiae Cdc25 protein essential for its function as an upstream regulator of Ras (Boguski and McCormick, 1993).

A variety of other domains present within the body of specific GEF family members (SH2, SH3, PH, PR, etc; Figure 4) allow for protein-protein and protein-lipid interactions which are important for the spatio-temporal regulation of the Ras activation process by different stimuli. Thus, Sos family members are involved in the coupling of growth factor receptors (specially tyrosine kinase receptors) to Ras-dependent mitogenic signaling pathways (Schlessinger, 1993). In contrast, Ras-GRFs are involved in Ca^{2+} influx/calmodulin-dependent activation of Ras (Farnsworth et al., 1995). Finally, the Ras-GRP proteins activate Ras through mechanisms regulated not only by Ca^{2+} but also by diacylglycerol (DAG) (Ebinu et al., 1998).

Crystallographic studies about Sos-Ras binding, which involves both switch 1 and switch 2 regions of Ras (Boriack-Sjodin et al., 1998), have suggested a model of Ras activation by Sos which would also be applicable to the rest of Ras-GEFs (Figure 5). Specifically, according to this model, the interaction of the REM domain with switch 2 mediates the anchoring of Ras to the Ras-GEF and the interaction of the CDC25-H domain (two β-sheets) with switch 1 leads to disruption of the nucleotide-binding site and GDP dissociation (Hall et al., 2001) (Figure 5).

3.2 Ras-GRP/CalDAG-GEF Proteins

The four known members of the Ras-GRP/CalDAG GEF family are expressed in various tissues including mainly nervous (Ras-GRP1, Ras GRP2 and Ras-GRP3) and hematopoietic cells (Ras-GRP1, Ras GRP3 and Ras-GRP4) (Ebinu et al., 1998; Reuther et al., 2002; Tognon et al., 1998). Besides the REM and CDC25-H domains, a pair of Ca^{2+}-binding EF-hands and the DAG-binding ZF (Zinc Finger) domain have been identified in Ras-GRP (Figure 4). Whereas deletion of the ZF domain eliminates the transforming activity of Ras-GRP, the EF-hands seem to be

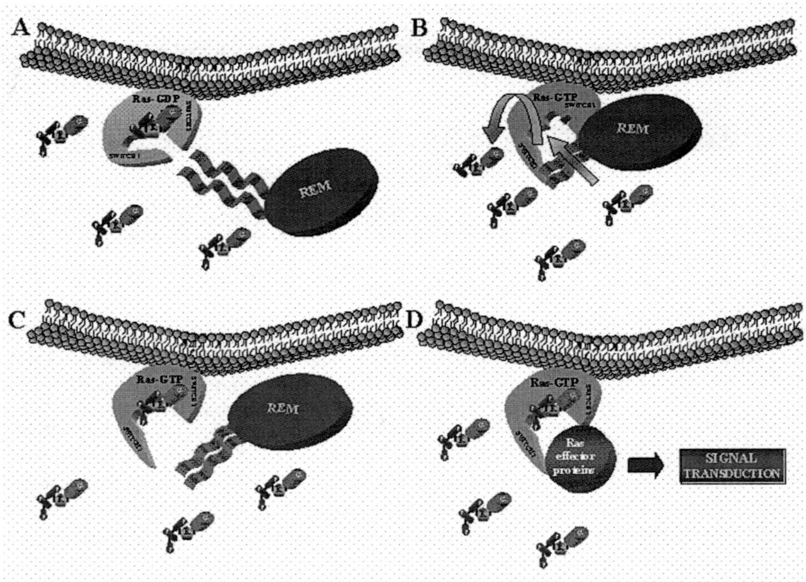

Figure 5. Model of GEF-dependent activation of Ras proteins. *A*, Ras bound to GDP is in the inactive conformation state due to the inaccessibility of Ras switch 1 to the downstream Ras effector proteins. Upon mitogenic stimulation, the Ras-GEFs travel to the plasma membrane, thus getting in close proximity to their target, Ras-GDP. *B*, Interaction of the REM domain with Switch 2, and of the two β-sheets of the CDC25-H domain with Switch 1 of Ras-GDP leads to disruption of the nucleotide-binding site and GDP dislodgement. Following Ras-GDP dissociation, intracellular GTP – with higher intracellular concentration than GDP – fills the Ras nucleotide-binding site pocket. *C*, The Ras-GTP structural conformation determines loss of binding affinity to the REM and CDC25-H domains of the Ras-GEF thus leading to disruption of the activation complex. *D*, As a consequence of Ras structural modification associated with GTP binding, the switch 1 becomes accessible to interaction with Ras effectors proteins leading to activation of different downstream signal transduction pathways

dispensable (Ebinu et al., 1998; Tognon et al., 1998). The activation of most Ras-GRP proteins is mediated by the phospholipase Cγ (PLC-γ)-dependent generation of DAG (Quilliam et al., 2002). This second messenger binds to the ZF domain of Ras-GRP proteins, thus facilitating the translocation of Ras-GRPs to membranes and their association with target GTPases (Quilliam et al., 2002). Consistent with this model, Ras-GTP formation is enhanced by a DAG analog in the presence of Ras-GRP, suggesting an involvement of these Ras-GEF in DAG-mediated activation of the Ras pathway (Ebinu et al., 1998; Kawasaki et al., 1998; Ohba et al., 2000; Tognon et al., 1998).

Ras-GRP1 (CalDAG-GEFII), Ras-GRP3 (CalDAG-GEFIII) and Ras-GRP4 have been described to activate Ras subfamily proteins (H, N, and K-Ras, R-Ras and TC21), but Ras-GRP3 can also activate Rap1 (Quilliam et al., 2002; Reuther et al., 2002; Yang et al., 2002). In contrast, Ras-GRP2 (CalDAG-GEFI) is mostly specific for Rap1 and apparently unable to activate any Ras isoforms (Clyde-Smith et al., 2000;

Matallanas et al., 2003). Ras-GRP3 is primarily active on H-Ras and exhibits weaker exchange activities on N-Ras and K-Ras (Matallanas et al., 2003). Ras-GRP1 also exhibits a stronger activation effect on H-Ras while displaying similar exchange activities on N-Ras and K-Ras (Matallanas et al., 2003).

Analysis of Ras-GRP1 KO mice indicates that this GEF plays an essential role on normal thymocyte differentiation, most likely in the process of positive selection (Dower et al., 2000). Although no KO mice are yet available for the other Ras-GRP family members, recent data using chicken DT40 B-cells indicate that Ras-GRP3 is important for the coupling of the B-cell receptor with Ras activation (Oh-hora et al., 2003).

Interestingly, and in sharp contrast to the rest of Ras-GEFs, Ras-GRP1 and Ras-GRP3 are present in the Golgi apparatus of exponentially growing cells (Bivona et al., 2003; Caloca et al., 2003a), thus mediating H-Ras and N-Ras activation in this endomembrane cell compartment (Figure 6) (Bivona et al., 2003; Caloca et al., 2003a). An interesting cross-talk between Vav/Rac1 and Ras signaling pathways, established through the stimulation of Ras-GRP1 has been demonstrated recently (Caloca et al., 2003b). This effect requires the convergence of two separate

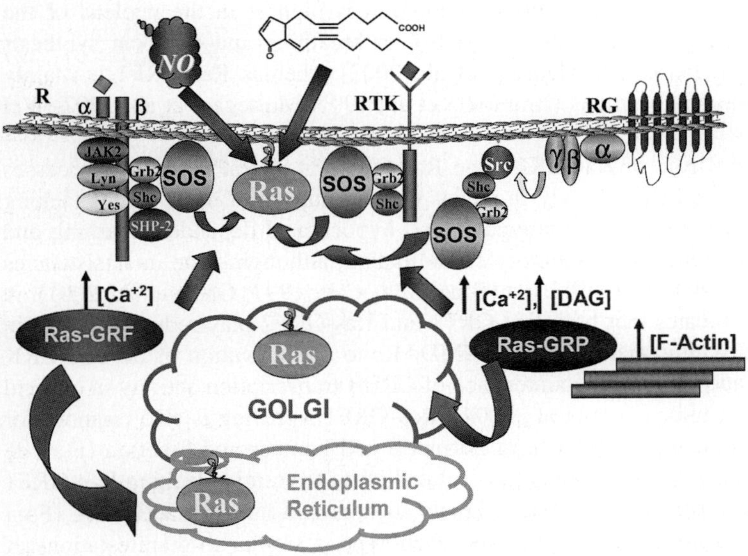

Figure 6. Different mechanisms of Ras protein activation. Ras proteins can become activated in the plasma membrane or in the Golgi apparatus (H-Ras) by the action of different stimuli. Sos proteins are able to mediate Ras activation upon ligand-stimulated activation of cell-surface receptors, receptor-associated tyrosine kinases, or agonist mediated through G-coupled receptors. Ras-GRF and Ras-GRP proteins are able to induce Ras activation upon increases of intracellular Ca^{2+} and DAG (Ras-GRP). Ras-GRP1 and Ras-GRP3 proteins can induce H-Ras and N-Ras activation in the Golgi apparatus, whereas Ras-GRF family proteins activate H-Ras in the endoplasmic reticulum. Finally, physical interaction of Ras proteins with some inorganic or lipid molecules, such as NO or cyclopentenone prostaglandins, may also induce their activation

Vav/Rac1-dependent steps: stimulation of PLC-γ (increasing the DAG levels) and F-actin polymerization (Bivona et al., 2003; Caloca et al., 2003b). This Vav/Ras-GRP signaling interaction is evolutionarily conserved and crucial for activation of the Ras pathway in lymphoid cells (Caloca et al., 2003b). A similar Vav/Rac connection occurs also in the case of Ras-GRP2, leading to regionalized activation of Rap1 in juxtamembrane areas of the cell (Caloca et al., 2004).

3.3 Ras-GRF Proteins

The Ras-GRF family encompasses two distinct members in mammalian cells, Ras-GRF1 (Cen et al., 1992; Martegani et al., 1992; Shou et al., 1992; Wei et al., 1992) and Ras-GRF2 (Fam et al., 1997). Ras-GRF1 – *hras-grf1* gene located on chromosome 15q24 (Guerrero et al., 1996) – is expressed mainly in nervous tissues (Brambilla et al., 1997; Martegani et al., 1992; Wei et al., 1993), although it is also detected in other organs (Guerrero et al., 1996) including pancreatic islets (Font de Mora et al., 2003). Ras-GRF2 transcripts are detected in a variety of tissues including brain, spleen and lung (Fam et al., 1997). Ras-GRF1 and Ras-GRF2 exhibit very different patterns of expression within the brain (Fernandez-Medarde et al., 2002). Thus, Ras-GRF2 brain expression is highest in the nucleus of the solitary tract, a region implicated in control of breathing and oxytocin synthesis during lactation (Fernandez-Medarde et al., 2002) whereas Ras-GRF1 is mainly expressed in the hippocampus (Brambilla et al., 1997; Martegani et al., 1992; Wei et al., 1993).

The functional role of Ras-GRF1 and Ras-GRF2 in neural signaling processes has been an active field of study in recent years. Analysis of mice strains lacking Ras-GRF1 show alterations in amygdala and hypocampal dependent learning and suggest an important role in memory consolidation, although some inconsistencies between published studies still exist (Brambilla et al., 1997; Giese et al., 2001). A recent report indicates that both Ras-GRF1 and Ras-GRF2 play redundant roles in coupling NMDA glutamate receptors NMDARs to the activation of the Ras-ERK signaling cascade and to the maintenance of CREB transcription activity in cortical neurons of adult mice (Tian et al., 2004). Ras-GRF1 signaling is also required for the development and maintenance of normal β-cell number and function (Font de Mora et al., 2003). IGF-I treatment of isolated islets from mice lacking Ras-GRF1 fails to activate critical downstream signals such as Akt and Erk activation (Font de Mora et al., 2003), and the observed phenotype is similar to manifestations of preclinical type 2 diabetes. In sharp contrast, animals lacking Ras-GRF2 show a normal phenotype and do not appear to develop any physical illness due to the null mutation, suggesting that Ras-GRF2 activity is dispensable for mouse growth, fertility, and somatic development to the adult stage (Fernandez-Medarde et al., 2002). Furthermore, the double-null *grf1/grf2* mice appear to be perfectly viable and fertile, keeping the same phenotype observed in single *grf1*-null mice and not showing any observable compensatory oversynthesis of the other Ras-GRF protein in the simple knockouts (Fernandez-Medarde et al., 2002).

In addition to the REM and CDC25-H domains, both Ras-GRF proteins possess a pleckstrin homology (PH) domain, an isoleucine-glutamine (IQ) motif, and tandem Dbl homology (DH)/PH domains (Fam et al., 1997; Fan et al., 1998) (Figure 4). The IQ motif is responsible for interaction with calmodulin, thus mediating signals triggered by increases of intracellular Ca^{2+} concentration (Buchsbaum et al., 1996; Farnsworth et al., 1995). Tandem DH/PH domains are conserved among Dbl family proteins, GEFs known to target Rho family GTP-binding proteins (Whitehead et al., 1997).

Regarding their regulation, it has been shown that both Ras-GRF1 and Ras-GRF2 are activated by G-protein-coupled receptors but are largely insensitive to receptors of the tyrosine kinase type (Fam et al., 1997; Mattingly and Macara, 1996; Shou et al., 1995; Zippel et al., 1996). Augmentation of intracellular calcium levels by calcium ionophores can also bring about the activation of Ras-GRF1 and Ras-GRF2. This is achieved through the IQ motif (Buchsbaum et al., 1996; Fam et al., 1997; Fan et al., 1998; Farnsworth et al., 1995). Interestingly, Ras-GRF1 exhibits GEF activity toward Rac as well in response to signals mediated by $G\beta\gamma$ subunits (Kiyono et al., 1999), while Ras-GRF2 shows constitutive and Ca^{2+}-stimulated Rac-GEF activity (Fan et al., 1998), for which the DH/PH domains are required. Phosphorylation of a serine residue is important for triggering Ras-GEF activation (Mattingly, 1999), whereas tyrosine phosphorylation is indispensable for Rac-GEF activation (Kiyono et al., 2000b; Kiyono et al., 1999). On the other hand, oligomerization of Ras-GRFs, mediated by their DH domains (Anborgh et al., 1999), may be required for biological function, because a mutation within the DH domain that abolishes oligomerization renders Ras-GRFs incapable of inducing transformation of NIH 3T3 cells. Furthermore, the ability of Ras-GRF1 to activate Ras is regulated by the Rho family GTPase Cdc42, through a mechanism entailing the translocation of Ras-GRF1 to the cell particulate fraction (Arozarena et al., 2000; Arozarena et al., 2001). Although it has been published that Ras-GEF activity of Ras-GRF1 is stimulated upon tyrosine phosphorylation by the Cdc42-regulated kinase ACK1 (Kiyono et al., 2000a), other authors report that Cdc42-GDP inhibits Ras-GRF-induced MAPK activation but neither Cdc42-GTP nor the Cdc42 downstream effectors (MLKS, ACK1, PAK1, and WASP) affect Ras-GRF performance (Arozarena et al., 2001).

Ras-GRF2 contains a PEST-like region (rich in proline, glutamic acid, serine, and threonine) containing a candidate destruction box (DB) – located between the REM and CDC25-H domains – (Figure 4) which is implicated in targeting proteins for ubiquitination and subsequent degradation via the 26S proteasome (Fam et al., 1997). Ras binding triggers the ubiquitination of Ras-GRF2 through conformational changes (induced by GTPase binding) exposing the DB and thereby targeting Ras-GRF2 for destruction (de Hoog et al., 2001). Moreover, the amino-terminal half (residues 1-625) of Ras-GRF1 binds (*in vitro* and *in vivo*) the deubiquitinating enzyme UBPy, and this interaction may play a role in regulating the levels of Ras-GRF1 (Gnesutta et al., 2001).

Different studies state that Ras-GRF1 activates H-Ras and R-Ras *in vivo*, but not N-Ras or K-Ras 4B (Gotoh et al., 2001; Jones and Jackson, 1998; Matallanas et al.,

2003). In contrast, Ras-GRF2 is reported capable of stimulating GDP/GTP exchange on N-Ras and on K-Ras, although at lower levels than H-Ras (Matallanas et al., 2003), but it seems unable to activate R-Ras *in vivo* (Gotoh et al., 2001). The inability of Ras-GRF2 to activate R-Ras appears to be the consequence of post-translational modification (geranylgeranylation) of this GTPase, since Ras-GRF2 has been reported to activate (*in vitro* or *in vivo*) unprocessed R-Ras (Gotoh et al., 2001). Analysis of chimeras between Ras-GRF1 and Ras-GRF2 demonstrates that a 30 amino acid segment embedded with their catalytic domains is responsible for recognizing the presence of different lipids bound to Ras (Gotoh et al., 2001). Finally, it has been shown that Ras-GRF family GEFs colocalize with H-Ras in the endoplasmic reticulum (ER) but not in the Golgi apparatus (Figure 6) and also that Ras-GRF1 and Ras-GRF2 can efficiently induce nucleotide exchange on reticular H-Ras (Arozarena et al., 2004). The Ras-GRF DH domain is required for the activation of H-Ras in the ER but not in the plasma membrane (Arozarena et al., 2004), supporting the notion that this DH domain harbors some targeting signal necessary to adequately position Ras-GRF proteins in the ER. Surprisingly, Ras-GRF mediation favors the activation of H-Ras in the ER by lysophosphatidic acid treatment, whereas ionomycin preferentially activates H-Ras at the plasma membrane (Arozarena et al., 2004) (Figure 6), although the effect of both stimuli is dependent on the Ras-GRF IQ motif (Innocenti et al., 1999; Zippel et al., 2000). The molecular factors inducing the selectivity of Ras-GRF proteins toward the ER and plasma membrane H-Ras pools under different stimuli are unknown. Likewise, the biological significance of H-Ras activation in endomembranes is still unclear.

3.4 Sos Proteins

The functional role of Sos in the activation of Ras downstream of a variety of receptors including tyrosine kinase-type (RTK), cytokine, and G protein-coupled receptors has been extensively clarified in recent years. Regarding RTK signalling in particular, recruitment of Sos (complexed with the adaptor protein Grb2) to the tyrosine-phosphorylated receptor at the plasma membrane is considered to be a crucial step for the onset of Ras activation (Schlessinger, 1993).

Sos family GEFs are known to be able to induce GDP/GTP exchange on all Ras isoforms, in the hierarchy H-Ras > N-Ras > K-Ras (Jaumot et al., 2002). They are also reportedly able to effectively induce GDP-GTP exchange on H-Ras at the plasma membrane and the ER (although with a more restrictive pattern than Ras-GRF proteins), but not on Golgi-associated H-Ras (Arozarena et al., 2004).

Son of sevenless (Sos) was initially discovered as a Ras activator by genetic studies of R7 cell development in the eye of *Drosophila melanogaster* (Simon et al., 1991). Ras-dependent RTK signalling pathways are well conserved at the mechanistic level throughout metazoan evolution, and homologues to dSos have been found from *Caenorhabditis elegans* (Chang et al., 2000), to mammalian cells (Bowtell et al., 1992; Chardin et al., 1993). Sos1 and Sos2, the two known members of the mammalian Sos family share 69% of overall homology (lower homology corresponding to carboxyl-terminal region) (Bowtell et al., 1992; Chardin et al., 1993). The *hsos1* and *hsos2* genes are located on chromosomes 2p22-p16 and 14q21-q22, respectively (Chardin and Mattei, 1994), and both proteins are ubiquitously expressed (Guerrero et al., 1996).

Despite their similarities in structure and expression pattern, the functional relevances of Sos1 and Sos2 are clearly different. Sos1 is essential for intrauterine development, with homozygous null mice dying in mid-gestation (Qian et al., 2000; Wang et al., 1997) due to impaired development of trophoblastic layers of the placenta (Qian et al., 2000), whereas mice lacking Sos2 are viable (Esteban et al., 2000). It appears that Sos1 can participate in both short-term and long-term activation of the Ras-MAPK pathway, while Sos2-dependent signals are predominantly short-term (Qian et al., 2000). Moreover, Sos2 proteins are degraded by the 26S proteasome following targeting by ubiquitination, whereas Sos1 proteins are not ubiquitinated and show longer half-life than Sos2 (Nielsen et al., 1997).

Sos proteins are involved in the coupling of growth factor receptors to Ras-dependent mitogenic signaling pathways (Boguski and McCormick, 1993; Downward, 1994; Feig, 1994; Quilliam et al., 1995; Schlessinger, 1993). A proposed mechanism suggests that recruitment of Sos to the plasma membrane via formation of a complex with Grb2 adapter proteins is responsible for activation of the mature, membrane-bound Ras proteins (Egan and Weinberg, 1993; McCormick, 1993; Pawson and Schlessinger, 1993). In this model, both the cytosolic and membrane-bound Sos forms are thought to exhibit similar nucleotide exchange activity, and no change of this exchange activity is supposed to occur as a consequence of relocation inside the cell. In support of this notion, constitutive or conditional membrane targeting of these exchange factors has been shown to potentiate Ras activation in transfected cells (Aronheim et al., 1994; Holsinger et al., 1995; Quilliam et al., 1994). However, other reports suggest that, irrespective of subcellular location, the intrinsic guanine nucleotide exchange activity of Sos may be different before and after stimulation of surface tyrosine kinase receptors (Li et al., 1996; Li et al., 1993a; Li et al., 1995; Rojas et al., 1999; Zarich et al., 2000). Crystallographic studies of Ras bound to the catalytic module of Sos have demonstrated a highly conserved specific Ras-GTP binding site on Sos (Margarit et al., 2003). The Ras-GTP binding stabilizes the active site of Sos allosterically, increasing the rate of GDP-release from Ras (Margarit et al., 2003).

In addition to the REM and CDC25-H domains, Sos proteins contain several defined domains, each involving a distinct function (Figure 4). The SH3 domains of Grb2 bind to specific proline-rich sequences located in the carboxyl-terminal region of Sos (Chardin et al., 1993; Li et al., 1993b; Rozakis-Adcock et al., 1993) (Figure 4) that adopt a left-handed polyproline type II helix conformation (Feng et al., 1994; Lim and Richards, 1994; Yu et al., 1994). Some reports suggest that the carboxyl-terminal portion of Sos may exert negative regulation over the activity of the whole Sos1 protein (Byrne et al., 1996; Corbalan-Garcia et al., 1998; Karlovich et al., 1995; Kim et al., 1998; McCollam et al., 1995; Wang et al., 1995; Zarich et al., 2000). Consistent with this notion, a mutation in the *hsos1* gene (segregating in a dominant manner) has been discovered that creates a premature stop codon abolishing the proline-rich SH3 binding domains present in the carboxyl-terminal region of the hSos1 protein (Hart et al., 2002). This mutation is associated to hereditary gingival fibromatosis, a rare, autosomal dominant form of gingival overgrowth. A transgenic mouse construct with a comparable Sos1 chimera produces a phenotype with skin hypertrophy (Hart et al., 2002). Furthermore, the proline-rich, carboxyl-terminal region of Sos contains a number of phosphorylation sites for MAPK and p90 RSK-2 (Cherniack et al., 1994; Corbalan-Garcia et al., 1996; Douville and Downward, 1997; Rozakis-Adcock et al., 1995), and this phosphorylation may play a negative feedback role on the Ras pathway. Several studies have demonstrated an interaction between Sos1 and Intersectin (ITSN), an adaptor protein consisting of multiple modular domains (including two amino-terminal Eps15 homology domains, a central coiled-coil domain, and five carboxyl-terminal SH3 domains) (Tong et al., 2000a; Tong et al., 2000b). Indeed, overexpression of the SH3 domains of ITSN blocked EGF activation of both Ras and MAPK (Adams et al., 2000; Tong et al., 2000a), and ITSN stimulates Ras-GTP levels and physically complexes with Ras *in vivo* on intracellular vesicles (Mohney et al., 2003) likely through Sos binding to the SH3 domains of ITSN.

The amino-terminal region of Sos is approximately 600 amino acids long and contains regions of homology to Dbl (DH) and pleckstrin (PH) domains (Figure 4), whose structures have also been determined (Koshiba et al., 1997; Soisson et al., 1998; Zheng et al., 1997) and are thought to be involved, respectively, in Rac1 activation (Nimnual et al., 1998; Scita et al., 1999; Scita et al., 2001), and phospholipid binding (Chen et al., 1997). The connection Sos-Rac1 is mediated by a complex of Sos proteins with the molecular adaptors Eps8 and E3b1-Abi-1 (Scita et al., 1999; Scita et al., 2001). Two distinct types of Sos complexes, endowed respectively with Ras- (Sos-Grb2) and Rac-specific (Sos-E3b1-Eps8) GEF activity have been described under physiological *in vivo* conditions (Innocenti et al., 2002; Scita et al., 1999; Scita et al., 2001; Scita et al., 2000). Therefore, it appears that Sos proteins can be engaged in dual interactions, each leading to the activation of a different biological response. Thus, the Sos-Grb2 complex is disrupted upon RTK activation (Cherniack et al., 1994; Douville and Downward, 1997; Innocenti et al., 2002), whereas the Sos-E3b1-Eps8 complex is not (Innocenti et al., 2002). Furthermore, the activation of Ras by growth factors is short-lived, whereas the activation of Rac is sustained (Innocenti et al., 2002). The Sos-dependent Rac1 activation is negatively regulated by the interaction of the Sos DH domain with LC3, a microtubule-associated small protein (Furuta et al., 2002). In addition to DH and PH domains, the amino-terminal region of Sos1 contains a HF (Histone Folds) motif (with homology to histone H2A) located upstream of the DH domain (residues 1-191) (Figure 4) with two tandem histone folds (Jorge et al., 2002; Sondermann et al., 2003), which may be responsible for negative control of Sos1, probably mediated by intramolecular binding with the PH domain (Jorge et al., 2002). The Sos PH domain shows approximately 5-fold higher affinity for PI-3,4,5-P_3 than for PI-4,5-P_2 (Rameh et al., 1997), and the specific phosphoinositide composition in a local environment could induce conformational changes of the PH domain by such an interaction that could reduce the HF-PH binding affinity.

The functional role of the amino-terminal region of Sos is not yet absolutely understood. Some reports have suggested that this region may be responsible for positive regulation of overall Sos1 activity (Byrne et al., 1996; Karlovich et al., 1995; Wang et al., 1995), basically by collaborating to the plasma membrane localization of Sos1 protein. However, it has also been described that stable membrane association of Sos by addition of a myristoylation signal to this protein, but lacking the amino terminus region, is not sufficient for Sos to be biologically active (Qian et al., 1998). In contrast, others reports propose that the amino-terminal region of Sos1 is involved in negative regulation of its catalytic activity by exerting negative allosteric control on the interaction of the Sos catalytic domain with Ras (Corbalan-Garcia et al., 1998; Hall et al., 2002). Using truncated mutants, a recent study has demonstrated that this amino-terminal region is absolutely necessary for the hSos1 activity (Jorge et al., 2002), suggesting that the first 200 amino acid residues of Sos1, (corresponding to the HF motif) exert a negative control on the overall Sos1 functional activity through its specific interaction with the PH domain (Jorge et al., 2002). However, ectopic overexpression of the PH domain in mouse fibroblasts has not shown any inhibitory effect on the Ras pathway (Jorge et al., 2002), an observation consistent with previously described effects of injected, isolated PH domain peptides on the insulin dependent induction of germinal vesicle breakdown in *Xenopus* oocytes (Font de Mora et al., 1996).

Two distinct human Sos1 isoforms are known (designated hSos1 Isf I and Isf II) which differ only by the presence, in hSos1 Isf II, of a 15 amino acid stretch located close the first proline-rich motif required for Grb2 binding (Rojas et al., 1996). Some human tissues express only one isoform (fetal brain, and adult skeletal muscle, liver, lung and pancreas) whereas others express different proportions of both in fetal and adult developmental stages. *In vitro* binding assays and yeast two-hybrid analysis showed that hSos1 Isf II exhibits higher Grb2 affinity than hSos1 Isf I (Rojas et al., 1996). Furthermore, direct Ras guanine nucleotide exchange activity assays in cellular lysates showed that hSos1 Isf II transfectants consistently exhibited higher activity than hSos1 Isf I transfectants under unstimulated conditions (Rojas et al., 1999). hSos1 Isf II is also significantly more effective than hSos1 Isf I to induce transforming phenotype of NIH3T3 cells when transfected alone or in conjunction with normal H-Ras (Rojas et al., 1999). The isoform II-specific 15 amino acid stretch contains a SH3-minimal binding site (SH3-MBS) responsible for the higher Grb2 affinity and Ras signaling activity (Zarich et al., 2000).

3.5 Alternative Mechanisms of Ras Activation

An alternative mechanism of Ras activation has been proposed whereby nitric oxide (NO) modification of cysteine 118 results in Ras biological activation by stimulating GDP release (Lander et al., 1997; Mott et al., 1997; Rojas and Santos, 2002)

(Figure 6). The GTPase activity, GDP dissociation rate and structure of a C118S mutant (insensitive to NO modification) appear to be similar to wild type Ras (Mott et al., 1997). Stable nitrosylation of Cys118 has little effect on the nucleotide exchange, effector recognition, or structure of Ras. However, transient nitrosylation of Cys118 increases guanine nucleotide dissociation of Ras, thus resulting in accumulation of the biologically active Ras-GTP complex *in vivo*. A hypothesis accounting for the mechanism of Ras activation by NO is the destabilization of a crucial interaction between amino acid residues in the GDP-binding pocket and the bound nucleotide. According to this model, transient denitrosylation would compete with guanine nucleotide substrate interactions thereby facilitating guanine nucleotide dissociation and GTP-loading of Ras. Such a model would suggest that Ras function may also be regulated directly by changes in the redox state in the cell.

The cyclopentenone 15-deoxy-$\Delta^{12,14}$-prostaglandin J_2 (15d-PGJ_2) has also been reported to induce cell proliferation (Chinery et al., 1999; Oliva et al., 2003; Rovin et al., 2002; Shahabi et al., 1987) and ERK activation (Oliva et al., 2003; Wilmer et al., 2001) through a process mediated by 15d-PGJ_2-elicited H-Ras activation (Oliva et al., 2003) (Figure 6). Such an activation pathway appears to be specific for H-Ras through the formation of a covalent adduct of 15d-PGJ_2 with Cys184 of H-Ras, but not with N-Ras or K-Ras (Oliva et al., 2003). Cyclopentenone prostaglandins (CyPG) are naturally occurring eicosanoids displaying varied biological activities including antiviral (Santoro, 1997) and antitumoral (Kato et al., 1986) effects, modulation of the heat shock response (Rossi et al., 1997) and induction of oxidative stress (Kondo et al., 2001) and apoptosis (Kim et al., 1993). The CyPG of the J_2 series, such as 15d-PGJ_2, arise from the spontaneous dehydration of PGD_2, whereas PGA_2 is produced by PGE_2 dehydration.

CyPG are reactive compounds possessing an α, β-unsaturated carbonyl group in the cyclopentenone ring which is able to react with sulfhydril groups of cysteine residues of proteins by Michael's addition (Chen et al., 1999; Narumiya et al., 1987; Parker, 1996). A recent report indicates that mutation of Cys184 inhibits H-Ras modification and activation by 15d-PGJ_2, whereas serum-elicited stimulation is not affected (Oliva et al., 2003). It is not clear whether the 15d-PGJ_2 effects require the contribution of a specific Ras-GEF. In any event, these observations suggest a novel mechanism for activation of the Ras signaling pathway as a consequence of chemical modification of H-Ras through formation of a covalent adduct with cyclopentenone prostaglandins. PGA_1, which also possesses a cyclopentenone ring, forms an adduct with H-Ras, but not with PGE_2 or cPGI_2, which are not cyclopentenones (Oliva et al., 2003).

NO and CyPG generation has been linked to inflammatory processes (Gilroy et al., 1999), whose resolution may require the contribution of multiple mechanisms. Unravelling the relative contribution of NO and CyPG to the promotion of cell survival and growth may provide new clues to understanding specific pathological processes such as chronic inflammation and colon carcinogenesis.

4. CONCLUSIONS AND PERSPECTIVES

The Ras proteins are conserved molecular switches which, along evolution, have shown a proven ability to mediate internalization of a great variety of extracellular signals and to also elicit (activating different downstream pathways) a great variety of cellular responses including differentiation, proliferation or even cell death (Gille and Downward, 1999) (Figure 6).

Interestingly, the same basic Ras cycle, composed of the same Ras isoforms (H, N, K-Ras) and the same positive and negative regulators (GAPs and GEFs), has been conserved and used along the evolutionary lines to manage such a diversity of extracellular signals and related cellular responses. Given the simultaneous occurrence in most cells of the different players and regulators of the Ras cycle, one of the most interesting questions concerning signal transduction is understanding how specific signals lead to a specific responses through the action of the same Ras proteins.

The last fifteen years have produced rapid advances in this regard. Ras-GEF and Ras-GAP research has generated a rich and complex body of knowledge, revealing a divergence of answers that reflect the intricate physiological aspects of these proteins. Still, much remains yet to be known regarding the questions of specificity and redundancy involved in Ras mediated signaling. The next period will be definitive to find the final answer to the above functional questions. New tools, such as proteomics or genomics, are going to be key instruments for this purpose and for opening new perspectives in the Ras field.

ACKNOWLEDGEMENTS

This work was supported by MCYT-SAF2003-02604 grant to JMR and MSC-FIS PI021570, MCYT-SAF2003-04177 and JC&L-SA013/02 grants to ES.

REFERENCES

Adams, A., Thorn, J.M., Yamabhai, M., Kay, B.K. and O'Bryan, J.P. (2000) Intersectin, an adaptor protein involved in clathrin-mediated endocytosis, activates mitogenic signaling pathways. J. Biol. Chem. *275*, 27414-27420.

Ahmadian, M.R., Kiel, C., Stege, P. and Scheffzek, K. (2003) Structural fingerprints of the Ras-GTPase activating proteins neurofibromin and p120GAP. J. Mol. Biol. *329*, 699-710.

Allen, M., Chu, S., Brill, S., Stotler, C. and Buckler, A. (1998) Restricted tissue expression pattern of a novel human rasGAP-related gene and its murine ortholog. Gene *218*, 17-25.

Anborgh, P.H., Qian, X., Papageorge, A.G., Vass, W.C., DeClue, J.E. and Lowy, D.R. (1999) Ras-specific exchange factor GRF: oligomerization through its Dbl homology domain and calcium-dependent activation of Raf. Mol. Cell. Biol. *19*, 4611-4622.

Aronheim, A., Engelberg, L., Al-Alawi, N., Schlessinger, J. and Karin, M. (1994) Membrane targeting of the nucleotide exchange factor Sos is sufficient for activating the Ras signaling pathway. Cell *78*, 949-961.

Arozarena, I., Aaronson, D.S., Matallanas, D., Sanz, V., Ajenjo, N., Tenbaum, S.P., Teramoto, H., Ighishi, T., Zabala, J.C., Gutkind, J.S. and Crespo, P. (2000) The Rho family GTPase Cdc42 regulates the activation of Ras/MAP kinase by the exchange factor Ras-GRF. J. Biol. Chem. *275*, 26441-26448.

Arozarena, I., Matallanas, D. and Crespo, P. (2001) Maintenance of CDC42 GDP-bound state by Rho-GDI inhibits MAP kinase activation by the exchange factor Ras-GRF evidence for Ras-GRF function being inhibited by Cdc42-GDP but unaffected by CDC42-GTP. J. Biol. Chem. *276*, 21878-21884.

Arozarena, I., Matallanas, D., Berciano, M.T., Sanz-Moreno, V., Calvo, F., Munoz, M.T., Egea, G., Lafarga, M. and Crespo, P. (2004) Activation of H-Ras in the endoplasmic reticulum by the RasGRF family guanine nucleotide exchange factors. Mol. Cell. Biol. *24*, 1516-1530.

Bajenaru, M.L., Donahoe, J., Corral, T., Reilly, K.M., Brophy, S., Pellicer, A. and Gutmann, D.H. (2001) Neurofibromatosis 1 (NF1) heterozygosity results in a cell-autonomous growth advantage for astrocytes. Glia *33*, 314-323.

Ballester, R., Marchuk, D., Boguski, M., Saulino, A., Letcher, R., Wigler, M. and Collins, F. (1990) The NF1 locus encodes a protein functionally related to mammalian GAP and yeast IRA proteins. Cell *63*, 851-859.

Bernards, A. (2003) GAPs galore! A survey of putative Ras superfamily GTPase activating proteins in man and Drosophila. Biochim. Biophys. Acta *1603*, 47-82.

Bivona, T.G. and Philips, M.R. (2003) Ras pathway signaling on endomembranes. Curr. Opin. Cell Biol. *15*, 136-142.

Bivona, T.G., PereZ De Castro, I., Ahearn, I.M., Grana, T.M., Chiu, V.K., Lockyer, P.J., Cullen, P.J., Pellicer, A., Cox, A.D. and Philips, M.R. (2003) Phospholipase Cgamma activates Ras on the Golgi apparatus by means of RasGRP1. Nature *424*, 694-698.

Boguski, M. and McCormick, F. (1993) Proteins regulating Ras and its relatives. Nature *366*, 643-654.

Bollag, G., Clapp, D.W., Shih, S., Adler, F., Zhang, Y.Y., Thompson, P., Lange, B.J., Freedman, M.H., McCormick, F., Jacks, T. and Shannon, K. (1996) Loss of NF1 results in activation of the Ras signaling pathway and leads to aberrant growth in haematopoietic cells. Nat. Genet. *12*, 144-148.

Boriack-Sjodin, P.A., Margarit, S.M., Bar-Sagi, D. and Kuriyan, J. (1998) The structural basis of the activation of Ras by Sos [see comments]. Nature *394*, 337-343.

Bowtell, D., Fu, P., Simon, M. and Senior, P. (1992) Identification of murine homologues of the Drosophila son of sevenless gene: potential activators of ras. Proc. Natl. Acad. Sci. U.S.A. *89*, 6511-6515.

Brambilla, R., Gnesutta, N., Minichiello, L., White, G., Roylance, A.J., Herron, C.E., Ramsey, M., Wolfer, D.P., Cestari, V., Rossi-Arnaud, C., et al. (1997) A role for the Ras signalling pathway in synaptic transmission and long-term memory. Nature *390*, 281-286.

Brownbridge, G.G., Lowe, P.N., Moore, K.J., Skinner, R.H. and Webb, M.R. (1993) Interaction of GTPase activating proteins (GAPs) with p21ras measured by a novel fluorescence anisotropy method. Essential role of Arg-903 of GAP in activation of GTP hydrolysis on p21ras. J. Biol. Chem. *268*, 10914-10919.

Buchsbaum, R., Telliez, J.B., Goonesekera, S. and Feig, L.A. (1996) The N-terminal pleckstrin, coiled-coil, and IQ domains of the exchange factor Ras-GRF act cooperatively to facilitate activation by calcium. Mol. Cell. Biol. *16*, 4888-4896.

Byrne, J.L., Paterson, H.F. and Marshall, C.J. (1996) p21Ras activation by the guanine nucleotide exchange factor Sos, requires the Sos/Grb2 interaction and a second ligand-dependent signal involving the Sos N-terminus. Oncogene *13*, 2055-2065.

Caloca, M.J., Zugaza, J.L. and Bustelo, X.R. (2003a) Exchange factors of the RasGRP family mediate Ras activation in the Golgi. J. Biol. Chem. *278*, 33465-33473.

Caloca, M.J., Zugaza, J.L., Matallanas, D., Crespo, P. and Bustelo, X.R. (2003b) Vav mediates Ras stimulation by direct activation of the GDP/GTP exchange factor Ras GRP1. EMBO J. *22*, 3326-3336.

Caloca, M.J., Zugaza, J.L., Vicente-Manzanares, M., Sanchez-Madrid, F. and Bustelo, X.R. (2004) F-actin-dependent translocation of the Rap1 GDP/GTP exchange factor RasGRP2. J Biol Chem. *279*, 20435-20446.

Casci., T., Vinos, J. and Freeman, M. (1999) Sprouty, an intracellular inhibitor of Ras signaling. Cell *96*, 655-665.

Cawthon, R.M., Weiss, R., Xu, G.F., Viskochil, D., Culver, M., Stevens, J., Robertson, M., Dunn, D., Gesteland, R., O'Connell, P. et al. (1990) A major segment of the neurofibromatosis type 1 gene: cDNA sequence, genomic structure, and point mutations. Cell *62*, 193-201.

Cen, H., Papageorge, A.G., Zippel, R., Lowy, D.R. and Zhang, K. (1992) Isolation of multiple mouse cDNAs with coding homology to Saccharomyces cerevisiae CDC25: identification of a region related to Bcr, Vav, Dbl and CDC24. EMBO J. *11*, 4007-4015.

Chang, C., Hopper, N.A. and Sternberg, P.W. (2000) Caenorhabditis elegans SOS-1 is necessary for multiple RAS-mediated developmental signals. EMBO J. *19*, 3283-3294.

Chardin, P., Camonis, J.H., Gale, N.W., van Aelst, L., Schlessinger, J., Wigler, M.H. and Bar-Sagi, D. (1993) Human Sos1: a guanine nucleotide exchange factor for Ras that binds to GRB2. Science *260*, 1338-1343.

Chardin, P. and Mattei, M.G. (1994) Chromosomal localization of two genes encoding human ras exchange factors: SOS1 maps to the 2p22 → p16 region and SOS2 to the 14q21 → q22 region of the human genome. Cytogenet. Cell Genet. *66*, 68-69.

Chen, R.H., Corbalan-Garcia, S. and Bar-Sagi, D. (1997) The role of the PH domain in the signal-dependent membrane targeting of Sos. EMBO J. *16*, 1351-1359.

Chen, Y., Morrow, J.D. and Roberts, L.J. 2nd (1999) Formation of reactive cyclopentenone compounds in vivo as products of the isoprostane pathway. J. Biol. Chem. *274*, 10863-10868.

Cherniack, A., Klarlund, J. and Czech, M. (1994) Phosphorylation of the Ras nucleotide exchange factor son of sevenless by mitogen-activated protein kinase. J. Biol. Chem. *269*, 4717-4720.

Chinery, R., Coffey, R.J., Graves-Deal, R., Kirkland, S.C., Sanchez, S.C., Zackert, W.E., Oates, J.A. and Morrow, J.D. (1999) Prostaglandin J2 and 15-deoxy-delta12,14-prostaglandin J2 induce proliferation of cyclooxygenase-depleted colorectal cancer cells. Cancer Res. *59*, 2739-2746.

Chow, A. and Gawler, D. (1999) Mapping the site of interaction between annexin VI and the p120GAP C2 domain. FEBS Lett. *460*, 166-172.

Chow, A., Davis, A.J. and Gawler, D.J. (1999) Investigating the role played by protein-lipid and protein-protein interactions in the membrane association of the p120GAP CaLB domain. Cell. Signal. *11*, 443-451.

Clark, J.D., Schievella, A.R., Nalefski, E.A. and Lin, L.L. (1995) Cytosolic phospholipase A2. J. Lipid Mediat. Cell. Signal. *12*, 83-117.

Clyde-Smith, J., Silins, G., Gartside, M., Grimmond, S., Etheridge, M., Apolloni, A., Hayward, N. and Hancock, J.F. (2000) Characterization of RasGRP2, a plasma membrane-targeted, dual specificity Ras/Rap exchange factor. J. Biol. Chem. *275*, 32260-32267.

Corbalan-Garcia, S., Yang, S.S., Degenhardt, K.R. and Bar-Sagi, D. (1996) Identification of the mitogen-activated protein kinase phosphorylation sites on human Sos1 that regulate interaction with Grb2. Mol. Cell. Biol. *16*, 5674-5682.

Corbalan-Garcia, S., Margarit, S.M., Galron, D., Yang, S.S. and Bar-Sagi, D. (1998) Regulation of Sos activity by intramolecular interactions. Mol. Cell. Biol. *18*, 880-886.

Costa, R.M., Yang, T., Huynh, D.P., Pulst, S.M., Viskochil, D.H., Silva, A.J. and Brannan, C.I. (2001) Learning deficits, but normal development and tumor predisposition, in mice lacking exon 23a of Nf1. Nat. Genet. *27*, 399-405.

Cozier, G.E., Lockyer, P.J., Reynolds, J.S., Kupzig, S., Bottomley, J.R., Millard, T.H., Banting, G. and Cullen, P.J. (2000) GAP1IP4BP contains a novel group I pleckstrin homology domain that directs constitutive plasma membrane association. J. Biol. Chem. *275*, 28261-28268.

Cullen, P.J., Hsuan, J.J., Truong, O., Letcher, A.J., Jackson, T.R., Dawson, A.P. and Irvine, R.F. (1995) Identification of a specific Ins (1,3,4,5) P4-binding protein as a member of the GAP1 family. Nature *376*, 527-530.

Cullen, P.J. and Lockyer, P.J. (2002) Integration of calcium and Ras signalling. Nat. Rev. Mol. Cell. Biol. *3*, 339-348.

Dasgupta, B. and Gutmann, D.H. (2003) Neurofibromatosis 1: closing the GAP between mice and men. Curr. Opin. Genet. Dev. *13*, 20-27.

de Hoog, C.L., Koehler, J.A., Goldstein, M.D., Taylor, P., Figeys, D. and Moran, M.F. (2001) Ras binding triggers ubiquitination of the Ras exchange factor Ras-GRF2. Mol. Cell. Biol. *21*, 2107-2117.

Douville, E. and Downward, J. (1997) EGF induced SOS phosphorylation in PC12 cells involves P90 RSK-2. Oncogene *15*, 373-383.

Dower, N.A., Stang, S.L., Bottorff, D.A., Ebinu, J.O., Dickie, P., Ostergaard, H.L. and Stone, J.C. (2000) RasGRP is essential for mouse thymocyte differentiation and TCR signaling. Nat. Immunol. *1*, 317-321.

Downward, J. (1994) The GRB2/Sem-5 adaptor protein. FEBS Lett. *338*, 113-117.

Downward, J. (1996) Control of ras activation. Cancer Surv. *27*, 87-100.

Downward, J. (2003) Targeting RAS signalling pathways in cancer therapy. Nat. Rev. Cancer *3*, 11-22.

Duchesne, M., Schweighoffer, F., Parker, F., Clerc, F., Frobert, Y., Thang, M.N. and Tocque, B. (1993) Identification of the SH3 domain of GAP as an essential sequence for Ras-GAP-mediated signaling. Science *259*, 525-528.

Ebinu, J.O., Bottorff, D.A., Chan, E.Y., Stang, S.L., Dunn, R.J. and Stone, J.C. (1998) RasGRP, a Ras guanyl nucleotide-releasing protein with calcium- and diacylglycerol-binding motifs. Science *280*, 1082-1086.

Egan, S. and Weinberg, R. (1993) The pathway to signal achievement. Nature *365*, 781-783.

Ellis, C., Measday, V. and Moran, M.F. (1995) Phosphorylation-dependent complexes of p120 Ras-specific GTPase-activating protein with p62 and p190. Methods Enzymol. *255*, 179-192.

Esteban, L.M., Fernandez-Medarde, A., Lopez, E., Yienger, K., Guerrero, C., Ward, J.M., Tessarollo, L. and Santos, E. (2000) Ras-guanine nucleotide exchange factor sos2 is dispensable for mouse growth and development. Mol. Cell. Biol. *20*, 6410-6413.

Fam, N.P., Fan, W.T., Wang, Z., Zhang, L.J., Chen, H. and Moran, M.F. (1997) Cloning and characterization of Ras-GRF2, a novel guanine nucleotide exchange factor for Ras. Mol. Cell. Biol. *17*, 1396-1406.

Fan, W.-T., Koch, C., de Hoog, C., Fam, N. and Moran, M. (1998) The exchange factor Ras-GRF2 activates Ras-dependent and Rac-dependent mitogen-activated protein kinase pathways. Curr. Biol. *8*, 935-938.

Farnsworth, C.L., Freshney, N.W., Rosen, L.B., Ghosh, A., Greenberg, M.E. and Feig, L.A. (1995) Calcium activation of Ras mediated by neuronal exchange factor Ras-GRF. Nature *376*, 524-527.

Feig, L. (1994) Guanine nucleotide exchange factors: a family of positive regulators of Ras and related GTPases. Curr. Opin. Cell. Biol. *6*, 204-211.

Feldmann, P., Eicher, E.N., Leevers, S.J., Hafen, E. and Hughes, D.A. (1999) Control of growth and differentiation by drosophila RasGAP, a homolog of p120 ras-GTPase-activating protein. Mol. Cell. Biol. *19*, 1928-1937.

Feng, S. and Chen, J.K., YuH., Simon, J.A. and Schreiber, S.L. (1994) Two binding orientations for peptides to the Src SH3 domain: development of a general model for SH3–ligand interactions. Science *266*, 1241-1247.

Fernandez-Medarde, A., Esteban, L.M., Nunez, A., Porteros, A., Tessarollo, L. and Santos, E. (2002) Targeted disruption of Ras-Grf2 shows its dispensability for mouse growth and development. Mol. Cell. Biol. *22*, 2498-2504.

Filvaroff, E., Calautti, E., McCormick, F. and Dotto, G.P. (1992) Specific changes of Ras GTPase-activating protein (GAP) and a GAP-associated p62 protein during calcium-induced keratinocyte differentiation. Mol. Cell. Biol. *12*, 5319-5328.

Font de Mora, J., Guerrero, C., Mahadevan, D., Coque, J., Rojas, J., Esteban, L., Rebecchi, M. and Santos, E. (1996) Isolated Sos1 PH domain exhibits germinal vesicle breakdown-inducing activity in Xenopus oocytes. JBC *271*, 18272-18276.

Font de Mora, J., Esteban, L.M., Burks, D.J., Nunez, A., Garces, C., Garcia-Barrado, M.J., Iglesias-Osma, M.C., Moratinos, J., Ward, J.M. and Santos, E. (2003) Ras-GRF1 signaling is required for normal beta-cell development and glucose homeostasis. EMBO J. *22*, 3039-3049.

Friedman, E., Gejman, P.V., Martin, G.A. and McCormick, F. (1993) Nonsense mutations in the C-terminal SH2 region of the GTPase activating protein (GAP) gene in human tumours. Nat. Genet. *5*, 242-247.

Furuta, S., Miura, K., Copeland, T., Shang, W.H., Oshima, A. and Kamata, T. (2002) Light Chain 3 associates with a Sos1 guanine nucleotide exchange factor: its significance in the Sos1-mediated Rac1 signaling leading to membrane ruffling. Oncogene *21*, 7060-7066.

Gaul, U., Mardon, G. and Rubin, G.M. (1992) A putative Ras GTPase activating protein acts as a negative regulator of signaling by the Sevenless receptor tyrosine kinase. Cell *68*, 1007-1019.

Gawler, D.J., Zhang, L.J. and Moran, M.F. (1995a) Mutation-deletion analysis of a Ca(2+)-dependent phospholipid binding (CaLB) domain within p120 GAP, a GTPase-activating protein for p21 ras. Biochem. J. *307* (Pt 2), 487-491.

Gawler, D.J., Zhang, L.J., Reedijk, M., Tung, P.S. and Moran, M.F. (1995b) CaLB: a 43 amino acid calcium-dependent membrane/phospholipid binding domain in p120 Ras GTPase-activating protein. Oncogene *10*, 817-825.

Giese, K.P., Friedman, E., Telliez, J.B., Fedorov, N.B., Wines, M., Feig, L.A. and Silva, A.J. (2001) Hippocampus-dependent learning and memory is impaired in mice lacking the Ras-guanine-nucleotide releasing factor 1 (Ras-GRF1). Neuropharmacology *41*, 791-800.

Gille, H. and Downward, J. (1999) Multiple ras effector pathways contribute to G(1) cell cycle progression. J. Biol. Chem. *274*, 22033-22040.

Gilroy, D.W., Colville-Nash, P.R., Willis, D., Chivers, J., Paul-Clark, M.J. and Willoughby, D.A. (1999) Inducible cyclooxygenase may have anti-inflammatory properties. Nat. Med. *5*, 698-701.

Glover, T.W., Stein, C.K., Legius, E., Andersen, L.B., Brereton, A. and Johnson, S. (1991) Molecular and cytogenetic analysis of tumors in von Recklinghausen neurofibromatosis. Genes Chromosomes Cancer *3*, 62-70.

Gnesutta, N., Ceriani, M., Innocenti, M., Mauri, I., Zippel, R., Sturani, E., Borgonovo, B., Berruti, G. and Martegani, E. (2001) Cloning and characterization of mouse UBPy, a deubiquitinating enzyme that interacts with the ras guanine nucleotide exchange factor CDC25(Mm)/Ras-GRF1. J. Biol. Chem. *276*, 39448-39454.

Gotoh, T., Tian, X. and Feig, L.A. (2001) Prenylation of target GTPases contributes to signaling specificity of Ras-guanine nucleotide exchange factors. J. Biol. Chem. *276*, 38029-38035.

Guerrero, C., Rojas, J.M., Chedid, M., Esteban, L.M., Zimonjic, D.B., Popescu, N.C., Font de Mora, J. and Santos, E. (1996) Expression of alternative forms of Ras exchange factors GRF and SOS1 in different human tissues and cell lines. Oncogene *12*, 1097-1107.

Guo, H.F., The, I., Hannan, F., Bernards, A. and Zhong, Y. (1997) Requirement of Drosophila NF1 for activation of adenylyl cyclase by PACAP38-like neuropeptides. Science *276*, 795-798.

Gutmann, D.H. (1998) Recent insights into neurofibromatosis type 1: clear genetic progress. Arch. Neurol. *55*, 778-780.

Gutmann, D.H. and Collins, F.S. (1992) Recent progress toward understanding the molecular biology of von Recklinghausen neurofibromatosis. Ann. Neurol. *31*, 555-561.

Gutmann, D.H. and Collins, F.S. (1993) The neurofibromatosis type 1 gene and its protein product, neurofibromin. Neuron *10*, 335-343.

Gutmann, D.H., Loehr, A., Zhang, Y., Kim, J., Henkemeyer, M. and Cashen, A. (1999) Haploinsufficiency for the neurofibromatosis 1 (NF1) tumor suppressor results in increased astrocyte proliferation. Oncogene *18*, 4450-4459.

Gutmann, D.H., Wu, Y.L., Hedrick, N.M., Zhu, Y., Guha, A. and Parada, L.F. (2001) Heterozygosity for the neurofibromatosis 1 (NF1) tumor suppressor results in abnormalities in cell attachment, spreading and motility in astrocytes. Hum. Mol. Genet. *10*, 3009-3016.

Hall, B.E., Yang, S.S., Boriack-Sjodin, P.A., Kuriyan, J. and Bar-Sagi, D. (2001) Structure-based mutagenesis reveals distinct functions for Ras switch 1 and switch 2 in Sos-catalyzed guanine nucleotide exchange. J. Biol. Chem. *276*, 27629-27637.

Hall, B.E., Yang, S.S. and Bar-Sagi, D. (2002) Autoinhibition of Sos by intramolecular interactions. Front Biosci. *7*, d288-d94.

Hancock, J.F. (2003) Ras proteins: different signals from different locations. Nat. Rev. Mol. Cell. Biol. *4*, 373-384.

Hart, T.C., Zhang, Y., Gorry, M.C., Hart, P.S., Cooper, M., Marazita, M.L., Marks, J.M., Cortelli, J.R. and Pallos, D. (2002) A mutation in the SOS1 gene causes hereditary gingival fibromatosis type 1. Am. J. Hum. Genet. *70*, 943-954.

Henkemeyer, M., Rossi, D.J., Holmyard, D.P., Puri, M.C., Mbamalu, G., Harpal, K., Shih, T.S., Jacks, T. and Pawson, T. (1995) Vascular system defects and neuronal apoptosis in mice lacking ras GTPase-activating protein. Nature *377*, 695-701.

Hiatt, K.K., Ingram, D.A., Zhang, Y., Bollag, G. and Clapp, D.W. (2001) Neurofibromin GTPase-activating protein-related domains restore normal growth in Nf1-/- cells. J. Biol. Chem. *276*, 7240-7245.

Holsinger, L.J., Spencer, D.M., Austin, D.J., Schreiber, S.L. and Crabtree, G.R. (1995) Signal transduction in T lymphocytes using a conditional allele of Sos. Proc. Natl. Acad. Sci. U.S.A. *92*, 9810-9814.

Innocenti, M., Zippel, R., Brambilla, R. and Sturani, E. (1999) CDC25 (Mm)/Ras-GRF1 regulates both Ras and Rac signaling pathways. FEBS Lett. *460*, 357-362.

Innocenti, M., Tenca, P., Frittoli, E., Faretta, M., Tocchetti, A., Di Fiore, P.P. and Scita, G. (2002) Mechanisms through which Sos-1 coordinates the activation of Ras and Rac. J. Cell Biol. *156*, 125-136.

Jaumot, M., Yan, J., Clyde-Smith, J., Sluimer, J. and Hancock, J.F. (2002) The linker domain of the Ha-Ras hypervariable region regulates interactions with exchange factors, Raf-1 and phosphoinositide 3-kinase. J. Biol. Chem. *277*, 272-278.

Jiang, Y., Ma, W., Wan, Y., Kozasa, T., Hattori, S. and Huang, X.Y. (1998) The G protein G alpha12 stimulates Bruton's tyrosine kinase and a rasGAP through a conserved PH/BM domain. Nature *395*, 808-813.

Johnson, M.R., DeClue, J.E., Felzmann, S., Vass, W.C., Xu, G., White, R. and Lowy, D.R. (1994) Neurofibromin can inhibit Ras-dependent growth by a mechanism independent of its GTPase-accelerating function. Mol. Cell. Biol. *14*, 641-645.

Jones, M.K. and Jackson, J.H. (1998) Ras-GRF activates Ha-Ras, but not N-Ras or K-Ras 4B, protein in vivo. J. Biol. Chem. *273*, 1782-1787.

Jorge, R., Zarich, N., Oliva, J.L., Azanedo, M., Martinez, N., de la Cruz, X. and Rojas, J.M. (2002) HSos1 contains a new amino-terminal regulatory motif with specific binding affinity for its pleckstrin homology domain. J. Biol. Chem. *277*, 44171-44179.

Karlovich, C.A., Bonfini, L., McCollam, L., Rogge, R.D., Daga, A., Czech, M.P. and Banerjee, U. (1995) In vivo functional analysis of the Ras exchange factor son of sevenless. Science *268*, 576-579.

Kato, T., Fukushima, M., Kurozumi, S. and Noyori, R. (1986) Antitumor activity of delta 7-prostaglandin A1 and delta 12-prostaglandin J2 in vitro and in vivo. Cancer Res. *46*, 3538-3542.

Kawasaki, H., Springett, G.M., Toki, S., Canales, J.J., Harlan, P., Blumenstiel, J.P., Chen, E.J., Bany, I.A., Mochizuki, N., Ashbacher, A., et al. (1998) A Rap guanine nucleotide exchange factor enriched highly in the basal ganglia. Proc. Natl. Acad. Sci. U.S.A. *95*, 13278-13283.

Kim, I.K., Lee, J.H., Sohn, H.W., Kim, H.S. and Kim, S.H. (1993) Prostaglandin A2 and delta 12-prostaglandin J2 induce apoptosis in L1210 cells. FEBS Lett. *321*, 209-214.

Kim, J.H., Shirouzu, M., Kataoka, T., Bowtell, D. and Yokoyama, S. (1998) Activation of Ras and its downstream extracellular signal-regulated protein kinases by the CDC25 homology domain of mouse Son-of-sevenless 1 (mSos1). Oncogene *16*, 2597-2607.

Kiyono, M., Satoh, T. and Kaziro, Y. (1999) G protein beta gamma subunit-dependent Rac-guanine nucleotide exchange activity of Ras-GRF1/CDC25(Mm). Proc. Natl. Acad. Sci. U.S.A. *96*, 4826-4831.

Kiyono, M., Kato, J., Kataoka, T., Kaziro, Y. and Satoh, T. (2000a) Stimulation of Ras guanine nucleotide exchange activity of Ras-GRF1/CDC25(Mm) upon tyrosine phosphorylation by the Cdc42-regulated kinase ACK1. J. Biol. Chem. *275*, 29788-29793.

Kiyono, M., Kaziro, Y. and Satoh, T. (2000b) Induction of rac-guanine nucleotide exchange activity of Ras-GRF1/CDC25(Mm) following phosphorylation by the nonreceptor tyrosine kinase Src. J. Biol. Chem. *275*, 5441-5446.

Koehler, J.A. and Moran, M.F. (2001a) RACK1, a protein kinase C scaffolding protein, interacts with the PH domain of p120GAP. Biochem. Biophys. Res. Commun. *283*, 888-895.

Koehler, J.A. and Moran, M.F. (2001b) Regulation of extracellular signal-regulated kinase activity by p120 RasGAP does not involve its pleckstrin homology or calcium-dependent lipid binding domains but does require these domains to regulate cell proliferation. Cell Growth Differ. *12*, 551-561.

Kondo, M., Oya-Ito, T., Kumagai, T., Osawa, T. and Uchida, K. (2001) Cyclopentenone prostaglandins as potential inducers of intracellular oxidative stress. J. Biol. Chem. *276*, 12076-12083.

Koshiba, S., Kigawa, T., Kim, J.H., Shirouzu, M., Bowtell, D. and Yokoyama, S. (1997) The solution structure of the pleckstrin homology domain of mouse Son-of-sevenless 1 (mSos1). J. Mol. Biol. *269*, 579-591.

Lander, H.M., Hajjar, D.P., Hempstead, B.L., Mirza, U.A., Chait, B.T., Campbell, S. and Quilliam, L.A. (1997) A molecular redox switch on p21(ras). Structural basis for the nitric oxide-p21(ras) interaction. J. Biol. Chem. *272*, 4323-4326.

Largaespada, D.A., Brannan, C.I., Jenkins, N.A. and Copeland, N.G. (1996) Nf1 deficiency causes Ras-mediated granulocyte/macrophage colony stimulating factor hypersensitivity and chronic myeloid leukaemia. Nat. Genet. *12*, 137-143.

Li, B., Subleski, M., Fusaki, N., Yamamoto, T., Copeland, T., Princler, G.L., Kung, H. and Kamata, T. (1996) Catalytic activity of the mouse guanine nucleotide exchanger mSOS is activated by Fyn tyrosine protein kinase and the T-cell antigen receptor in T cells. Proc. Natl. Acad. Sci. U.S.A. *93*, 1001-1005.

Li, B.Q., Subleski, M., Shalloway, D., Kung, H.F. and Kamata, T. (1993a) Mitogenic activation of the Ras guanine nucleotide exchange factor in NIH 3T3 cells involves protein tyrosine phosphorylation. Proc. Natl. Acad. Sci. U.S.A. *90*, 8504-8508.

Li, B.Q., Wang, M.H., Kung, H.F., Ronsin, C., Breathnach, R., Leonard, E.J. and Kamata, T. (1995) Macrophage-stimulating protein activates Ras by both activation and translocation of SOS nucleotide exchange factor. Biochem. Biophys. Res. Commun. *216*, 110-118.

Li, N., Batzer, A., Daly, R., Yajnik, V., Skolnik, E., Chardin, P., Bar-Sagi, D., Margolis, B. and Schlessinger, J. (1993b) Guanine-nucleotide-releasing factor hSos1 binds to Grb2 and links receptor tyrosine kinases to Ras signalling. Nature *363*, 85-88.

Lim, W.A. and Richards, F.M. (1994) Critical residues in an SH3 domain from Sem-5 suggest a mechanism for proline-rich peptide recognition. Nat. Struct. Biol. *1*, 221-225.

Lockyer, P.J., Bottomley, J.R., Reynolds, J.S., McNulty, T.J., Venkateswarlu, K., Potter, B.V., Dempsey, C.E. and Cullen, P.J. (1997) Distinct subcellular localisations of the putative inositol 1,3,4,5-tetrakisphosphate receptors GAP1IP4BP: and GAP1m result from the GAP1IP4BP PH domain directing plasma membrane targeting. Curr. Biol. *7*, 1007-1010.

Lockyer, P.J., Wennstrom, S., Kupzig, S., Venkateswarlu, K., Downward, J. and Cullen, P.J. (1999) Identification of the ras GTPase-activating protein GAP1(m) as a phosphatidylinositol-3,4,5-trisphosphate-binding protein in vivo. Curr. Biol. *9*, 265-268.

Lockyer, P.J., Kupzig, S. and Cullen, P.J. (2001) CAPRI regulates Ca(2+)-dependent inactivation of the Ras-MAPK pathway. Curr. Biol. *11*, 981-986.

Maekawa, M., Nakamura, S. and Hattori, S. (1993) Purification of a novel ras GTPase-activating protein from rat brain. J. Biol. Chem. *268*, 22948-22952.

Maekawa, M., Li, S., Iwamatsu, A., Morishita, T., Yokota, K., Imai, Y., Kohsaka, S., Nakamura, S. and Hattori, S. (1994) A novel mammalian Ras GTPase-activating protein which has phospholipid-binding and Btk homology regions. Mol. Cell. Biol. *14*, 6879-6885.

Mangues, R., Corral, T., Lu, S., Symmans, W.F., Liu, L. and Pellicer, A. (1998) NF1 inactivation cooperates with N-ras in in vivo lymphogenesis activating Erk by a mechanism independent of its Ras-GTPase accelerating activity. Oncogene *17*, 1705-1716.

Margarit, S.M., Sondermann, H., Hall, B.E., Nagar, B., Hoelz, A., Pirruccello, M., Bar-Sagi, D. and Kuriyan, J. (2003) Structural evidence for feedback activation by Ras.GTP of the Ras-specific nucleotide exchange factor SOS. Cell *112*, 685-695.

Martegani, E., Vanoni, M., Zippel, R., Coccetti, P., Brambilla, R., Ferrari, C., Sturani, E. and Alberghina, L. (1992) Cloning by functional complementation of a mouse cDNA encoding a homologue of CDC25, a Saccharomyces cerevisiae RAS activator. EMBO J. *11*, 2151-2157.

Martin, G.A., Viskochil, D., Bollag, G., McCabe, P.C., Crosier, W.J., Haubruck, H., Conroy, L., Clark, R., O'Connell, P., Cawthon, R.M. et al. (1990) The GAP-related domain of the neurofibromatosis type 1 gene product interacts with ras p21. Cell *63*, 843-849.

Matallanas, D., Arozarena, I., Berciano, M.T., Aaronson, D.S., Pellicer, A., Lafarga, M. and Crespo, P. (2003) Differences on the inhibitory specificities of H-Ras, K-Ras, and N-Ras (N17) dominant negative mutants are related to their membrane microlocalization. J. Biol. Chem. *278*, 4572-4581.

Mattingly, R.R. (1999) Phosphorylation of serine 916 of Ras-GRF1 contributes to the activation of exchange factor activity by muscarinic receptors. J. Biol. Chem. *274*, 37379-37384.

Mattingly, R.R. and Macara, I.G. (1996) Phosphorylation-dependent activation of the Ras-GRF/CDC25Mm exchange factor by muscarinic receptors and G-protein beta gamma subunits. Nature *382*, 268-272.

McCollam, L., Bonfini, L., Karlovich, C.A., Conway, B.R., Kozma, L.M., Banerjee, U. and Czech, M.P. (1995) Functional roles for the pleckstrin and Dbl homology regions in the Ras exchange factor Son-of-sevenless. J. Biol. Chem. *270*, 15954-15957.

McCormick, F. (1993) How receptors turn Ras on. Nature *363*, 15-16.

McCormick, F. (1998) Going for the GAP. Curr. Biol. *8*, R673-R674.

Mohney, R.P., Das, M., Bivona, T.G., Hanes, R., Adams, A.G., Philips, M.R. and O'Bryan, J.P. (2003) Intersectin activates Ras but stimulates transcription through an independent pathway involving JNK. J. Biol. Chem. *278*, 47038-47045.

Mott, H.R., Carpenter, J.W. and Campbell, S.L. (1997) Structural and functional analysis of a mutant Ras protein that is insensitive to nitric oxide activation. Biochemistry *36*, 3640-3644.

Narumiya, S., Ohno, K., Fukushima, M. and Fujiwara, M. (1987) Site and mechanism of growth inhibition by prostaglandins. III. Distribution and binding of prostaglandin A2 and delta 12-prostaglandin J2 in nuclei. J. Pharmacol. Exp. Ther. *242*, 306-311.

Nielsen, K.H., Papageorge, A.G., Vass, W.C., Willumsen, B.M. and Lowy, D.R. (1997) The Ras-specific exchange factors mouse Sos1 (mSos1) and mSos2 are regulated differently: mSos2 contains ubiquitination signals absent in mSos1. Mol. Cell. Biol. *17*, 7132-7138.

Nimnual, A.S., Yatsula, B.A. and Bar-Sagi, D. (1998) Coupling of Ras and Rac guanosine triphosphatases through the Ras exchanger Sos. Science *279*, 560-563.

Ohba, Y., Mochizuki, N., Yamashita, S., Chan, A.M., Schrader, J.W., Hattori, S., Nagashima, K. and Matsuda, M. (2000) Regulatory proteins of R-Ras, TC21/R-Ras2, and M-Ras/R-Ras3. J. Biol. Chem. *275*, 20020-20026.

Oh-hora, M., Johmura, S., Hashimoto, A., Hikida, M. and Kurosaki, T. (2003) Requirement for Ras guanine nucleotide releasing protein 3 in coupling phospholipase C-gamma2 to Ras in B cell receptor signaling. J. Exp. Med. *198*, 1841-1851.

Oliva, J.L., Perez-Sala, D., Castrillo, A., Martinez, N., Canada, F.J., Bosca, L. and Rojas, J.M. (2003) The cyclopentenone 15-deoxy-delta 12,14-prostaglandin J2 binds to and activates H-Ras. Proc. Natl. Acad. Sci. U.S.A. *100*, 4772-4777.

Parker, J. (1996) Prostaglandin A_2 protein interactions and inhibition of cellular proliferation. Prostaglandins *50*, 359-375.

Pawson, T. and Schlessinger, J. (1993) SH2 and SH3 domains. Curr. Biol. *3*, 434-442.

Qian, X., Vass, W.C., Papageorge, A.G., Anborgh, P.H. and Lowy, D.R. (1998) N terminus of Sos1 Ras exchange factor: critical roles for the Dbl and pleckstrin homology domains. Mol. Cell. Biol. *18*, 771-778.

Qian, X., Esteban, L., Vass, W.C., Upadhyaya, C., Papageorge, A.G., Yienger, K., Ward, J.M., Lowy, D.R. and Santos, E. (2000) The Sos1 and Sos2 Ras-specific exchange factors: differences in placental expression and signaling properties. EMBO J. *19*, 642-654.

Quilliam, L.A., Huff, S.Y., Rabun, K.M., Wei, W., Park, W., Broek, D. and Der, C.J. (1994) Membrane-targeting potentiates guanine nucleotide exchange factor CDC25 and SOS1 activation of Ras transforming activity. Proc. Natl. Acad. Sci. U.S.A. *91*, 8512-8516.

Quilliam, L.A., Khosravi-Far, R., Huff, S.Y., and Der, C.J. (1995) Guanine nucleotide exchange factors: activators of the Ras superfamily of proteins. Bioessays *17*, 395-404.

Quilliam, L.A., Rebhun, J.F. and Castro, A.F. (2002) A growing family of guanine nucleotide exchange factors is responsible for activation of Ras-family GTPases. Prog. Nucleic Acid Res. Mol. Biol. *71*, 391-444.

Rameh, L.E., Arvidsson, A., Carraway, K.L., 3rd, Couvillon, A.D., Rathbun, G., Crompton, A., VanRenterghem, B., Czech, M.P., Ravichandran, K.S., Burakoff, S.J. et al. (1997) A comparative analysis of the phosphoinositide binding specificity of pleckstrin homology domains. J. Biol. Chem. *272*, 22059-22066.

Resat, H., Straatsma, T.P., Dixon, D.A. and Miller, J.H. (2001) The arginine finger of RasGAP helps Gln-61 align the nucleophilic water in GAP-stimulated hydrolysis of GTP. Proc. Natl. Acad. Sci. U.S.A. *98*, 6033-6038.

Reuther, G.W., Lambert, Q.T., Rebhun, J.F., Caligiuri, M.A., Quilliam, L.A. and Der, C.J. (2002) RasGRP4 is a novel Ras activator isolated from acute myeloid leukemia. J. Biol. Chem. *277*, 30508-30514.

Rojas, J.M., Coque, J.J.R., Guerrero, C., Aroca, P., Font de Mora, J., de la Cruz, X., Lorenzi, M.V., Esteban, L.M. and Santos, E. (1996) A 15 amino acid stretch close to the Grb2-binding domain defines two differentially expressed hSos1 isoforms with markedly different Grb2 binding affinity and biological activity. Oncogene *12*, 2291-2300.

Rojas, J.M., Subleski, M., Coque, J.J., Guerrero, C., Saez, R., Li, B.Q., Lopez, E., Zarich, N., Aroca, P., Kamata, T. and Santos, E. (1999) Isoform-specific insertion near the Grb2-binding domain modulates the intrinsic guanine nucleotide exchange activity of hSos1. Oncogene *18*, 1651-1661.

Rojas, J.M. and Santos, E. (2002) Ras genes and human cancer: different implications and different roles. Curr. Genomics *3*, 295-311.

Rossi, A., Elia, G. and Santoro, G. (1997) Inhibition of nuclear factor kappa B by prostaglandin A_1: an effect associated with heat shock transcription factor activation. Proc. Natl. Acad. Sci. U.S.A. *94*, 746-750.

Rovin, B.H., Wilmer, W.A., Lu, L., Doseff, A.I., Dixon, C., Kotur, M. and Hilbelink, T. (2002) 15-Deoxy-delta12,14-prostaglandin J2 regulates mesangial cell proliferation and death. Kidney Int. *61*, 1293-1302.

Rozakis-Adcock, M., Fernley, R., Wade, J., Pawson, T. and Bowtell, D. (1993) The SH2 and SH3 domains of mammalian Grb2 couple the EGF receptor to the Ras activator mSos1 [see comments]. Nature *363*, 83-85.

Rozakis-Adcock, M., van der Geer, P., Mbamalu, G. and Pawson, T. (1995) MAP kinase phosphorylation of mSos1 promotes dissociation of mSos1-Shc and mSos1-EGF receptor complexes. Oncogene *11*, 1417-1426.

Sanchez-Margalet, V. and Najib, S. (2001) Sam68 is a docking protein linking GAP and PI3K in insulin receptor signaling. Mol. Cell. Endocrinol. *183*, 113-121.

Santoro, M.G. (1997) Antiviral activity of cyclopentenone prostanoids. Trends Microbiol. *5*, 276-281.

Scheffzek, K., Ahmadian, M.R., Wiesmuller, L., Kabsch, W., Stege, P., Schmitz, F. and Wittinghofer, A. (1998) Structural analysis of the GAP-related domain from neurofibromin and its implications. EMBO J. *17*, 4313-4327.

Schlessinger, J. (1993) How receptor tyrosine kinases activate Ras. Trends Biochem. Sci. *18*, 273-276.

Scita, G., Nordstrom, J., Carbone, R., Tenca, P., Giardina, G., Gutkind, S., Bjarnegard, M., Betsholtz, C. and Di Fiore, P.P. (1999) EPS8 and E3B1 transduce signals from Ras to Rac. Nature *401*, 290-293.

Scita, G., Tenca, P., Frittoli, E., Tocchetti, A., Innocenti, M., Giardina, G. and Di Fiore, P.P. (2000) Signaling from Ras to Rac and beyond: not just a matter of GEFs. EMBO J. *19*, 2393-2398.

Scita, G., Tenca, P., Areces, L.B., Tocchetti, A., Frittoli, E., Giardina, G., Ponzanelli, I., Sini, P., Innocenti, M. and Di Fiore, P.P. (2001) An effector region in Eps8 is responsible for the activation of the Rac-specific GEF activity of Sos-1 and for the proper localization of the Rac-based actin-polymerizing machine. J. Cell Biol. *27*, 27.

Seizinger, B.R. (1993) NF1: a prevalent cause of tumorigenesis in human cancers? Nat. Genet. *3*, 97-99.

Sermon, B.A., Eccleston, J.F., Skinner, R.H. and Lowe, P.N. (1996) Mechanism of inhibition by arachidonic acid of the catalytic activity of Ras GTPase-activating proteins. J. Biol. Chem. *271*, 1566-1572.

Shahabi, N.A., Chengini, N. and Wittliff, J.L. (1987) Alterations of MCF-7 human breast cancer cell after prostaglandins PGA_1 and PGF_{2a} treatment. Exp. Cell Biol. *55*, 18-27.

Shou, C., Farnsworth, C.L., Neel, B.G. and Feig, L.A. (1992) Molecular cloning of cDNAs encoding a guanine-nucleotide-releasing factor for Ras p21. Nature *358*, 351-354.

Shou, C., Wurmser, A., Suen, K.L., Barbacid, M., Feig, L.A. and Ling, K. (1995) Differential response of the Ras exchange factor, Ras-GRF to tyrosine kinase and G protein mediated signals. Oncogene *10*, 1887-1893.

Simon, M.A., Bowtell, D.D., Dodson, G.S., Laverty, T.R. and Rubin, G.M. (1991) Ras1 and a putative guanine nucleotide exchange factor perform crucial steps in signaling by the sevenless protein tyrosine kinase. Cell *67*, 701-716.

Soisson, S.M., Nimnual, A.S., Uy, M., Bar-Sagi, D. and Kuriyan, J. (1998) Crystal structure of the Dbl and pleckstrin homology domains from the human Son of sevenless protein. Cell *95*, 259-268.

Sondermann, H., Soisson, S.M., Bar-Sagi, D. and Kuriyan, J. (2003) Tandem histone folds in the structure of the N-terminal segment of the ras activator Son of Sevenless. Structure (Camb.) *11*, 1583-1593.

The, I., Hannigan, G.E., Cowley, G.S., Reginald, S., Zhong, Y., Gusella, J.F., Hariharan, I.K. and Bernards, A. (1997) Rescue of a Drosophila NF1 mutant phenotype by protein kinase A. Science *276*, 791-794.

Tian, X., Gotoh, T., Tsuji, K., Lo, E.H., Huang, S. and Feig, L.A. (2004) Developmentally regulated role for Ras-GRFs in coupling NMDA glutamate receptors to Ras, Erk and CREB. EMBO J. *23*, 1567-1575.

Tocque, B., Delumeau, I., Parker, F., Maurier, F., Multon, M.C. and Schweighoffer, F. (1997) Ras-GTPase activating protein (GAP): a putative effector for Ras. Cell. Signal. *9*, 153-158.

Tognon, C.E., Kirk, H.E., Passmore, L.A., Whitehead, I.P., Der, C.J. and Kay, R.J. (1998) Regulation of RasGRP via a phorbol ester-responsive C1 domain. Mol. Cell. Biol. *18*, 6995-7008.

Tong, J., Hannan, F., Zhu, Y., Bernards, A. and Zhong, Y. (2002) Neurofibromin regulates G protein-stimulated adenylyl cyclase activity. Nat. Neurosci. *5*, 95-96.

Tong, X.K., Hussain, N.K., Adams, A.G., O'Bryan, J.P. and McPherson, P.S. (2000a) Intersectin can regulate the Ras/MAP kinase pathway independent of its role in endocytosis. J. Biol. Chem. *275*, 29894-29899.

Tong, X.-K., Hussain, N.K., de Heuvel, E., Kurakin, A., Abi-Jaoude, E., Quinn, C.C., Olson, M.F., Marais, R., Baranes, D., Kay, B.K. and McPherson, P.S. (2000b) The endocytic protein intersectin is a major binding partner for the Ras exchange factor mSos1 in rat brain. EMBO J. *19*, 1263-1271.

Trahey, M., Wong, G., Halenbeck, R., Rubinfeld, B., Martin, G.A., Ladner, M., Long, C.M., Crosier, W.J., Watt, K., Koths, K. and et al. (1988) Molecular cloning of two types of GAP complementary DNA from human placenta. Science *242*, 1697-1700.

van der Geer, P., Henkemeyer, M., Jacks, T. and Pawson, T. (1997) Aberrant Ras regulation and reduced p190 tyrosine phosphorylation in cells lacking p120-Gap. Mol. Cell. Biol. *17*, 1840-1847.

Walker, S.A., Kupzig, S., Lockyer, P.J., Bilu, S., Zharhary, D. and Cullen, P.J. (2002) Analyzing the role of the putative inositol 1,3,4,5-tetrakisphosphate receptor GAP1IP4BP in intracellular Ca2+ homeostasis. J. Biol. Chem. *277*, 48779-48785.

Walker, S.A., Kupzig, S., Bouyoucef, D., Davies, L.C., Tsuboi, T., Bivona, T.G., Cozier, G.E., Lockyer, P.J., Buckler, A., Rutter, G.A. et al. (2004) Identification of a Ras GTPase-activating protein regulated by receptor-mediated Ca(2+) oscillations. EMBO J. *23*, 1749-1760.

Wang, D.Z., Hammond, V.E., Abud, H.E., Bertoncello, I., McAvoy, J.W. and Bowtell, D.D. (1997) Mutation in Sos1 dominantly enhances a weak allele of the EGFR, demonstrating a requirement for Sos1 in EGFR signaling and development. Genes Dev. *11*, 309-320.

Wang, W., Fisher, E., Jia, Q., Dunn, J., Porfiri, E., Downward, J. and Egan, S. (1995) The Grb2 binding domain of mSos1 is not required for downstream signal transduction. Nat. Gen. *10*, 294-300.

Wei, W., Mosteller, R.D., Sanyal, P., Gonzales, E., McKinney, D., Dasgupta, C., Li, P., Liu, B.X. and Broek, D. (1992) Identification of a mammalian gene structurally and functionally related to the CDC25 gene of Saccharomyces cerevisiae. Proc. Natl. Acad. Sci. U.S.A. *89*, 7100-7104.

Wei, W., Schreiber, S.S., Baudry, M., Tocco, G. and Broek, D. (1993) Localization of the cellular expression pattern of cdc25NEF and ras in the juvenile rat brain. Brain Res. Mol. Brain Res. *19*, 339-344.

Whitehead, I.P., Campbell, S., Rossman, K.L. and Der, C.J. (1997) Dbl family proteins. Biochim. Biophys. Acta *1332*, F1-F23.

Williams, J.A., Su, H.S., Bernards, A., Field, J. and Sehgal, A. (2001) A circadian output in Drosophila mediated by neurofibromatosis-1 and Ras/MAPK. Science *293*, 2251-2256.

Wilmer, W.A., Dixon, C., Lu, L., Hilbelink, T. and Rovin, B.H. (2001) A cyclopentenone prostaglandin activates mesangial MAP kinase independently of PPARgamma. Biochem. Biophys. Res. Commun. *281*, 57-62.

Wittinghofer, A., Scheffzek, K. and Ahmadian, M.R. (1997) The interaction of Ras with GTPase-activating proteins. FEBS Lett. *410*, 63-67.

Yamamoto, T., Matsui, T., Nakafuku, M., Iwamatsu, A. and Kaibuchi, K. (1995) A novel GTPase-activating protein for R-Ras. J. Biol. Chem. *270*, 30557-30561.

Yang, Y., Li, L., Wong, G.W., Krilis, S.A., Madhusudhan, M.S., Sali, A. and Stevens, R.L. (2002) RasGRP4, a new mast cell-restricted Ras guanine nucleotide-releasing protein with calcium- and diacylglycerol-binding motifs. Identification of defective variants of this signaling protein in asthma, mastocytosis, and mast cell leukemia patients and demonstration of the importance of RasGRP4 in mast cell development and function. J. Biol. Chem. *277*, 25756-25774.

Yu, H., Chen, J.K., Feng, S., Dalgarno, D.C., Brauer, A.W. and Schreiber, S.L. (1994) Structural basis for the binding of proline-rich peptides to SH3 domains. Cell *76*, 933-945.

Zarich, N., Oliva, J.L., Jorge, R., Santos, E. and Rojas, J.M. (2000) The isoform-specific stretch of hSos1 defines a new Grb2-binding domain. Oncogene *19*, 5872-5883.

Zhang, Y.Y., Vik, T.A., Ryder, J.W., Srour, E.F., Jacks, T., Shannon, K. and Clapp, D.W. (1998) Nf1 regulates hematopoietic progenitor cell growth and ras signaling in response to multiple cytokines. J. Exp. Med. *187*, 1893-1902.

Zheng, J., Chen, R.H., Corblan-Garcia, S., Cahill, S.M., Bar-Sagi, D. and Cowburn, D. (1997) The solution structure of the pleckstrin homology domain of human SOS1. A possible structural role for the sequential association of diffuse B cell lymphoma and pleckstrin homology domains. J. Biol. Chem. *272*, 30340-30344.

Zippel, R., Orecchia, S., Sturani, E. and Martegani, E. (1996) The brain specific Ras exchange factor CDC25 Mm: modulation of its activity through Gi-protein-mediated signals. Oncogene *12*, 2697-2703.

Zippel, R., Balestrini, M., Lomazzi, M. and Sturani, E. (2000) Calcium and calmodulin are essential for Ras-GRF1-mediated activation of the Ras pathway by lysophosphatidic acid. Exp. Cell Res. *258*, 403-408.

CHAPTER 3

STRUCTURAL PRINCIPLES OF RAS INTERACTION WITH REGULATORS AND EFFECTORS

DENNIS FIEGEN, RADOVAN DVORSKY, AND
MOHAMMAD REZA AHMADIAN

Max-Planck-Institute of Molecular Physiology, Department Structural Biology, Otto-Hahn-Strasse 11, 44227 Dortmund, Germany. Tel.: #49 231 133 21 05; Fax: #49 231 133 21 99; e-mail: reza.ahmadian@mpi-dortmund.mpg.de

Abstract: In the past years, a considerable progress has been made in the fundamental understanding of the functions and underlying mechanisms of the proto-oncogen Ras. Comprehensive structural studies resulted in determination of more than 50 structures, which provided a deep insight into the three-dimensional folds, the consequences of ligand binding and hydrolysis, the principles of regulation and the specificity of effector binding. This chapter is concerned with the structural aspects of a molecular switch as a pivotal component of the signal transduction machinery

Keywords: Ras, structure, GTPase, G-domain, universal switch, GAP, GEF, effector

Abbreviations: cagedGTP: P3-1[S/R]-(2-nitrophenyl)-ethyl-guanosine-5′-triphosphate (S/R-diastereomer); DABP-GTP: guanosine-5′-(phosphoramidate-N-[[dihydroxyphosphanyl]-imine-N-methane]); GAP: GTPase activating protein; GDP: guanosine 5′-diphosphate; GEF: guanine nucleotide exchange factor; GppCp: guanosine 5′-[β,γ-methylene]triphosphate; GppNHp: guanosine 5′-[β,γ-imido]triphosphate; GTP: guanosine 5′-triphosphate; mant-GppNHp: 3′-O-(N-methyl-anthraniloyl)-2′-deoxy-GppNHp; RBD: Ras binding domain

1. INTRODUCTION

Since the discovery of the Ras proto-oncogene (25 years ago) more than 100 different small GTPases have been identified. Most of them act as regulatory GTP hydrolases controlling a wide variety of processes within the cell, ranging from gene expression, cytoskeletal reorganization and microtubule organization to vesicular and nuclear transport.

C. Der (ed.), RAS Family GTPases, 45–66.

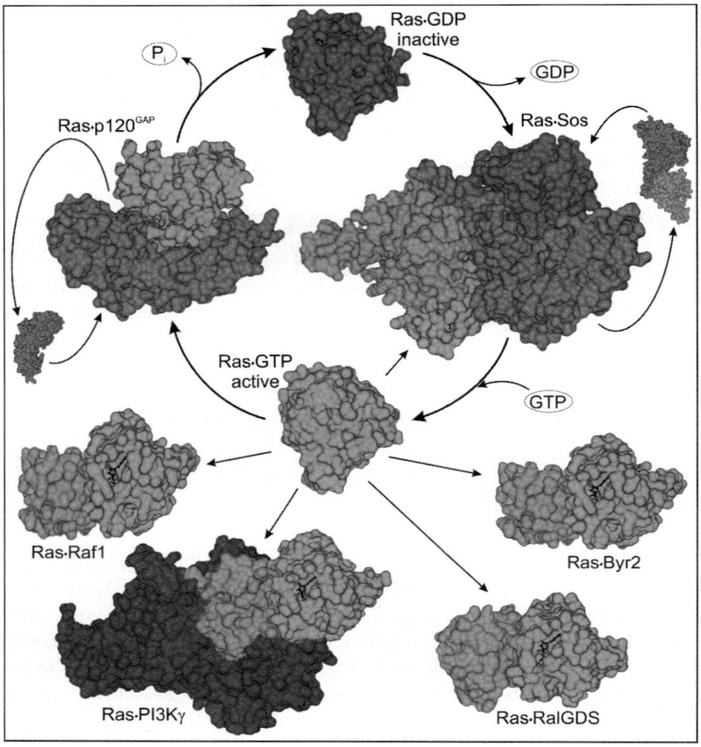

Figure 1. The Ras cycle. Binding of the catalytic domain of the RasGEF Sos (orange-cyan) to the inactive Ras•GDP (indianred) results in accelerated GDP-dissociation and formation of the nucleotide-free Ras-SOS complex (see 3.1). GTP-binding to Ras dissociates Sos from its complex with Ras and induces an active conformation in Ras shown as the Ras•GppNHp complex (green). Signal transduction proceeds through complex formation of activated Ras with downstream effectors such as Raf1, PI3Kγ, RalGDS and Byr2 (see 3.2). The Ras-binding domain (RBD) of the respective effectors is shown in cyan. The catalytic domain of Sos consists of a second subdomain (cyan), which also binds Ras in a GTP-dependent manner (see 3.1). Association of the catalytic domain of the RasGAP (p120GAP, orange) with Ras leads to the formation of the transition state complex, the stimulation of the GTP-hydrolysis reaction and finally to the conversion of Ras to its inactive GDP-bound state (see 3.3). For better comparison, the orientation of the given structures is retained in the following figures

The Ras protein is the prototype of the Ras superfamily of small GTPases, which can be subdivided into the Ras, Rho, Rab, Arf, Rad, Ran, Rheb, Rit and Rag families. All these proteins share a high degree of sequence similarity and a common three-dimensional structure, called the GTP-binding (G) domain. This domain enables them to act as molecular switches cycling between two defined conformational states: an inactive guanosine diphosphate (GDP)-bound and an active guanosine triphosphate (GTP)-bound state. Small GTPases are inefficient enzymes because their intrinsic functions, the exchange of the bound GDP to the cellular abundant GTP as well as the GTP-hydrolysis (GTPase) reaction are extremely low. Thus, nature invented two classes of proteins that tightly regulate the cycle between the active and inactive forms of the GTPase (Boguski and McCormick, 1993; Vetter and Wittinghofer, 2001). Guanosine nucleotide exchange factors (GEFs) accelerate the exchange of bound GDP for GTP, whereas GTPase-activating proteins (GAPs) stimulate the GTP-hydrolysis reaction. In its active GTP-bound conformation the GTPase can interact with and regulate a spectrum of functionally diverse effector proteins, participating in a network of signaling cascades (Figure 1).

In this chapter, we focus on the structural aspects of the Ras protein, comprising the fundamental principles of its intrinsic functions, the regulation provided by GEFs and GAPs, and the interaction with downstream effectors. A large amount of the available information is based on the fascinating structural investigations summarized in three Tables.

2. THREE-DIMENSIONAL STRUCTURES OF RAS AND THEIR IMPLICATIONS

2.1 Crystallization of Ras

The first crystallization trials on the wildtype Ras protein have not been successful. It is known that mobile elements of a protein can prevent it from being packed into crystals. The highly variable carboxy-terminal part of the (H-, K- and N-) Ras proteins, which acts as a signal for the posttranslational modification and membrane localization, turned out to be highly flexible and unfavorable for crystallization of this protein. Truncation of the C-terminal amino acids consequently resulted in successful crystallization and X-ray analysis of the G domain (Morikawa et al., 1988; Jancarik et al., 1988; Scherer et al., 1989). Additional studies showed that truncation did not impair the biochemical properties of Ras (John et al., 1989). Accordingly, a large number of crystal structures of Ras in the active and inactive state have been determined under various conditions (Table 1). An important aspect obtained from time-resolved X-ray crystallography has been the structural investigation of the GTPase reaction of Ras in the crystalline state (Schlichting et al., 1989; Schlichting et al., 1990). These studies provided new insights into the GTP/GDP transition and the resulting structural changes at two regions, called switch I and II (see section 2.3).

In addition to X-ray crystallography nuclear magnetic resonance (NMR) spectroscopy has been used not only to demonstrate the similar fold of the Ras structure in solution, but also to monitor crucial dynamic elements of the structure which exists in two or more slowly interconverting conformations termed as regional polysterism (Kraulis et al., 1994; Ito et al., 1997; Hu et al., 1997).

2.2 Overall Structure

The central building block of the Ras protein is an 18.5 kDa domain that carries out the basic functions of nucleotide binding and hydrolysis. This core domain reveals a conserved overall structural fold consisting of a central six-stranded β-sheet and

Table 1. Overview of Ras-Structures

Ras	PDB	Resolution (Å)	References
GppNHp	5p21	1.35	Pai et al., 1990
GppCp	121p	1.54	Wittinghofer et al., 1991
D38E/G12R/G12V/Q61 H/Q61L·GppNHp	221p/421p/521p/ 621p/721p	2.3, 2.2, 2.6, 2.4, 2.0	Krengel et al., 1990
GDP/G12V·GDP	1q21/2q21	2.2, 2.2	Tong et al., 1991
GDP/GppCp	4q21/6q21	2.0/1.95	Milburn et al., 1990
G12D/G12P·GppNHp	1agp/821p	2.3/1.5	Franken et al., 1993
GDP	1crr	NMR	Kraulis et al., 1994
CagedGTP/GTP/GDP	1plj/1plk/1pll	2.8/2.8/2.8	Scheidig et al., 1994
mantGppNHp/cagedGTP/ cagedGTP	1gnp/1gnq/1gnr	2.7/2.5/1.85	Scheidig et al., 1995
G12P·GppCp·Mg^{2+} and Mn^{2+}	1jah/1jai	1.8/1.8	Schweins et al., 1997
GDP	1aa9	NMR	Ito et al., 1997
DABP-GppNHp/ G12P·DABP-GTP	1clu/1rvd	1.7/1.9	Ahmadian et al., 1999
GTP/GppNHp	1qra/1ctq	1.6/1.26	Scheidig et al., 1999
T35S·GppNHp	1iaq	2.9	Spoerner et al., 2001
GDP	1ioz	2.0	Kigawa et al., 2002
A59G·GTP and GDP	1lf0/1lf5	1.7/1.7	Hall et al., 2002
GppNHp, various conditions	1p2s/1p2t/1p2u/1p2v	2.45/2.0/2.0/2.3	Buhrman et al., 2003

five α-helices connected by ten peptide loops. Five of these loops (G1-G5) are responsible for the specificity and high affinity (10^{11}-10^{12} M^{-1}) of nucleotide (Figure 2) (John et al., 1990; Bourne et al., 1991; Schmidt et al., 1996; Via et al., 2000).

The most important contribution to high-affinity binding of the nucleotide is provided by the G1 or L1 with the consensus sequence ^{10}GxxxxGK(S/T)^{17}. It is involved in the binding of nucleotide phosphate groups (also known as the phosphate binding loop or P-loop) (Saraste et al., 1990) and contains three important residues: codon 12 encoding for Gly12 is the most frequently mutated Ras codon in human tumors (Barbacid et al., 1987; Bos, 1989); Lys16 forms a ring-like structure wrapping around the β-phosphate and creating a positively charged environment; Ser17 in Ras coordinates both, the important Mg^{2+}-ion and the β-phosphate. Its substitution for Asn (S17N) renders Ras dominant-negative and inhibits Ras signaling pathways. The reason for this effect is a dramatic reduction in nucleotide binding affinity and consequently a higher affinity for guanine nucleotide exchange factors (Farnsworth et al., 1991; John et al., 1993; Feig, 1999). G2 or L2 is an integral part of effector-binding loop (see section 3.2) containing the invariant Thr35. This residue is absolutely necessary for the functional dynamics of the switch I region and is therefore crucial for the interaction with effector proteins (Spoerner et al., 2001). G3 or L4 contains the ^{57}DxxG^{60}-motif. The Asp57 side chain is involved in Mg^{2+} binding, whereas Gly60 coordinates the γ-phosphate by a main chain hydrogen bond and is an important sensor for the conformational change of the

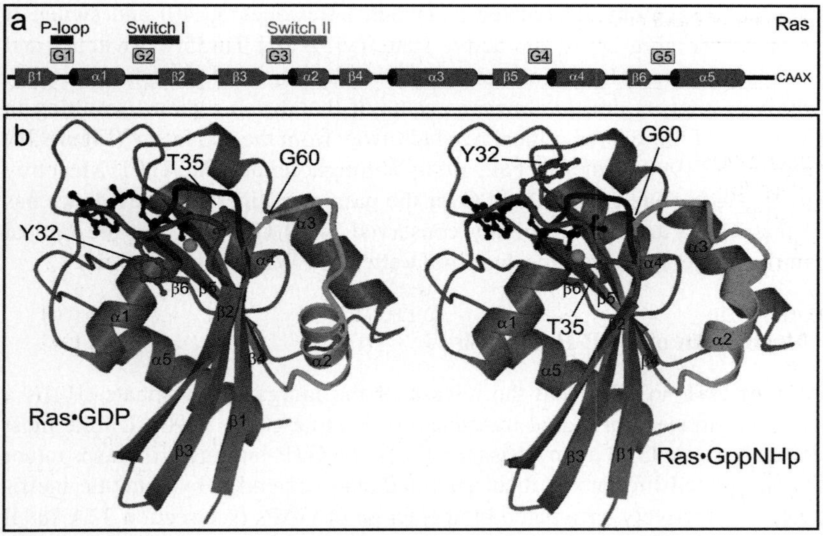

Figure 2. Structural overview of the Ras protein. (a) Secondary structure elements are illustrated as cylinders (α-helices) and arrows (β-strands). The guanine nucleotide-binding site is formed by five peptide loops (G1-G5; yellow boxes), which are highly conserved throughout all GTPases. The bold lines indicate the position of GTPase signatures such as P-loop (blue), switch I (red) and switch II (green). The isoprenylation site (called CAAX-box) is pointed out at the C-terminus. (b) Ribbon structures of the C-terminal truncated Ras·GDP (left panel) (Pai et al., 1990) and Ras·GppNHp (right panel) (Milburn et al., 1990) are shown with the P-loop in blue, switch I in red, switch II in green, the nucleotides as ball-and-stick in black and the Mg^{2+} ion as a cyan sphere. To emphasize the nucleotide-dependent conformational changes, critical residues like Tyr32, Thr35 and Gln61 are highlighted

switch II region (Wittinghofer et al., 1993). G4 and G5 are responsible for the guanine base recognition. Important in conferring the specificity is the ^{116}NKxD^{119}-motif (G4-loop) that tightly interacts with the guanine base. Mutation of Asp119 in Ras has been shown to change the nucleotide specificity from guanosine to xanthosine nucleotides (Schmidt et al., 1996; Cool et al., 1999). The ^{145}SAK^{147}-motif (G5-loop) provides Ser145 that stabilizes Asp119. Ala146 binds the guanine base and is another determinant for the guanine-binding ability of Ras. The G1, G2 and G3 motifs are centered around the γ-phosphate of the nucleotide and comprise the active site of the universal switch (Bourne et al., 1991; Wittinghofer and Pai, 1991; Sprang, 1997a).

2.3 The molecular switch function

The transition between the GTP- and the GDP-bound form of Ras is accompanied by a conformational changes that dramatically affects its affinity for downstream signaling molecules. The crystal and NMR structures of Ras (Table 1), in the active and inactive state, revealed that structural differences are primarily confined to two highly mobile regions, designated as switch I (residues 30-40) and switch II (residues 60-76) (Figure 2). In the active state Tyr32 and Thr35 in switch I and Gly60 in switch II form a hydrogen bonding network with the γ-phosphate of GTP. GTP hydrolysis triggers drastic rearrangements of the switch regions, resulting in the reorientation of these three critical residues away from the active site (Figure 2b) (Schlichting et al., 1990; Milburn et al., 1990; Wittinghofer and Pai, 1991; Menetrey and Cherfils, 1999; Diaz et al., 2000). Over the past years, increasing evidence has emerged that Thr35 and Gly60, highly conserved in all GTPases, are the critical players in the universal switch mechanism (Vetter and Wittinghofer, 2001).

2.4 Mechanism of GTP-Hydrolysis

Conversion of GTP to GDP and the release of the inorganic phosphate (P_i) is a key process in intracellular signal transduction (Bourne et al., 1990). Unlike most enzymes, the intrinsic GTP hydrolysis reaction of the GTP-binding proteins is rather slow. The prolonged lifetime of their activated state, extended by catalytic inefficiency, can be effectively terminated by the action of GAPs (see section 3.3). In all Ras structures (Table 1) the γ-phosphate of the GTP is coordinated by interactions with the Mg^{2+} ion, five conserved residues (Lys16, Tyr32, Thr35, Gly60 and Gln61) and in particular two water molecules (Pai et al., 1990; Scheidig et al., 1999).

In the proposed catalytic mechanism GTP itself plays a central role in the hydrolysis reaction by acting as a general base (substrate-assisted catalysis; Schweins et al., 1995). Accordingly, the γ-phosphate of the GTP abstracts a proton from the catalytic water molecule yielding a nucleophilic hydroxyl ion. This ion subsequently attacks the γ-phosphate and produces a trigonal bipyramidal transition state, which then dissociates into the reaction products P_i and GDP. Moreover, the role of a second water molecule has been discussed on the basis of a GTP-bound Ras structure, determined at 1.26 Å resolution, proposing a proton shuffling mechanism between two attacking water molecules and one oxygen of the γ-phosphate (Scheidig et al., 1999).

Based on mutational and crystallographic data, both functional groups of the critical Gln61 side chain have been suggested to position the nucleophilic water and stabilize the transition state of the reaction (Pai et al., 1990; Wittinghofer and Pai, 1991; French et al., 1994). The carbonyl group makes a hydrogen bond to the catalytic water whereas the amide group interacts directly with one of the γ-phosphate oxygen atoms (Wittinghofer et al., 1993; Schweins and Warshel, 1996; Scheidig et al., 1999). Replacement of Gln61 by virtually any other amino acid significantly reduces the intrinsic hydrolysis rate, prevents the GAP-mediated inactivation and thus induces oncogenic transformation by constitutive activation of the GTPase (Der et al., 1986; Vogel et al., 1988; Barbacid, M., 1987; Krengel et al., 1990; Sprang, 1997c; Ahmadian et al., 1999). In addition, oncogenic mutations of Gly12 perturb the conformation of Gln61 or its interaction with the catalytic water (Krengel et al., 1990; Franken et al., 1993; Scheffzek et al., 1997) and consequently the GTP-hydrolysis reaction (Ahmadian et al., 1999).

3. INTERACTION OF RAS WITH REGULATORS AND EFFECTORS

The vast amount of uncomplexed Ras structures has provided a deep insight into guanine nucleotide binding, the mechanism of the GTP hydrolysis reaction and the molecular switch function. In addition, a detailed picture of functionally important sites and residues of Ras, which is an absolute prerequisite for understanding its mode of action and particularly its signaling specificity at the molecular level, has emerged from structural investigations of Ras in complex with GEFs, GAPs and effectors. Here we address structural features of the interacting partners, their binding characteristics as well as the molecular mechanism behind the GEF and GAP function.

3.1 Activation by GEFs

The intrinsic guanine nucleotide release from Ras is very slow but can be accelerated by several orders of magnitude by RasGEFs (Lenzen et al., 1998). The *Saccharomyces cerevisiae* Ras-GEF Cdc25 ($Cdc25^{Sc}$) was the first Ras-GEF to be identified (Broek et al., 1987). In higher eukaryotes there are two different classes of Ras-specific $Cdc25^{Sc}$ homologues: Sos proteins (Bowtell et al., 1992) and $Cdc25^{Mm}$ (Martegani et al., 1992), also known as RasGRF.

A reaction scheme for the GEF-catalyzed exchange has been proposed on the basis of detailed kinetic studies of $Cdc25^{Mm}$ (Lenzen et al., 1998). This includes the formation of a low affinity GEF·GTPase·GDP·Mg^{2+} quaternary complex that rapidly converts to a high affinity GEF·GTPase binary complex with the concomitant expulsion of GDP and Mg^{2+} (Cherfils and Chardin, 1999; Mori et al., 2002). On the cellular level, the exchange process is driven by the $\sim$30-fold higher concentration of GTP compared to that of GDP (Pan and Wessling-Resnick, 1998). The association with GTP·Mg^{2+} leads again to an unstable quaternary complex of GEF·GTPase·GTP·Mg^{2+}, followed by the dissociation of the GEF from the GTP-bound GTPase.

The molecular architecture of RasGEFs comprises several functionally and structurally diverse domains (Figure 3a). The central catalytic domain, which is conserved among different RasGEFs, consists of a REM (Ras exchanger motif) and the catalytically active Cdc25 module (Chardin et al., 1993; Margarit et al., 2003). The crystal structures of Ras in complex with the catalytic domain of human Sos (Table 2; Figure 3b) (Boriack-Sjodin et al., 1998; Margarit et al., 2003) have shed light on the determinants of their intermolecular interaction, the molecular mechanism of the accelerated exchange reaction and the feedback regulation of Sos by Ras·GTP.

The first structure of the nucleotide-free binary Ras-Sos complex revealed, that from the two distinct α-helical Sos domains, only the Cdc25 domain directly interacts with Ras via the P-loop, switch I, II and the α3 helix. The structure demonstrates that the disruption of the P-loop and the Mg^{2+}-binding site is crucial for the exchange mechanism. One of the most conspicuous effects of Sos binding

Table 2. Overview of Ras-regulator structures and complexes

Regulator	PDB	Resolution (Å)	References
p120GAP	1wer	1.6	Scheffzek et al., 1996
NF1	1nf1	2.5	Scheffzek et al., 1998b
Complexes			
Ras-p120GAP	1wq1	2.5	Scheffzek et al., 1997
Ras-Sos1	1bkd	2.8	Boriack-Sjodin et al., 1998
Ras-Sos1·GTP/ GppNHp/GppNHp/GTP	1nvu/1nvv/1nvw/1nvx	2.2/2.18/2.7/3.2	Margarit et al., 2003

to Ras is the induction of an open switch I conformation. Sos inserts an α-helix into the nucleotide-binding site and thereby induces a displacement of the switch I region. In this open conformation, critical residues like Phe28 and Thr35, normally contacting the nucleotide, are stabilized by a protruding α-helical hairpin of Sos (Figure 3b and 3c). Other critical features, contributing to the dramatic alteration of the chemical environment of the Mg^{2+}- and phosphate-binding sites, arise from a distorted conformation of the switch II region of Ras but particularly from the α-helical hairpin of Sos itself. Thereby, Ala59 of switch II and Leu938 and Glu942 of Sos impinge the Mg^{2+}-binding site and lead to the expulsion of the Mg^{2+}-cofactor (Figure 3c). In addition, the complete rearrangement of the loop connecting the β2-β3 sheets leads to a tight electrostatic interaction of Lys16 of the P-loop with Glu62 of the switch II region. Lys16, which is invariant in all phosphate binding proteins, contributes significantly to the stabilization of the phosphate moieties (Saraste et al., 1990). Its mutation leads to a reduced nucleotide affinity (Sigal et al., 1986a). Glu62 on the other hand has been shown to be important for the exchange activity of the GEF proteins (Mistou et al., 1992).

Taken together, the GEF action leads to a perturbation of the nucleotide-binding site by structural rearrangement of P-loop, switch I and switch II and consequently to an overall 10^5-fold rate enhancement of the nucleotide release. The binary complex of the intermediate nucleotide-free state is quickly dissociated in the presence of guanine nucleotides (Lenzen et al., 1998). The association of GTP is thought to function on the same reaction pathway as the nucleotide release, based on the principle of microreversibility. At present little is known about the intermediate ternary complexes, which await further structural and biochemical investigation.

A puzzling aspect of the Sos-mediated exchange mechanism is the role of the REM domain. The structure of the nucleotide-free Ras-Sos complex showed that the REM domain does not directly contribute to Ras binding but rather structurally stabilizes a helical hairpin of the catalytic Cdc25 domain (Boriack-Sjodin et al., 1998). It has been shown that this intramolecular interaction is critical for Sos function, since mutations at this interface severely compromise catalytic activity (Hall et al., 2001). The breakthrough towards understanding the effect of the REM domain on Sos activity has been provided by a second set of crystallographic investigations (Margarit et al., 2003). This study showed that the REM domain binds

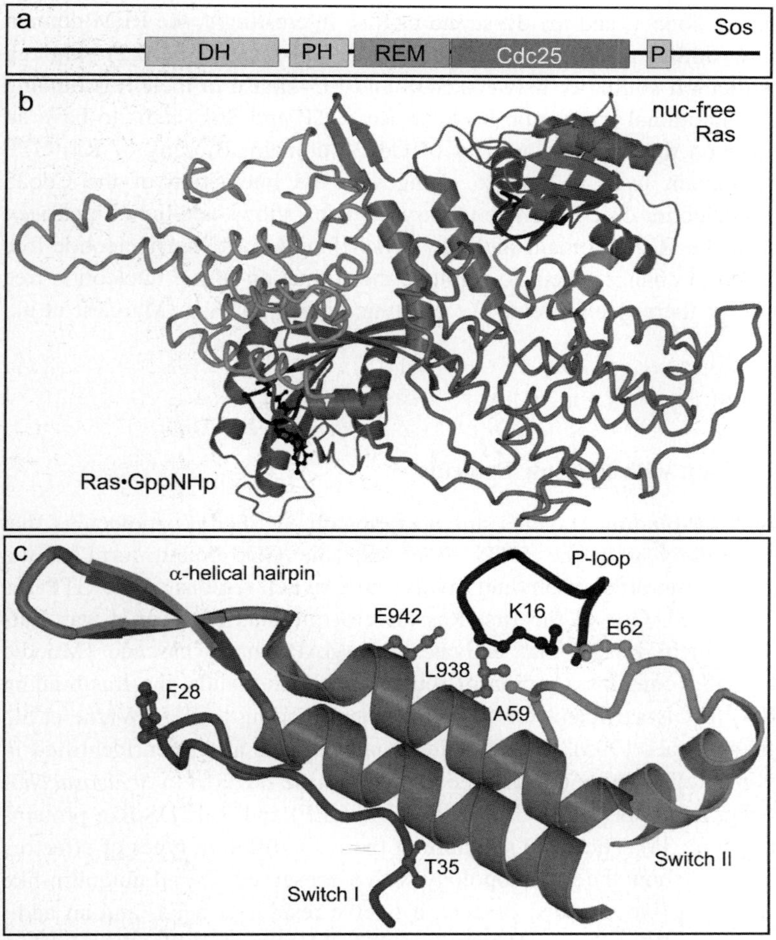

Figure 3. Structural determinants for the GEF-catalyzed nucleotide release from Ras. (a) Schematic domain organization of Sos highlights REM in goldengreen and Cdc25 in orange DH stands for Dbl homology, P for proline-rich and PH for pleckstrin homology domains. (b) Crystal structure of the nucleotide-free Ras·Sos·Ras·GppNHp complex (Margarit et al., 2003) suggests a positive feedback worm representation) consists of the REM domain (goldengreen) that binds Ras in a GTP-dependent manner and the Cdc25 domain (orange), which catalyzes the nucleotide release from Ras and forms a stable complex with the nucleotide-free Ras. The Ras molecules are shown in the same colors as in Figure 2b. The α-helical hairpin of the Cdc25 domain is shown as ribbon. (c) Structural changes of the P-loop, switch I and switch II of Ras, induced by α-helical hairpin of the Cdc25 domain, are highlighted

a second Ras molecule in a GTP-dependent manner and provides a positive feedback mechanism for the activation of Ras by Sos. This new distal Ras binding site is located between the Cdc25 and the REM domain, just opposite to the active site (Figure 3b). The intermolecular interactions mainly involve switch I and switch II but also helix α1, loop 1, and the β2-strand of Ras. Interestingly, the REM domain binds Ras in a similar manner as the Ras effector PI3-kinase (see 3.2) although there is no structural similarity between SOS and PI3-kinase in their Ras binding elements. The additional interaction between Ras·GTP and Sos seems to have an allosteric effect on the catalytic activity of Cdc25 domain. Binding of Ras·GTP to the REM domain leads in turn to changes in the interaction of the Cdc25 domain with nucleotide-free Ras. The ternary complex shows additional hydrogen bonds between the REM domain and the switch I region of the nucleotide-free Ras. The structural changes seem to stabilize the formation of the nucleotide-free Ras·Sos complex thereby enhancing the catalytic activity of Sos (Margarit et al., 2003).

3.2 Interaction with Effector Proteins

Effectors for GTP-binding proteins are operationally defined as molecules that interact more tightly with the GTP- than with the GDP-bound form of the GTPase implying that effector binding involves the switch regions of the GTPases (Wittinghofer, 2003). One of the first Ras effector proteins discovered was Raf-1, a Ser/Thr protein kinase that activates the MAP kinase cascade (Moodie et al., 1993). Raf-1 contains a region of roughly 80 amino acids, the Ras-binding domain (RBD) that is sufficient for GTP dependent binding to Ras (Warne et al., 1993; Herrmann et al., 1995). Such binding domains have also been identified in phosphatidylinositol-3-kinase (PI3K), the Ser/Thr kinase Byr2 from *Schizosaccharomyces pombe*, RalGDS (a Ral GTPase specific GEF) and RalGDS-like proteins Rgl and Rlf (Figure 3a). The RBD structure of these six different types of effectors (Table 3; Figure 1) show the same topology with a conserved overall ubiquitin-like fold consisting of a five-stranded β-sheet, a twelve residue α-helix, and an additional one-turn helix (ββαββαβ) (Herrmann, 2003). Most remarkably, the way they interact with the Ras protein is very similar. The crystal structures of the Rap·Raf-RBD (Nassar et al., 1995; Nassar et al., 1996), Ras·RalGDS-RBD (Huang et al., 1998) and Ras·Byr2 (Scheffzek et al., 2001) complexes (Table 3; Figure 1) have shown that RBD binding to the switch I region is conserved and creates an interface involving a continuous antiparallel β-sheet (Figure 4b and 4c). Among other variable contacts at the interface, seven residues (Asp33, Ile36, Glu37, Asp38, Ser39, Tyr40 and Arg41) in the switch I region are involved in the formation of all interfaces between Ras and its effectors, as shown in Figure 4c for the Rap1·Raf1 complex. These so-called effector residues play a critical role in the selective effector binding and activation, and have thus served as useful tools for unraveling the underlying networks of signal transduction pathways (Polakis and McCormick, 1993; White et al., 1995; Rodriguez-Viciana et al., 1997; Kinashi et al., 2000; Herrmann, 2003).

Both, Ras and Rap bind to the same set of effectors in vitro which is not surprising considering the similar amino acid composition of the switch I region, although Rap

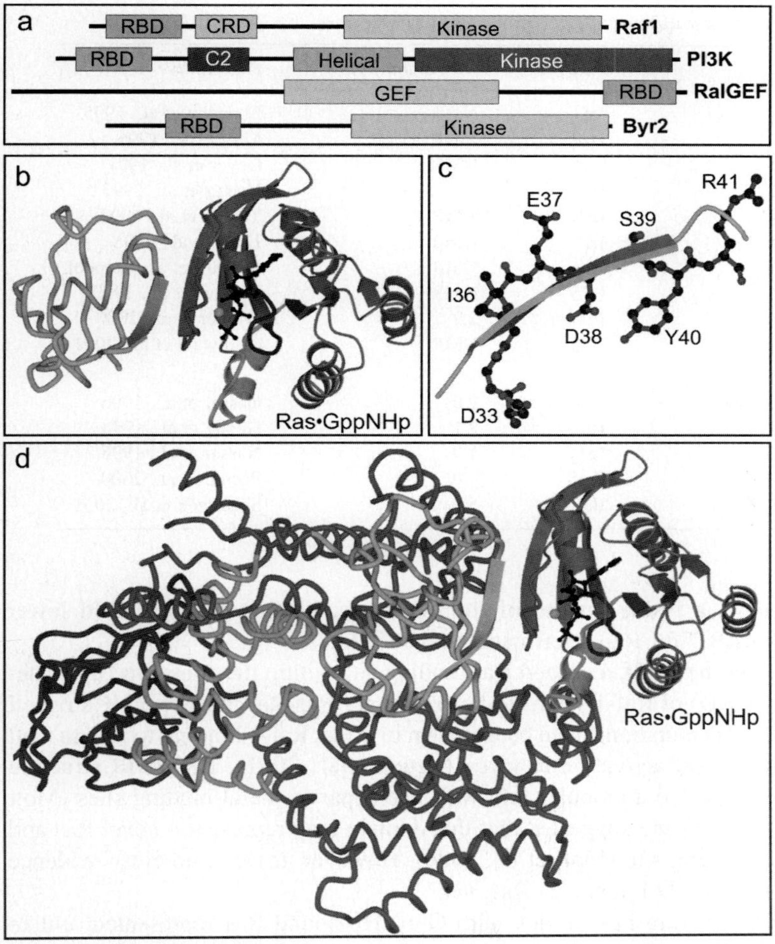

Figure 4. Structural features of the Ras interaction with effectors. (a) Schematic domain organization of four different effectors is shown. The RBD (Ras binding domain) is colored goldengreen, C2 darkgreen, the helical domain of PI3K lightgreen and the kinase domain indianred. CRD, cysteine-rich domain; GEF, Ral GRPase-specific exchange domain. (b) Crystal structure of the Ras-related GTPase Rapl·GppNHp (Ribbon) in complex with the Rafl-RBD (worm except for the interacting β-sheet) (Nassar et al., 1995) represents the characteristic Ras·RBD binding mode as shown for RalGDS-RBD (Huang et al., 1998) and Byr2-RBD (Scheffzek et al., 2001). Rap l is colored as Ras in Figure 2b. The Ras contacting β2-sheet of Rafl-RBD is represented as ribbon. (c) The conserved effector-binding residues of the switch I region of Ras are shown as ball-and-stick with of the β2-sheet of the RBD in front. (d) Crystal structure of Ras·GppNHp·PI3Kγ complex (Pacold et al., 2000) is shown with PI3Kγ in the same colors as in (a) and Ras as in Figure 2b

appears to have a different biological function in vivo (Bos, 1998; Stork, 2003). The major determinant concerning specificity is provided by the Glu31 residue of Ras, which creates a favorable complementary interface for the Ras-Raf interaction (Nassar et al., 1995; Nassar et al., 1996). Rap on the other hand contains a positively

Table 3. Overview of Ras-regulator structures and complexes

Effector	PDB	Resolution (Å)	References
Raf1-RBD	1rfa	NMR	Emerson et al., 1995
Raf1-CRD	1faq	NMR	Mott et al., 1996
Ral-RBD	2rgf	NMR	Geyer et al., 1997
Ral-RBD	1lxd	2.4	Huang et al., 1997
Raf1-RBD	1rrb	NMR	Terada et al., 1999
Rlf-RBD	1rlf	NMR	Esser et al., 1998
Ral-RBD	1rax	NMR	Mueller et al., (unpubl.)
Rgl-RBD	1ef5	NMR	Kigawa et al., 1998
PI3Kγ	1e8x	2.2	Walker et al., 2000
Byr2-RBD	1i35	NMR	Gronwald et al., 2001
Complexes			
Rap1a-RafRBD	1gua	2.0	Nassar et al., 1996
Ras-RalRBD	1lfd	2.1	Huang et al., 1998
Rap-RafRBD	1cly	1.9	Nassar et al., 1995
Ras G12V-PI3Kγ	1he8	3.0	Pacold et al., 2000
Ras-Byr2RBD	1k8r	3.0	Scheffzek et al., 2001

charged residue, a lysine instead of the glutamate, leading to a 100-fold lower affinity of RafRBD for Rap (Herrmann et al., 1996).

In addition to the RBD, a second Ras binding site within the conserved cysteine-rich domain (CRD) of Raf-1 seems to be critical for Ras-Raf interaction (Campbell et al., 1998). CRD mutations have been shown to affect Raf-1 binding to Ras as well as Ras-mediated Raf activation in vivo (Avruch et al., 2001). The NMR structure of Raf1-CRD revealed a globular fold with two separate metal-binding sites (Mott et al., 1996). It has been suggested that this domain may represent a novel Ras and phospholipid-binding site (Mott et al., 1996). However, there is no clear evidence for a direct Raf1-CRD binding to Ras, yet.

The PI3Kγ structure in complex with GppNHp-bound Ras represented, unlike the isolated RBD structures, the first view of a GTPase interacting with an essentially complete effector molecule (Figure 1 and 4d) (Pacold et al., 2000). PI3Kγ consists of four domains, an N-terminal RBD, a C2 domain, a helical domain and a catalytic domain (Figure 4a). The catalytic domain consists of two lobes: a smaller N-terminal lobe (residues 726-883) and a larger C-terminal lobe (884-1092). The overall organization of the uncomplexed PI3Kγ catalytic subunit is preserved in the structure of the Ras·PI3Kγ complex (Pacold et al., 2000; Walker et al., 2000). Contacts between PI3Kγ and Ras are made primarily via the PI3Kγ RBD, which adopts a topology homologous to the RafRBD-like structures, and the switch I region of Ras (Figure 4d). Thus formation of an intermolecular β-sheet seems to be a common mode of Ras interaction with its effectors (Geyer et al., 1996; Pacold et al., 2000; Herrmann, 2003). The complex structure of Ras·PI3Kγ revealed for the first time that a Ras effector makes essential intermolecular contacts also with the switch II region of Ras (Pacold et al., 2000). These interactions, which are mediated by the RBD but also by the C-terminal lobe, induce structural changes in the kinase domain and presumably affect phospholipid binding and catalytic activity of PI3Kγ (Figure 4d) (Pacold et al., 2000). Whereas allosteric regulation of PI3Kγ has been demonstrated in vitro and in vivo, the mechanism of activation for other Ras effectors remains unclear.

3.3 Inactivation by GAPs

Hydrolysis of the bound GTP is the timing mechanism that returns Ras to the GDP bound inactive state and thereby completes the GTPase cycle. The intrinsic GTP hydrolysis reaction is very slow but can be accelerated efficiently by several orders of magnitude upon interaction with GAPs (Gideon et al., 1992; Ahmadian et al., 1997a; Vetter and Wittinghofer, 2001; Scheffzek et al., 1998a). Two models have been discussed for the mechanism of GAP action (Scheffzek et al., 1998a; Phillips et al., 2003). In the fist scenario, the GTPase itself can be an efficient GTPase machine when it adopts an enzymatically competent conformation upon interaction with the GAP ('isomerization' hypothesis). In the second model the GAP molecule participates actively in the process of GTP hydrolysis by supplying a catalytic residue to the active site, which stabilizes the transition state of the reaction ('arginine finger' hypothesis). A major breakthrough to explore the mechanism of the GAP-stimulated GTP hydrolysis reaction was provided by the observation that small GTPases are able to bind aluminum fluoride only in the presence of the catalytic domain of the GAPs (Mittal et al., 1996; Ahmadian et al., 1997b). It is generally appreciated that the bound aluminium fluoride mimics the γ-phosphate in its pentavalent transition state of the GAP-GTPase complex during GTP-hydrolysis (Wittinghofer, 1997; Scheffzek et al., 1998a; Sprang, 2000).

Several mammalian Ras specific GAPs of various sizes and modular architectures have been described to date with $p120^{GAP}$ and neurofibromin (NF-1) as the currently best-characterized ones (Figure 5a) (McCormick, 1998; Scheffzek et al., 1998b; Cichowski and Jacks, 2001; Donovan et al., 2002; Bernards, 2002). They have a high sequence similarity in their catalytic GAP domains (Scheffzek et al., 1998a, 1998b). p120GAP is the prototype of this class since it was the first to be isolated (Trahey et al., 1987, 1988; Vogel et al., 1988). Neurofibromin is the product of neurofibromatosis type 1 gene (Xu et al., 1990a, 1990b; Martin et al., 1990; Ballester et al., 1990), which is has been found frequently mutated in patients with the disease neurofibromatosis type I (Cichowski and Jacks, 2001).

The crystal structures of the catalytic domain of p120GAP and neurofibromin show a helical elongated protein (Table 2; Fig 5b) that contains all residues conserved among RasGAPs (Ahmadian et al., 1996; Scheffzek et al., 1996; Scheffzek et al., 1998b). The structure of the Ras-RasGAP complex, formed by Ras·GDP and GAP-334 in the presence of aluminium fluoride (AlF_3) has shown that GAP binds and stabilizes the switch I and II regions of Ras and supplies an arginine residue (Arg789 in p120GAP or Arg1276 in neurofibromin) into the catalytic machinery of Ras that stabilizes the transition state of the GTP-hydrolysis reaction (Figure 5c) (Scheffzek et al., 1997, 1998a). Mutational analysis has shown, that the substitution of the 'arginine finger' R1276 by lysine, glutamine, asparagine and alanine or even its deletion dramatically reduces the GAP-stimulated GTP-hydrolysis reaction of Ras (Ahmadian et al., 1997c; Ahmadian et al., 2003). The R1276P mutation, found in patients with type 1 neurofibromatosis, has been shown to nearly abolish the GAP-function of neurofibromin (Klose et al., 1998).

Three prominent regions in RasGAPs, the arginine-finger loop, the phenylalanine-leucine-arginine (FLR-) region and the α7/variable loop have been shown to contain structural fingerprints governing the GAP function (Ahmadian et al., 2003). The finger loop provides the catalytic arginine ('arginine finger') that neutralizes developing charges during the transition state of the reaction and additionally stabilizes the critical Gln61 (Figure 5c). Gln61, itself, also contacts AlF_3 and a water molecule that corresponds to the attacking nucleophile. The invariant FLR region, a hallmark for RasGAPs, indirectly contributes to GTPase stimulation by forming a scaffold, which stabilizes the Ras switch regions. The α7/variable loop uses several conserved residues including two lysines, which are involved in numerous interactions with the switch I region of Ras. Structural and biochemical studies on the GTPase-GAP interactions which basically confirmed both, 'arginine finger' and 'isomerization' hypotheses, have revealed that the basic mechanism of the GTPase stimulation relies on two major features: firstly, GAPs stabilize flexible residues in the switch I and II regions of the GTPase such as the essential catalytic residue Gln61; secondly, GAPs supply a catalytically critical arginine to the GTPase active site that stabilizes the transition state of the GTP hydrolysis reaction (Ahmadian et al., 1997c; Scheffzek et al., 1997; Scheffzek et al., 1998a).

Impaired GTPase activity, particularly in the presence of RasGAPs, has been found to be the biochemical reason for the oncogenicity of the Gly12 and Gln61 mutations, preventing Ras from being switched off (McCormick, 1998; Scheffzek et al., 1997). Therefore, the mechanism of the GAP-stimulated GTPase reaction was of major medical importance to disclose the mystery behind these molecular defects. The structure of the Ras·GAP complex (Scheffzek et al., 1997) has provided an explanation why these mutants are insensitive to GAP (Figure 5c): Glycine 12 is in close proximity to the finger loop such that even the smallest possible change (e.g. alanine) would sterically interfere with the geometry of the transition state. The apparent involvement of Gln61 in stabilizing the transition state (Mittal et al., 1996) confirms the notion that Gln61 plays a vital role in catalysis.

Understanding why oncogenic mutants of Ras cannot be switched off by GAP has invoked the concept of restoring the GTPase activity of oncogenic Ras mutants as a therapeutic approach for Ras-directed cancer therapy (Wittinghofer and Waldmann, 2000; Ahmadian, 2002). This concept has gained impact by recent reports showing that oncogenic Ras mutants can be chemically inactivated and are not irreversibly damaged in their capability to act as molecular switches. The defective GTPase reaction of different oncogenic mutants of Ras could be

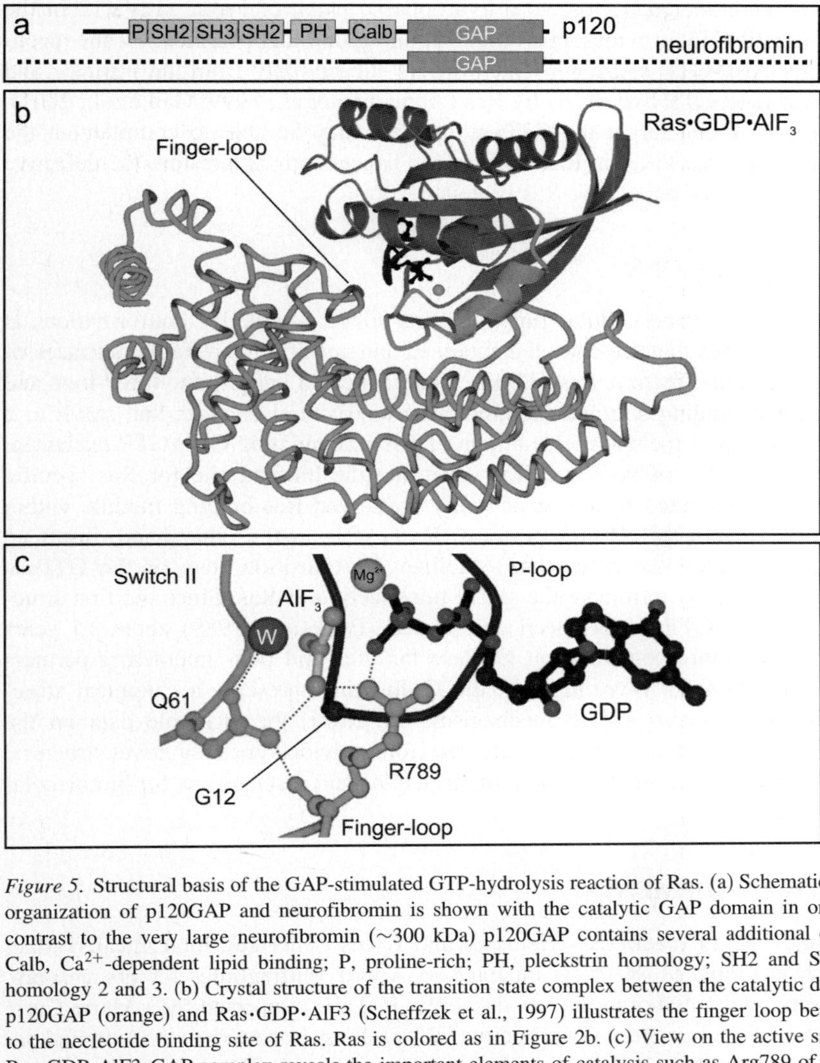

Figure 5. Structural basis of the GAP-stimulated GTP-hydrolysis reaction of Ras. (a) Schematic domain organization of p120GAP and neurofibromin is shown with the catalytic GAP domain in orange. In contrast to the very large neurofibromin (~300 kDa) p120GAP contains several additional domains: Calb, Ca^{2+}-dependent lipid binding; P, proline-rich; PH, pleckstrin homology; SH2 and SH3, Src-homology 2 and 3. (b) Crystal structure of the transition state complex between the catalytic domain of p120GAP (orange) and Ras·GDP·AlF3 (Scheffzek et al., 1997) illustrates the finger loop being close to the necleotide binding site of Ras. Ras is colored as in Figure 2b. (c) View on the active site of the Ras·GDP·AlF3·GAP complex reveals the important elements of catalysis such as Arg789 of the GAP finger-loop and Gln61 of Ras. AlF3 is supposed to mimic the γ-phosphate of the GTP in the pentavalent transition state. The nucleophilic water (w) molecule is shown as a red sphere and the Mg^{2+}-ion as a cyan sphere

increased up to three orders of magnitude by using a modified GTP analogue, 3,4-diaminobenzophenone-phosphoramidate-GTP (DABP-GTP), instead of GTP (Ahmadian et al., 1999). The structures of DABP-GppNHp bound to Pro12 and Val12 mutants of Ras, respectively (Table 1), that could be helpful in designing the proper scaffold for a GTPase directed lead compound, have shown that the DABP moiety is accommodated close to a hydrophobic patch of Pro12 or Val12 in the P-loop. DABP-GTP provides an aromatic amino group that is critical for the mechanism of DABP-GTP cleavage, which differs substantially from the intrinsic and GAP-stimulated GTP hydrolysis by Ras (Ahmadian et al., 1999; Gail et al., 2001). Catalytic drugs that target the GTPase reaction may be able to complement the insensitive GAP activities in Ras transformed cancer cells and restore the defective GTPase reaction of oncogenic Ras proteins.

4. CONCLUSIONS

To perform its myriad cellular functions Ras adopts alternative conformations at the switch regions that are complementary to and specific for different surfaces of its regulators and effectors. RasGEF inserts an α-helical hairpin into the P-loop and the Mg^{2+}-ion binding site and thereby induces structural changes that result in a drastic reduction of the nucleotide binding affinity facilitating GDP/GTP-exchange. The switch regions of activated Ras constitute the binding site for Ras specific effectors, characterized by the structurally conserved Ras-binding module within the otherwise variable effector proteins. RasGAPs, on the other hand, insert an arginine finger into the active site, stabilize the transition state of the GTPase reaction and thereby terminate the signal transduction of Ras. Since the first structure of a small GTPase has been determined (Pai et al., 1989) about 15 years ago, many structures on different GTPase families and their interacting partners have been published, revealing that the G-domain represents a canonical structure with a conserved switch mechanism. Moreover, the available data on the GTPase complexes with regulators and effectors provided not only novel structural features but also shed light on common principles and mechanisms for bimolecular interaction.

ACKNOWLEGDEMENTS

We would like to thank Lars Hemsath and Klaus Scheffzek for critical reading of the manuscript. M. R. A. is indebted to Alfred Wittinghofer for his wisdom, collegiality, and collaboration over the years. R. D. is supported by a Marie Curie Fellowship of the European Community (EC). D. F. is supported by the Verband der chemischen Industrie (VCI) and the Bundesministerium für Bildung und Forschung (BMBF).

REFERENCES

Ahmadian, M.R., Wiesmuller, L., Lautwein, A., Bischoff, F.R. and Wittinghofer, A. (1996) Structural differences in the minimal catalytic domains of the GTPase-activating proteins p120GAP and neurofibromin. J. Biol. Chem. *271*, 16409-16415.

Ahmadian, M.R., Hoffmann, U., Goody, R.S. and Wittinghofer, A. (1997a) Individual rate constants for the interaction of Ras proteins with GTPase-activating proteins determined by fluorescence spectroscopy. Biochemistry *36*, 4535-4541.

Ahmadian, M.R., Mittal, R., Hall, A. and Wittinghofer, A. (1997b) Aluminum fluoride associates with the small guanine nucleotide binding proteins. FEBS Lett. *408*, 315-318.

Ahmadian, M.R., Stege, P., Scheffzek, K. and Wittinghofer, A. (1997c) Confirmation of the arginine-finger hypothesis for the GAP-stimulated GTP-hydrolysis reaction of Ras. Nat. Struct. Biol. *4*, 686-689.

Ahmadian, M.R., Zor, T., Vogt, D., Kabsch, W., Selinger, Z., Wittinghofer, A. and Scheffzek, K. (1999) Guanosine triphosphatase stimulation of oncogenic Ras mutants. Proc. Natl. Acad. Sci. U.S.A. *96*, 7065-7070.

Ahmadian, M.R. (2002) Prospects for anti-ras drugs. Br. J. Haematol. *116*, 511-518.

Ahmadian, M.R., Kiel, C., Stege, P. and Scheffzek, K. (2003) Structural fingerprints of the Ras-GTPase activating proteins neurofibromin and p120GAP. J. Mol. Biol. *329*, 699-710.

Avruch, J., Khokhlatchev, A., Kyriakis, J.M., Luo, Z., Tzivion, G., Vavvas, D. and Zhang, X.F. (2001) Ras activation of the Raf kinase: tyrosine kinase recruitment of the MAP kinase cascade. Recent Prog. Horm. Res. *56*, 127-155.

Ballester, R., Marchuk, D., Boguski, M., Saulino, A., Letcher, R., Wigler, M. and Collins, F. (1990) The NF1 locus encodes a protein functionally related to mammalian GAP and yeast IRA proteins. Cell *63*, 851-859.

Barbacid, M. (1987) ras genes. Annu. Rev. Biochem. *56*, 779-827.

Bernards, A. (2002) GAPs galore! A survey of putative Ras superfamily GTPase activating proteins in man and Drosophila. Biochim. Biophys. Acta *1603*, 47-82.

Boguski, M.S. and McCormick, F. (1993) Proteins regulating Ras and its relatives. Nature *366*, 643-654.

Bollag, G. and McCormick, F. (1991) Differential Regulation of rasGAP and neurofibromatosis gene product activities. Nature *351*, 576-579.

Boriack-Sjodin, P.A., Margarit, S.M., Bar-Sagi, D. and Kuriyan, J. (1998) The structural basis of the activation of Ras by Sos. Nature *394*, 337-343.

Bos, J.L. (1989) Ras oncogenes in human cancer: a review. Cancer Res. *49*, 4682-4689.

Bos, J.L. (1998) All in the family? New insights and questions regarding interconnectivity of Ras, Rap1 and Ral. EMBO J. *17*, 6776-6782.

Bourne, H.R., Sanders, D.A. and McCormick, F. (1990) The GTPase superfamily: a conserved switch for diverse cell functions. Nature *348*, 125-132.

Bourne, H.R., Sanders, D.A. and McCormick, F. (1991) The GTPase superfamily: conserved structure and molecular mechanism. Nature *349*, 117-127.

Bowtell, D., Fu, P., Simon, M. and Senior, P. (1992) Identification of murine homologues of the Drosophila son of sevenless gene: potential activators of ras. Proc. Natl. Acad. Sci. U.S.A. *89*, 6511-6515.

Broek, D., Toda, T., Michaeli, T., Levin, L., Birchmeier, C., Zoller, M., Powers, S. and Wigler, M. (1987) The S. cerevisiae CDC25 gene product regulates the RAS/adenylate cyclase pathway. Cell *48*, 789-799.

Buhrman, G., de Serrano, V. and Mattos, C. (2003) Organic solvents order the dynamic switch II in Ras crystals. Structure *11*, 747-751.

Campbell, S.L., Khosravi-Far, R., Rossman, K.L., Clark, G.J., and Der, C.J. (1998) Increasing complexity of Ras signaling. *Oncogene 17*, 1395-1413.

Chardin, P., Camonis, J.H., Gale, N.W., van Aelst, L., Schlessinger, J., Wigler, M.H. and Bar-Sagi, D. (1993) Human Sos1: a guanine nucleotide exchange factor for Ras that binds to GRB2. Science *260*, 1338-1343.

Cherfils, J. and Chardin, P. (1999) GEFs: structural basis for their activation of small GTP-binding proteins. Trends Biochem. Sci. *24*, 306-311.

Cichowski, K. and Jacks, T. (2001) NF1 tumor suppressor gene function: narrowing the GAP. Cell *104*, 593-604.

Cool, R.H., Schmidt, G., Lenzen, C.U., Prinz, H., Vogt, D. and Wittinghofer, A. (1999) The Ras mutant D119N is both dominant negative and activated. Mol. Cell. Biol. *19*, 6297-6305.

Der, C.J., Finkel, T. and Cooper, G.M. (1986) Biological and biochemical properties of human rasH genes mutated at codon 61. Cell *44*, 167-176.

Diaz, J.F., Escalona, M.M., Kuppens, S. and Engelborghs, Y. (2000) Role of the switch II region in the conformational transition of activation of Ha-ras-p21. Protein Sci. *9*, 361-368.

Donovan, S., Shannon, K.M. and Bollag, G. (2002) GTPase activating proteins: critical regulators of intracellular signaling. Biochim. Biophys. Acta *1602*, 23-45.

Emerson, S.D., Madison, V.S., Palermo, R.E., Waugh, D.S., Scheffler, J.E., Tsao, K.L., Kiefer, S.E., Liu, S.P. and Fry, D.C. (1995) Solution structure of the Ras-binding domain of c-Raf-1 and identification of its Ras interaction surface. Biochemistry *34*, 6911-6918.

Esser, D., Bauer, B., Wolthuis, R.M., Wittinghofer, A., Cool, R.H. and Bayer, P. (1998) Structure determination of the Ras-binding domain of the Ral-specific guanine nucleotide exchange factor Rlf. Biochemistry *37*, 13453-13462.

Farnsworth, C.L. and Feig, L.A. (1991) Dominant inhibitory mutations in the Mg(2+)-binding site of RasH prevent its activation by GTP. Mol. Cell. Biol. *11*, 4822-4829.

Feig, L.A. (1999) Tools of the trade: use of dominant-inhibitory mutants of Ras-family GTPases. Nat. Cell Biol. *1*, E25-E27.

Franken, S.M., Scheidig, A.J., Krengel, U., Rensland, H., Lautwein, A., Geyer, M., Scheffzek, K., Goody, R.S., Kalbitzer, H.R. and Pai, E.F. (1993) Three-dimensional structures and properties of a transforming and a nontransforming glycine-12 mutant of p21H-ras. Biochemistry *32*, 8411-8420.

French, J.E., Libbus, B.L., Hansen, L., Spalding, J., Tice, R.R., Mahler, J. and Tennant, R.W. (1994) Cytogenetic analysis of malignant skin tumors induced in chemically treated TG-AC transgenic mice. Mol. Carcinog. *11*, 215-226.

Gail, R., Costisella, B., Ahmadian, M.R. and Wittinghofer, A. (2001) Ras-mediated cleavage of a GTP analogue by a novel mechanism. Chembiochem *2*, 570-575.

Geyer, M., Schweins, T., Herrmann, C., Prisner, T., Wittinghofer, A. and Kalbitzer, H.R. (1996) Conformational transitions in p21ras and in its complexes with the effector protein Raf-RBD and the GTPase activating protein GAP. Biochemistry *35*, 10308-10320.

Geyer, M., Herrmann, C., Wohlgemuth, S., Wittinghofer, A. and Kalbitzer, H.R. (1997) Structure of the Ras-binding domain of RalGEF and implications for Ras binding and signalling. Nat. Struct. Biol. *4*, 694-699.

Gideon, P., John, J., Frech, M., Lautwein, A., Clark, R., Scheffler, J.E. and Wittinghofer, A. (1992) Mutational and kinetic analyses of the GTPase-activating protein (GAP)-p21 interaction: the C-terminal domain of GAP is not sufficient for full activity. Mol. Cell. Biol. *12*, 2050-2056.

Gronwald, W., Huber, F., Grunewald, P., Sporner, M., Wohlgemuth, S., Herrmann, C. and Kalbitzer, H.R. (2001) Solution structure of the Ras binding domain of the protein kinase Byr2 from Schizosaccharomyces pombe. Structure *9*, 1029-1041.

Hall, B.E., Yang, S.S., Boriack-Sjodin, P.A., Kuriyan, J. and Bar-Sagi, D. (2001) Structure-based mutagenesis reveals distinct functions for Ras switch 1 and switch 2 in Sos-catalyzed guanine nucleotide exchange. J. Biol. Chem. *276*, 27629-27637.

Hall, B.E., Bar-Sagi, D. and Nassar, N. (2002) The structural basis for the transition from Ras-GTP to Ras-GDP. Proc. Natl. Acad. Sci. U.S.A. *99*, 12138-12142.

Herrmann, C. (2003) Ras–effector interactions: after one decade. Curr. Opin. Struct. Biol. *13*, 122-129.

Herrmann, C., Martin, G.A. and Wittinghofer, A. (1995) Quantitative analysis of the complex between p21ras and the Ras-binding domain of the human Raf-1 protein kinase. J. Biol. Chem. *270*, 2901-2905.

Herrmann, C., Horn, G., Spaargaren, M. and Wittinghofer, A. (1996) Differential interaction of the ras family GTP-binding proteins H-Ras, Rap1A, and R.-Ras with the putative effector molecules Raf kinase and Ral-guanine nucleotide exchange factor. J. Biol. Chem. *271*, 6794-6800.

Hu, J.S. and Redfield, A.G. (1997) Conformational and dynamic differences between N-ras P21 bound to GTPgammaS and to GMPPNP as studied by NMR. Biochemistry *36*, 5045-5052.

Huang, L., Weng, X., Hofer, F., Martin, G.S. and Kim, S.H. (1997) Three-dimensional structure of the Ras-interacting domain of RalGDS. Nat. Struct. Biol. *4*, 609-615.

Huang, L., Hofer, F., Martin, G.S. and Kim, S.H. (1998) Structural basis for the interaction of Ras with RalGDS. Nat. Struct. Biol. *5*, 422-426.

Ito, Y., Yamasaki, K., Iwahara, J., Terada, T., Kamiya, A., Shirouzu, M., Muto, Y., Kawai, G., Yokoyama, S., Laue, E.D., Walchli, M., Shibata, T., Nishimura, S. and Miyazawa, T. (1997) Regional polysterism in the GTP-bound form of the human c-Ha-Ras protein. Biochemistry *36*, 9109-9119.

Jancarik, J., de Vos, A., Kim, S.H., Miura, K., Ohtsuka, E., Noguchi, S. and Nishimura, S. (1988) Crystallization of human c-H-ras oncogene products. J. Mol. Biol. *200*, 205-207.

John, J., Schlichting, I., Schiltz, E., Rosch, P. and Wittinghofer, A. (1989) C-terminal truncation of p21H preserves crucial kinetic and structural properties. J. Biol. Chem. *264*, 13086-13092.

John, J., Sohmen, R., Feuerstein, J., Linke, R., Wittinghofer, A. and Goody, R.S. (1990) Kinetics of interaction of nucleotides with nucleotide-free H-ras p21. Biochemistry *29*, 6058-6065.

John, J., Rensland, H., Schlichting, I., Vetter, I., Borasio, G.D., Goody, R.S. and Wittinghofer, A. (1993) Kinetic and structural analysis of the Mg(2+)-binding site of the guanine nucleotide-binding protein p21H-ras. J. Biol. Chem. *268*, 923-929.

Kigawa, T., Endo, M., Ito, Y., Shirouzu, M., Kikuchi, A. and Yokoyama, S. (1998) Solution structure of the Ras-binding domain of RGL. FEBS Lett. *441*, 413-418.

Kigawa, T., Yamaguchi-Nunokawa, E., Kodama, K., Matsuda, T., Yabuki, T., Matsuda, N., Ishitani, R., Nureki, O. and Yokoyama, S. (2002) Selenomethionine incorporation into a protein by cell-free synthesis. J. Struct. Funct. Genomics *2*, 29-35.

Klose, A., Ahmadian, M.R., Schuelke, M., Scheffzek, K., Hoffmeyer, S., Gewies, A., Schmitz, F., Kaufmann, D., Peters, H., Wittinghofer, A. and Nurnberg, P. (1998) Selective disactivation of neurofibromin GAP activity in neurofibromatosis type 1. Hum. Mol. Genet. *7*, 1261-1268.

Kraulis, P.J., Domaille, P.J., Campbell-Burk, S.L., Van Aken, T. and Laue, E.D. (1994) Solution structure and dynamics of ras p21.GDP determined by heteronuclear three- and four-dimensional NMR spectroscopy. Biochemistry *33*, 3515-3531.

Krengel, U., Schlichting, L., Scherer, A., Schumann, R., Frech, M., John, J., Kabsch, W., Pai, E.F. and Wittinghofer, A. (1990) Three-dimensional structures of H-ras p21 mutants: molecular basis for their inability to function as signal switch molecules. Cell *62*, 539-548.

Lenzen, C., Cool, R.H., Prinz, H., Kuhlmann, J. and Wittinghofer, A. (1998) Kinetic analysis by fluorescence of the interaction between Ras and the catalytic domain of the guanine nucleotide exchange factor Cdc25Mm. Biochemistry *37*, 7420-7430.

Margarit, S.M., Sondermann, H., Hall, B.E., Nagar, B., Hoelz, A., Pirruccello, M., Bar-Sagi, D. and Kuriyan, J. (2003) Structural evidence for feedback activation by Ras. GTP of the Ras-specific nucleotide exchange factor SOS. Cell *112*, 685-695.

Martegani, E., Vanoni, M., Zippel, R., Coccetti, P., Brambilla, R., Ferrari, C., Sturani, E. and Alberghina, L. (1992) Cloning by functional complementation of a mouse cDNA encoding a homologue of CDC25, a Saccharomyces cerevisiae RAS activator. EMBO J. *11*, 2151-2157.

Martin, G.A., Viskochil, D., Bollag, G., McCabe, P.C., Crosier, W.J., Haubruck, H., Conroy, L., Clark, R., O'Connell, P., Cawthon, R.M., Innis, M.A. and McCormick, F. (1990) The GAP-related domain of the neurofibromatosis type 1 gene product interacts with ras p21. Cell *63*, 843-849.

McCormick, F. (1998) Going for the GAP. Curr. Biol. *8*, R673-R674.

Menetrey, J. and Cherfils, J. (1999) Structure of the small G protein Rap2 in a non-catalytic complex with GTP. Proteins *37*, 465-473.

Milburn, M.V. and Tong, L. deVos, A.M., Brunger, A., Yamaizumi, Z., Nishimura, S. and Kim, S.H. (1990) Molecular switch for signal transduction: structural differences between active and inactive forms of protooncogenic ras proteins. Science *247*, 939-945.

Mistou, M.Y., Jacquet, E., Poullet, P., Rensland, H., Gideon, P., Schlichting, I., Wittinghofer, A. and Parmeggiani, A. (1992) Mutations of Ha-ras p21 that define important regions for the molecular mechanism of the SDC25 C-domain, a guanine nucleotide dissociation stimulator. EMBO J. *11*, 2391-2397.

Mittal, R., Ahmadian, M.R., Goody, R.S. and Wittinghofer, A. (1996) Formation of a transition-state analog of the Ras GTPase reaction by Ras-GDP, tetrafluoroaluminate, and GTPase-activating proteins. Science *273*, 115-117.

Moodie, S.A., Willumsen, B.M., Weber, M.J. and Wolfman, A. (1993) Complexes of Ras. GTP with Raf-1 and mitogen-activated protein kinase kinase. Science *260*, 1658-1661.

Mori, K., Hata, M., Neya, S. and Hoshino, T. (2002) A study on the role of Mg2+ in a Ras protein by MD simulation, Chem.-Bio Inform. J. *2* (*4*), 147-155.

Morikawa, K., Tsujimoto, M., Higashi, T., Matsumoto, O., Miura, K., Ohtsuka, E., Noguchi, S. and Nishimura, S. (1988) Crystallization and preliminary crystallographic data of a truncated derivative from the human c-Ha-ras protein. J. Mol. Biol. *201*, 237-238.

Mott, H.R., Carpenter, J.W., Zhong, S., Ghosh, S., Bell, R.M. and Campbell, S.L. (1996) The solution structure of the Raf-1 cysteine-rich domain: a novel ras and phospholipid binding site. Proc. Natl. Acad. Sci. U.S.A. *93*, 8312-8317.

Nassar, N., Horn, G., Herrmann, C., Scherer, A., McCormick, F. and Wittinghofer, A. (1995) The 2.2 A crystal structure of the Ras-binding domain of the serine/threonine kinase c-Raf1 in complex with Rap1A and a GTP analogue. Nature *375*, 554-560.

Nassar, N., Horn, G., Herrmann, C., Block, C., Janknecht, R. and Wittinghofer, A. (1996) Ras/Rap effector specificity determined by charge reversal. Nat. Struct. Biol. *3*, 723-729.

Pacold, M.E., Suire, S., Perisic, O., Lara-Gonzalez, S., Davis, C.T., Walker, E.H., Hawkins, P.T., Stephens, L., Eccleston, J.F. and Williams, R.L. (2000) Crystal structure and functional analysis of Ras binding to its effector phosphoinositide 3-kinase gamma. Cell *103*, 931-943.

Pai, E.F., Kabsch, W., Krengel, U., Holmes, K.C., John, J. and Wittinghofer, A. (1989) Structure of the guanine-nucleotide-binding domain of the Ha-ras oncogene product p21 in the triphosphate conformation. Nature *341*, 209-214.

Pai, E.F., Krengel, U., Petsko, G.A., Goody, R.S., Kabsch, W. and Wittinghofer, A. (1990) Refined crystal structure of the triphosphate conformation of H-ras p21 at 1.35 A resolution. implications for the mechanism of GTP hydrolysis. EMBO J. *9*, 2351-2359.

Pan, J.Y., Wessling-Resnick, M. (1998) GEF-mediated GDP/GTP exchange by monomeric GTPases: a regulatory role for Mg2+? Bioessays, June;20(6):516-521.

Phillips, R.A., Hunter, J.L., Eccleston, J.F. and Webb, M.R. (2003) The mechanism of Ras GTPase activation by neurofibromin. Biochemistry *42*, 3956-3965.

Polakis, P. and McCormick, F. (1993) Structural requirements for the interaction of p21ras with GAP, exchange factors, and its biological effector target. J. Biol. Chem. *268*, 9157-9160.

Rodriguez-Viciana, P., Warne, P.H., Khwaja, A., Marte, B.M., Pappin, D., Das, P., Waterfield, M.D., Ridley, A., Downward, J. (1997) Role of phosphoinositide 3-OH kinase in cell transformation and control of the actin cytoskeleton by Ras. Cell. May 2;89(3):457-467.

Saraste, M., Sibbald, P.R. and Wittinghofer, A. (1990) The P-loop – a common motif in ATP- and GTP-binding proteins. Trends Biochem. Sci. *15*, 430-434.

Scheffzek, K., Lautwein, A., Kabsch, W., Ahmadian, M.R. and Wittinghofer, A. (1996) Crystal structure of the GTPase-activating domain of human p120GAP and implications for the interaction with Ras. Nature *384*, 591-596.

Scheffzek, K., Ahmadian, M.R., Kabsch, W., Wiesmuller, L., Lautwein, A., Schmitz, F. and Wittinghofer, A. (1997) The Ras-RasGAP complex: structural basis for GTPase activation and its loss in oncogenic Ras mutants. Science *277*, 333-338.

Scheffzek, K., Ahmadian, M.R. and Wittinghofer, A. (1998b) GTPase-activating proteins: helping hands to complement an active site. Trends Biochem. Sci. *23*, 257-262.

Scheffzek, K., Ahmadian, M.R., Wiesmuller, L., Kabsch, W., Stege, P., Schmitz, F. and Wittinghofer, A. (1998b) Structural analysis of the GAP-related domain from neurofibromin and its implications. EMBO J. *17*, 4313-4327.

Scheffzek, K., Grunewald, P., Wohlgemuth, S., Kabsch, W., Tu, H., Wigler, M., Wittinghofer, A. and Herrmann, C. (2001) The Ras-Byr2RBD complex: structural basis for Ras effector recognition in yeast. Structure *9*, 1043-1050.

Scheidig, A.J., Sanchez-Llorente, A., Lautwein, A., Pai, E.F., Corrie, J.E., Reid, G.P., Wittinghofer, A. and Goody, R.S. (1994) Crystallographic studies on p21(H-ras) using the synchrotron Laue method: improvement of crystal quality and monitoring of the GTPase reaction at different time points. Acta Crystallogr. D. Biol. Crystallogr. *50*, 512-520.

Scheidig, A.J., Franken, S.M., Corrie, J.E., Reid, G.P., Wittinghofer, A., Pai, E.F. and Goody, R.S. (1995) X-ray crystal structure analysis of the catalytic domain of the oncogene product p21H-ras complexed with caged GTP and mant dGppNHp. J. Mol. Biol. *253*, 132-150.

Scheidig, A.J., Burmester, C. and Goody, R.S. (1999) The pre-hydrolysis state of p21(ras) in complex with GTP: new insights into the role of water molecules in the GTP hydrolysis reaction of ras-like proteins. Structure *7*, 1311-1324.

Scherer, A., John, J., Linke, R., Goody, R.S., Wittinghofer, A., Pai, E.F. and Homes, K.C. (1989) Crystallization and preliminary X-ray analysis of the human c-H-ras-oncogene product p21 complexed with GTP analogues. J. Mol. Biol. *206*, 257-259.

Schlichting, I., Rapp, G., John, J., Wittinghofer, A., Pai, E.F. and Goody, R.S. (1989) Biochemical and crystallographic characterization of a complex of c-Ha-ras p21 and caged GTP with flash photolysis. *Proceedings* Natl. Acad. Sci. U.S.A. *86*, 7687-7690.

Schlichting, I., Almo, S.C., Rapp, G., Wilson, K., Petratos, K., Lentfer, A., Wittinghofer, A., Kabsch, W., Pai, E.F., Petsko, G.A. and Good, R.S. (1990) Time-resolved X-ray crystallographic study of the conformational change in Ha-Ras p21 protein on GTP hydrolysis. Nature *345*, 309-315.

Schmidt, G., Lenzen, C., Simon, I., Deuter, R., Cool, R.H., Goody, R.S. and Wittinghofer, A. (1996) Biochemical and biological consequences of changing the specificity of p21ras from guanosine to xanthosine nucleotides. Oncogene *12*, 87-96.

Schweins, T. and Wittinghofer, A. (1994) GTP-binding proteins. Structures, interactions and relationships. Curr. Biol. *4*, 547-550.

Schweins, T., Geyer, M., Scheffzek, K., Warshel, A., Kalbitzer, H.R. and Wittinghofer, A. (1995) Substrate-assisted catalysis as a mechanism for GTP hydrolysis of p21ras and other GTP-binding proteins. Nat. Struct. Biol. *2*, 36-44.

Schweins, T. and Warshel, A. (1996) Mechanistic analysis of the observed linear free energy relationships in p21ras and related systems. Biochemistry *35*, 14232-14243.

Schweins, T., Scheffzek, K., Asshéuer, R. and Wittinghofer, A. (1997) The role of the metal ion in the p21ras catalysed GTP-hydrolysis: Mn^{2+} versus Mg^{2+}. J. Mol. Biol. *266*, 847-856.

Sigal, I.S., Gibbs, J.B., D'Alonzo, J.S., Temeles, G.L., Wolanski, B.S., Socher, S.H. and Scolnick, E.M. (1986) Mutant ras-encoded proteins with altered nucleotide binding exert dominant biological effects. Proc. Natl. Acad. Sci. U.S.A. *83*, 952-956.

Spoerner, M., Herrmann, C., Vetter, I.R., Kalbitzer, H.R. and Wittinghofer, A. (2001) Dynamic properties of the Ras switch I region and its importance for binding to effectors. Proc. Natl. Acad. Sci. U.S.A. *98*, 4944-4949.

Sprang, S.R. (1997) G protein mechanisms: insights from structural analysis. Annu. Rev. Biochem. *66*, 639-678.

Sprang, S.R. (1997) G proteins, effectors and GAPs: structure and mechanism. Curr. Opin. Struct. Biol. *7*, 849-856.

Sprang, S.R. (2000) Conformational display: a role for switch polymorphism in the superfamily of regulatory GTPases, Sci. STKE 2000, PE1.

Stork, P.J. (2003) Does Rap1 deserve a bad Rap? Trends Biochem. Sci. *28*, 267-275.

Terada, T., Ito, Y., Shirouzu, M., Tateno, M., Hashimoto, K., Kigawa, T., Ebisuzaki, T., Takio, K., Shibata, T., Yokoyama, S., Smith, B.O., Laue, E.D. and Cooper, J.A. (1999) Nuclear magnetic resonance and molecular dynamics studies on the interactions of the Ras-binding domain of Raf-1 with wild-type and mutant Ras proteins. J. Mol. Biol. *286*, 219-232.

Tong, L.A., de Vos, A.M., Milburn, M.V. and Kim, S.H. (1991) Crystal structures at 2.2 A resolution of the catalytic domains of normal ras protein and an oncogenic mutant complexed with GDP. J. Mol. Biol. *217*, 503-516.

Trahey, M. and McCormick, F. (1987) A cytoplasmic protein stimulates normal N-ras p21 GTPase, but does not affect oncogenic mutants. Science *238*, 542-545.

Trahey, M., Wong, G., Halenbeck, R., Rubinfeld, B., Martin, G.A., Ladner, M., Long, C.M., Crosier, W.J., Watt, K., Koths, K. and McCormick, F. (1988) Molecular cloning of two types of GAP complementary DNA from human placenta. Science *242*, 1697-1700.

Vetter, I.R. and Wittinghofer, A. (2001) The guanine nucleotide-binding switch in three dimensions. Science *294*, 1299-1304.

Via, A., Ferre, F., Brannetti, B., Valencia, A. and Helmer-Citterich, M. (2000) Three-dimensional view of the surface motif associated with the P-loop structure: cis and trans cases of convergent evolution. J. Mol. Biol. *303*, 455-465.

Vogel, U.S., Dixon, R.A., Schaber, M.D., Diehl, R.E., Marshall, M.S., Scolnick, E.M., Sigal, I.S. and Gibbs, J.B. (1988) Cloning of bovine GAP and its interaction with oncogenic ras p21. Nature *335*, 90-93.

Walker, E.H., Pacold, M.E., Perisic, O., Stephens, L., Hawkins, P.T., Wymann, M.P. and Williams, R.L. (2000) Structural determinants of phosphoinositide 3-kinase inhibition by wortmannin, LY294002, quercetin, myricetin, and staurosporine. Mol. Cell *6*, 909-919.

Warne, P.H., Viciana, P.R. and Downward, J. (1993) Direct interaction of Ras and the amino-terminal region of Raf-1 in vitro. Nature *364*, 352-355.

Wittinghofer, A. (2003) Structural Considerations of Small GTP-Binding Proteins. In: Handbook of Cell Signaling, 1st edn. Elsevier Science (USA), pp. 757-761.

Wittinghofer, A. and Pai, E.F. (1991) The structure of Ras protein: a model for a universal molecular switch. Trends Biochem. Sci. *16*, 382-387.

Wittinghofer, A. and Waldmann, H. (2000) Ras, a molecular switch involved in tumor formation. Angew. Chem. *39*, 4192-4214.

Wittinghofer, A., Franken, S.M., Scheidig, A.J., Rensland, H., Lautwein, A., Pai, E.F. and Goody, R.S. (1993) Three-dimensional structure and properties of wild-type and mutant H-ras-encoded p21. Ciba Found. Symp. *176*, 6-21.

Wittinghofer, A., Scheffzek, K. and Ahmadian, M.R. (1997) The interaction of Ras with GTPase-activating proteins. FEBS Lett. *410*, 63-67.

Xu, G., O'Connel, P., Viskochil, D., Cawthon, R., Robertson, M., Culver, M., Dunn, D., Stevens, J., Gesteland, R., White, R. and Weiss, R. (1990a) The neurofibromatosis type 1 gene encodes a protein related to GAP. Cell *62*, 599-608.

Xu, G.F., Lin, B., Tanaka, K., Dunn, D., Wood, D., Gesteland, R., White, R., Weiss, R. and Tamanoi, F. (1990b) The catalytic domain of the neurofibromatosis type 1 gene product stimulates ras GTPase and complements ira mutants of S. cerevisiae. Cell *63*, 835-841.

CHAPTER 4

RAS AND THE RAF/MEK/ERK CASCADE

DEBORAH K. MORRISON AND IRA O. DAAR
Laboratory of Protein Dynamics and Signaling, NCI-Frederick, Frederick, MD 21701

Abstract: The Raf/MEK/ERK protein kinases constitute a key effector cascade used by Ras to relay signals regulating cell growth, survival, proliferation, and differentiation. These kinases are activated in a sequential manner through direct phosphorylation. Raf is the initiating kinase that interacts with membrane-localized GTP-bound Ras. The signal is then transduced from Raf to MEK and from MEK to ERK, ultimately resulting in the phosphorylation of critical cellular targets by activated ERK. In addition to the core enzymes of the cascade, various scaffolding proteins and signaling modulators have been identified that affect the efficiency and level of signaling through this important kinase cascade. An emerging concept is that these factors contribute to the spatiotemporal control of Ras/ERK signaling, allowing sensitive activation and deactivation of the pathway in response to diverse extracellular cues

Keywords: Signal transduction, protein kinase, phosphorylation

1. INTRODUCTION

Signal transduction is the process whereby cells translate extracellular signals into specific biological responses. In many cell types, modules of sequentially activating protein kinases, such as the MAPK cascades, are essential for this process, functioning as a relay route from the cell surface to the nucleus and as central integrators of the signaling inputs. Signal transduction mediated by the RasGTPase is no exception and in higher eukaryotic organisms, the kinase module used by Ras is the MAPK cascade comprised of the Raf/MEK/ERK kinases, also known as the ERK module (Pearson et al., 2001). The first kinase in this module, Raf, is a direct effector of Ras that binds specifically to active GTP-bound Ras. This interaction recruits the cytoplasmic Raf protein to the plasma membrane where it becomes activated. Raf then phosphorylates and activates MEK, which in turn phosphorylates and activates ERK. The cascade culminates when activated ERK

C. Der (ed.), RAS Family GTPases, 67–93.

phosphorylates critical cytoplasmic and nuclear substrates required for a specific cellular response.

In this chapter, we will examine how signal transmission through the Raf/MEK/ERK cascade is regulated. First, we will review the molecular mechanisms that control the activity of the core kinase components Raf, MEK, and ERK. Then we will discuss the scaffolding proteins and signaling modulators that affect the efficiency, spatiotemporal dynamics, and level of signaling through this major kinase cascade.

2. THE RAF KINASES

Members of the Raf serine/threonine kinase family are the initiating enzymes in the three-tiered ERK kinase cascade. In mammalian cells, there are three Raf proteins, Raf-1, A-Raf, and B-Raf (Hagemann and Rapp, 1999). Invertebrate organisms such as *Drosophila melanogaster* and *Caenorhabditis elegans* encode a single Raf kinase, but no homolog is present in yeast. The Raf kinases were first discovered when the *raf-1* gene was identified as the cellular counterpart of the murine retroviral oncogene, *v-raf* (Rapp et al., 1983). Raf-1 is the most widely expressed member of the three mammalian Raf kinases with significant protein levels detected in all cell types examined (Storm et al., 1990). Expression of the other family members is more limited with A-Raf expression highest in urogenital tissues and B-Raf expression highest in neuronal tissues, testis, and haematopoietic cells. Determining how Raf kinase activity is regulated has been a daunting task that has challenged investigators for years – due largely to the complexity of the process. In this section, we will examine the molecular mechanisms involved in Raf regulation. Our discussion will focus primarily on Raf-1, the most extensively studied Raf protein; however, distinct regulatory features of the other Raf family members will also be described.

2.1 Regulation of Raf by Autoinhibition

All Raf proteins contain three conserved regions, CR1, CR2, and CR3, and can be divided into two functional domains – an N-terminal regulatory domain and a C-terminal catalytic domain (Figure 1). The N-terminal regulatory domain contains both CR1, which consists of a Ras binding domain (RBD) and a cysteine-rich domain (CRD), and CR2, a region rich in serine and threonine residues, whereas the C-terminal catalytic domain comprises the CR3 (Daum et al., 1994). The first indication that the Raf N-terminus serves a regulatory role came from the observation that this domain is absent in the oncogenic v-Raf protein (Rapp et al., 1983). Subsequently, it was shown that deletion of this domain converts any of the mammalian Raf proteins into constitutively active kinases capable of inducing cell transformation (Heidecker et al., 1990; Stanton et al., 1989). The N-terminus thus functions as a repressor, inhibiting the activity of the catalytic domain through intramolecular interactions (Chong and Guan, 2003; Cutler et al., 1998), and for Raf

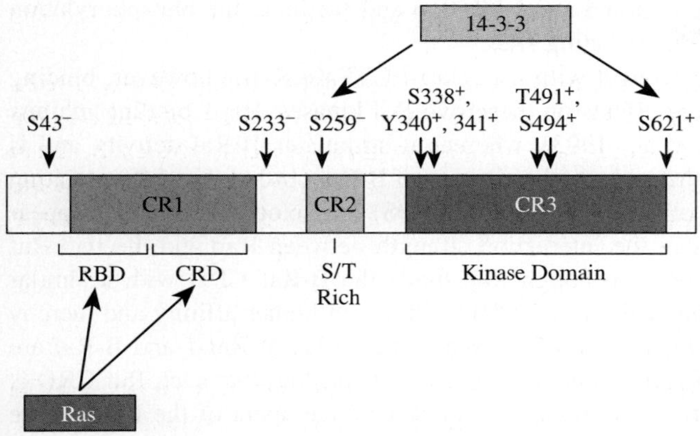

Figure 1. Schematic diagram of Raf-1. The three regions that are highly conserved among Raf proteins are shown, namely CR1, CR2, and CR3. CR1 contains the Ras binding domain (RBD) and the cysteine-rich domain (CRD), both of which interact with activated Ras; CR2 is a region rich in serine/threonine residues; and CR3 is the kinase domain. Sites of Raf-1 phosphorylation are also shown (the single-letter amino acid code is used); and sites that positively (+) and negatively (−) affect Raf-1 are indicated. S259 and S621 function as phosphorylation-dependent 14-3-3 binding sites; S388 and Y340/341 are activating phosphorylation sites found in the negative charge regulatory region (N-region); and T491 and S494 are activating (+) phosphorylation sites found in the activation segment

to become activated, this autoinhibition must be relieved. The cell accomplishes this task through an intricate series of events that includes a change in subcellular localization, protein and lipid interactions, as well as phosphorylation events.

2.2 Regulation of Raf by Protein Interactions

2.2.1 Ras and Ras-related proteins

A major breakthrough in understanding Raf activation came from the discovery that all Raf proteins interact with activated Ras (Van Aelst et al., 1993; Vojtek et al., 1993; Zhang et al., 1993). Ras binding does not stimulate Raf enzymatic activity directly, but instead localizes the normally cytoplasmic Raf to the plasma membrane. This change in localization is a critical step in the Raf activation process, and Raf-1 proteins artificially targeted to the plasma membrane are constitutively active in a Ras-independent manner (Leevers et al., 1994; Stokoe et al., 1994). The interaction with Ras is mediated by the Raf RBD and CRD, and binding of Ras to both regions is required for full Raf activation (Luo et al., 1997; Roy et al., 1997). The initial contact is made by the RBD, which binds with high affinity (18 nM) to the Ras effector domain in its GTP-bound state (Herrmann et al., 1995). This interaction then allows the CRD to make distinct contacts with lipid-modified Ras (Hu et al., 1995; Williams et al., 2000). In addition to altering Raf subcellular localization, the interaction with Ras is thought to induce conformational changes in Raf that disrupt its N-terminal autoinhibition and facilitate the phosphorylation of the catalytic domain on activating sites.

Raf-1 and B-Raf also interact with the related GTPase Rap1; however, binding to Rap1 has very different effects on these two Raf kinases. Rap1 binding inhibits Raf-1 activation (Cook et al., 1993), whereas it stimulates B-Raf activity and is thought to contribute to the prolonged activation of B-Raf observed in differentiating PC-12 cells (Ohtsuka et al., 1996; York et al., 1998). These opposing effects appear to be due to differences in the interaction strength between Rap and the two Raf CRDs (Okada et al., 1999). Although Rap binds the B-Raf CRD with a similar affinity as does Ras, it binds the Raf-1 CRD with much higher affinity and thereby inhibits Raf-1 from interacting with Ras. When the CRDs of Raf-1 and B-Raf are exchanged, the effect of Rap binding is reversed, suggesting that each Raf CRD is an important determinant in the regulation of Raf by members of the RasGTPase family.

2.2.2 14-3-3

Another regulator of Raf activity is the phosphoserine/phosphothreonine binding protein, 14-3-3. All Raf proteins contain two high-affinity, phosphorylation-dependent 14-3-3 binding sites, one in the CR2 domain (S259 of Raf-1) and one following the kinase domain (S621 of Raf-1) (Muslin et al., 1996). 14-3-3 molecules exist as dimers with two binding pockets and are capable of interacting with both Raf sites simultaneously. The functional consequence of the Raf/14-3-3 interaction appears to be twofold. First, binding of 14-3-3 to the S259 and S621 sites of Raf-1 (both of which are highly phosphorylated under quiescent conditions in many cell types) appears to stabilize the autoinhibited Raf conformation (Michaud et al., 1995). In addition, binding of the 14-3-3 dimer to the S259 site and other contacts within the Raf-1 N-terminal domain, such as with the CRD, apparently mask N-terminal regions required for Raf activation, including the RBD as well as the CRD (Clark et al., 1997; McPherson et al., 1999).

Following signal activation, 14-3-3 appears to play a second role in facilitating and stabilizing the active Raf conformation. Cell stimulation disrupts 14-3-3 binding to the Raf-1 N-terminus due to the Ras/Raf interaction (Rommel et al., 1996) and dephosphorylation of the S259 site (Dhillon et al., 2002). However, binding to the C-terminal site persists and displacement of 14-3-3 from this site by peptide competition prevents Raf-1 activation *in vivo* (Tzivion et al., 1998). Binding of the 14-3-3 dimer to the C-terminal site may serve to localize Raf with an upstream activator(s), facilitate Raf oligomerization, and/or protect the activated Raf catalytic domain from phosphatase attack.

2.2.3 Protein phosphatase 2A and the Pin1 prolyl-isomerase

The serine/threonine protein phosphatase 2A (PP2A) also interacts with Raf-1, and recent studies indicate that PP2A contributes to the Raf-1 activation process as well as the recycling of attenuated Raf-1 following mitogen-induced activation.

PP2A is a heterotrimeric phosphatase composed of a dimeric core enzyme (the structural A and the catalytic C subunits) and a regulatory subunit (B-type subunits). Coimmunoprecipitation experiments and mass spectroscopy analyses indicate that the dimeric core subunits of PP2A interact constitutively with Raf-1, whereas association of the regulatory B subunit is induced by stimulus treatment (Abraham et al., 2000; Ory et al., 2003). The underlying mechanism(s) for why binding of the B subunit is induced following growth factor treatment is currently unknown; however, incorporation of the B subunit into the complex would be expected to increase the catalytic activity of PP2A towards Raf. In the Raf-1 activation process, PP2A functions by dephosphorylating the inhibitory phospho-S259 site that mediates 14-3-3 binding (Ory et al., 2003). An issue yet to be resolved is whether dephosphorylation of this site occurs prior to, coincident with, or after Ras binding.

Following Raf-1 activation, PP2A appears to play a second role by dephosphorylating feedback phosphorylation sites that downmodulate Raf-1 activity (Dougherty et al., 2005). Interestingly, five of the six feedback phosphorylation sites (discussed below) occur on serine residues followed by a proline (SP), and for PP2A to efficiently dephosphorylate these sites, the phospho-SP peptide bond must be in the *trans*-conformation (Zhou et al., 2000). The cellular protein that specifically isomerizes phospho-(S/T)P bonds is Pin1 (Yaffe et al., 1997), and Pin1 has recently been shown to associate with Raf-1 (Dougherty et al., 2005). The interaction with Raf appears to be mediated by the Pin1 WW domain and is selective for Raf-1 hyperphosphorylated on the feedback phospho-SP sites. Confirmation that Pin1 and PP2A are important for the recycling of downregulated Raf-1 comes from the observation that Raf-1 persists in a hyperphosphorylated/desensitized state in mouse embryo fibroblasts (MEFs) that lack Pin1 and in cells treated with the PP2A inhibitor okadaic acid. Moreover, mitogen-induced activation of Raf-1 is strongly suppressed in Pin1-deficient MEFs, indicating that in the absence of Pin1, the conversion of the pSP sites to a *trans*-conformation occurs more slowly, thereby reducing PP2A-mediated dephosphorylation of these sites and the recycling of Raf-1 to its signaling competent state.

2.2.4 *Molecular chaperones, scaffolds, and signaling modulators*

Various molecular chaperones, scaffolding proteins and signaling modulators have also been reported to associate with the Raf kinases. The molecular chaperones include Hsp90, p50/cdc37, and the Hsp70 binding protein Bag1. Hsp90 and p50 interact with the Raf catalytic domain and are required for protein stability (Schulte et al., 1996). Bag1 binding has been reported to stimulate Raf-1 kinase activity; however, the activating mechanism is unclear as are the conditions that promote this interaction (Song et al., 2001). The scaffolding proteins and signaling modulators identified as Raf-interacting molecules include kinase suppressor of Ras (KSR), connector enhancer of KSR (CNK), suppressor of Ras 8 (Sur-8), Raf kinase inhibitor protein (RKIP), Spry, Spred, and Erbin. The effect of these interactions on Raf function and ERK signaling will be discussed in sections below.

2.3 Regulation of Raf by Oligomerization

The activation of numerous protein kinases requires dimerization or oligomerization, and a growing body of evidence indicates that oligomerization contributes to Raf-1 activation. Previously, it had been shown that forced oligomerization of Raf-1 results in constitutive kinase activation (Farrar et al., 1996; Luo et al., 1996). In addition, Raf-1 and B-Raf have been observed to heterodimerize (Weber et al., 2001). More recently, studies investigating the mechanisms of B-Raf-mediated oncogenesis have revealed that oncogenic B-Raf proteins are complexed with Raf-1 and that mutational activation of B-Raf results in the transactivation of Raf-1 (Wan et al., 2004). The exact mechanism for how oligomerization promotes kinase activation is unknown, but potential mechanisms include disruption of the Raf autoinhibited state, promotion of its active kinase conformation, and/or increased transphosphorylation on activating sites.

2.4 Regulation of Raf by Lipid Interactions

Lipids have also been implicated in Raf activation, and both Raf-1 and B-Raf have been reported to bind phosphatidyl serine (PS), phosphatidic acid (PA), ceramide, and cholesterol. PS binding is mediated by the Raf CRD and is thought to occur following membrane recruitment (Ghosh et al., 1996). Contact with PS has been shown to disrupt interactions occurring between the CRD and bound 14-3-3 dimers and may help relieve the autoinhibition of the N-terminal domain as well as stabilize the membrane localization of Raf (McPherson et al., 1999). Surprisingly, PA binding is mediated by the Raf catalytic domain (residues 389 to 423 of Raf-1; Ghosh et al., 1996), and studies indicate that this interaction may direct the translocation of Raf to the appropriate membrane environment for kinase activation (Rizzo et al., 1999). The region(s) of Raf required for ceramide and cholesterol binding have not been characterized; however, given that both of these molecules are prominent components of lipid rafts, their binding has been proposed to localize Raf to these membrane microdomains (Hekman et al., 2002). Further studies are needed to clarify these interactions and their functional importance.

2.5 Regulation of Raf by Phosphorylation

All Raf kinases are phosphoproteins and phosphorylation is essential for Raf regulation. In addition to the phosphorylation sites that mediate 14-3-3 binding (S259 and S621 of Raf-1), Raf proteins are phosphorylated on numerous other residues that modulate its function. In particular, sites required for enzyme activation are found in two distinct regions of the kinase domain: the negative-charge regulatory region (N-Region), located prior to kinase subdomain I, and the activation segment, located between kinase subdomains VII and VIII. The N-region of all Raf proteins contains either negatively charged amino acids or residues that become negatively charged as a result of phosphorylation (Mason et al., 1999). Negative charges in this region are important for relieving the autoinhibition of the N-terminus and for stabilizing the active conformation of the kinase domain (Cutler et al., 1998; Wan et al., 2004). The N-region of both Raf-1 and A-Raf are comprised of phosphoacceptor residues that only become phosphorylated in response to signaling stimuli (S338 and Y340/Y341 for Raf-1; Fabian et al., 1993; King et al., 1998; Marais et al., 1995). In contrast, the N-region of B-Raf contains two negatively charged aspartic acid residues and a serine residue (S445) that is constitutively phosphorylated (Mason et al., 1999). Phosphorylation of the B-Raf S445 site appears to be mediated by numerous cellular kinases, including members of the AGC kinase family, whereas the inducible phosphorylation of Raf-1 on the N-region sites has been found to be mediated by members of the Src family of tyrosine kinases, PAK kinases, and protein kinase C (PKC) family members (Chaudhary et al., 2000; Fabian et al., 1993; Hamilton et al., 2001; King et al., 1998; Marais et al., 1995; Mason et al., 1999).

Similar to other protein kinases, all Raf members require phosphorylation of the activation segment for full kinase activity (Chong et al., 2001; Chong et al., 2003). Phosphorylation of this region is critical for exposing the catalytic cleft and for stabilizing the active kinase conformation. The activation loop residues phosphorylated for Raf-1 are T491 and S494, and for B-Raf, they are T598 and S601. Whether phosphorylation of these sites is mediated in *trans* by oligomerized Raf molecules or is the result of another kinase activity is yet to be determined.

It is important to note that of the mammalian Raf proteins B-Raf exhibits the highest basal activity *in vitro* and the strongest activity towards MEK *in vivo* (Marais et al., 1997). The elevated activity of B-Raf is thought to be due to the charged nature of the B-Raf N-region. Moreover, because the N-region of B-Raf is constitutively phosphorylated, phosphorylation of the activation segment alone is sufficient to turn the enzyme on, perhaps explaining why B-Raf is the only Raf family member that has been found to be mutationally activated in human cancers. Interestingly, the vast majority of oncogenic B-Raf mutations induce conformational changes in the activation segment, with the most prevalent mutation (V600E) mimicking phosphorylation of the loop (Davies et al., 2002; Wan et al., 2004).

Raf function can also be negatively affected by phosphorylation. As mentioned above, S259 phosphorylation inhibits Raf-1 activity by mediating 14-3-3 binding. This site is conserved in all Raf proteins, and both protein kinase A (PKA) and AKT are kinases that phosphorylate Raf-1 on this site (Dumaz and Marais, 2003; Rommel et al., 1999). Similarly, AKT phosphorylates B-Raf on the S259-equivalent site (S364) as well as on two unique sites (S428 and T439) that contribute to kinase inhibition (Guan et al., 2000). In addition, other Raf-1-specific sites are phosphorylated by PKA—S43 and S233 (Dumaz and Marais, 2003). Phosphorylation of S43 has been suggested to interfere with the Ras/Raf interaction (Wu et al., 1993), whereas S233 serves as an additional 14-3-3 binding site (Dumaz and Marais, 2003). By mutational analysis, phosphorylation of all three sites, S43, S233, and S259, has been shown to contribute to Raf-1 inhibition induced by cAMP (Dumaz and Marais, 2003).

Finally, phosphorylation sites have recently been identified that mediate the downregulation of Raf-1 activity following mitogen-induced activation (Dougherty et al., 2005). These sites include S43, as well as five novel SP sites: S29, S289, S296, S301, and S642. Hyperphosphorylation of these six sites inhibits the Ras/Raf interaction and desensitizes Raf-1 to subsequent activational events. Moreover, phosphorylation of all six sites is dependent on downstream MEK signaling, indicating a negative feedback regulatory mechanism.

2.6 Model for Mitogen-induced Raf-1 Activation and Inactivation

After almost two decades of research, the following model for Raf-1 activation has evolved. In a quiescent cell, Raf-1 exists in an inactive state in the cytosol. The inactive conformation of Raf-1 is maintained by autoinhibitory interactions occurring between the N-terminal regulatory and the C-terminal catalytic domains, and by the binding of a 14-3-3 dimer that contacts the N-terminal phospho-S259 and C-terminal phospho-S621 sites. In response to signaling events, Ras becomes activated and GTP-loaded. Binding of the Raf-1 RBD to Ras, together with the PP2A-mediated dephosphorylation of S259, disrupts the interaction of 14-3-3 with the Raf-1 N-terminus, thereby allowing the CRD to contact Ras and membrane phospholipids. Together, these events relieve the autoinhibition of the regulatory domain and induce conformational changes that facilitate phosphorylation of the catalytic domain on activating sites, including those located in the N-region (S338 and Y340/341) and the activation loop (S491 and T494). Once activated, Raf-1 then propagates the signal by phosphorylating and activating the MEKs, which in turn phosphorylate and activate the ERKs. Activation of MEK/ERK signaling induces the hyperphosphorylation of Raf-1 on six sites that contributed to the downregulation and desensitization of Raf-1. Raf-1 is subsequently recycled to a signaling competent state through dephosphorylation events requiring the activities of both PP2A and Pin1. As this model illustrates, Raf-1 is subject to highly complex regulatory mechanisms that contribute to the precise control of its activity under physiological conditions.

3. THE MEKs

The second class of enzymes in the ERK cascade are the MEKs. These proteins belong to a family of dual specificity kinases that phosphorylate threonine and tyrosine residues. MEK1 and MEK2 are the MEK proteins that participate in ERK signaling in mammalian cells (Zheng and Guan, 1993); however, only one MEK kinase is required in worms, flies, and frogs (Kosako et al., 1993). The MEKs are ~45 kDa in size and consist of a C-terminal catalytic domain that is preceded by a short N-terminal region (Figure 2). MEK1 and MEK2 are activated by members of the Raf kinase family and their only known substrates are the ERKs.

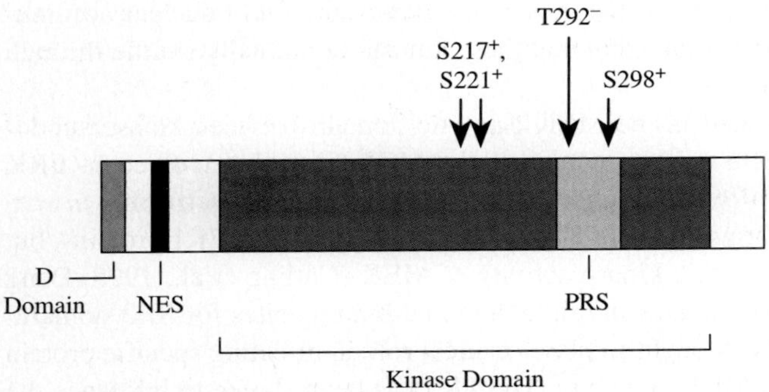

Figure 2. Schematic diagram of MEK1. The docking or D domain, nuclear export sequence (NES), the proline-rich sequence (PRS), and kinase domain are indicated. Activating phosphorylation sites (S218 and S222) are located in the activation segment and modulatory phosphorylation sites are found in the PRS

3.1 Enzymatic Activation of the MEKs

Converting MEK from an inactive to an active enzyme is achieved by the phosphorylation of two highly conserved residues in the activation segment of the kinase domain, namely S217 and S221 for MEK1 and S222 and S226 for MEK2 (Alessi et al., 1994; Mansour et al., 1994b; Zheng and Guan, 1994). Phosphorylation of either site partially increases MEK activity, and a constitutively active MEK kinase can be generated by mimicking the phosphorylation of these sites using negatively charged amino acids (Cowley et al., 1994; Mansour et al., 1994a).

3.2 MEK Regulatory Domains

In addition to the catalytic domain, the MEKs contain three other domains important for their function: an ERK binding site, a nuclear export sequence (NES) and a proline-rich sequence (PRS) insert. Like several other protein kinases, MEK can complex with its substrate via a binding site that is distinct from the catalytic domain active site. The binding site for ERK is located near the MEK N-terminus and is known as a docking or D domain (Tanoue et al., 2000). The D domain is composed of basic and hydrophobic amino acids and is present in other members of this dual specificity kinase family as well as in numerous substrates of the MAPK family. Mutations that modify or delete the D-domain interfere with the ability of MEK to activate ERK *in vivo* (Xu et al., 1999). Further demonstrating the importance of the D domain in MEK/ERK signaling is the finding that a key mechanism of action for anthrax lethal toxins resides in its ability to cleave the D domain from MEK, thereby inhibiting ERK activation (Duesbery et al., 1998).

The MEK NES motif is also found in the N-terminal region and mediates the nuclear export of MEK via CRM-dependent mechanisms (Fukuda et al., 1996). In quiescent cells, MEK is found primarily in the cytoplasm. However, when leptomycin B is used to inhibit CRM-dependent export, the rapid nuclear accumulation of MEK is observed, indicating that these kinases continually shuttle through the nucleus.

The PRS insert is located in the MEK catalytic domain between kinase subdomains IX and X, and is a unique feature of the MEK proteins involved in ERK signaling (Catling et al., 1995). Deletion of the PRS inhibits ERK activation *in vivo* and abolishes the transforming potential of constitutively active MEK1 proteins, but has little effect on the intrinsic kinase activity of MEK (Catling et al., 1995; Dang et al., 1998). This region contains multiple potential binding sites for SH3 domain-containing proteins and is thought to play a critical role in directing specific protein interactions required for MEK signaling. The PRS has been shown to influence the Raf/MEK association even though Raf does not bind directly to the insert (Catling et al., 1995; Dang et al., 1998), and the MP1 scaffolding protein has been found to selectively bind the PRS of MEK1 (Schaeffer et al., 1998). In addition, the MEK1 PRS contains several phosphorylation sites not present in the MEK2 PRS, and these sites have been implicated in the regulation of MEK1 protein interactions and activation kinetics (Frost et al., 1997; Gardner et al., 1994; Mansour et al., 1994b). For example, phosphorylation of S298 in the MEK1 PRS by PAK1 enhances the interaction between MEK1 and ERK2 under conditions of cell adhesion (Eblen et al., 2002). ERK2, in turn, can phosphorylate MEK1 on T292, which acts as a negative feedback mechanism to prevent the PAK-mediated phosphorylation of S298, thus limiting adhesion-mediated MEK1/ERK signaling (Eblen et al., 2004).

3.3 MEK Regulation by Scaffolding Proteins and Signaling Modulators

Another important aspect of MEK regulation is modulating the ability of MEK to interact with its upstream activator Raf and downstream target ERK. Although MEK can associate directly with these proteins, scaffolding molecules and signaling modulators such as KSR, MP1, and RKIP have been found to play a critical role in regulating these interactions as well as modulating MEK's intracellular localization. A more detailed description of these molecules and their role in ERK cascade signaling will be discussed in sections below.

4. THE ERKs

Signaling through the mammalian ERK module culminates with the activation of ERK1 and ERK2 (Boulton et al., 1991; Boulton et al., 1990). These proteins have ~83% amino acid identity and are expressed to various extents in all tissues. The ERKs consist of a central kinase domain that is flanked by short N- and C-terminal noncatalytic regions (Figure 3). Once activated by the MEKs, the ERKs phosphorylate a diverse range of substrates throughout the cell.

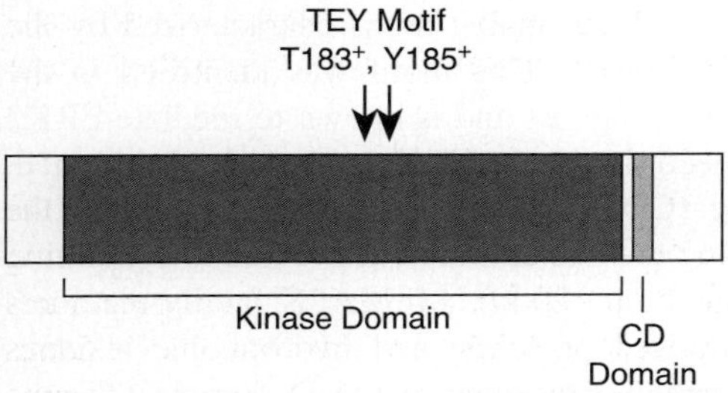

Figure 3. Schematic diagram of ERK2. The common docking (CD) domain and the kinase domain are indicated. Activating phosphorylation sites in the conserved TEY motif of the activation segment are indicated

4.1 Enzymatic Activation of the ERKs

As with the MEKs, full activation of the ERKs is achieved by phosphorylation of two conserved residues in the activation loop of the catalytic domain (Payne et al., 1991). However, in contrast to the MEKs, phosphorylation of one site does not increase ERK activity, and the phosphorylation of these sites cannot be mimicked by negatively charged amino acids (Zhang et al., 1995). The two phosphoacceptor sites in the ERK1/ERK2 activation loop are separated by a glutamic acid residue to give the motif TEY. The tyrosine residue is phosphorylated first, followed by the threonine (Haystead et al., 1992; Robbins et al., 1992), and phosphorylation of these sites is non-processive, apparently requiring two separate interactions with MEK (Ferrell Jr. and Bhatt, 1997).

4.2 Docking Sites for Substrates and Regulators

Activated ERKs phosphorylate serine/threonine residues most often followed by a proline (Gonzalez et al., 1991). ERK substrates are found in various subcellular compartments and include membrane proteins, receptors, cytoskeletal proteins, protein kinases, and nuclear transcription factors. Many of these substrates contain specific interaction motifs that facilitate their phosphorylation by the ERKs (Tanoue and Nishida, 2003). The docking or D domain is a well-characterized binding domain found on numerous substrates as well as regulators of ERK activity, including the MEKs and specific protein phosphatases (Tanoue et al., 2000). The D domain is comprised of basic amino acid residues flanked on one or both sites by hydrophobic residues. These domains can be found upstream or downstream of the phosphoacceptor site, and they can be recognized by more than one class of MAPK.

A second interaction motif found in some ERK substrates consists of the sequence FxF that is often followed by a proline residue. This motif has been named the DEF domain (docking site for ERK1/2, FxFP) based upon preferential binding to the ERKs (Galanis et al., 2001; Jacobs et al., 1999). DEF domains usually lie downstream of the phosphoacceptor sites, and they can occur in combination with

D domains. A third docking domain is a hydrophobic motif characterized by the sequence LxLxxxF (Seidel and Graves, 2002). This motif was identified in the Pointed domain of certain Ets transcription factors and is known to mediate ERK2 binding. Whether this motif exhibits selectivity for the ERKs is yet to be determined.

On the ERKs, a common docking (CD) domain that lies C-terminal to the catalytic domain is involved in mediating the interaction with D domain–containing proteins (Rubinfeld et al., 1999; Tanoue et al., 2000). All MAPK family members possess a CD domain, which is comprised of acidic and hydrophobic residues that interact with the basic and hydrophobic residues of the D domain (Tanoue et al., 2000; Xu et al., 2001). The CD domain is positioned in what has been described as a docking groove, located opposite the catalytic active site in the steric structure of MAPKs (Tanoue et al., 2001). Hydrogen exchange mass spectrometry and mutational analyses have identified other sequence motifs in this docking groove that impact D domain interactions (Lee et al., 2004; Tanoue et al., 2001; Tarrega et al., 2002; Zhang et al., 2003), and it is currently thought that residues throughout the groove contribute to the binding specificity of individual MAPK family members.

Although ERK residues required for binding to the LxLxxxF motif have not been identified, recent experiments have defined a hydrophobic binding pocket that is involved in DEF domain interactions (Lee et al., 2004). Residues proximal to the ERK activation segment contribute to this binding pocket, and conformational changes induced by phosphorylation of the tyrosine residue in the TEY motif are required for full exposure of this binding site (Lee et al., 2004).

4.3 Subcellular Localization of the ERKs

ERK localization can vary depending on the cell type and signaling condition. For example, in quiescent fibroblasts, ERK1 and ERK2 are found primarily in the cytoplasm. However, upon cell stimulation, a significant portion of these molecules accumulates in the nucleus (Chen et al., 1992; Gonzalez et al., 1993; Lenormand et al., 1993). The nuclear localization of ERK contributes to its function and is required for growth factor-induced DNA replication, fibroblast transformation, and PC12 cell differentiation (Brunet et al., 1999; Cowley et al., 1994; Robinson et al., 1998). ERKs do not contain classical nuclear localization or nuclear export sequences (NLS or NES), and exactly how ERK import and export occurs is not fully understood. To date, mechanisms that have been shown to contribute to the regulation of ERK localization include cytoplasmic anchoring, phosphorylation-induced dimerization, nuclear entry by diffusion, active transport across the nuclear membrane, and nuclear retention (Adachi et al., 1999; Fukuda et al., 1997; Khokhlatchev et al., 1998; Lenormand et al., 1998; Matsubayashi et al., 2001; Whitehurst et al., 2002). In particular, the interaction with MEK appears to play a prominent role in keeping ERK localized in the cytosol of unstimulated cells and in mediating its export following signal termination (Fukuda et al., 1997). The MEKs are as abundant as the ERKs and contain a *bona fide* NES. When the ERK binding site on MEK is deleted or the nuclear export of MEK is compromised, ERK rapidly accumulates in the nucleus (Fukuda et al., 1997; Volmat et al., 2001). During normal signaling events, ERK is released from the MEK/ERK complex as a result of its phosphorylation and activation, allowing ERK to then translocate to the nucleus (Fukuda et al., 1997).

The spatial localization of ERK is also influenced by interactions with other cellular proteins. Additional molecules that act as cytoplasmic anchors include calponin (Menice et al., 1997), phosphoprotein enhanced in astrocytes 15 kDa (PEA15) (Formstecher et al., 2001), beta-arrestin (Tohgo et al., 2002), Sef (Toril et al., 2004), and the tyrosine phosphatase PTP-SL (Blanco-Aparicio et al., 1999; Zuniga et al., 1999). Unlike MEK binding, however, the interaction with PEA15, beta arrestin or Sef retains active ERK in the cytosol, thereby directing ERK activity towards cytoplasmic substrates. Protein interactions also contribute to the nuclear accumulation of ERK. In this case, the ERK-dependent transcriptional induction of short-lived nuclear anchoring proteins is required (Lenormand et al., 1998). Candidates for these molecules include the dual specificity phosphatases MKP1 and MKP2.

4.4 ERK Inactivation

Inactivation of the ERKs can be achieved by dephosphorylation of either the phosphothreonine or phosphotyrosine site. As a result, ERKs can be inactivated by all three major classes of protein phosphatases: serine/threonine, tyrosine, and dual specificity phosphatases. A group of dual specificity phosphatases largely dedicated to the inactivation of MAPKs is known as the MAP kinase phosphatases (MKPs; Camps et al., 2000; Keyse, 2000). The MKPs consist of an N-terminal regulatory domain and a C-terminal catalytic domain, and their expression can be restricted to specific subcellular compartments. The substrate specificity of each MKP has not been fully elucidated; however, experimental evidence has implicated MKP1, 2, and 3 in the inactivation of the ERKs. Moreover, ERK signaling has been shown to contribute to the positive regulation of these MKPs. Expression of MKP1 and MKP2 is induced by activation of the ERK cascade, and these proteins are stabilized as a result of ERK phosphorylation (Brondello et al., 1997; Brondello, 1999 #71). In addition, binding of ERK to D domains found in the N-terminal regulatory regions of MKP1 and MKP3 dramatically increases the catalytic activity of these phosphatases (Camps et al., 1998; Slack et al., 2001).

Also implicated in ERK inactivation are the tyrosine phosphatases PTP-SL, STEP, and He-PTP/LC-PTP (Oh-hora et al., 1999; Pettiford and Herbst, 2000; Tarrega et al., 2002) and the serine/threonine phosphatase PP2A (Alessi et al., 1995; Sontag et al., 1993). Which phosphatases are *bona fide* ERK inactivators *in vivo* has been difficult to determine, and it is likely that the specific phosphatases involved may vary depending on the cell type and signaling conditions as well as the subcellular localization of the activated ERK molecules.

5. ERK SCAFFOLDS

Although the kinase components of the ERK module may interact via a series of sequential binary interactions to create a protein kinase cascade, these kinases can also be organized in complexes by scaffolding molecules. The importance of scaffolds to MAPK signaling originates from studies of the Ste5p protein in budding yeast, where Ste5p functions as an essential docking platform for the MAPK components during pheromone-induced mating (Elion, 2001). Scaffolding proteins facilitate kinase activation by colocalizing the components of the module. Moreover, they provide specificity to MAPK signaling by insulating the module from irrelevant stimuli and by regulating the module's intracellular localization. Two molecules that serve as ERK scaffolds for the Ras pathway are KSR and MP1.

5.1 KSR

KSR was discovered to be a positive effector of Ras signaling though genetic studies performed in *Drosophila* and *C. elegans* [Kornfeld, 1995 #307; Sundaram, 1995 #308 (Therrien et al., 1995). KSR homologs are found in vertebrates and invertebrates but not yeast, and two KSR proteins (KSR1 and KSR2) are present in mammalian cells. Following its discovery, it was initially thought that KSR might function as a protein kinase due to the presence of a C-terminal kinase-like domain. However, the fact that the mammalian KSR proteins lack a critical lysine residue normally required for the phosphotransfer reaction (Therrien et al., 1995) together with the finding that mutagenesis of residues predicted to be required for kinase activity do not impair the biological activity of *C. elegans* or *Drosophila* KSR (Roy et al., 2002; Stewart et al., 1999) suggests that KSR is not a functional protein kinase.

The role of KSR as a scaffold began to emerge when KSR1 was found to associate with numerous signaling molecules, including the three kinase components of the ERK cascade. Both MEK1 and MEK2 constitutively interact with the KSR kinase-like domain (Denouel-Galy et al., 1998; Yu et al., 1998), and the function of KSR as an ERK scaffold is dependent on MEK binding as well as the translocation of KSR to the plasma membrane following cell stimulation (Michaud et al., 1997; Müller et al., 2001; Stewart et al., 1999). In quiescent cells, KSR1 is localized predominantly in the cytosol, where it constitutively associates with MEK, a dimer of 14-3-3, Cdc25C-associated kinase 1 (C-TAK1), and the catalytic core subunits of PP2A (Cacace et al., 1999; Müller et al., 2001; Ory et al., 2003). C-TAK1 functions to maintain the phosphorylation status of one (S392) of two 14-3-3 binding sites (S297 and S392), and binding of the 14-3-3 dimer to these sites is critical for retaining the KSR1 complex in the cytosol (Müller et al., 2001). In response to signal activation, the B regulatory subunit of PP2A binds to the KSR-associated PP2A catalytic core complex and stimulates the dephosphorylation of S392, resulting in the release of 14-3-3 from this site and the rapid translocation of the KSR1 complex to the plasma membrane (Ory et al., 2003). The S392/14-3-3 binding site is in close proximity to the KSR1 cysteine-rich C1 domain, and mutational analysis indicates that the C1 domain is required for the translocation of KSR1 to the plasma membrane (Michaud et al., 1997; Zhou et al., 2002). Therefore, it is likely that release of 14-3-3 from the S392 site exposes this region, thereby facilitating membrane localization. In addition, KSR1 proteins mutated at S392 display stronger binding to ERK, indicating that release of 14-3-3 may also expose the FxFP site required for ERK binding (Müller et al., 2001). The end result is that KSR1 localizes MEK together with activated Raf-1 at the plasma membrane and provides a docking platform for ERK, thereby facilitating the sequential phosphorylation events required for ERK cascade signaling.

5.2 MP1

Another protein that has been demonstrated to have ERK scaffolding properties is MEK-partner 1 (MP1). MP1 is a small 13.5 kDa protein that was first isolated in a yeast two-hybrid screen for MEK1-interacting proteins. MP1 binds MEK1 and ERK1 to the exclusion of MEK2 and ERK2, and selectively promotes the activation of ERK1 (Schaeffer et al., 1998). MP1 localizes to late endosomes through a constitutive interaction with a highly conserved 14 kDa adaptor protein, known as p14 (Wunderlich et al., 2001). Studies using RNAi to address the functional importance of the MP1/p14 complex in signal transmission indicate that the endosomal localization of this complex is required for full ERK activation in response to EGF stimulation (Teis et al., 2002). Whether the localization of ERK1 signaling to endosomes by the p14/MP1 complex reflects the completion or extension of a signal transduction process that is initiated at the plasma membrane or whether it is qualitatively different from ERK-mediated signals emanating from the plasma membrane has yet to be determined.

6. SIGNALING MODULATORS

Because diverse extracellular cues are transduced through the Ras/ERK cascade, qualitative differences in properties such as duration, amplitude and subcellular localization of the ERK signal are important for achieving the appropriate biological response. These aspects of ERK signaling are influenced by the activities of cellular proteins that modulate the activation and/or function of the core pathway components (Figure 4). Several such signaling modulators have now been identified through genetic, molecular, and biochemical approaches. Genetics screens in model organisms such as *Drosophila* and *C. elegans*, for instance, have led to the identification of the KSR scaffold as well as Sur-8 and CNK, two modulators that augment Ras/ERK signaling. Proteins that antagonize signal transmission have also been identified and include RIN1, Erbin, Spry, Spred, RKIP, and IMP.

6.1 Sur-8

The Sur-8 protein is comprised of 18 tandem leucine-rich repeats (LRR) and is found in all multicellular organisms. Sur-8 was initially discovered in *C. elegans*, where genetic epistasis experiments positioned it to function downstream of Ras

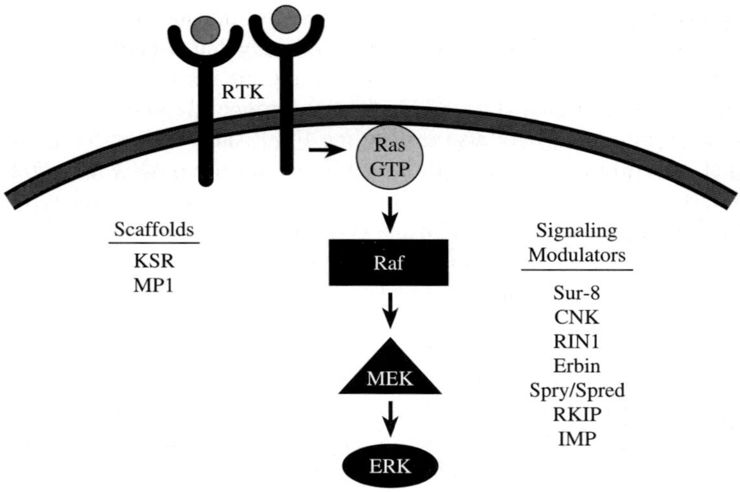

Figure 4. The Ras/Raf/MEK/ERK signaling cascade. The Raf/MEK/ERK protein kinases constitute a key effector cascade used by Ras to relay many diverse extracellular cues, such as those that are received at the cell surface by receptor tyrosine kinases (RTK). Raf is the initiating kinase that interacts with membrane-localized GTP-bound Ras. The signal is then transduced from RAF to MEK and from MEK to ERK. In addition to the core enzymes of the cascade, various scaffolding proteins (KSR and MP1) and signaling modulators (Sur-8, CNK, RIN1, Erbin, Spry/Spred, RKIP, and IMP) have been identified that affect the efficiency and level of signaling through this important kinase cascade

but upstream or in parallel to Raf (Sieburth et al., 1998). By two-hybrid analysis, Sur-8 interacts with Ras, but not other related small GTPases, and shows preferential binding to K-Ras (Li et al., 2000). Sur-8 has also been found to complex with Ras and Raf-1 in mammalian cells and, when overexpressed, to increase the Ras-dependent activation of Raf-1 (Li et al., 2000). These findings have led to the model that Sur-8 acts to enhance the signaling strength of Ras by promoting the Ras/Raf interaction.

6.2 CNK

CNK is also found in all multicellular organisms and contains several protein interaction motifs (Therrien et al., 1998). Members of the CNK family, which includes the mammalian membrane-associated guanylate kinase interacting (MAGUIN) proteins (Yao et al., 1999), possess a sterileα motif (SAM), a conserved region in CNK (CRIC) domain, a PSD-95/DLG-1/ZO-1 (PDZ) domain, and a pleckstrin homology (PH) domain. Both CNK and the MAGUIN proteins have been reported to interact with the C-terminal kinase domain of Raf and have been shown to partially localize to the plasma membrane (Therrien et al., 1998; Yao et al., 2000). The biological importance of CNK in ERK signaling has been addressed in *Drosophila* S2 tissue culture cells by depleting CNK protein levels. In these studies, CNK was found to be required for activated Ras-, insulin-, and phorbol ester-mediated Raf activation and to be responsible for the compartmentalization of a pool of Raf at the plasma membrane (Anselmo et al., 2002). These results suggest that CNK may facilitate the localization of Raf to membrane-bound signaling complexes, thereby resulting in efficient Raf activation. Interestingly, CNK has also been reported to interact with components of the Ral and Rho signaling pathways (Jaffe et al., 2004; Lanigan et al., 2003) and with the human tumor suppressor RASSF1A (Rabizadeh et al., 2004). These findings suggest that CNK functions in multiple Ras-dependent pathways and that it may act as a signal integrator, mediating or regulating crosstalk between these various effector cascades.

6.3 RIN1

RIN1 is 84 kDa protein that was discovered in a genetic screen for mammalian proteins that antagonize Ras signaling in *S. cerevisiae* (Han and Colicelli, 1995). RIN1 is a classical Ras effector that contains an RBD and interacts with the Ras effector domain in a GTP-dependent manner. RIN1 binds Ras with affinities similar to that of Raf (22 nM for the RIN1 RBD and 18 nM for the Raf-1 RBD) and can act as a competitive inhibitor of the Ras/Raf interaction (Wang et al., 2002). RIN1 is expressed at low to undetectable levels in most tissues, but in cells where it is highly expressed, such as a subset of brain neurons, RIN1 has been suggested to redirect Ras signaling from the Raf/MEK/ERK cascade to other downstream effectors (Dhaka et al., 2003).

6.4 Erbin

Another negative regulator of Ras/ERK cascade signaling is Erbin. The 180 kDa Erbin protein was originally identified as a ErbB2 binding partner and is a member of the LRR and PDZ domain-containing protein (LAP) family (Borg et al., 2000). Like RIN1, Erbin appears to inhibit ERK signaling by directly interfering with the binding of Raf to activated Ras (Huang et al., 2003). Overexpression of Erbin blocks ERK activation induced by stimulated RTKs and activated Ras but not that induced by activated Raf. Erbin has also been found to coimmunoprecipitate with activated Ras and, when overexpressed, to reduce the amount of Raf bound to Ras. Erbin does not contain a classical RBD, but does possess LRR sequences similar to those found in the Ras-interacting Sur-8 protein, and it is likely that this is the region of Erbin that mediates the interaction with Ras, given that the inhibitory effects of Erbin are dependent on the LRR sequences.

6.5 Sprouty (Spry) and Spred Proteins

The Sprouty (Spry) proteins are conserved inducible inhibitors of RTK/Ras/ERK cascade signaling. The first member of this family was discovered genetically as an antagonist of FGF receptor signaling in *Drosophila* (Hacohen et al., 1998).

Subsequent characterization of this protein family has revealed that the Sprys specifically target RTK/ERK signaling, with no inhibitory effect on JNK, p38 MAPK, or phosphoinositide 3-kinase pathways (Yusoff et al., 2002). Four Spry proteins have been identified in humans, all of which contain a conserved C-terminal, cysteine-rich region that mediates the oligomerization and plasma membrane localization of this protein family.

Spry proteins have been found to influence RTK signaling at multiple points in the cascade depending on the biological context. When cells are stimulated with various growth factors, a conserved tyrosine residue in the N-terminal region of Sprys (Y55 in human Spry2) becomes phosphorylated (Fong et al., 2003; Hanafusa et al., 2002; Rubin 2003 #49). In the context of FGF signaling, this phosphotyrosine residue binds the SH2 domain of Grb2, an adaptor which couples the stimulation of the FGF receptor to Ras activation via the recruitment of the guanylnucleotide exchange factor Sos to either the FGF-receptor substrate 2 (FRS2) or SH2 domain-containing protein tyrosine phosphatase 2 (SHP2) (Hanafusa et al., 2002). Thus, by sequestering Grb2, Spry proteins inhibit Ras activation and ERK cascade signaling. Interestingly, in response to EGF receptor activation, the same Spry phosphotyrosine residue competes with activated EGF receptors for binding to the SH2 domain of c-Cbl, an E3 ubiquitin ligase (Fong et al., 2003; Rubin et al., 2003). As a result, Sprys are ubiquitinated and degraded by the proteosome (Hall et al., 2003; Rubin et al., 2003), whereas expression of the EGF receptor on the cell surface is prolonged (Rubin et al., 2003; Wong et al., 2002), resulting in sustained RTK/Ras/ERK cascade signaling.

The Spry proteins also interact with the catalytic domain of Raf-1 through a conserved Ras binding motif present in their cysteine-rich domains (Sasaki et al., 2003). This interaction blocks the activation of Raf-1 mediated by PKC-δ, but appears to have no effect on Ras-dependent Raf-1 activation (Sasaki et al., 2003). In contrast, the Sprouty-related proteins with EVH1 domains (Spreds) are capable of blocking Ras-dependent Raf-1 activation (Wakioka et al., 2001). Spreds have been found to associate with Ras, but this interaction does not prevent growth factor-induced Ras activation or inhibit the Ras/Raf interaction. Strikingly, however, phosphorylation of Raf-1 on the activating S338 site is blocked, suggesting that when Raf-1 is associated with the Spred/Ras complex, it may not be accessible to its upstream activating kinase.

6.6 RKIP

RKIP is a 20 kDa protein that antagonizes Ras signaling by inhibiting the Raf/MEK interaction. RKIP belongs to a highly conserved family of phosphotidylethanolamine binding proteins and was identified as a Raf-interacting protein by yeast two-hybrid analysis (Yeung et al., 1999). RKIP was subsequently shown to interact with the kinase domain of both MEK and Raf, but due to overlapping binding sites, binds Raf and MEK in a mutually exclusive manner (Yeung et al., 2000). *In vitro*, RKIP is able to dissociate Raf/MEK complexes and behaves as a competitive inhibitor of MEK phosphorylation. In cells, overexpression of RKIP blocks Ras-dependent ERK signaling, whereas downregulation of RKIP has the opposite effect. PKC-mediated phosphorylation of RKIP on S153 disrupts the RKIP/Raf interaction and appears to account for the release of RKIP from Raf following signaling events (Corbit et al., 2003). Based on these findings, RKIP is thought to play an important role in preventing aberrant MEK activation in the absence of signaling events.

6.7 IMP

IMP is an E3 ubiquitin ligase that has been proposed to modulate ERK signaling by regulating the ERK scaffolding activity of KSR1 (Matheny et al., 2004). IMP was isolated in a yeast two-hybrid screen for proteins that bind activated Ras. IMP is a ubiquitously expressed, highly conserved protein that contains three structural motifs: a RING-H2 domain followed by an ubiquitin-protease-like zinc finger (UBP-ZnF; also known as a PAZ domain) and leucine heptad repeats. As would be expected for a Ras effector, IMP binds Ras in a GTP- and stimulus-dependent manner; however, IMP does not possess a classical RBD and instead has been reported to interact with Ras through a region encompassing the UBP-ZnF domain. IMP also interacts with KSR1, and this association apparently promotes the mislocalization and sequestration of KSR1 in a cellular compartment that is inaccessible to upstream activators of KSR1 function. Ras activation appears to relieve the IMP-mediated repression of KSR1 by recruiting IMP to the cell surface and inducing its autoubiquitination. The proposed model for IMP is that it acts as a negative regulator to keep Ras signaling in check until cells receive an activating stimulus. In addition, because depletion of IMP protein levels has been found to allow cells to respond to suboptimal doses of an extracellular stimulus, IMP may modulate the threshold sensitivity of cells to stimulus, permitting them to adapt to chronic or complex signaling conditions.

7. CONCLUDING REMARKS

The past two decades have witnessed tremendous advances in our understanding of the Ras/ERK signaling cascade. Following the initial discoveries of the individual components of the pathway, such as Ras, Raf, MEK, and ERK, many years of study were devoted to the elucidation of their biochemical functions and enzyme-substrate relationships within the cascade. In recent years, the accelerating pace of discovery has identified numerous auxiliary factors, such as scaffolding proteins and signaling modulators, that play key roles in the spatiotemporal control of the cascade. While much remains to be done, it has already become apparent that Ras/Raf/MEK/ERK signal transduction is controlled by highly dynamic and complex regulatory mechanisms. Understanding the details of these events will be crucial in comprehending how this important signaling pathway mediates such diverse biological responses.

REFERENCES

Abraham, D., Podar, K., Pacher, M., Kubicek, M., Welzel, N., Hemmings, B.A., Dilworth, S.M., Mischak, H., Kolch, W. and Baccarini, M. (2000) Raf-1-associated protein phosphatase 2A as a positive regulator of kinase activation. J. Biol. Chem. *275*, 22300-22304.

Adachi, M., Fukuda, M. and Nishida, E. (1999) Two co-existing mechanisms for nuclear import of MAP kinase: passive diffusion of a monomer and active transport of a dimer. EMBO J. *18*, 5347-5358.

Alessi, D.R., Saito, Y., Campbell, D.G., Cohen, P., Sithanandam, G., Rapp, U., Ashworth, A., Marshall, C.J. and Cowley, S. (1994) Identification of the sites in MAP kinase kinase-1 phosphorylated by p74raf-1. EMBO J. *13*, 1610-1619.

Alessi, D.R., Gomez, N., Moorhead, G., Lewis, T., Keyse, S.M. and Cohen, P. (1995) Inactivation of p42 MAP kinase by protein phosphatase 2A and a protein tyrosine phosphatase, but not CL100, in various cell lines. Curr. Biol. *5*, 283-295.

Anselmo, A.N., Bumeister, R., Thomas, J.M. and White, M.A. (2002) Critical contribution of linker proteins to Raf kinase activation. J. Biol. Chem. *277*, 5940-5943.

Blanco-Aparicio, C., Torres, J. and Pulido, R. (1999) A novel regulatory mechanism of MAP kinases activation and nuclear translocation mediated by PKA and the PTP-SL tyrosine phosphatase. J. Cell Biol. *147*, 1129-1136.

Borg, J.P., Marchetto, S., Le Bivic, A., Ollendorff, V., Jaulin-Bastard, F., Saito, H., Fournier, E., Adelaide, J., Margolis, B. and Birnbaum, D. (2000) ERBIN: a basolateral PDZ protein that interacts with the mammalian ERBB2/HER2 receptor. Nat. Cell Biol. *2*, 407-414.

Boulton, T.G., Yancopoulos, G.D., Gregory, J.S., Slaughter, C., Moomaw, C., Hsu, J. and Cobb, M.H. (1990) An insulin-stimulated protein kinase similar to yeast kinases involved in cell cycle control. Science *249*, 64-67.

Boulton, T.G., Nye, S.H., Robbins, D.J., Ip, N.Y., Radziejewska, E., Morgenbesser, S.D., DePinho, R.A., Panayotatos, N., Cobb, M.H. and Yancopoulos, G.D. (1991) ERKs: a family of protein-serine/threonine kinases that are activated and tyrosine phosphorylated in response to insulin and NGF. Cell *65*, 663-675.

Brondello, J.M., Brunet, A., Pouyssegur, J. and McKenzie, F.R. (1997) The dual specificity mitogen-activated protein kinase phosphatase-1 and -2 are induced by the p42/p44 MAPK cascade. J. Biol. Chem. *272*, 1368-1376.

Brondello, J.M., Pouyssegur, J. and Mckenzie, F.R. (1999) Reduced MAP kinase phosphatase-1 degradation after p42/p44 MAPK-dependent phosphorylation. Science *286*, 2514-2517.

Brunet, A., Roux, D., Lenormand, P., Dowd, S., Keyse, S. and Pouyssegur, J. (1999) Nuclear translocation of p42/p44 mitogen-activated protein kinase is required for growth factor-induced gene expression and cell cycle entry. EMBO J. *18*, 664-674.

Cacace, A.M., Michaud, N.R., Therrien, M., Mathes, K., Copeland, T., Rubin, G.M. and Morrison, D.K. (1999) Identification of constitutive and ras-inducible phosphorylation sites of KSR: implications for 14-3-3 binding, mitogen-activated protein kinase binding, and KSR overexpression. Mol. Cell. Biol. *19*, 229-240.

Camps, M., Nichols, A., Gillieron, C., Antonsson, B., Muda, M., Chabert, C., Boschert, U. and Arkinstall, S. (1998) Catalytic activation of the phosphatase MKP-3 by ERK2 mitogen-activated protein kinase. Science *280*, 1262-1265.

Camps, M., Nichols, A. and Arkinstall, S. (2000) Dual specificity phosphatases: a gene family for control of MAP kinase function. FASEB J. *14*, 6-16.

Catling, A.D., Schaeffer, H.J., Reuter, C.W., Reddy, G.R. and Weber, M.J. (1995) A proline-rich sequence unique to MEK1 and MEK2 is required for raf binding and regulates MEK function. Mol. Cell. Biol. *15*, 5214-5225.

Chaudhary, A., King, W.G., Mattaliano, M.D., Frost, J.A., Diaz, B., Morrison, D.K., Cobb, M.H., Marshall, M.S. and Brugge, J.S. (2000) Phosphatidylinositol 3-kinase regulates Raf1 through Pak phosphorylation of serine 338. Curr. Biol. *10*, 551-554.

Chen, R.H., Sarnecki, C. and Blenis, J. (1992) Nuclear localization and regulation of erk- and rsk-encoded protein kinases. Mol. Cell. Biol. *12*, 915-927.

Chong, H., Lee, J. and Guan, K.L. (2001) Positive and negative regulation of Raf kinase activity and function by phosphorylation. EMBO J. *20*, 3716-3727.

Chong, H. and Guan, K.-L. (2003) Regulation of Raf through phosphorylation and N-terminus–C terminus interaction. J. Biol. Chem. *278*, 36269-36276.

Chong, H., Vikis, H.G. and Guan, K.L. (2003) Mechanisms of regulating the Raf kinase family. Cell. Signal. *15*, 463-469.

Clark, G.J., Drugan, J.K., Rossman, K.L., Carpenter, J.W., Rogers-Graham, K., Fu, H., Der, C.J. and Campbell, S.L. (1997) 14-3-3 zeta negatively regulates raf-1 activity by interactions with the Raf-1 cysteine-rich domain. J. Biol. Chem. *272*, 20990-20993.

Cook, S.J., Rubinfeld, B., Albert, I. and McCormick, F. (1993) RapV12 antagonizes Ras-dependent activation of ERK1 and ERK2 by LPA and EGF in Rat-1 fibroblasts. EMBO J. *12*, 3475-3485.

Corbit, K.C., Trakul, N., Eves, E.M., Diaz, B., Marshall, M. and Rosner, M.R. (2003) Activation of Raf-1 signaling by protein Kinase C through a mechanism Involving Raf kinase inhibitory protein. J. Biol. Chem. *278*, 13061-13068.

Cowley, S., Paterson, H., Kemp, P. and Marshall, C.J. (1994) Activation of MAP kinase kinase is necessary and sufficient for PC12 differentiation and for transformation of NIH 3T3 cells. Cell *77*, 841-852.

Cutler, R.E.J., Stephens, R.M., Saracino, M.R. and Morrison, D.K. (1998) Autoregulation of the Raf-1 serine/threonine kinase. Proc. Natl. Acad. Sci. U.S.A. *95*, 9214-9219.

Dang, A., Frost, J.A. and Cobb, M.H. (1998) The MEK1 proline-rich insert is required for efficient activation of the mitogen-activated protein kinases ERK1 and ERK2 in mammalian cells. J. Biol. Chem. *273*, 19909-19913.

Daum, G., Eisenmann-Tappe, I., Fries, H.W., Troppmair, J. and Rapp, U.R. (1994) The ins and outs of Raf kinases. Trends Biochem. Sci. *19*, 474-480.

Davies, H., Bignell, G.R., Cox, C., Stephens, P., Edkins, S., Clegg, S., Teague, J., Woffendin, H., Garnett, M.J., Bottomley, W., Davis, N., Dicks, E., Ewing, R., Floyd, Y., Gray, K., Hall, S., Hawes, R., Hughes, J., Kosmidou, V., Menzies, A., Mould, C., Parker, A., Stevens, C., Watt, S., Hooper, S., Wilson, R., Jayatilake, H., Gusterson, B.A., Cooper, C., Shipley, J., Hargrave, D., Pritchard-Jones, K., Maitland, N., Chenevix-Trench, G., Riggins, G.J., Bigner, D.D., Palmieri, G., Cossu, A., Flanagan, A., Nicholson, A., Ho, J.W., Leung, S.Y., Yuen, S.T., Weber, B.L., Seigler, H.F., Darrow, T.L. and Paterson, H. (2002) Mutations of the BRAF gene in human cancer. Nature *417*, 949-954.

Denouel-Galy, A., Douville, E.M., Warne, P.H., Papin, C., Laugier, D., Calothy, G., Downward, J. and Eychene, A. (1998) Murine Ksr interacts with MEK and inhibits Ras-induced transformation. Curr. Biol. *8*, 46-55.

Dhaka, A., Costa, R.M., Hu, H., Irvin, D.K., Patel, A., Kornblum, H.I., Silva, A.J., O'Dell, T.J. and Colicelli, J. (2003) The RAS effector RIN1 modulates the formation of aversive memories. J. Neurosci. *23*, 748-757.

Dhillon, A.S., Meikle, S., Yazici, Z., Eulitz, M. and Kolch, W. (2002) Regulation of Raf-1 activation and signalling by dephosphorylation. EMBO J. *21*, 64-71.

Dougherty, M.K., Müller, J., Ritt, D.A., Zhou, M., Zhou, X.Z., Copeland, T.D., Conrads, T.P., Veenstra, T.D., Lu, K.P. and Morrison, D.K. (2005) Regulation of Raf-1 by direct feedback phosphorylation. Mol. Cell *17*, 215-224.

Duesbery, N.S., Webb, C.P., Leppla, S.H., Gordon, V.M., Klimpel, K.R., Copeland, T.D., Ahn, N.G., Oskarsson, M.K., Fukasawa, K., Paull, K.D. and Vande Woude, G.F. (1998) Proteolytic inactivation of MAP-kinase-kinase by anthrax lethal factor. Science *280*, 734-737.

Dumaz, N. and Marais, R. (2003) Protein kinase A blocks Raf-1 activity by stimulating 14-3-3 binding and blocking Raf-1 interaction with Ras. J. Biol. Chem. *278*, 29819-29823.

Eblen, S.T., Slack, J.K., Weber, M.J. and Catling, A.D. (2002) Rac-PAK signaling stimulates extracellular signal-regulated kinase (ERK) activation by regulating formation of MEK1-ERK complexes. Mol. Cell. Biol. *22*, 6023-6033.

Eblen, S.T., Slack-Davis, J.K., Tarcsafalvi, A., Parsons, J.T., Weber, M.J. and Catling, A.D. (2004) Mitogen-activated protein kinase feedback phosphorylation regulates MEK1 complex formation and activation during cellular adhesion. Mol. Cell. Biol. *24*, 2308-2317.

Elion, E.A. (2001) The Ste5p scaffold. J. Cell Sci. *114*, 3967-3978.
Fabian, J.R., Daar, I.O. and Morrison, D.K. (1993) Critical tyrosine residues regulate the enzymatic and biological activity of Raf-1 kinase. Mol. Cell. Biol. *13*, 7170-7179.
Farrar, M.A., Alberol, I. and Perlmutter, R.M. (1996) Activation of the Raf-1 kinase cascade by coumermycin-induced dimerization. Nature *383*, 178-181.
Ferrell Jr., J.E. and Bhatt, R.R. (1997) Mechanistic studies of the dual phosphorylation of mitogen-activated protein kinase. J. Biol. Chem. *272*, 19008-19016.
Fong, C.W., Leong, H.F., Wong, E.S.M., Lim, J., Yusoff, P. and Guy, G.R. (2003) Tyrosine phosphorylation of Sprouty2 enhances its interaction with c-Cbl and is crucial for its function. J. Biol. Chem. *278*, 33456-33464.
Formstecher, E., Ramos, J.W., Fauquet, M., Calderwood, D.A., Hsieh, J.C., Canton, B., Nguyen, X.T., Barnier, J.V., Camonis, J., Ginsberg, M.H. and Chneiweiss, H. (2001) PEA-15 mediates cytoplasmic sequestration of ERK MAP kinase. Dev. Cell *1*, 239-250.
Frost, J.A., Steen, H., Shapiro, P., Lewis, T., Ahn, N., Shaw, P.E. and Cobb, M.H. (1997) Cross-cascade activation of ERKs and ternary complex factors by Rho family proteins. EMBO J. *16*, 6426-6438.
Fukuda, M., Gotoh, I., Gotoh, Y. and Nishida, E. (1996) Cytoplasmic localization of mitogen-activated protein kinase kinase directed by its NH2-terminal, leucine-rich short amino acid sequence, which acts as a nuclear export signal. J. Biol. Chem. *271*, 20024-20028.
Fukuda, M., Gotoh, Y. and Nishida, E. (1997) Interaction of MAP kinase with MAP kinase kinase: its possible role in the control of nucleocytoplasmic transport of MAP kinase. EMBO J. *16*, 1901-1908.
Galanis, A., Yang, S.H. and Sharrocks, A.D. (2001) Selective targeting of MAPKs to the ETS domain transcription factor SAP-1. J. Biol. Chem. *276*, 965-973.
Gardner, A.M., Vaillancourt, R.R., Lange-Carter, C.A. and Johnson, G.L. (1994) MEK-1 phosphorylation by MEK kinase, Raf, and mitogen-activated protein kinase: analysis of phosphopeptides and regulation of activity. Mol. Biol. Cell *5*, 193-201.
Ghosh, S., Strum, J.C., Sciorra, V.A., Daniel, L. and Bell, R.M. (1996) Raf-1 kinase possesses distinct binding domains for phosphatidylserine and phosphatidic acid. J. Biol. Chem. *271*, 8472-8480.
Gonzalez, F., Seth, A., Raden, D., Bowman, D., Fay, F. and Davis, R. (1993) Serum-induced translocation of mitogen-activated protein kinase to the cell surface ruffling membrane and the nucleus. J. Cell Biol. *122*, 1089-1101.
Gonzalez, F.A., Raden, D.L. and Davis, R.J. (1991) Identification of substrate recognition determinants for human ERK1 and ERK2 protein kinases. J. Biol. Chem. *266*, 22159-22163.
Guan, K.-L., Figueroa, C., Brtva, T.R., Zhu, T., Taylor, J., Barber, T.D. and Vojtek, A.B. (2000) Negative regulation of the serine/threonine kinase B-Raf by Akt. J. Biol. Chem. *275*, 27354-27359.
Hacohen, N., Kramer, S., Sutherland, D., Hiromi, Y. and Krasnow, M.A. (1998) Sprouty encodes a novel antagonist of FGF signaling that patterns apical branching of the Drosophila airways. Cell *92*, 253-263.
Hagemann, C. and Rapp, U.R. (1999) Isotype-specific functions of Raf kinases. Exp. Cell Res. *253*, 34-46.
Hall, A.B., Jura, N., DaSilva, J., Jang, Y.J., Gong, D. and Bar-Sagi, D. (2003) hSpry2 Is targeted to the ubiquitin-dependent proteasome pathway by c-Cbl. Curr. Biol. *13*, 308-314.
Hamilton, M., Liao, J., Cathcart, M.K. and Wolfman, A. (2001) Constitutive association of c-N-Ras with c-Raf-1 and protein kinase C epsilon in latent signaling modules. J. Biol. Chem. *276*, 29079-29090.
Han, L. and Colicelli, J. (1995) A human protein selected for interference with Ras function interacts directly with Ras and competes with Raf1. Mol. Cell. Biol. *15*, 1318-1323.
Hanafusa, H., Torii, S., Yasunaga, T. and Nishida, E. (2002) Sprouty1 and Sprouty2 provide a control mechanism for the Ras/MAPK signalling pathway. Nat. Cell Biol. *4*, 850-858.
Haystead, T.A., Dent, P., Wu, J., Haystead, C.M. and Sturgill, T.W. (1992) Ordered phosphorylation of p42mapk by MAP kinase kinase. FEBS Lett. *306*, 17-22.
Heidecker, G., Huleihel, M., Cleveland, J.L., Kolch, W., Beck, T.W., Lloyd, P., Pawson, T. and Rapp, U.R. (1990) Mutational activation of c-raf-1 and definition of the minimal transforming sequence. Mol. Cell. Biol. *10*, 2503-2512.

Hekman, M., Hamm, H., Villar, A.V., Bader, B., Kuhlmann, J., Nickel, J. and Rapp, U.R. (2002) Associations of B- and C-Raf with cholesterol, phosphatidylserine, and lipid second messengers. J. Biol. Chem. *277*, 24090-24102.

Herrmann, C., Martin, G.A. and Wittinghofer, A. (1995) Quantitative analysis of the complex between p21ras and the Ras-binding domain of the human Raf-1 protein kinase. J. Biol. Chem. *270*, 2901-2905.

Hu, C.D., Kariya, K., Tamada, M., Akasaka, K., Shirouzu, M., Yokoyama, S. and Kataoka, T. (1995) Cysteine-rich region of Raf-1 interacts with activator domain of post-translationally modified Ha-Ras. J. Biol. Chem. *270*, 30274-30277.

Huang, Y.Z., Zang, M., Xiong, W.C., Luo, Z. and Mei, L. (2003) Erbin suppresses the MAP kinase pathway. J. Biol. Chem. *278*, 1108-1114.

Jacobs, D., Glossip, D., Xing, H., Muslin, A.J. and Kornfeld, K. (1999) Multiple docking sites on substrate proteins form a modular system that mediates recognition by ERK MAP kinase. Genes Dev. *13*, 163-175.

Jaffe, A.B., Aspenstrom, P. and Hall, A. (2004) Human CNK1 acts as a scaffold protein, linking Rho and Ras signal transduction pathways. Mol. Cell. Biol. *24*, 1736-1746.

Keyse, S.M. (2000) Protein phosphatases and the regulation of mitogen-activated protein kinase signalling. Curr. Opin. Cell Biol. *12*, 186-192.

Khokhlatchev, A.V., Canagarajah, B., Wilsbacher, J., Robinson, M., Atkinson, M., Goldsmith, E. and Cobb, M.H. (1998) Phosphorylation of the MAP kinase ERK2 promotes its homodimerization and nuclear translocation. Cell *93*, 605-615.

King, A.J., Sun, H., Diaz, B., Barnard, D., Miao, W., Bagrodia, S. and Marshall, M.S. (1998) The protein kinase Pak3 positively regulates Raf-1 activity through phosphorylation of serine 338. Nature *396*, 180-183.

Kornfeld, K., Hom, D.B. and Horvitz, H.R. (1995) The Ksr-1 gene encodes a novel protein kinase involved in Ras-mediated signaling in C. elegans. Cell *83*, 903-913.

Kosako, H., Nishida, E. and Gotoh, Y. (1993) cDNA cloning of MAP kinase kinase reveals kinase cascade pathways in yeasts to vertebrates. EMBO J. *12*, 787-794.

Lanigan, T.M., Liu, A., Huang, Y.Z., Mei, L., Margolis, B. and Guan, K.L. (2003) Human homologue of Drosophila CNK interacts with Ras effector proteins Raf and Rlf. FASEB J. *17*, 2048-2060.

Lee, T., Hoofnagle, A.N., Kabuyama, Y., Stroud, J., Min, X., Goldsmith, E.J., Chen, L., Resing, K.A. and Ahn, N.G. (2004) Docking motif interactions in MAP kinases revealed by hydrogen exchange mass spectrometry. Mol. Cell *14*, 43-55.

Leevers, S.J., Paterson, H.F. and Marshall, C.J. (1994) Requirement for Ras in Raf activation is overcome by targeting Raf to the plasma membrane. Nature *369*, 411-414.

Lenormand, P., Sardet, C., Pages, G., L'Allemain, G., Brunet, A. and Pouyssegur, J. (1993) Growth factors induce nuclear translocation of MAP kinases (p42mapk and p44mapk) but not of their activator MAP kinase kinase (p45mapkk) in fibroblasts. J. Cell Biol. *122*, 1079-1088.

Lenormand, P., Brondello, J.-M., Brunet, A. and Pouyssegur, J. (1998) Growth factor-induced p42/p44 MAPK nuclear translocation and retention requires both MAPK activation and neosynthesis of nuclear anchoring proteins. J. Cell Biol. *142*, 625-633.

Li, W., Han, M. and Guan, K.-L. (2000) The leucine-rich repeat protein SUR-8 enhances MAP kinase activation and forms a complex with Ras and Raf. Genes Dev. *14*, 895-900.

Luo, Z., Tzivion, G., Belshaw, P.J., Vavvas, D., Marshall, M. and Avruch, J. (1996) Oligomerization activates c-Raf-1 through a Ras-dependent mechanism. Nature *383*, 181-185.

Luo, Z., Diaz, B., Marshall, M.S. and Avruch, J. (1997) An intact Raf zinc finger is required for optimal binding to processed Ras and for Ras-dependent Raf activation in situ. Mol. Cell. Biol. *17*, 46-53.

Mansour, S.J., Matten, W.T., Hermann, A.S., Candia, J.M., Rong, S., Fukasawa, K., Vande Woude, G.F. and Ahn, N.G. (1994a) Transformation of mammalian cells by constitutively active MAP kinase kinase. Science *265*, 966-970.

Mansour, S.J., Resing, K.A., Candi, J.M., Hermann, A.S., Gloor, J.W., Herskind, K.R., Wartmann, M., Davis, R.J. and Ahn, N.G. (1994b) Mitogen-activated protein (MAP) kinase phosphorylation of MAP kinase kinase: determination of phosphorylation sites by mass spectrometry and site-directed mutagenesis. J. Biochem. *116*, 304-314.

Marais, R., Light, Y., Paterson, H.F. and Marshall, C.J. (1995) Ras recruits Raf-1 to the plasma membrane for activation by tyrosine phosphorylation. EMBO J. *14*, 3136-3145.

Marais, R., Light, Y., Paterson, H.F., Mason, C.S. and Marshall, C.J. (1997) Differential regulation of Raf-1, A-Raf, and B-Raf by oncogenic ras and tyrosine kinases. J. Biol. Chem. *272*, 4378-4383.

Mason, C.S., Springer, C.J., Cooper, R.G., Superti-Furga, G., Marshall, C.J. and Marais, R. (1999) Serine and tyrosine phosphorylations cooperate in Raf-1, but not B-Raf activation. EMBO J. *18*, 2137-2148.

Matheny, S.A., Chen, C., Kortum, R.L., Razidlo, G.L., Lewis, R.E. and White, M.A. (2004) Ras regulates assembly of mitogenic signalling complexes through the effector protein IMP. Nature *427*, 256-260.

Matsubayashi, Y., Fukuda, M. and Nishida, E. (2001) Evidence for existence of a nuclear pore complex-mediated, cytosol-independent pathway of nuclear translocation of ERK MAP kinase in permeabilized cells. J. Biol. Chem. *276*, 41755-41760.

McPherson, R.A., Harding, A., Roy, S., Lane, A. and Hancock, J.F. (1999) Interactions of c-Raf-1 with phosphatidylserine and 14-3-3. Oncogene *18*, 3862-3869.

Menice, C.B., Hulvershorn, J., Adam, L.P., Wang, C.-L.A. and Morgan, K.G. (1997) Calponin and mitogen-activated protein kinase signaling in differentiated vascular smooth muscle. J. Biol. Chem. *272*, 25157-25161.

Michaud, N.R., Fabian, J.R., Mathes, K.D. and Morrison, D.K. (1995) 14-3-3 is not essential for Raf-1 function: Identification of Raf-1 proteins that are biologically activated in a 14-3-3 and Ras-independent manner. Mol. Cell. Biol. *15*, 3390-3397.

Michaud, N.R., Therrien, M., Cacace, A., Edsall, L.C., Spiegel, S., Rubin, G.M. and Morrison, D.K. (1997) KSR stimulates Raf-1 activity in a kinase-independent manner. Proc. Natl. Acad. Sci. U.S.A. *94*, 12792-12796.

Müller, J., Ory, S., Copeland, T., Piwnica-Worms, H. and Morrison, D.K. (2001) C-TAK1 regulates Ras signaling by phosphorylating the MAPK scaffold, KSR1. Mol. Cell *8*, 983-993.

Muslin, A.J., Tanner, J.W., Allen, P.M. and Shaw, A.S. (1996) Interaction of 14-3-3 with signaling proteins is mediated by the recognition of phosphoserine. Cell *84*, 889-897.

Oh-hora, M., Ogata, M., Mori, Y., Adachi, M., Imai, K., Kosugi, A. and Hamaoka, T. (1999) Direct suppression of TCR-mediated activation of extracellular signal-regulated kinase by leukocyte protein tyrosine phosphatase, a tyrosine-specific phosphatase. J. Immunol. *163*, 1282-1288.

Ohtsuka, T., Shimizu, K., Yamamori, B., Kuroda, S. and Takai, Y. (1996) Activation of brain B-Raf protein kinase by Rap1B small GTP-binding protein. J. Biol. Chem. *271*, 1258-1261.

Okada, T., Hu, C.-D., Jin, T.-G., Kariya, K., Yamawaki-Kataoka, Y. and Kataoka, T. (1999) The strength of interaction at the Raf cysteine-rich domain is a critical determinant of response of Raf to Ras family small GTPases. Mol. Cell. Biol. *19*, 6057-6064.

Ory, S., Zhou, M., Conrads, T.P., Veenstra, T.D. and Morrison, D.K. (2003) Protein phosphatase 2A positively regulates Ras signaling by dephosphorylating KSR1 and Raf-1 on critical 14-3-3 binding sites. Curr. Biol. *13*, 1356-1364.

Payne, D.M., Rossomando, A.J., Martino, P., Erickson, A.K., Her, J.H., Shabanowitz, J., Hunt, D.F., Weber, M.J. and Sturgill, T.W. (1991) Identification of the regulatory phosphorylation sites in pp42/mitogen-activated protein kinase (MAP kinase). EMBO J. *10*, 885-892.

Pearson, G., Robinson, F., Beers Gibson, T., Xu, B.E., Karandikar, M., Berman, K. and Cobb, M.H. (2001) Mitogen-activated protein (MAP) kinase pathways: regulation and physiological functions. Endocr. Rev. *22*, 153-183.

Pettiford, S.M. and Herbst, R. (2000) The MAP-kinase ERK2 is a specific substrate of the protein tyrosine phosphatase HePTP. Oncogene *19*, 858-869.

Rabizadeh, S., Xavier, R.J., Ishiguro, K., Bernabeortiz, J., Lopez-Ilasaca, M., Khokhlatchev, A., Mollahan, P., Pfeifer, G.P., Avruch, J. and Seed, B. (2004) The scaffold protein CNK1 interacts with the tumor suppressor RASSF1A and augments RASSF1A-induced cell death. J. Biol. Chem. *279*, 29247-29254.

Rapp, U.R., Goldsborough, M.D., Mark, G.E., Bonner, T.I., Groffen, J., Reynolds, F.H. Jr., and Stephenson, J.R. (1983) Structure and biological activity of v-raf, a unique oncogene transduced by a retrovirus. Proc. Natl. Acad. Sci. U.S.A. *80*, 4218-4222.

Rizzo, M.A., Shome, K., Vasudevan, C., Stolz, D.B., Sung, T.-C., Frohman, M.A., Watkins, S.C. and Romero, G. (1999) Phospholipase D and its product, phosphatidic acid, mediate agonist-dependent Raf-1 translocation to the plasma membrane and the activation of the mitogen-activated protein kinase pathway. J. Biol. Chem. *274*, 1131-1139.

Robbins, D.J., Cheng, M., Zhen, E., Vanderbilt, C.A., Feig, L.A. and Cobb, M.H. (1992) Evidence for a Ras-dependent extracellular signal-regulated protein kinase (ERK) cascade. Proc. Natl. Acad. Sci. U.S.A. *89*, 6924-6928.

Robinson, M.J., Stippec, S.A., Goldsmith, E., White, M.A. and Cobb, M.H. (1998) A constitutively active and nuclear form of the MAP kinase ERK2 is sufficient for neurite outgrowth and cell transformation. Curr. Biol. *8*, 1141-1150.

Rommel, C., Radziwill, G., Lovric, J., Noeldeke, J., Heinicke, T., Jones, D., Aitken, A. and Moelling, K. (1996) Activated Ras displaces 14-3-3 protein from the amino terminus of c-Raf-1. Oncogene *12*, 609-619.

Rommel, C., Clarke, B.A., Zimmermann, S., Nunez, L., Rossman, R., Reid, K., Moelling, K., Yancopoulos, G.D. and Glass, D.J. (1999) Differentiation stage-specific inhibition of the Raf-MEK-ERK pathway by Akt. Science *286*, 1738-1741.

Roy, F., Laberge, G., Douziech, M., Ferland-McCollough, D. and Therrien, M. (2002) KSR is a scaffold required for activation of the ERK/MAPK module. Genes Dev. *16*, 427-438.

Roy, S., Lane, A., Yan, J., McPherson, R. and Hancock, J.F. (1997) Activity of plasma membrane-recruited Raf-1 is regulated by Ras via the Raf zinc finger. J. Biol. Chem. *272*, 20139-20145.

Rubin, C., Litvak, V., Medvedovsky, H., Zwang, Y., Lev, S. and Yarden, Y. (2003) Sprouty fine-tunes EGF signaling through interlinked positive and negative feedback loops. Curr. Biol. *13*, 297-307.

Rubinfeld, H., Hanoch, T. and Seger, R. (1999) Identification of a cytoplasmic-retention sequence in ERK2. J. Biol. Chem. *274*, 30349-30352.

Sasaki, A., Taketomi, T., Kato, R., Saeki, K., Nonami, A., Sasaki, M., Kuriyama, M., Saito, N., Shibuya, M. and Yoshimura, A. (2003) Mammalian Sprouty4 suppresses Ras-independent ERK activation by binding to Raf1. Nat. Cell Biol. *5*, 427-432.

Schaeffer, H.J., Catling, A.D., Eblen, S.T., Collier, L.S., Krauss, A. and Weber, M.J. (1998) MP1: a MEK binding partner that enhances enzymatic activation of the MAP kinase cascade. Science *281*, 1668-1671.

Schulte, T.W., Blagosklonny, M.V., Romanova, L., Mushinski, J.F., Monia, B.P., Johnston, J.F., Nguyen, P., Trepel, J. and Neckers, L.M. (1996) Destabilization of Raf-1 by geldanamycin leads to disruption of the Raf-1-MEK-mitogen-activated protein kinase signalling pathway. Mol. Cell. Biol. *16*, 5839-5845.

Seidel, J.J. and Graves, B.J. (2002) An ERK2 docking site in the pointed domain distinguishes a subset of ETS transcription factors. Genes Dev. *16*, 127-137.

Sieburth, D.S., Sun, Q. and Han, M. (1998) SUR-8, a conserved Ras-binding protein with leucine-rich repeats, positively regulates Ras-mediated signaling in C. elegans. Cell *94*, 119-130.

Slack, D.N., Seternes, O.M., Gabrielsen, M. and Keyse, S.M. (2001) Distinct binding determinants for ERK2/p38alpha and JNK MAP kinases mediate catalytic activation and substrate selectivity of MAP kinase phosphatase-1. J. Biol. Chem. *276*, 16491-16500.

Song, J., Takeda, M. and Morimoto, R.I. (2001) Bag1-Hsp70 mediates a physiological stress signalling pathway that regulates Raf-1/ERK and cell growth. Nat. Cell Biol. *3*, 276-282.

Sontag, E., Fedorov, S., Kamibayashi, C., Robbins, D., Cobb, M. and Mumby, M. (1993) The interaction of SV40 small tumor antigen with protein phosphatase 2A stimulates the MAP kinase pathway and induces cell proliferation. Cell *75*, 887-897.

Stanton, V.P., Nichols, D.W., Laudano, A.P. and Cooper, G.M. (1989) Definition of the human raf amino-terminal regulatory region by deletion mutagenesis. Mol. Cell. Biol. *9*, 639-647.

Stewart, S., Sundaram, M., Zhang, Y., Lee, J., Han, M. and Guan, K.L. (1999) Kinase suppressor of Ras forms a multiprotein signaling complex and modulates MEK localization. Mol. Cell. Biol. *19*, 5523-5534.

Stokoe, D., Macdonald, S.G., Cadwallader, K., Symons, M. and Hancock, J.F. (1994) Activation of Raf as a result of recruitment to the plasma membrane. Science *264*, 1463-1467.

Storm, S.M., Cleveland, J.L. and Rapp, U.R. (1990) Expression of raf family proto-oncogenes in normal mouse tissues. Oncogene *5*, 345-351.

Sundaram, M. and Han, M. (1995) The C. elegans Ksr-1 gene encodes a novel Raf-related kinase involved in Ras-mediated signal transduction. Cell *83*, 889-901.

Tanoue, T., Adachi, M., Moriguchi, T. and Nishida, E. (2000) A conserved docking motif in MAP kinases common to substrates, activators and regulators. Nat. Cell Biol. *2*, 110-116.

Tanoue, T., Maeda, R., Adachi, M. and Nishida, E. (2001) Identification of a docking groove on ERK and p38 MAP kinases that regulates the specificity of docking interactions. EMBO J. *20*, 466-479.

Tanoue, T. and Nishida, E. (2003) Molecular recognitions in the MAP kinase cascades. Cell. Signal. *15*, 455-462.

Tarrega, C., Blanco-Aparicio, C., Munoz, J.J. and Pulido, R. (2002) Two clusters of residues at the docking groove of mitogen-activated protein kinases differentially mediate their functional interaction with the tyrosine phosphatases PTP-SL and STEP. J. Biol. Chem. *277*, 2629-2636.

Teis, D., Wunderlich, W. and Huber, L.A. (2002) Localization of the MP1-MAPK scaffold complex to endosomes is mediated by p14 and required for signal transduction. Dev. Cell *3*, 803-814.

Therrien, M., Chang, H.C., Solomon, N.M., Karim, F.D., Wassarman, D.A. and Rubin, G.M. (1995) KSR, a novel protein kinase required for RAS signal transduction. Cell *83*, 879-888.

Therrien, M., Wong, A.M. and Rubin, G.M. (1998) CNK, a RAF-binding multidomain protein required for RAS signaling. Cell *95*, 343-353.

Tohgo, A., Pierce, K.L., Choy, E.W., Lefkowitz, R.J. and Luttrell, L.M. (2002) Beta-arrestin scaffolding of the ERK cascade enhances cytosolic ERK activity but inhibits ERK-mediated transcription following angiotensin AT1a receptor stimulation. J. Biol. Chem. *277*, 9429-9436.

Toril, S., Kusakabe, M., Yamamoto, T., Maekawa, M. and Nishida, E. (2004) Sef is a spatial regulator for Ras/MAP kinase signaling. Dev. Cell *7*, 33-44.

Tzivion, G., Luo, Z. and Avruch, J. (1998) A dimeric 14-3-3 protein is an essential cofactor for Raf kinase activity. Nature *394*, 88-92.

Van Aelst, L., Barr, M., Marcus, S., Polverino, A. and Wigler, M. (1993) Complex formation between Ras and Raf and other protein kinases. Proc. Natl. Acad. Sci. U.S.A. *90*, 6213-6217.

Vojtek, A.B., Hollenberg, S.M. and Cooper, J.A. (1993) Mammalian Ras interacts directly with the serine/threonine kinase Raf. Cell *74*, 205-214.

Volmat, V., Camps, M., Arkinstall, S., Pouyssegur, J. and Lenormand, P. (2001) The nucleus, a site for signal termination by sequestration and inactivation of p42/p44 MAP kinases. J. Cell Sci. *114*, 3433-3443.

Wakioka, T., Sasaki, A., Kato, R., Shouda, T., Matsumoto, A., Miyoshi, K., Tsuneoka, M., Komiya, S., Baron, R. and Yoshimura, A. (2001) Spred is a Sprouty-related suppressor of Ras signalling. Nature *412*, 647-651.

Wan, P.T., Garnett, M.J., Roe, S.M., Lee, S., Niculescu-Duvaz, D., Good, V.M., Jones, C.M., Marshall, C.J., Springer, C.J., Barford, D. and Marais, R. (2004) Mechanism of activation of the RAF-ERK signaling pathway by oncogenic mutations of B-RAF. Cell *116*, 855-867.

Wang, Y., Waldron, R.T., Dhaka, A., Patel, A., Riley, M.M., Rozengurt, E. and Colicelli, J. (2002) The RAS effector RIN1 directly competes with RAF and is regulated by 14-3-3 proteins. Mol. Cell. Biol. *22*, 916-926.

Weber, C.K., Slupsky, J.R., Kalmes, H.A. and Rapp, U.R. (2001) Active Ras induces heterodimerization of cRaf and BRaf. Cancer Res. *61*, 3595-3598.

Whitehurst, A.W., Wilsbacher, J.L., You, Y., Luby-Phelps, K., Moore, M.S. and Cobb, M.H. (2002) ERK2 enters the nucleus by a carrier-independent mechanism. Proc. Natl. Acad. Sci. U.S.A. *99*, 7496-7501.

Williams, J.G., Drugan, J.K., Yi, G.S., Clark, G.J., Der, C.J. and Campbell, S.L. (2000) Elucidation of binding determinants and functional consequences of Ras/Raf-cysteine-rich domain interactions. J. Biol. Chem. *275*, 22172-22179.

Wong, E.S., Fong, C.W., Lim, J., Yusoff, P., Low, B.C., Langdon, W.Y. and Guy, G.R. (2002) Sprouty2 attenuates epidermal growth factor receptor ubiquitylation and endocytosis, and consequently enhances Ras/ERK signalling. EMBO J. *21*, 4796-4808.

Wu, J., Dent, P., Jelinek, T., Wolfman, A., Weber, M.J. and Sturgill, T.W. (1993) Inhibition of the EGF-activated MAP kinase signaling pathway by adenosine 3′,5′-monophosphate. Science *262*, 1065-1069.

Wunderlich, W., Fialka, I., Teis, D., Alpi, A., Pfeifer, A., Parton, R.G., Lottspeich, F. and Huber, L.A. (2001) A novel 14-kilodalton protein interacts with the mitogen-activated protein kinase scaffold mp1 on a late endosomal/lysosomal compartment. J. Cell Biol. *152*, 765-776.

Xu, B., Stippec, S., Robinson, F.L. and Cobb, M.H. (2001) Hydrophobic as well as charged residues in both MEK1 and ERK2 are important for their proper docking. J. Biol. Chem. *276*, 26509-26515.

Xu, B., Wilsbacher, J.L., Collisson, T. and Cobb, M.H. (1999) The N-terminal ERK-binding site of MEK1 is required for efficient feedback phosphorylation by ERK2 in vitro and ERK activation in vivo. J. Biol. Chem. *274*, 34029-34035.

Yaffe, M.B., Schutkowski, M., Shen, M., Zhou, X.Z., Stukenberg, P.T., Rahfeld, J.U., Xu, J., Kuang, J., Kirschner, M.W., Fischer, G., Cantley, L.C. and Lu, K.P. (1997) Sequence-specific and phosphorylation-dependent proline isomerization: a potential mitotic regulatory mechanism. Science *278*, 1957-1960.

Yao, I., Hata, Y., Ide, N., Hirao, K., Deguchi, M., Nishioka, H., Mizoguchi, A. and Takai, Y. (1999) MAGUIN, a novel neuronal membrane-associated guanylate kinase-interacting protein. J. Biol. Chem. *274*, 11889-11896.

Yao, I., Ohtsuka, T., Kawabe, H., Matsuura, Y., Takai, Y. and Hata, Y. (2000) Association of membrane-associated guanylate kinase-interacting protein-1 with Raf-1. Biochem. Biophys. Res. Commun. *270*, 538-542.

Yeung, K., Seitz, T., Li, S., Janosch, P., McFerran, B., Kaiser, C., Fee, F., Katasanakis, K.D., Rose, D.W., Mischak, H., Sedivy, J.M. and Kolch, W. (1999) Suppression of Raf-1 kinase activity and MAP kinase signalling by RKIP. Nature *401*, 173-177.

Yeung, K., Janosch, P., McFerran, B., Rose, D.W., Mischak, H., Sedivy, J.M. and Kolch, W. (2000) Mechanism of suppression of the Raf/MEK/extracellular signal-regulated kinase pathway by the Raf kinase inhibitor protein. Mol. Cell. Biol. *20*, 3079-3085.

York, R.D., Yao, H., Dillon, T., Ellig, C.L., Eckert, S.P., McCleskey, E.W. and Stork, P.J. (1998) Rap1 mediates sustained MAP kinase activation induced by nerve growth factor. Nature *392*, 622-626.

Yu, W., Fantl, W.J., Harrowe, G. and Williams, L.T. (1998) Regulation of the MAP kinase pathway by mammalian Ksr through direct interaction with MEK and ERK. Curr. Biol. *8*, 56-64.

Yusoff, P., Lao, D.H., Ong, S.H., Wong, E.S., Lim, J., Lo, T.L., Leong, H.F., Fong, C.W. and Guy, G.R. (2002) Sprouty2 inhibits the Ras/MAP kinase pathway by inhibiting the activation of Raf. J. Biol. Chem. *277*, 3195-3201.

Zhang, J., Zhang, F., Ebert, D., Cobb, M.H. and Goldsmith, E.J. (1995) Activity of the MAP kinase ERK2 is controlled by a flexible surface loop. Structure *3*, 299-307.

Zhang, J., Zhou, B., Zheng, C.F. and Zhang, Z.Y. (2003) A bipartite mechanism for ERK2 recognition by its cognate regulators and substrates. J. Biol. Chem. *278*, 29901-29912.

Zhang, X.F., Settleman, J., Kyriakis, J.M., Takeuchi-Suzuki, E., Elledge, S.J., Marshall, M.S., Bruder, J.T., Rapp, U.R. and Avruch, J. (1993) Normal and oncogenic $p21^{ras}$ proteins bind to the amino-terminal regulatory domain of c-Raf-1. Nature *364*, 308-313.

Zheng, C. and Guan, K. (1993) Cloning and characterization of two distinct human extracellular signal-regulated kinase activator kinases, MEK1 and MEK2. J. Biol. Chem. *268*, 11435-11439.

Zheng, C.F. and Guan, K.L. (1994) Activation of MEK family kinases requires phosphorylation of two conserved Ser/Thr residues. EMBO J. *13*, 1123-1131.

Zhou, M., Horita, D.A., Waugh, D.S., Byrd, R.A. and Morrison, D.K. (2002) Solution structure and functional analysis of the cysteine-rich C1 domain of kinase suppressor of Ras (KSR). J. Mol. Biol. *315*, 435-446.

Zhou, X.Z., Kops, O., Werner, A., Lu, P.J., Shen, M., Stoller, G., Kullertz, G., Stark, M., Fischer, G. and Lu, K.P. (2000) Pin1-dependent prolyl isomerization regulates dephosphorylation of Cdc25C and tau proteins. Mol. Cell *6*, 873-883.

Zuniga, A., Torres, J., Ubeda, J. and Pulido, R. (1999) Interaction of mitogen-activated protein kinases with the kinase interaction motif of the tyrosine phosphatase PTP-SL provides substrate specificity and retains ERK2 in the cytoplasm. J. Biol. Chem. *274*, 21900-21907.

CHAPTER 5

NONCANONICAL EFFECTOR TARGETS OF ONCOGENIC RAS PROTEINS

An overview of alternative Ras-binding partners and their potential significance

BENJAMIN BOETTNER AND LINDA VAN AELST

Affiliation Cold Spring Harbor Laboratory, 1 Bungtown Road, Cold Spring Harbor, NY11724, USA

Abstract: The understanding of the Raf, PI3-kinase and RalGDS mediated pathways that relay physiological signals from oncogenic Ras-proteins has been consistently improved over the recent years. The proliferative, anti-apoptotic and some of the more cell-type and tumor-idiosyncratic effects of Ras GTPases in various scenarios could be ascribed to one of these effectors. However, individual tumor cells undergo drastic changes in their cell fates and differentiation states which likely require the activation of other than Raf-, PI3-kinase- and RalGDS-initiated signaling mechanisms. In addition, Ras GTPases participate in a multitude of developmental processes that entail growth, proliferative, differentiative and migratory programs. Proteins such as AF-6, Nore1, certain protein kinase C (PLC) isoforms, Tiam1, Rin1 and a few others have been identified as candidate Ras-effectors mostly by virtue of their physical interaction properties in various affinity-based protocols but also as a result of genetic and computational approaches. This selection of alternative binding partners for oncogenic Ras-proteins can thus serve as a source for more in depth investigations of particular Ras-related phenomena. The following chapter will scrutinize these molecules with respect to their functions and biochemical properties

Keywords: Ras-signaling, Ras-effectors, Rap, Nore, RASSF, Tiam1, *Drosophila* Still life, Rin1, *Drosophila* sprint, PLCε, AF-6, Afadin, *Drosophila* Canoe

1. INTRODUCTION

All members of the Ras family of small GTPases are cycling between a GDP-bound inactive state and a GTP-bound active state. GDP·Ras is recognized by specific guanine nucleotide exchange factors (GEFs) that activate their cognate GTPase(s) by exchanging GDP for GTP. GTP·Ras then is competent to interact with a set of downstream effectors thereby relaying the signal down the pathway. This

C. Der (ed.), RAS Family GTPases, 95–120.

action, however, is a transient one, since Ras GTPases possess a weak hydrolytical activity that converts GTP to GDP. GTPase activating proteins (GAPs) markedly enhance this de-activating mechanism thus, in concert with GEFs, allowing for a tight regulation of a particular Ras GTPase in a particular process. Oncogenic or constitutively active forms of Ras are mutationally locked in the GTP-bound state and insensitive to extrinsic GTPase activating activities.

A typical feature of Ras effectors is the presence of an RA (Ras association) domain, also referred to as RBD (Ras binding) domain. The canonical effectors of Ras, namely the Raf kinase, PI3 kinase and RalGEF families, whose functions and properties are discussed elsewhere in this volume are examples for RA-domain containing proteins but a growing number of alternative or non-canonical effectors is being revealed. In this chapter we will discuss those non-canonical effectors. In particular, we will focus on the Nore/RASSF, Rin1, Tiam1, PLCε and AF-6/Afadin effectors and the conceptualization of their roles in various processes. It should be mentioned that other candidates such as Rain (Ras interacting protein; Mitin et al., 2004) and IMP (Impedes Mitogenic signal Propagation; Matheny et al., 2004) have recently been identified and studies on them begin to illuminate particular aspects of Ras regulation and signaling. This summary merges data obtained from biochemical approaches and developmental systems such as mouse and *Drosophila* to point out proven or potentially conserved functions of effectors.

2. NONCANONICAL RAS EFFECTORS

2.1 Nore1: Triggering Apoptosis Downstream of Ras

Activated Ras proteins can exert an anti- as well as a pro-apoptotic effect depending on the cellular context they are acting in. It is widely understood that in addition to their proliferative effects Ras proteins also help their host cells to counteract apoptosis during the oncogenic transformation of a target cell. The latter aspect has been convincingly shown to rely on Ras' activation of PI3-kinase signaling (Kauffmann-Zeh et al., 1997; Khwaja et al., 1997; Marte et al., 1997). Also an NF-κB-triggering mechanism by Ras contributes to the protection of cells against apoptosis (Mayo et al., 1997). Conversely, however, Ras, when introduced as a constitutively active mutant into primary cells rather provokes them to arrest their cell cycles or even to enter an apoptotic program.

How does Ras achieve these opposite effects? One part of the answer to this question comes from the finding that halting the cell cycle in G1 as a downstream effect of Ras often involves the activation of $p14^{ARF}$ and consequently inactivation of the p53 inhibitor MDM2 (Bates et al., 1998; Palmero et al., 1998). As a result p53 activity is enhanced and the p21 checkpoint activated (Lloyd et al., 1997; Serrano et al., 1997). Thus, inactivation of the p53 gene severely compromises the ability of primary mouse embryo fibroblasts (MEFs) to induce apoptosis upon introduction of oncogenic Ras. p53 activation by Ras requires Raf/MAP kinase signaling (Kauffmann-Zeh et al., 1997; Lin et al., 1998).

Employing a super-repressor form of IκBα, a potent inhibitor of NF-κB, Mayo et al. could show that constitutively active Ras can still bring about an apoptotic response in p53 deficient MEFs with suppressed NF-κB activity (Mayo et al., 1997). This strongly suggested that a p53-independent pro-apoptotic Ras-triggered pathway was operational.

More recently, a novel mechanism that involves the pro-apoptotic protein Nore1 as an immediate Ras-effector has begun to emerge. Mouse Nore1 was initially isolated as a Ras-interacting non-catalytic protein and the analysis of its sequence revealed a C-terminally located RA-domain that confers affinity to Ras (Figure 1; Vavvas et al., 1998). In addition, Nore1 harbors several PXXP motifs in its amino-terminal portion and a more centrally located C1 zinc finger domain. The work of several laboratories has revealed the existence of several Nore1-related polypeptides, the most closely related of which is RASSF1A (for Ras association domain family 1). Work by Ortiz-Vega and colleagues has established that RASSF1A possesses the ability to homodimerize and heterodimerize with Nore1 and that the latter association likely brings RASSF1A into a complex with Ras (Ortiz-Vega et al., 2002). Another close relative of Nore1, namely RASSF2, was recently shown to interact with oncogenic Ras proteins. Like Nore1 and RASSF1A, RASSF2 can also trigger apoptosis and cell cycle arrest (Vos et al., 2003). Of particular relevance with respect to oncogenesis are insights into the methylation status of the Nore1/RASSF1 genes in cancer cells. Chromosomal modifications in form of hypermethylations have been observed in the Nore1 gene-regulatory region (Hesson et al., 2003) and also in that of its family sibling RASSF1A (Spugnardi et al., 2003). The Nore1 encoding locus in humans is located at chromosomal region 1q32.1 (Tommasi et al., 2002). Two adjacent coding units give rise to Nore1A and Nore1B, respectively. Nore1A is subject to alternative splicing resulting in 2 isoforms, Nore1Aα and Nore1β. Whereas the former is the homolog of the mouse Nore1 geneproduct encoding the central DAG (di-acyl-glycerol) and the C-terminal RA (Ras association)-domain mentioned above, the latter constitutes a truncated variant of Nore1Aα that lacks the C-terminal RA-domain. Nore1B is transcribed/translated into a protein with an RA- but without a DAG domain (Tommasi et al., 2002). Hesson and colleagues have reported that CpG islands in the Nore1A promotor are hypermethylated in a considerable proportion of tumor cell lines (Hesson et al., 2003). In an interesting study Nore1 has been positionally cloned as a breakpoint spanning gene involved in clear cell renal carcinomas (CCRCC; Chen et al., 2003). Some hereditary CCRCCs are associated with balanced chromosomal translocations and breakpoint regions have been proposed to harbor disease-relevant genes. In all CCRCC families that have been described to date, translocations involving a region of chromosome 3 are underlying the predisposition (Bodmer et al., 1998; van Kessel et al., 1999). Chen an co-workers succeeded in defining the chromosomal breakpoints involved in the t(1;3) (q32.1; q13.3) translocation present in a Japanese family inflicted with CCRCC (Kanayama et al., 2001) and found that one of the breakpoints lies in the Nore1 gene thereby causing its disruption (Chen et al., 2003). Subsequent analysis of the Nore1 promotor revealed a high degree

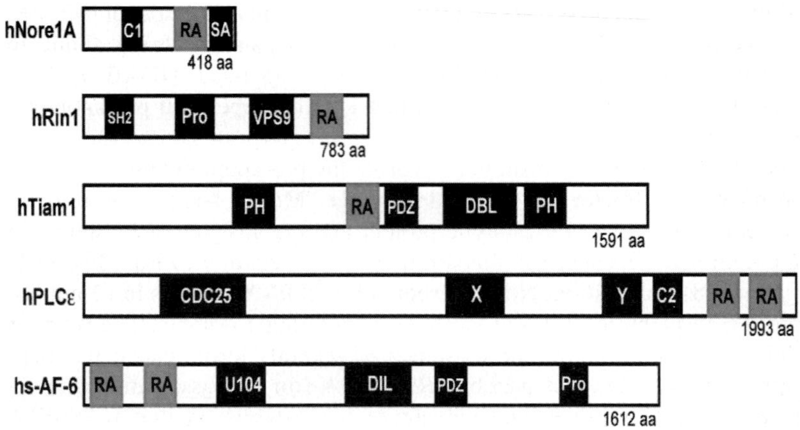

Figure 1. Schematic overviews of alternative Ras effector molecules and their domain structures. RA = Ras associating domain; C1 = DAG-PE phorbol ester, diacylglycerol binding domain; C = Ca^{2+}-dependent membrane targetingmodule; CDC25 = RAS GEF domain; DBL = Dbl homology domain with Rho GEF activity; DIL = myosin homology domain; PDZ = PSD-95, Discs large, ZO-1 homology domain; PH = Pleckstrin homology domain tethering activity to phospholipids-enriched membranes; Pro = Proline-rich sequence; SA = SARAH, Sav/Rassf/Hpo motif; SH2 = Src homology 2 domain with phosphotyrosine affinity; U104 = kinesin homology domain; VPS9 = Vacuolar sorting protein 9 domain involved in Rab GTPase activation; X and Y = phospholipase catalytic activity

of hypermethylation resulting in a markedly decreased expression of the gene in a number of RCC cell lines and sporadic CCRCCs (Chen et al., 2003).

Hypermethylation of promotor sequences has also been associated with the gene coding for human RASSF1A on chromosome 3p21.3 in cases of lung, breast, kidney and thyroid carcinomas and malignant cutaneous melanomas (Dreijerink et al., 2001; Morrissey et al., 2001; Yoon et al., 2001; Tommasi et al., 2002; Dammann et al., 2003). Vos et al. reported, that also RASSF2 is downregulated in lung tumor cell lines although it is not clear yet by which means (Vos et al., 2003). Like some other *bona fide* tumor suppressors, members of the Nore1/RASSF gene family appear to be epigenetically silenced under cancerogenic circumstances, which in the light of the findings described above probably will compromise the proapoptotic abilities of oncogenic Ras and thus assist oncogenic progression.

Direct biochemical evidence obtained in cell culture studies and more indirect evidence from genetic research in *Drosophila* are supporting a role of Nore1/RASSF as a potential tumor suppressor. Khohlatchev and his coworkers showed that both Nore1 and RASSF1A are constitutively associated with the proapoptotic kinase Mst1 (mammalian STE20-like kinase 1) and the Nore1/Mst1 complex binds to Ras in a serum-dependent manner (Khokhlatchev et al., 2002). Mst1 has previously been associated with an apoptotic potential (Graves et al., 1998). Interestingly, a myristoylated version of Mst1 can function as a potent inducer of apoptosis. This is indicative of a mechanism in which membrane recruitment of Mst1 by activated Ras through mediation by Nore1 contributes to kinase activation. Similarly, Nore1 carrying a K-Ras-derived CAAX-box at its C-terminus displays a significantly increased apoptotic efficacy. Khokhlatchev et al. further expanded on these observations and compared the ability of H- and K-Ras to evoke an apoptic phenotype. Constitutively active H-Ras^{G12V} is considerably weaker in its apoptotic potential than K-Ras^{G12V}. When endowed with effector loop mutation E37G that eliminates the ability of Ras to interact with the Raf- or PI3-kinases but that preserves its affinity to Nore1, H-Ras$^{G12V/E37G}$ retains a markedly elevated apoptotic potential (Khokhlatchev et al., 2002). Taken together these findings reveal the first components and mechanistic details of a Ras-specific pathway that can potently induce an apoptotic cell fate. Hence, Nore1/RASSF family members and Mst1 kinases are interesting molecules for further explorations into eminent questions, such as the search for substrate(s) of Mst1 kinase activity and higher-order epistatic elements and the relevance of this pathway *in vivo*.

In addition to being stimulated by Ras, Mst1 can be activated by Caspase3-dependent proteolytic cleavage. The latter eliminates the inhibitory N-terminal dimerization domain (Graves et al., 1998; Lee et al., 1998). More recently, caspase-cleaved Mst1 has been identified as the agent that in cells driven into apoptosis phosphorylates histone H2B at serine 14 which is a modification thought to be associated with apoptotic chromatin (Cheung et al., 2003). Whether these findings also relate to the Ras-dependent activation of Mst1 is unknown. As an alternative, Ras through Nore1 could propagate a conformational change that helps to build up the full phosphorylation potential towards one or more alternative cytoplasmic substrates.

A growing body of evidence implicating Nore1/Mst1 into one or more cell death and cell cycle arresting pathways comes from studies in *Drosophila*. The mammalian Mst1 (and Mst2) kinase has a counterpart, namely the product of the *hippo* gene in flies (Harvey et al., 2003; Pantalacci et al., 2003; Udan et al., 2003; Wu et al., 2003). Hippo appears to function together with 2 other proteins, Warts (also known as Lats) and Salvador (also known as Shar-pei) to inhibit cell cycle progression and to promote cell death. *hippo* loss-of-function induces cells in larval imaginal discs to overproliferate and to evade apoptosis when they normally should die. The same phenotypes also result from loss of *salvador* and *warts*. Very strikingly, human Mst2 can compensate for the absence of Hippo (Wu et al., 2003), which strongly indicates a conservation that exceeds the mere similarity between the proteins. The *Salvador* gene product is a scaffold protein, whereas the *warts* gene encodes a Ser/Thr kinase. Salvador has been found to directly associate with Hippo and Warts and like Hippo, both proteins restrict proliferation and promote cell death (Kango-Singh et al., 2002; Tapon et al., 2002). It is noteworthy, that mutations in the human homolog of the *salvador* gene, *hWW45*, have been detected in cancer cell lines (Tapon et al., 2002). Also the *warts* gene appears to be represented in the human genome in form of two orthologs, *Lats-1* and *Lats-2*, the latter of which has been associated with tumor suppressor activities (Li et al., 2003). Paralleling these biochemical interactions, several laboratories demonstrated genetic interactions between all three genes (Tapon et al., 2002; Harvey et al., 2003; Udan et al., 2003; Wu et al., 2003). Taken together, these studies indicate that hippo, savador and warts operate in a common pathway arresting the cell cycle and triggering apoptosis.

The biochemical consequences of these interactions have not been sufficiently unraveled, however, some insights have emerged that further the case. Hippo can phosphorylate Salvador (Harvey et al., 2003; Pantalacci et al., 2003; Wu et al., 2003) and Warts (Wu et al., 2003) *in vitro*. *In vivo*, the situation is more complex. Here, the phosphorylation of the latter two proteins depends on Hippo and its association with Salvador. Experiments with a kinase-dead Hippo mutant have shown that while Hippo's kinase activity is absolutely required for it to trigger cell-cycle and apoptosis related events (Udan et al., 2003; Wu et al., 2003), Salvador under these conditions is still phosphorylated (Pantalacci et al., 2003). This suggests the possibility that Hippo needs to associate with Salvador to recruit yet another kinase to modify Salvador. However, Warts so far remains the best candidate for Hippo's own kinase activity.

How do Hippo, Salvadore and Warts impinge on the cell cycle and apoptosis machineries? Independently conducted studies showed that all three genes and their products negatively regulate expression of Cyclin E (Kango-Singh et al., 2002; Tapon et al., 2002), which is a limiting factor for entry into the S phase of the cell cycle in *Drosophila* imaginal discs (Richardson et al., 1995; Neufeld et al., 1998). Curiously, however, this cannot be the only answer for the following reasons. Overexpression of Cyclin E does accelerate G1 to S progression but in normal cells this is compensated for by elongation of the S phase. In contrast, in cells mutant for any of the *hippo*, *savador* or *warts* genes every phase of the cell cycle is shortened and, very strikingly, mutant cells reach a normal size. This argues for a scenario where Hippo, Salvador and Warts in concert delicately regulate more than one target with relevance not only to single cell cycle check-points but also to growth regulation. To enhance apoptosis, Hippo, Salvador and Warts were shown to decrease the levels of the apoptosis inhibitor DIAP1 and they may do so by different means. Work from three different laboratories demonstrated, that (1) DIAP1 expression was hampered and (2), that existing DIAP1 was destabilized by phosphorylation (Tapon et al., 2002; Udan et al., 2003; Wu et al., 2003).

In summary, Hippo, Salvador and Warts are part of a mechanism that coordinately regulates proliferation and cell death. Such regulation is vital for tissue development and homeostasis and its dysregulation is likely to give rise to cancer-like behaviour. In conjunction with the loss-of-function data associated with Nore1/RASSF in cancer cells one may postulate vital tumor suppressive properties of the pathways. It will be fascinating to follow the development in either paradigm and to see whether there is a Ras/Nore1 regulatory input into Hippo/Salvador/Warts function in *Drosophila*. The *Drosophila* genome offers a candidate, CG4546 that may fulfill Nore1-like properties. Reciprocally, it is an intriguing idea that a Hippo/Salvador/Warts-based mechanism may translate a Ras stimulus into cell cycle regulatory and apoptotic responses. These possibilities are strengthened by the existence of a complement of all genes in both genomes. *hWW45* and *Lats-1/2* are *salvadore* and *warts* orthologs in the human genome and *CG4656* is a *Nore1/Rassf*-like gene in *Drosophila*. This is further underscored by the recent computational prediction of a common coiled-coil motif, termed SARAH (for Sav/Rassf/Hpo), that is present in the C-termini of the hWW45/Salvador, Nore1/Rassf/CG4656 and Mst1/Mst2/Hippo families of proteins (Scheel and Hofmann, 2003) and that may enable the formation of dimeric or even trimeric complexes among representatives of the three groups of signal transducers.

2.2 Rin1: Connecting Ras to Endocytic Events

Rin1 was originally isolated as a human cDNA capable of suppressing the lethality phenotype produced by a dominant active Ras2^{V19} mutant in *Saccharomyces cerevisiae*. Hence, Rin was chosen as an acronym for Ras interaction/interference (Colicelli et al., 1991). Subsequently, Han and Collicelli determined that Rin1 binds preferentially to the GTP-loaded form of H-Ras suggesting that it may function as an effector (Han and Colicelli, 1995). The organization of the Rin1 protein, as depicted in Figure 1, features an RA-domain at the C-terminus that enables Ras-association. An SH2-domain at the N-terminus followed by a Pro-rich sequence commonly used to anchor proteins containing SH3 domains likely endow Rin1 with further signaling functions. More recently a Vsp9 (Vacuolar sorting protein 9) homology region has been ascribed to Rin1 and, as will be described below, presents a functional feature with catalytic GDP/GTP-exchange activity towards Rab GTPases.

In contrast to the more established oncogenic Ras effectors Raf, PI3K and RalGEFs, overexpression of Rin1 in classical transformation assays blocks Ras-induced transformation. This observation substantiates the Ras-inhibitory function found in yeast (Han and Colicelli, 1995). As found for the Raf-family kinases, also Rin1 associates with 14-3-3 proteins in a phosphorylation-dependent manner (Wang et al., 2002). The critical residue for the formation of this complex is Ser-351 in the peptide sequence of Rin1. A mutant in which Ser-351 has been converted to Ala fails to complex with 14-3-3 proteins and impedes oncogenic Ras-signaling more efficiently. S-351 is part of a consensus phosphorylation site for protein kinase D (PKD; identical with PKCμ) and Wang et al. have shown that PKD in fact uses Rin1 as a substrate *in vitro* and *in vivo* (Wang et al., 2002). In light of these data, one can envision a model in which a still to be identified mechanism activates PKD that in turn phosphorylates Rin1 at Ser-351 to release it out of its complex with 14-3-3 proteins. Rin1 then becomes available to compete with Raf for Ras-binding and thus exerts its negative influence on cell transformation. In fact, the reported affinity (K(d) values) of Rin1 for Ras with 22 nM, which is comparable to that of Raf to Ras, would be in favour of such a competition model.

In contrast to its antagonistic effect in Ras-dependent transformation of fibroblasts, Rin1 has been assigned a pro-oncogenic role in BCR-ABL signaling during leukemogenesis (Afar et al., 1997). BCR-ABL is a compound protein that is expressed in hematopoetic cells whose genome has been subjected to a reciprocal chromosomal translocation t(9;22)(q34;q11). In the fusion product BCR sequences are C-terminally joined with c-abl sequences rendering the ABL tyrosine function constitutively and aberrantly active. This activity gives rise to the pathological and ultimately lethal features of acute lymphocytic and chronic myelogenous leukemias.

Mutants of the BCR-ABL protein with a weakened transformation capacity have been used to delineate downstream signaling events with the rationale that overexpression of epistatic downstream-elements in a particular pathway can compensate for a mutation in one of the crucial BCR-ABL domains (Afar et al., 1997). In this scenario, ectopic amounts of e.g. Myc or Cyclin D1 could overcome the deficiency of a BCR-ABL mutant harboring a mutation in the SH2-binding region of the Abl portion. Similarly, overexpression of Shc can compensate for a mutation in the Grb-binding site of BCR-ABL. Employing this assay, Afar and his colleagues demonstrated, that elevated Rin1 levels in cells expressing the Grb-binding deficient BCR-ABL mutant can re-install and even inforce the oncogenic effects of BCR-ABL. Using its N-terminus, Rin1 directly associates with the Abl-SH2 and SH3 domains and becomes Tyr-phosphorylated (Afar et al., 1997). At first glance, this appears to be paradoxical: on one hand, forced expression of Rin1 can rescue deficient BCR-ABL and on the other hand Rin1 can block Ras-signaling, the latter considered to be obligatory for BCR-ABL's transforming effect. According to Afar et al., Rin1 rescues the same mutation in BCR-ABL that likely impairs signaling to Ras, namely the Grb-binding function. One explanation could be that Rin1 performs two completely unrelated tasks in both pathways. It should be interesting to see whether the signal flow through the Ras/MAPK pathway is disturbed under conditions in which BCR-ABL co-operates with an elevated dose of Rin1 or whether Raf/MAPK signaling occurs with normal dynamics.

What function could endogenous Rin1 perform in the cell? An interesting study by Tall et al. began to shed light on this question by identifying an enzymatic activity in Rin1 that is associated with its Vps9p domain (Tall et al., 2001). The Vps9p peptide in Rin1 exerts a GDP/GTP nucleotide exchange activity towards Rab5, a small GTPase of the Rab subfamily of small GTPases that is intimately involved in endosome fusion and EGF receptor mediated endocytosis. This was examined by performing endosome fusion assays and measurements of [^{125}I]-EGF uptake, respectively. In the former experiment the fusion-ability of endosomes was significantly increased by the addition of purified Rin1 and, importantly, even more so by adding GTP-loaded H-Ras (Tall et al., 2001). This synergism between Ras and Rin1 was equally effective in the examination of EGF receptor mediated endocytosis. Here, cells that have been transduced with combinations of Ras, Rin1 and Rab5 proteins were exposed to radioactively labeled EGF ligand and the uptake was monitored over time. Viral introduction of Rab5, Rin1 and Ras alone markedly elevated the amount of internalized EGF receptors and co-introduction of Rin1 and Ras further enhanced this effect (Tall et al., 2001). Consistent with these findings, on an ultrastructural level, elevated levels of Rin1 resulted in an intracellular enlargement of endosomal bodies. Noteworthy, Tall et al. reported that a naturally occurring Rin1 splice variant that is lacking a 62 amino acid peptide overlapping the Vps9p exchange domain (Rin1Δ) can not perform these endocytosis-related functions and in some cases may even act as a dominant negative version of Rin1. Rin1Δ when overexpressed in fibroblasts inhibits EGF receptor mediated endocytosis and endosome/endosome fusion. By demonstrating that the

SH2-domain of Rin1 directly associates with Tyr-phosphorylated EGF receptor upon stimulation, Barbieri and co-workers revealed the structural basis of these observations (Barbieri et al., 2003). Earlier observations already had indicated that Rin1 partially redistributes to the plasma membrane following EGF treatment. In line with these data, when comparing MAPK activation in cells transfected with Rin1 or Rin1Δ, Tall et al. described opposite effects of the two splice variants on the pathway. While Rin1, as to be expected, blunts MAPK activation after EGF treatment of serum starved NR6 fibroblasts, Rin1Δ significantly amplifies MAPK activity in the same situation (Tall et al., 2001).

Putting these observations into perspective, a model can be envisioned, in which the stimulation of EGF receptors with EGF ligand recruits and activates Ras via a combination of adaptor and activator proteins. GTP-loaded Ras in turn binds to Rin1, which exchanges GDP for GTP on Rab5-like proteins that subsequently trigger endocytic events resulting in the uptake of EGF receptors and downregulation of its various effector pathways whereby a negative feedback is implemented (Tall et al., 2001). It remains to be seen, however, whether and how Rin1's stimulation of endosome fusion and EGF receptor mediated endocytosis contributes to its negative effect on Ras induced transformation.

More recently, in an attempt to define the developmental function of Rin1, Dhaka et al. disrupted the gene in mice by targeted recombination (Dhaka et al., 2003). Surprisingly, homozygous mutant mice were born according to Mendelian predictions and fully viable without displaying morphological abnormalities. The same study also assessed Rin1 in more detail and showed that expression is particularly high in brain and low or undetectable in other tissues. Examining brain regions using *in situ* hybridization techniques, they found that the Rin1 message is maximally noticeable in the forebrain in areas such as the hippocampus, the amygdala, the striatum, and the cortex at day P21 but not during embryonic nervous development and early postnatal development (Dhaka et al., 2003). This peak of expression timely coincides with massive synaptogenesis in the brain and the manifestation of behavioral patterns. Given this, the authors were prompted to examine Rin1 -/- mice in a series of learning paradigms. It turned out that specifically aversive memory formation was impaired in Rin1 -/- animals as became evident in auditory fear conditioning and taste aversion protocols. Loss of Rin1 appears to render the affected animals in both situations hypersensitive. Very interestingly, these findings correlated with an elevated long-term potentiation (LTP) in the amygdala, the brain region responsible for processing emotional cues into memory. In contrast, LTP in hippocampal neurons being associated with other types of learning and memory was unaffected. Although LTP and synaptic plasticity have been found to rely on Ras signaling in the hippocampus (Zhu et al., 2002) and Dhaka et al. have traced Rin1 in complexes with Ras using mouse forebrain tissue (Dhaka et al., 2003), a mechanistic dependency of Rin1-modified LTP on a potential input by Ras still remains to be investigated.

It is noteworthy, that the mouse expression data resemble those obtained for the Rin1 homolog in *Drosophila*. Szabo et al. have identified a *Drosophila* homologue of Rin1, which they called *sprint* (Szabo et al., 2001). The *sprint* gene is expressed in a number of tissues during fly embryogenesis including a subset of differentiating neurons in the central nervous system (CNS). *sprint* loss-of-function alleles are not yet available but it will be exciting to see whether there are commonalities between the neuronal functions of Rin1 in mice and Sprint in the fly and in which pathways Sprint will be integrated in *Drosophila* nervous system development and/or function.

2.3 Tiam1: Linking Ras to Rac Signaling

A recent addition to the list of Ras-effectors is the Rac-specific GTP-GDP-exchange protein Tiam1 (= T-cell invasion and metastasis gene product). Tiam1 was initially isolated in a screen in which the potential of virally tagged T-cells to adopt an invasive phenotype was assayed (Habets et al., 1994). Subsequent studies revealed its nature as a Rac-specific exchange factor and the necessity to drive Rac-activation in order to confer invasiveness on host cells (Michiels et al., 1995).

As a structural underpinning for this function, Tiam1 harbors the canonical Dbl/PH (DH/PH) domain that is the hallmark and catalytical portion of Rho-family directed exchange factors (see Figure 1). Preceeding the DH/PH domain, are a second N-terminal PH-domain, an RA-domain and a central PDZ-domain. It was known that Rac-stimulation is one of the many consequences of enhanced Ras activity (Bar-Sagi and Feramisco, 1986; Ridley et al., 1992; Rodriguez-Viciana et al., 1997). More recently, the RA domain of Tiam1 has been validated as a Ras-associating entity in a series of biochemical experiments (Ponting, 1999; Lambert et al., 2002). Lambert and colleagues demonstrated that the RA-domain directly binds to activated Ras triggering GTP/GDP exchange on Rac. Moreover, ectopic expression of a constitutively active Ras^{Q61L} mutant or of Ras-GRP4, a Ras-specific exchanger in NIH 3T3 cells, in conjunction with elevated Tiam1 levels, result in Rac-activation in a synergistic manner. Several pieces of evidence produced by the authors exclude that this effect is dependent on PI3-kinase, but must rather rely on a direct interface between Ras and TIAM1. First, a Ras-effector loop mutant that does not associate with the catalytic subunit of PI3-kinase still efficiently stimulates the Tiam1/Rac module. Secondly, exposure of HEK 293T cells to the PI3-kinase inhibitor LY294002 does not abolish the increased Rac GTP/GDP ratio. Lastly, co-expression of p110-CAAX, an activated mutant form of PI3-kinase cannot substitute for the presence of constitutively active Ras. Nevertheless, this mutant can induce Akt-stimulation to expected levels (Lambert et al., 2002).

It should be emphasized here that Rac-activation by Tiam1 has been reported to involve phospho-inositides as a stimulatory means by exerting an effect on one or both of Tiam1's PH-domains (Sander et al., 1998; Fleming et al., 2000). Fleming et al. found that the N-terminal PH-domain is crucial for phospho-lipid binding and that co-expression of constitutively active PI3-kinase with Tiam1 elevated the levels of GTP-bound Rac (Fleming et al., 2000). These findings correlated with studies in epithelial MDCK cells where, dependent on the composition of the extracellular matrix, Tiam1 may either promote cell-cell adhesion or cell-migration (Sander et al., 1998). In cells that are immobile, Tiam1 resides in adherens junctions whereas in cells that adopt migratory behaviour, it locates to leading, lamellipodial structures. On fibronectin and laminin1 as an extracellular matrix (ECM), TIAM1/Rac signaling in MDCK cells appears to favour a cell-cell adhesive phenotype whereas Tiam1 transfected MDCK cells on various collagens are rendered motile. Both activities are inhibited by reduced PI3-kinase activity (Sander et al., 1998).

In a recent cancer-directed *in vivo* approach, the interface between Ras and Tiam1 signaling has been convincingly corroborated. Malliri and co-workers have disrupted the *Tiam1* locus in mice. Although, widely expressed with predominance in brain and testis, Tiam1 is not required for proper mouse development (Malliri et al., 2002). This is most likely due to compensatory mechanisms involving other Rac specific exchange factors such as a closer homologue of Tiam1, namely Tiam2 (or Stef). Tiam2 shares essentially the same domain structure with Tiam1 including a predicted RA-domain, a centrally located PDZ motif and the PH/DH region. Yet, a more in depth analysis of the *in vivo* functions of Tiam2 still needs to be carried out (for an overview of Rho GTPase activating proteins see (Schmidt and Hall, 2002).

Malliri et al. have found that the mutational status of the *Tiam1* gene critically influences tumor development in mice whose skin is subjected to a combined 7,12-dimethylbenzanthracene (DMBA)/10-o-tetradecanophorbol-13-acetate (TPA) treatment. DMBA-exposure as a tumor-initiating event is known to invariably induce mutations in c-Ha-Ras, that render the protein constitutively active (H-Ras^{Q61L}) and this was confirmed by the study under discussion here. TPA then promotes growth and development of the tumor in the skin of animals subjected to the treatment. Of interest for this process, Tiam1 appears to interfere with both optimal tumor initiation and promotion since *Tiam*1-/- mice showed significantly lower numbers of tumors that in addition grew at slower rates. In fact, tumor incidence and growth correlated more tightly with the gene dose of *Tiam1* when compared between *Tiam*1-/- Tiam1-/+ and wild type animals. The authors could recapitulate this relation in focus forming assays performed with *Tiam*1-/- mouse embryo fibroblasts (MEFs) that displayed a marked resistance to Ras-induced transformation. However, further complexity was added. The simple hypothesis that Tiam1 inhibits Ras-induced tumor formation per se is incorrect, since the tumors induced by DMBA/TPA in the skin of *Tiam*1-/- mice exert a significantly elevated tendency to progress to malignancy. Etiologically, benign tumors that mostly have a papilloma phenotype on their path to malignancy adopt an invasive carcinoma-like phenotype. Thus, Malliri and co-workers infer a biphasic model from their data in which the initial stages of tumor development are favored by the presence of Tiam1, whereas later on during the papilloma to carcinoma transition Tiam1-function is rather of inhibitory nature (Malliri et al., 2002). In analogy to the latter aspect of the model, continuous activation of Ras in MDCK cells, that undergo an epithelial-mesenchymal transition in culture, ensues a decrease in Tiam1 expression and concomitantly lowered GTP-Rac levels. Moreover, forced expression of TIAM1 from a heterologous promotor negatively interferes with this transition and stabilizes cell-cell contacts (Zondag et al., 2000).

Apart from cell-cell adhesion effects, which other aspects could be a consequence of Tiam1-activation in this scenario? One correlation that Malliri et al. found is the one between the described Tiam1-dependent tumor phenotype and an increased apoptotic index in tumor tissue of *Tiam1-/-* mice, suggesting that Ras signals through Tiam1 to counteract apoptotic clearance. *In vitro*, Rac can activate NF-κB to suppress apoptosis triggered by oncogenic Ras (Joneson and Bar-Sagi, 1999) and it will be interesting to see whether this pathway is operative in TPA/DMBA-treated keratinocytes. Concerning the growth promoting effects of TIAM, Rac can also elicit cyclin D1 expression as a means to accelerate G1/S phase progression of the cell cycle (Bar-Sagi and Hall, 2000; Mettouchi et al., 2001; Welsh et al., 2001) and, moreover, cyclin D1 is required for Ras to provoke skin tumors in mice (Robles et al., 1998).

Developmental studies give yet another clue to Tiam1 function. A transposon insertion in the *still life* gene, the *Drosophila* homolog of *Tiam1*, causes neurological locomotion defects and male sterility. In their initial study, Sone et al. showed, that the fly homologue is present in the periactive zones of the synaptic terminals of neuromuscular junctions (Sone et al., 1997). In a second study the authors isolated further mutations in the *still life* gene and found the *still life* locus to genetically interact with *fasciclin 2* (*Fas2*) in the regulation of synaptic growth. In addition, the Still life protein depends on the cell-adhesion promoting Fas2 geneproduct for its specific localization (Sone et al., 2000). Since Fas2, Still life and also Discs large (Dlg) all regulate synaptic growth and localize to the same structure, the authors speculate that a protein network involving these proteins in the periactive zone regulates growth processes, while the adjacent proximal active zone in synaptic terminals performs its distinct neurotransmitting functions (Sone et al., 2000). To date, no genetic interaction between *still life* and *Ras* has been demonstrated in *Drosophila*, but as Tiam1, Still life harbors a computationally predictable RA-domain (Ponting, 1999), making it a potential Ras-effector target in invertebrate development. A role for Ras in synaptic plasticity has been demonstrated (Zhu et al., 2002) and, biochemically, Still life acts as a GTP-GDP exchange factor for *Drosophila* Rac (Sone et al., 2000). Clearly, many more pieces of the puzzle need to be assembled but an interesting picture is beginning to emerge in which Ras/TIAM/Rac signaling might be of broader biological relevance.

2.4 PLCε: A Ras Effector/Activator with Phospholipase Activity

As another potential effector protein for activated Ras, several research groups have independently identified a novel member of the phospholipase C (PLC) family, which subsequently has been termed PLCε (Shibatohge et al., 1998; Kelley et al., 2001; Lopez et al., 2001; Song et al., 2001). The lipid-modifying function, like in other PLC isoforms is encoded by a central catalytic domain which consists of the highly conserved X and Y motifs (see Figure 1). C-terminally adjacent to the catalytic domain all PLCs harbor a C2 domain. However, PLCε is distinguishable from the other family members by the lack of a PH-domain and EF-hands that in other PLC proteins flank the catalytic core N-terminally and by the presence of a tandem RA-repeat at its C-terminus. In addition, a Ras GTP exchange domain (CDC25 catalytic domain) has been predicted to reside in the very N-terminus of the protein. These latter two features are clearly suggestive of a role as a Ras-family effector and/or activator.

Several studies have provided initial evidence for the notion that PLCε fulfills the criteria of a genuine Ras-effector molecule. The second of the C-terminal RA-domains (RA2) can associate with high affinity to GTP Ras, as has been demonstrated biochemically in pull-down assays with GST- PLCεRA2 and H-Ras that was bound to the non-hydrolysable GTP-analog GTPγS or the constitutively GTP-loaded H-Ras^{Q61L} mutant (Kelley et al., 2001). Furthermore, when Kelley and co-workers mutated a critical Lys residue of the RA2 domain (K2150 in the rat homolog) to Glu thereby disrupting the interaction with Ras, the binding affinity of PLCε to Ras was drastically impaired (Kelley et al., 2001). They further investigated the ability of PLCε to bind to the previously characterized Ras effector loop mutants H-Ras^{T35S}, H-Ras^{Y40C} and H-Ras^{E37G}. H-Ras carrying a T35S mutation was shown to bind to Raf, but not to RalGDS or PI3K, while H-Ras^{Y40C} is specific for PI3K but does not associate with Raf or RalGDS. The H-Ras^{E37G} mutant displays selectivity towards RalGDS and at the same time incapacitates Ras' ability to activate Raf and PI3K. Interestingly, Kelley et al. found the latter mutant, Ras^{E37G} to be the only one of the three to interact with PLCε. This finding raises the question as to whether some of the effects in the various assays that have been performed with those mutants to address the function of RalGDS could be attributable to or at least modifiable by endogenous PLCε. As an example, the recently demonstrated ability of the H-Ras^{E37G} mutant to transform human cells could be fine-tuned by the activity of PLCε (Hamad et al., 2002). This issue merits some future consideration.

This leads us to a central question: does the association of PLCε with Ras stimulate its phospholipase activity? PLC catalytic domains in general hydrolyse phosphatidylinositol 4,5-bisphosphate (PIP2) at the inner plasma membrane leaflet. Thereby 1,4,5-triphosphate (IP3) and diacylglycerol (DAG) are generated, both of which act as second messengers: IP3 induces intracellular Ca^{2+} stores to open up and release Ca^{2+} into the cytosol, whereas DAG stimulates protein kinase C (PKC) isoforms setting of other signaling cascades. By measuring the IP3-release from radioactively labeled inositol in transfected COS cells, Kelley et al. found that PLCε, when co-expressed with constitutively active Ras, displays a significantly elevated catalytic activity. Of note is the observation that the above described K2150E RA-domain mutation that disrupted binding of PLCε to Ras also in this cell culture assay was eliminating the stimulatory effect of Ras. Similarly, the effector loop mutation E37G was deleterious to the capacity of Ras to stimulate PLCε's hydrolytic potential. These data could not be reproduced in an *in vitro* assay with purified proteins, suggesting that additional factors are required for PLCε activation in a cellular context (Kelley et al., 2001). This would be reminiscent of Ras' activation of the Raf-kinases where membrane related phosphorylation events critically contribute to the ensuing kinase-activity. In a liposome-based reconstitution assay, Song et al. managed to demonstrate a stimulatory effect of GTP Ras on the hydrolytic activity of PLCε (Song et al., 2001). Therefore it seems, that PLCε needs to be in close proximity to a membranous environment to exert its activity. Song et al. have investigated a GFP-fused version of PLCε in transfection experiments and they indeed found that Ras could recruit PLCε to the plasma membrane. This recruitment was also detected in starved cells that have been exposed to EGF, whose receptor in great part acts via Ras. Dominant negative Ras in the same experiment could efficiently prevent this re-localization behaviour (Song et al., 2001). In a subsequent study the authors examined the activation of PLCε by PDGF (Song et al., 2002). They found that PLCε's relocalization is a compound effect relying on Ras and Rap1, a close relative of the oncogenic Ras proteins that shares an identical effector loop region. Whereas Ras rapidly activates PLCε in an acute manner, Rap1 appears to be required for the sustained activity of PLCε (Song et al., 2002). A dual activation mode involving Ras and Rap also could be translating an EGF signal into a PLCε response (Kelley et al., 2004). Thus, it appears that growth factor triggered pathways that operate through Ras-family GTPases can recruit PLCε to specific membrane domains where it would meet its natural substrate(s).

Besides being stimulated downstream of receptor tyrosine kinases (RTKs) and Ras as described above, PLCε can mediate the action of heterotrimeric G-proteins that are linked to G-protein coupled receptors (GPCRs). Lopez et al. observed, that Gα12 subunits of heterotrimeric G-proteins, when present in a constitutively active form in TSA201 cells stimulated the hydrolytic activity of PLCε with great efficacy. This in turn promotes the formation of GTP-bound Ras and the activation of MAPK further downstream. Interestingly, the latter can also be achieved in the same transient transfection assay using a variant of PLCε in which due to a point mutation (H1144L in human PLCε) phospholipase activity is obliterated. Since the isolated CDC25 domain of PLCε was also able to evoke these responses, the authors inferred that a subclass of heterotrimeric G-proteins induces PLCε to activate the Ras/MAPK pathway as an upstream mediator and independently of its lipid-directed activity (Lopez et al., 2001). In contrast to its role in linking heterotrimeric G-proteins with Ras, the CDC25 domain of PLCε in hematopoietic BaF3 cells stimulated with PDGF displays GTP/GDP exchange activity towards Rap1, but not Ras (Song et al., 2002). Both studies, however, did not analyze the exchange activity of PLCε *in vitro*, so the observed levels of GTP-Ras and GTP-Rap could still be explicable by indirect effects. As another twist, signaling from some GPCR does not only elevate exchange factor activity of intracellular PLCε, but also stimulates its phospholipase activity.

In search of phospholipase stimulating factors, Kelley and colleagues scrutinized a series of hormones that trigger both RTK and GPCR signaling in COS-7 cells transfected with PLCε (Kelley et al., 2004). Apart from the earlier investigated growth factor EGF as an RTK-ligand, they identified LPA (lysophosphatidic acid),

S1P (sphingosine 1-phosphate) and thrombin as GPCR agonists. Moreover, overexpression of the constitutively active Gα12 and Gα13 subunits of heterotrimeric G-proteins could mimic these effects and it turned out that they did so independently of the RA2 domain of PLCε. RA2-independent phospholipase activity, besides, was also provoked by dominant active versions of the small Rho, Rac and Ral GTPases. However, when co-introduced as constitutively active forms together with PLCε into COS-7 cells, TC21 and Rap small GTPases appeared to trigger phospholipase activity via interaction with the PLCε's RA2 domain (Kelley et al., 2004). It remains obscure, however, how RA2-independent activation might occur and the responsive elements of PLCε will have to be carefully mapped. An equally if not more relevant issue is the one of how the different outputs, namely exchange activity on Ras GTPases and phospholipase activity on phospholipids are regulated upon a given stimulus. As alluded to earlier, some stimuli can achieve both and there could be an underlying interplay in form of a feedback mechanism.

Taken together, the potential of PLCε to be activated as a lipid-metabolizing enzyme by Ras GTPases and alternatively their ability to activate Ras GTPases in probably many physiological situations make PLCε a multifunctional molecule in Ras-dependent signaling pathways.

2.5 AF-6/Afadin: Ras GTPases and Cell Adhesion

The AF-6 protein has been identified as a potential Ras-interactor in different approaches. Using constitutively GTP-associated H-Ras as a probe in the yeast-2-hybrid system and an affinity purification protocol, Van Aelst et al. and Kuriyama et al., respectively, described the AF-6 protein as a Ras-binding molecule (Van Aelst et al., 1994; Kuriyama et al., 1996). This interaction was subsequently extended to K-Ras and N-Ras as further members of the subfamily of oncogenic Ras-proteins (Boettner et al., 2001). Also M-Ras, which at least in the context of the classical focus formation assay is endowed with oncogenic properties was included in the group of AF-6 interacting Ras proteins (Quilliam et al., 1999). In contrast to the established Ras-effectors Raf, PI3K and RalGDS, overexpression of AF-6 does not affect the focus forming activity of H-Ras (Van Aelst et al., 1994) or M-Ras (Quilliam et al., 1999). Prior to its identification as a Ras-interacting molecule, a fusion protein between the N-terminal part of ALL-1 and a C-terminal portion of AF-6 had been pinpointed as a cause for acute myeloid leukemia, hence the gene's name AF-6 for ALL-1 fused gene on chromosome 6. Despite this finding, however, a potential contribution of the AF-6 sequence to the leukemic phenotype so far has not been documented (Prasad et al., 1993).

In the domain composition of AF-6 (Figure 1), two RA-domains that are responsible for the interaction with Ras-family GTPases reside in the N-terminus (Ponting and Benjamin, 1996). This tandem RA motif is followed by U104 and DIL (dilute) motifs, the former being a feature of actin-dependent and the latter occurring in microtubule-dependent motor proteins, respectively (Ponting, 1995). Further

C-terminally located are a PDZ-domain (Ponting and Phillips, 1995) and a long C-terminal sequence without any predicable domains but Proline-rich patches, which can potentially serve as docking sites for binding partners. A larger splice variant of AF-6 has been purified in experiments aimed to identify F-actin binding proteins from rat brain lysates. A smaller splice variant corresponding to human AF-6, which lacks the C-terminal F-actin binding domain was co-purified. Both proteins were localized to adherens junctions (AJs) of various tissues and hence the authors of these studies coined the alternative name afadin for a splice variant of AF-6 in adherens junctions (Mandai et al., 1997). These localization data are slightly conflicting with another report in which AF-6 as a ZO-1 interactor localized to tight junctional structures in canine MDCK cells, although AF-6 together with ZO-1 concentrated at cell-cell contact sides in cells that intrinsically lack tight junctional structures such as Rat-1 fibroblasts and PC12 rat pheochromocytoma cells (Yamamoto et al., 1997). Together with the finding that AF-6 also tethers Profilin as an actin cytoskeletal regulator (Boettner et al., 2000), the notion that AF-6 serves to link specific membrane proteins with the actin cytoskeletal machinery is very plausible. The PDZ domain of AF-6 has been found to capture several membrane proteins. Among these are members of the Ephrin-related receptor tyrosine kinase family. Upon association with the Ephrin receptor, AF-6 can be phosphorylated in a ligand-dependent manner (Hock et al., 1998) as well as induce clustering of Ephrin receptors in heterologous cells when co-expressed (Buchert et al., 1999). Investigating AF-6 in the nervous system, Buchert et al. have traced the protein together with several Ephrin receptors to post-synaptic densities using electronmicroscopical techniques (Buchert et al., 1999). Ephrin signaling has been primarily associated with developmental aspects of the nervous system but is also gaining broader relevance in other areas of development. The consequences of the AF-6/Ephrin receptor interaction for aspects of brain development and biology, however, remain elusive. Another class of specifically interacting membrane proteins are members of the nectin family originally known as poliovirus receptor-related protein PRR (Takahashi et al., 1999). However, lateron it turned out that Nectins do not serve as poliovirus receptor but instead mediate entry and spreading of α–Herpes virus (Cocchi et al., 2000; Menotti et al., 2000). Nectin isoforms as Ca^{2+}-independent immunoglobulin-like cell-adhesion proteins undergo homophilic and heterophilic trans-interaction on the extracellular surface of neighboring cells and localize to epithelial cell-cell contact regions (Takahashi et al., 1999).

In early polarizing epithelial cells, Nectin transdimers have been proposed to form with higher velocity at initial cell-cell contact sites than Cadherin trans-dimers. Both, Nectin and Cadherin conjugates, however, are converted into micro-clusters that mix with each other to form primordial "spot"-like AJs. "Spot"-like AJs in turn mature into more elaborated "line"-like AJs that will further condense into the adherens junctional belt. Despite the difference in initial appearance in cell-cell contact regions, it is presently not clear whether nectins are absolutely required for AJ formation. Nevertheless, AF-6/Afadin may possibly bridge the Nectin and Cadherin adhesion systems. A Proline stretch in the longer form of AF-6 associates with Ponsin (Mandai et al., 1999) and its DIL domain binds ADIP (Afadin DIL domain interacting protein, Asada et al., 2003), both partners being adaptors. Whereas Ponsin may indirectly interact with α-Catenin through the Vinculin protein, ADIP may do so with the help of Actinin.

Following the appearance of AJs, Claudin-based tight junctions (TJs) are formed apically. Already the early spot-like AJs contain molecules such as ZO-1 (Suzuki et al., 2002) and JAM (junctional adhesion molecule, Ebnet et al., 2001), that lateron are crucial for TJ assembly, architecture and/or function. JAM proteins constitute a growing subfamily of the Ig-superfamily that homodimerize across the cleft of juxtaposed cells and are expressed in a variety of leukocytes, platelets, endothelial and epithelial cells (for review see Ebnet et al., 2004). In particular JAM-A marks the membrane region apically of AJs that is to develop into the elaborate tight junctional structure. JAM-A, via its exposed cytoplasmic tail can engage a number of PDZ-domain containing proteins (Ebnet et al., 2004). One of its partners with special relevance for polarization processes is the PAR-3 component of the PAR-3/PAR-6/aPKC complex. In an attractive model, Cadherin-mediated activation of the Rho GTPases Rac1 or CDC42 recruits the PAR-3/PAR-6/aPKC complex, PAR-6 acting as an immediate effector for the Rho GTPases. JAM-A docking to PAR-3 in this context could locally direct the ensuing kinase activity of aPKC (atypical protein kinase C) and thereby designate membrane domains for downstream events that require further TJ-building elements and processes. Interestingly, JAM-A also binds to AF-6/Afadin again tethering the PDZ-domain of the latter with its C-terminus (Ebnet et al., 2001).

Although clear functional evidence for a role of AF-6/Afadin in any of these polarizing steps is still missing, it is intriguing that a single molecule can interact with a whole spectrum of polarity-promoting factors. In line with these localization and interaction studies, targeted disruption of the AF-6/Afadin locus resulted in severe defects in embryonic morphogenesis. Deficient embryos are severely compromised in their ability to properly polarize cells of ectodermal origin and do not display the regular organization of tight and adherens junctional structures that demarcate the lateral membrane (Ikeda et al., 1999; Zhadanov et al., 1999). As a consequence of these and probably also other malfunctions, deficient embryos die at around day E10 with gross morphological abnormalities. It remains to be seen yet whether the problems arising in null embryos relate to the function of Ras-type GTPases in these processes.

In several respects, the effects elicited by oncogenic Ras in epithelial transformation processes constitute a converse scenario of embryonic epithelial polarization and differentiation. Ras-mediated transformation among other features is distinguished by a disruption of cell-cell contacts and subsequent contact-independent growth. Thus, AF-6/Afadin would be an attractive candidate that might serve as a direct entry point into cell-cell adhesion complexes for aberrant oncogenic Ras-proteins. Despite successful efforts to demonstrate the formation of a complex between ectopically expressed H-Ras and endogenous AF-6/Afadin by co-immunoprecipitation (Yamamoto et al., 1999) the actual relevance of

AF-6/Afadin as a Ras-effector and the consequences of this interaction for either mammalian development or specific aspects of tumorigenesis so far remain obscure.

Given the technical difficulties inherent to a comprehensive functional analysis of AF-6, genetically more tractable systems are more likely to provide insights into the relationship between Ras-GTPases and AF-6/Afadin. In fact, the AF-6/Afadin gene is conserved in model organisms such as *Drosophila melanogaster* (Miyamoto et al., 1995) and *Caenorhabditis elegans* (Watari et al., 1998) and in particular studies in *Drosophila* are yielding first valuable concepts highlighting the relevance of AF-6/Afadin as a Ras-target. Miyamoto and co-workers revealed the molecular identity of the canoe gene (cno) and found it to encode a protein with a domain composition that is identical to that of AF-6/Afadin (Miyamoto et al., 1995). Subsequently, Matsuo and colleagues confirmed in yeast-two-hybrid tests that both N-terminal RA-domains of Canoe could interact with activated Ras1, the fly counterpart of the mammalian oncogenic Ras-proteins (Matsuo et al., 1997). As AF-6/Afadin in vertebrate cells, also Canoe is an element of cell-cell adhesion complexes. In analogy to AF-6/Afadin, it directly binds to the *Drosophila* ZO-1 homolog Polychaetoid (Takahashi et al., 1998) and immunolocalization with different antibodies demonstrated that it resides in cell-cell AJs (Takahashi et al., 1998; Boettner et al., 2003) in cells with epithelial character.

Matsuo et al. examined a genetic interaction between *Ras1* and *canoe* in a specific aspect of eye differentiation, namely the generation of cone cells (Matsuo et al., 1997). The *Drosophila* compound eye represents a neuroepithelium with remarkable complexities. One of roughly 750 ommatidia in a fully differentiated eye is composed of 20 cells, 8 of which are photoreceptors and the remaining 12 being associated pigment cells and lens-secreting cone cells. During eye differentiation each cell fate is strictly determined by cell-cell interactive mechanisms regardless of their lineage.

Based on studies by M. Freemann, EGF-receptor (EGFR) signaling turned out to be crucial for the proper differentiation of all retinal cells and over-activation of EGFR-function results in the formation of supernumerary cells, whereas a dominant negative form of the receptor halts the differentiation program in the so-called R7 equivalence group (Freeman, 1996). Since Ras is an intrinsic element of the EGFR signaling pathway, Matsuo and co-workers successfully demonstrated, that constitutively active $Ras1^{V12}$, when expressed at the same stage, significantly elevated the number of cone cells in ommatidia. Conversely, expression of a dominant negative $Ras1^{N17}$ transgene caused a reduction of cone cells (Matsuo et al., 1997). Interestingly, a hypomorphic allele of the *canoe* gene, cno^{mis1}, when homozygously present, also induces a loss of cone cells. Prompted by these similarities, Matsuo and colleagues set out to more closely examine a possible interaction between *Ras1* and *canoe* in this context and came to unexpected findings. In these experiments, a lowered gene dosage at the *canoe* locus markedly amplified the inhibiting potential of dominant negative $Ras1^{N17}$ on cone cell formation, while the same manipulation in conjunction with a constitutively active $Ras1^{V12}$ transgene also had a stimulatory effect. Here, cone cells were formed in even greater excess (Matsuo et al., 1997). These data, on first sight, are difficult to consolidate and will require more in depth experimentation. The authors formulate a model in which the Canoe protein serves to coordinate crosstalk of EGFR/Ras1 signaling with another yet undefined pathway. A candidate for the latter is the Notch pathway since it crucially participates in cell fate decisions in the eye by conferring what is known as lateral inhibition between neighboring cells in a group of diversely differentiating cells. In support of this, *canoe* has previously been shown to genetically interact with Notch signaling components during wing development (Miyamoto et al., 1995).

More recently, in another paradigm, Gaengel and Mlodzik discovered a role for *canoe* during the ommatidial rotation process in the *Drosophila* compound eye (Gaengel and Mlodzik, 2003). Ommatidial rotation follows the cellular specification of the R3/R4 photoreceptor fates (Adler, 2002; Mlodzik, 2002) and entails the rotation of an entire but not yet fully differentiated ommatidial precluster, independent of its surrounding environment in a concerted fashion. Preclusters in the dorsal half of the eye imaginal disc rotate 90° counterclockwise and clusters in the ventral half rotate 90° clockwise towards the dorsalventral midline, which is also termed equator. Thereby, a very specific form of epithelial planar polarity (PCP) is established in which preclusters in the dorsal and ventral halves of the differentiating eye take on opposite chirality (reviewed in Reifegerste and Moses, 1999; Adler, 2002; Mlodzik, 2002).

Few mutations have been described, that impact the rotation process without affecting either R3/R4 photoreceptor specification or the actual direction of rotation. Gaengel and Mlodzik were able to molecularly unravel the *roulette* allele as a mutation specifically impairing ommatidial rotation (Gaengel and Mlodzik, 2003). *Roulette* represents an allele of the *argos* gene, a previously described secreted inhibitor of EGFR signaling (Freeman et al., 1992). The *roulette* mutation causes a rotation-specific phenotype in which ommatidia tend to rotate at random, some more and some less than 90° toward the equator (Gaengel and Mlodzik, 2003). The investigators showed a genetic interaction between *roulette* and *Ras1* and in addition demonstrated that also *pointed*, the gene encoding the transcription factor that is activated downstream of Ras1 by rolled/MAPK. This strongly suggested that the activated EGFR relays its signal through the Ras/MAPK pathway to regulate ommatidial rotation. The same study reported, that also *canoe* interacts genetically with elements of the EGFR-pathway. The rotation defects triggered by a heterozygous null mutation in the *star* gene, which encodes a positive regulator of EGFR signaling, are further aggravated when also the *canoe* gene dosage was halved. The fact that the subviable combination of the *canoe* hypomorph, cno^{mis1}, with a complete *canoe* loss-of-function allele also gave rise to strong rotation defects, reinforce the significance of *canoe* in the control of this process (Gaengel and Mlodzik, 2003). The evidence, however, that Canoe's function is that of a direct Ras-effector in this context still needs to be consolidated.

It should be mentioned, that such evidence has been provided for the physical interaction between Canoe and the Ras1 sister GTPase Rap1 in the embryonic dorsal closure process. Rap1, harboring an effector loop region that is identical to Ras, strongly interacts with Canoe and this holds true also for the mammalian counterparts Rap1A and AF-6/afadin (Van Aelst et al., 1994; Boettner et al., 2000; Boettner et al., 2003). Canoe actually was isolated in a yeast-two-hybrid screen that was conducted with a constitutively active *Drosophila* $Rap1^{V12}$ mutant as bait and it turned out that both N-terminal RA-domains in the Canoe protein apart from Ras1 could also associate with Rap1 (Boettner et al., 2003). Thermodynamic and other binding studies have shown that AF-6/Afadin binds even stronger to Rap1 than to oncogenic Ras proteins (Linnemann et al., 1999; Boettner et al., 2000). In dorsal closure, as part of the gastrulation process in the fly embryo, the lateral ectoderm stretches dorsalward to close the embryo at the dorsal midline, thereby replacing a transient epithelium also known as amnioserosa. In fact, the *canoe* locus was originally identified because a mutation in it caused a "dorsal open" phenotype (Jurgens et al., 1984). More recently, it was reported that *canoe* and *Rap1* mutations genetically interact to produce a synergistic phenotype (Boettner et al., 2003). Since altering Ras1 activity in the embryonic ectoderm at dorsal closure stages does not influence the completion of the process in any way, this experimental paradigm provided the advantage to assess potential signaling from Rap1 to canoe on a molecular level without any interference of Ras1 which in other cases can pose an obstacle to the interpretability of interaction data obtained *in vivo*. In a dominant negative approach, expression of dominant negative $Rap1^{N17}$ in the embryonic ectoderm provokes a severe dorsal closure defect, which can be significantly rescued by concomitant overexpression of a wild type *canoe* transgene. However, a deletion mutant of Canoe that is lacking both N-terminal RA-domains is unable to alleviate the strong defect caused by $Rap1^{N17}$, whereas it still possessed marked biological activity since it can be integrated into cell-cell adhesion complexes and more importantly, it partially rescues the *canoe*-specific dorsal closure defect. Canoe, in a Rap1-independent manner feeds into Jun-kinase signaling during DC, again indicating participation in more than one pathway in a given biological process (Boettner et al., 2003).

These data taken together suggest that the Canoe protein may perceive inputs from different Ras-type GTPases to orchestrate events in multiple processes only a few of which have been looked at in more detail. Very likely it will only be a matter of time until analogous correlations centering on the AF-6/Afadin protein will be revealed in mammalian systems. Although more information on the signaling pathways and networks that AF-6/Afadin/Canoe is involved in will be essential to comprehensively picture its function, already now a molecule with multifaceted regulatory and functional implications can be anticipated.

REFERENCES

Adler, P.N. (2002) Planar signaling and morphogenesis in Drosophila. Dev. Cell *2*, 525-535.

Afar, D.E., Han, L., McLaughlin, J., Wong, S., Dhaka, A., Parmar, K., Rosenberg, N., Witte, O.N. and Colicelli, J. (1997) Regulation of the oncogenic activity of BCR-ABL by a tightly bound substrate protein RIN1. Immunity *6*, 773-782.

Asada, M., Irie, K., Morimoto, K., Yamada, A., Ikeda, W., Takeuchi, M. and Takai, Y. (2003) ADIP, a novel Afadin- and alpha-actinin-binding protein localized at cell-cell adherens junctions. J. Biol. Chem. *278*, 4103-4111.

Bar-Sagi, D. and Feramisco, J.R. (1986) Induction of membrane ruffling and fluid-phase pinocytosis in quiescent fibroblasts by ras proteins. Science *233*, 1061-1068.

Bar-Sagi, D. and Hall, A. (2000) Ras and Rho GTPases: a family reunion. Cell *103*, 227-238.

Barbieri, M.A., Kong, C., Chen, P.I., Horazdovsky, B.F. and Stahl, P.D. (2003) The SRC homology 2 domain of Rin1 mediates its binding to the epidermal growth factor receptor and regulates receptor endocytosis. J. Biol. Chem. *278*, 32027-32036.

Bates, S., Phillips, A.C., Clark, P.A., Stott, F., Peters, G., Ludwig, R.L. and Vousden, K.H. (1998) p14ARF links the tumour suppressors RB and p53. Nature *395*, 124-125.

Bodmer, D., Eleveld, M.J., Ligtenberg, M.J., Weterman, M.A., Janssen, B.A., Smeets, D.F., de Wit, P.E., van den Berg, A., van den Berg, E., Koolen, M.I. and Geurts van Kessel, A. (1998) An alternative route for multistep tumorigenesis in a novel case of hereditary renal cell cancer and a t (2;3) (q35;q21) chromosome translocation. Am. J. Hum. Genet. *62*, 1475-1483.

Boettner, B., Govek, E.E., Cross, J. and Van Aelst, L. (2000) The junctional multidomain protein AF-6 is a binding partner of the Rap1A GTPase and associates with the actin cytoskeletal regulator profilin. Proc. Natl. Acad. Sci. U.S.A. *97*, 9064-9069.

Boettner, B., Herrmann, C. and Van Aelst, L. (2001) Ras and Rap1 interaction with AF-6 effector target. Methods Enzymol. *332*, 151-168.

Boettner, B., Harjes, P., Ishimaru, S., Heke, M., Fan, H.Q., Qin, Y., Van Aelst, L. and Gaul, U. (2003) The AF-6 homolog canoe acts as a Rap1 effector during dorsal closure of the Drosophila embryo. Genetics *165*, 159-169.

Buchert, M., Schneider, S., Meskenaite, V., Adams, M.T., Canaani, E., Baechi, T., Moelling, K. and Hovens, C.M. (1999) The junction-associated protein AF-6 interacts and clusters with specific Eph receptor tyrosine kinases at specialized sites of cell-cell contact in the brain. J. Cell Biol. *144*, 361-371.

Chen, J., Lui, W.O., Vos, M.D., Clark, G.J., Takahashi, M., Schoumans, J., Khoo, S.K., Petillo, D., Lavery, T., Sugimura, J., Astuti, D., Zhang, C., Kagawa, S., Maher, E.R., Larsson, C., Alberts, A.S., Kanayama, H.O. and Teh, B.T. (2003) The t(1;3) breakpoint-spanning genes LSAMP and NORE1 are involved in clear cell renal cell carcinomas. Cancer Cell *4*, 405-413.

Cheung, W.L., Ajiro, K., Samejima, K., Kloc, M., Cheung, P., Mizzen, C.A., Beeser, A., Etkin, L.D., Chernoff, J., Earnshaw, W.C. and Allis, C.D. (2003) Apoptotic phosphorylation of histone H2B is mediated by mammalian sterile twenty kinase. Cell *113*, 507-517.

Cocchi, F., Menotti, L., Dubreuil, P., Lopez, M. and Campadelli-Fiume, G. (2000) Cell-to-cell spread of wild-type herpes simplex virus type 1, but not of syncytial strains, is mediated by the immunoglobulin-like receptors that mediate virion entry, nectin1 (PRR1/HveC/HIgR) and nectin2 (PRR2/HveB). J. Virol. *74*, 3909-3917.

Colicelli, J., Nicolette, C., Birchmeier, C., Rodgers, L., Riggs, M. and Wigler, M. (1991) Expression of three mammalian cDNAs that interfere with RAS function in Saccharomyces cerevisiae. Proc. Natl. Acad. Sci. U.S.A. *88*, 2913-2917.

Dammann, R., Schagdarsurengin, U., Liu, L., Otto, N., Gimm, O., Dralle, H., Boehm, B.O., Pfeifer, G.P. and Hoang-Vu, C. (2003) Frequent RASSF1A promoter hypermethylation and K-ras mutations in pancreatic carcinoma. Oncogene *22*, 3806-3812.

Dhaka, A., Costa, R.M., Hu, H., Irvin, D.K., Patel, A., Kornblum, H.I., Silva, A.J., O'Dell, T.J. and Colicelli, J. (2003) The RAS effector RIN1 modulates the formation of aversive memories. J. Neurosci. *23*, 748-757.

Dreijerink, K., Braga, E., Kuzmin, I., Geil, L., Duh, F.M., Angeloni, D., Zbar, B., Lerman, M.I., Stanbridge, E.J., Minna, J.D., Protopopov, A., Li, J., Kashuba, V., Klein, G. and Zabarovsky, E.R. (2001) The candidate tumor suppressor gene, RASSF1A, from human chromosome 3p21.3 is involved in kidney tumorigenesis. Proc. Natl. Acad. Sci. U.S.A. *98*, 7504-7509.

Ebnet, K., Suzuki, A., Horikoshi, Y., Hirose, T., Meyer Zu Brickwedde, M.K., Ohno, S. and Vestweber, D. (2001) The cell polarity protein ASIP/PAR-3 directly associates with junctional adhesion molecule (JAM). EMBO J. *20*, 3738-3748.

Ebnet, K., Suzuki, A., Ohno, S. and Vestweber, D. (2004) Junctional adhesion molecules (JAMs): more molecules with dual functions?. J. Cell Sci. *117*, 19-29.

Fleming, I.N., Gray, A. and Downes, C.P. (2000) Regulation of the Rac1-specific exchange factor Tiam1 involves both phosphoinositide 3-kinase-dependent and -independent components. Biochem. J. *351*, 173-182.

Freeman, M., Klambt, C., Goodman, C.S. and Rubin, G.M. (1992) The argos gene encodes a diffusible factor that regulates cell fate decisions in the Drosophila eye. Cell. *69*, 963-975.

Freeman, M. (1996) Reiterative use of the EGF receptor triggers differentiation of all cell types in the Drosophila eye. Cell *87*, 651-660.

Gaengel, K. and Mlodzik, M. (2003) Egfr signaling regulates ommatidial rotation and cell motility in the Drosophila eye via MAPK/Pnt signaling and the Ras effector Canoe/AF6. Development, *130*, 5413-5423.

Graves, J.D., Gotoh, Y., Draves, K.E., Ambrose, D., Han, D.K., Wright, M., Chernoff, J., Clark, E.A. and Krebs, E.G. (1998) Caspase-mediated activation and induction of apoptosis by the mammalian Ste20-like kinase Mst1. EMBO J. *17*, 2224-2234.

Habets, G.G., Scholtes, E.H., Zuydgeest, D., van der Kammen, R.A., Stam, J.C., Berns, A. and Collard, J.G. (1994) Identification of an invasion-inducing gene, Tiam-1, that encodes a protein with homology to GDP-GTP exchangers for Rho-like proteins. Cell. *77*, 537-549.

Hamad, N.M., Elconin, J.H., Karnoub, A.E., Bai, W., Rich, J.N., Abraham, R.T., Der, C.J. and Counter C.M. (2002) Distinct requirements for Ras oncogenesis in human versus mouse cells. Genes Dev. *16*, 2045-2057.

Han, L. and Colicelli, J. (1995) A human protein selected for interference with Ras function interacts directly with Ras and competes with Raf1. Mol. Cell. Biol. *15*, 1318-1323.

Harvey, K.F., Pfleger, C.M. and Hariharan, I.K. (2003) The Drosophila Mst ortholog, hippo, restricts growth and cell proliferation and promotes apoptosis. Cell *114*, 457-467.

Hesson, L., Dallol, A., Minna, J.D., Maher, E.R. and Latif, F. (2003) NORE1A, a homologue of RASSF1A tumour suppressor gene is inactivated in human cancers. Oncogene *22*, 947-954.

Hock, B., Bohme, B., Karn, T., Yamamoto, T., Kaibuchi, K., Holtrich, U., Holland, S., Pawson, T., Rubsamen-Waigmann, H. and Strebhardt, K. (1998) PDZ-domain-mediated interaction of the Eph-related receptor tyrosine kinase EphB3 and the ras-binding protein AF6 depends on the kinase activity of the receptor. Proc. Natl. Acad. Sci. U.S.A. *95*, 9779-9784.

Ikeda, W., Nakanishi, H., Miyoshi, J., Mandai, K., Ishizaki, H., Tanaka, M., Togawa, A., Takahashi, K., Nishioka, H., Yoshida, H., Mizoguchi, A., Nishikawa, S. and Takai, Y. (1999) Afadin: a key molecule essential for structural organization of cell-cell junctions of polarized epithelia during embryogenesis. J. Cell Biol. *146*, 1117-1132.

Joneson, T. and Bar-Sagi, D. (1999) Suppression of Ras-induced apoptosis by the Rac GTPase. Mol. Cell. Biol. *19*, 5892-5901.

Jurgens, G.E., Wieschaus, C., Nusslein-Volhard, C. and Kluding, H. (1984) Mutations affecting the pattern of the larval cuticle in Drosophila melanogaster. II. Zygotic loci on the third chromosome. Roux's Arch. Dev. Biol. *193*, 28-44.

Kanayama, H., Lui, W.O., Takahashi, M., Naroda, T., Kedra, D., Wong, F.K., Kuroki, Y., Nakahori, Y., Larsson, C., Kagawa, S. and Teh, B.T. (2001) Association of a novel constitutional translocation t(1q;3q) with familial renal cell carcinoma. J. Med. Genet. *38*, 165-170.

Kango-Singh, M., Nolo, R., Tao, C., Verstreken, P., Hiesinger, P.R., Bellen, H.J. and Halder, G. (2002) Shar-pei mediates cell proliferation arrest during imaginal disc growth in Drosophila. Development *129*, 5719-5730.

Kauffmann-Zeh, A., Rodriguez-Viciana, P., Ulrich, E., Gilbert, C., Coffer, P., Downward, J. and Evan, G. (1997) Suppression of c-Myc-induced apoptosis by Ras signalling through PI(3)K and PKB. Nature *385*, 544-548.

Kelley, G.G., Reks, S.E., Ondrako, J.M. and Smrcka, A.V. (2001) Phospholipase C (epsilon): a novel Ras effector. EMBO J. *20*, 743-754.

Kelley, G.G., Reks, S.E. and Smrcka, A.V. (2004) Hormonal regulation of phospholipase Cepsilon through distinct and overlapping pathways involving G12 and Ras family G-proteins. Biochem. J. *378*, 129-139.

Khokhlatchev, A., Rabizadeh, S., Xavier, R., Nedwidek, M., Chen, T., Zhang, X.F., Seed, B. and Avruch, J. (2002) Identification of a novel Ras-regulated proapoptotic pathway. Curr. Biol. *12*, 253-265.

Khwaja, A., Rodriguez-Viciana, P., Wennstrom, S., Warne, P.H. and Downward, J. (1997) Matrix adhesion and Ras transformation both activate a phosphoinositide 3-OH kinase and protein kinase B/Akt cellular survival pathway. EMBO J. *16*, 2783-2793.

Kuriyama, M., Harada, N., Kuroda, S., Yamamoto, T., Nakafuku, M., Iwamatsu, A., Yamamoto, D., Prasad, R., Croce, C., Canaani, E. and Kaibuchi, K. (1996) Identification of AF-6 and canoe as putative targets for Ras. J. Biol. Chem. *271*, 607-610.

Lambert, J.M., Lambert, Q.T., Reuther, G.W., Malliri, A., Siderovski, D.P., Sondek, J., Collard, J.G. and Der, C.J. (2002) Tiam1 mediates Ras activation of Rac by a PI(3)K-independent mechanism. Nat. Cell Biol. *4*, 621-625.

Lee, K.K., Murakawa, M., Nishida, E., Tsubuki, S., Kawashima, S., Sakamaki, K. and Yonehara, S. (1998) Proteolytic activation of MST/Krs, STE20-related protein kinase, by caspase during apoptosis. Oncogene *16*, 3029-3037.

Li, Y., Pei, J., Xia, H., Ke, H., Wang, H. and Tao, W. (2003) Lats2, a putative tumor suppressor, inhibits G1/S transition. Oncogene *22*, 4398-4405.

Lin, A.W., Barradas, M., Stone, J.C., van Aelst, L., Serrano, M. and Lowe, S.W. (1998) Premature senescence involving p53 and p16 is activated in response to constitutive MEK/MAPK mitogenic signaling. Genes Dev. *12*, 3008-3019.

Linnemann, T., Geyer, M., Jaitner, B.K., Block, C., Kalbitzer, H.R., Wittinghofer, A. and Herrmann, C. (1999) Thermodynamic and kinetic characterization of the interaction between the Ras binding domain of AF6 and members of the Ras subfamily. J. Biol. Chem. *274*, 13556-13562.

Lloyd, A.C., Obermuller, F., Staddon, S., Barth, C.F., McMahon, M. and Land, H. (1997) Cooperating oncogenes converge to regulate cyclin/cdk complexes. Genes Dev. *11*, 663-677.

Lopez, I., Mak, E.C., Ding, J., Hamm, H.E. and Lomasney, J.W. (2001) A novel bifunctional phospholipase c that is regulated by Galpha 12 and stimulates the Ras/mitogen-activated protein kinase pathway. J. Biol. Chem. *276*, 2758-2765.

Malliri, A., van der Kammen, R.A., Clark, K., van der Valk, M., Michiels, F. and Collard, J.G. (2002) Mice deficient in the Rac activator Tiam1 are resistant to Ras-induced skin tumours. Nature *417*, 867-871.

Mandai, K., Nakanishi, H., Satoh, A., Obaishi, H., Wada, M., Nishioka, H., Itoh, M., Mizoguchi, A., Aoki, T., Fujimoto, T., Matsuda, Y., Tsukita, S. and Takai, Y. (1997) Afadin: a novel actin filament-binding protein with one PDZ domain localized at cadherin-based cell-to-cell adherens junction. J. Cell Biol. *139*, 517-528.

Mandai, K., Nakanishi, H., Satoh, A., Takahashi, K., Satoh, K., Nishioka, H., Mizoguchi, A. and Takai, Y. (1999) Ponsin/SH3P12: an 1-afadin- and vinculin-binding protein localized at cell-cell and cell-matrix adherens junctions. J. Cell Biol. *144*, 1001-1017.

Marte, B.M., Rodriguez-Viciana, P., Wennstrom, S., Warne, P.H. and Downward, J. (1997) R-Ras can activate the phosphoinositide 3-kinase but not the MAP kinase arm of the Ras effector pathways. Curr. Biol. *7*, 63-70.

Matheny, S.A., Chen, C., Kortum, R.L., Razidlo, G.L., Lewis, R.E. and White, M.A. (2004) Ras regulates assembly of mitogenic signalling complexes through the effector protein IMP. Nature. *427*, 256-260.

Matsuo, T., Takahashi, K., Kondo, S., Kaibuchi, K. and Yamamoto, D. (1997) Regulation of cone cell formation by Canoe and Ras in the developing Drosophila eye. Development. *124*, 2671-2680.

Mayo, M.W., Wang, C.Y., Cogswell, P.C., Rogers-Graham, K.S., Lowe, S.W., Der, C.J. and Baldwin, Jr. A.S. (1997) Requirement of NF-kappaB activation to suppress p53-independent apoptosis induced by oncogenic Ras. Science *278*, 1812-1815.

Menotti, L., Lopez, M., Avitabile, E., Stefan, A., Cocchi, F., Adelaide, J., Lecocq, E., Dubreuil, P. and Campadelli-Fiume, G. (2000) The murine homolog of human Nectin1delta serves as a species nonspecific mediator for entry of human and animal alpha herpesviruses in a pathway independent of a detectable binding to gD. Proc. Natl. Acad. Sci. U.S.A. *97*, 4867-4872.

Mettouchi, A., Klein, S., Guo, W., Lopez-Lago, M., Lemichez, E., Westwick, J.K. and Giancotti, F.G. (2001) Integrin-specific activation of Rac controls progression through the G(1) phase of the cell cycle. Mol. Cell *8*, 115-127.

Michiels, F., Habets, G.G., Stam, J.C., van der Kammen, R.A. and Collard, J.G. (1995) A role for Rac in Tiam1-induced membrane ruffling and invasion. Nature. *375*, 338-340.

Mitin, N.Y., Ramocki, M.B., Zullo, A.J., Der, C.J., Konieczny S.F. and Taparowsky, E.J. (2004) Identification and characterization of rain, a novel Ras-interacting protein with a unique subcellular localization, J. Biol. Chem. *279*, 22353-22361.

Miyamoto, H., Nihonmatsu, I., Kondo, S., Ueda, R., Togashi, S., Hirata, K., Ikegami, Y. and Yamamoto, D. (1995) Canoe encodes a novel protein containing a GLGF/DHR motif and functions with Notch and scabrous in common developmental pathways in Drosophila. Genes Dev. *9*, 612-625.

Mlodzik, M. (2002) Planar cell polarization: do the same mechanisms regulate Drosophila tissue polarity and vertebrate gastrulation? Trends Genet. *18*, 564-571.

Morrissey, C., Martinez, A., Zatyka, M., Agathanggelou, A., Honorio, S., Astuti, D., Morgan, N.V., Moch, H., Richards, F.M., Kishida, T., Yao, M., Schraml, P., Latif, F. and Maher, E.R. (2001) Epigenetic inactivation of the RASSF1A 3p21.3 tumor suppressor gene in both clear cell and papillary renal cell carcinoma. Cancer Res. *61*, 7277-7281.

Neufeld, T.P., de la Cruz, A.F., Johnston, L.A. and Edgar, B.A. (1998) Coordination of growth and cell division in the Drosophila wing. Cell *93*, 1183-1193.

Ortiz-Vega, S., Khokhlatchev, A., Nedwidek, M., Zhang, X.F., Dammann, R., Pfeifer, G.P. and Avruch, J. (2002) The putative tumor suppressor RASSF1A homodimerizes and heterodimerizes with the Ras-GTP binding protein Nore1. Oncogene *21*, 1381-1390.

Palmero, I., Pantoja, C. and Serrano, M. (1998) p19ARF links the tumour suppressor p53 to Ras. Nature *395*, 125-126.

Pantalacci, S., Tapon, N. and Leopold, P. (2003) The Salvador partner Hippo promotes apoptosis and cell-cycle exit in Drosophila. Nat. Cell Biol. *5*, 921-927.

Ponting, C.P. (1995) AF-6/cno: neither a kinesin nor a myosin, but a bit of both. Trends Biochem. Sci. *20*, 265-266.

Ponting, C.P. (1999) Raf-like Ras/Rap-binding domains in RGS12- and still-life-like signalling proteins. J. Mol. Med. *77*, 695-698.

Ponting, C.P. and Phillips, C. (1995) DHR domains in syntrophins, neuronal NO synthases and other intracellular proteins. Trends Biochem. Sci. *20*, 102-103.

Ponting, C.P. and Benjamin, D.R. (1996) A novel family of Ras-binding domains. Trends Biochem. Sci. *21*, 422-425.

Prasad, R., Gu, Y., Alder, H., Nakamura, T., Canaani, O., Saito, H., Huebner, K., Gale, R.P., Nowell, P.C., Kuriyama, K., et al. (1993) Cloning of the ALL-1 fusion partner, the AF-6 gene, involved in acute myeloid leukemias with the t(6;11) chromosome translocation. Cancer Res. *53*, 5624-5628.

Quilliam, L.A., Castro, A.F., Rogers-Graham, K.S., Martin, C.B., Der, C.J. and Bi, C. (1999) M-Ras/R-Ras3, a transforming ras protein regulated by Sos1, GRF1, and p120 Ras GTPase-activating protein, interacts with the putative Ras effector AF6. J. Biol. Chem. *274*, 23850-23857.

Reifegerste, R. and Moses, K. (1999) Genetics of epithelial polarity and pattern in the Drosophila retina. Bioessays *21*, 275-285.

Richardson, H., O'Keefe, L.V., Marty, T. and Saint, R. (1995) Ectopic cyclin E expression induces premature entry into S phase and disrupts pattern formation in the Drosophila eye imaginal disc. Development *121*, 3371-3379.

Ridley, A.J., Paterson, H.F., Johnston, C.L., Diekmann, D. and Hall, A. (1992) The small GTP-binding protein rac regulates growth factor-induced membrane ruffling. Cell *70*, 401-410.

Robles, A.I., Rodriguez-Puebla, M.L., Glick, A.B., Trempus, C., Hansen, L., Sicinski, P., Tennant, R.W., Weinberg, R.A., Yuspa, S.H. and Conti, C.J. (1998) Reduced skin tumor development in cyclin D1-deficient mice highlights the oncogenic ras pathway in vivo. Genes Dev. *12*, 2469-2474.

Rodriguez-Viciana, P., Warne, P.H., Khwaja, A., Marte, B.M., Pappin, D., Das, P., Waterfield, M.D., Ridley, A. and Downward, J. (1997) Role of phosphoinositide 3-OH kinase in cell transformation and control of the actin cytoskeleton by Ras. Cell *89*, 457-467.

Sander, E.E., van Delft, S., ten Klooster, J.P., Reid, T., van der Kammen, R.A., Michiels, F. and Collard, J.G. (1998) Matrix-dependent Tiam1/Rac signaling in epithelial cells promotes either cell-cell adhesion or cell migration and is regulated by phosphatidylinositol 3-kinase. J. Cell Biol. *143*, 1385-1398.

Scheel, H. and Hofmann, K. (2003) A novel interaction motif, SARAH, connects three classes of tumor suppressor. Curr. Biol. *13*, R899-R900.

Schmidt, A. and Hall, A. (2002) Guanine nucleotide exchange factors for Rho GTPases: turning on the switch. Genes Dev. *16*, 1587-1609.

Serrano, M., Lin, A.W., McCurrach, M.E., Beach, D. and Lowe, S.W. (1997) Oncogenic ras provokes premature cell senescence associated with accumulation of p53 and p16INK4a. Cell *88*, 593-602.

Shibatohge, M., Kariya, K., Liao, Y., Hu, C.D., Watari, Y., Goshima, M., Shima, F. and Kataoka, T. (1998) Identification of PLC210, a Caenorhabditis elegans phospholipase C, as a putative effector of Ras. J. Biol. Chem. *273*, 6218-6222.

Sone, M., Hoshino, M., Suzuki, E., Kuroda, S., Kaibuchi, K., Nakagoshi, H., Saigo, K., Nabeshima, Y. and Hama, C. (1997) Still life, a protein in synaptic terminals of Drosophila homologous to GDP-GTP exchangers. Science *275*, 543-547.

Sone, M., Suzuki, E., Hoshino, M., Hou, D., Kuromi, H., Fukata, M., Kuroda, S., Kaibuchi, K., Nabeshima, Y. and Hama, C. (2000) Synaptic development is controlled in the periactive zones of Drosophila synapses. Development *127*, 4157-4168.

Song, C., Hu, C.D., Masago, M., Kariyai, K., Yamawaki-Kataoka, Y., Shibatohge, M., Wu, D., Satoh, T. and Kataoka, T. (2001) Regulation of a novel human phospholipase C, PLCepsilon, through membrane targeting by Ras. J. Biol. Chem. *276*, 2752-2757.

Song, C., Satoh, T., Edamatsu, H., Wu, D., Tadano, M., Gao, X. and Kataoka, T. (2002) Differential roles of Ras and Rap1 in growth factor-dependent activation of phospholipase C epsilon. Oncogene *21*, 8105-8113.

Spugnardi, M., Tommasi, S., Dammann, R., Pfeifer, G.P. and Hoon, D.S. (2003) Epigenetic inactivation of RAS association domain family protein 1 (RASSF1A) in malignant cutaneous melanoma. Cancer Res. *63*, 1639-1643.

Suzuki, A., Ishiyama, C., Hashiba, K., Shimizu, M., Ebnet, K. and Ohno, S. (2002) aPKC kinase activity is required for the asymmetric differentiation of the premature junctional complex during epithelial cell polarization. J. Cell Sci. *115*, 3565-3573.

Szabo, K., Jekely, G. and Rorth, P. (2001) Cloning and expression of sprint, a Drosophila homologue of RIN1. Mech. Dev. *101*, 259-262.

Takahashi, K., Matsuo, T., Katsube, T., Ueda, R. and Yamamoto, D. (1998) Direct binding between two PDZ domain proteins Canoe and ZO-1 and their roles in regulation of the jun N-terminal kinase pathway in Drosophila morphogenesis. Mech. Dev. *78*, 97-111.

Takahashi, K., Nakanishi, H., Miyahara, M., Mandai, K., Satoh, K., Satoh, A., Nishioka, H., Aoki, J., Nomoto, A., Mizoguchi, A. and Takai, Y. (1999) Nectin/PRR: an immunoglobulin-like cell adhesion molecule recruited to cadherin-based adherens junctions through interaction with Afadin, a PDZ domain-containing protein. J. Cell Biol. *145*, 539-549.

Tall, G.G., Barbieri, M.A., Stahl, P.D. and Horazdovsky, B.F. (2001) Ras-activated endocytosis is mediated by the Rab5 guanine nucleotide exchange activity of RIN1. Dev. Cell. *1*, 73-82.

Tapon, N., Harvey, K.F., Bell, D.W., Wahrer, D.C., Schiripo, T.A., Haber, D.A. and Hariharan, I.K. (2002) salvador Promotes both cell cycle exit and apoptosis in Drosophila and is mutated in human cancer cell lines. Cell *110*, 467-478.

Tommasi, S., Dammann, R., Jin, S.G., Zhang, X.F., Avruch, J. and Pfeifer, G.P. (2002) RASSF3 and NORE1: identification and cloning of two human homologues of the putative tumor suppressor gene RASSF1. Oncogene *21*, 2713-2720.

Udan, R.S., Kango-Singh, M., Nolo, R., Tao, C. and Halder, G. (2003) Hippo promotes proliferation arrest and apoptosis in the Salvador/Warts pathway. Nat. Cell Biol. *5*, 914-920.

Van Aelst, L., White, M.A. and Wigler, M.H. (1994) Ras partners. Cold Spring Harb. Symp. Quant. Biol. *59*, 181-186.

van Kessel, A.G., Wijnhoven, H., Bodmer, D., Eleveld, M., Kiemeney, L., Mulders, P., Weterman, M., Ligtenberg, M., Smeets, D. and Smits, A. (1999) Renal cell cancer: chromosome 3 translocations as risk factors. J. Natl. Cancer Inst. *91*, 1159-1160.

Vavvas, D., Li, X., Avruch, J. and Zhang, X.F. (1998) Identification of Nore1 as a potential Ras effector. J. Biol. Chem. *273*, 5439-5442.

Vos, M.D., Ellis, C.A., Elam, C., Ulku, A.S., Taylor, B.J. and Clark, G.J. (2003) RASSF2 is a novel K-Ras-specific effector and potential tumor suppressor. J. Biol. Chem. *278*, 28045-28051.

Wang, Y., Waldron, R.T., Dhaka, A., Patel, A., Riley, M.M., Rozengurt, E. and Colicelli, J. (2002) The RAS effector RIN1 directly competes with RAF and is regulated by 14-3-3 proteins. Mol. Cell. Biol. *22*, 916-926.

Watari, Y., Kariya, K., Shibatohge, M., Liao, Y., Hu, C.D., Goshima, M., Tamada, M., Kikuchi, A. and Kataoka, T. (1998) Identification of Ce-AF-6, a novel Caenorhabditis elegans protein, as a putative Ras effector. Gene *224*, 53-58.

Welsh, C.F., Roovers, K., Villanueva, J., Liu, Y., Schwartz, M.A. and Assoian, R.K. (2001) Timing of cyclin D1 expression within G1 phase is controlled by Rho. Nat. Cell Biol. *3*, 950-957.

Wu, S., Huang, J., Dong, J. and Pan, D. (2003) hippo encodes a Ste-20 family protein kinase that restricts cell proliferation and promotes apoptosis in conjunction with salvador and warts. Cell *114*, 445-456.

Yamamoto, T., Harada, N., Kano, K., Taya, S., Canaani, E., Matsuura, Y., Mizoguchi, A., Ide, C. and Kaibuchi, K. (1997) The Ras target AF-6 interacts with ZO-1 and serves as a peripheral component of tight junctions in epithelial cells. J. Cell Biol. *139*, 785-795.

Yamamoto, T., Harada, N., Kawano, Y., Taya, S. and Kaibuchi, K. (1999) In vivo interaction of AF-6 with activated Ras and ZO-1. Biochem. Biophys. Res. Commun. *259*, 103-107.

Yoon, J.H., Dammann, R. and Pfeifer, G.P. (2001) Hypermethylation of the CpG island of the RASSF1A gene in ovarian and renal cell carcinomas. Int. J. Cancer *94*, 212-217.

Zhadanov, A.B., Provance, Jr. D.W., Speer, C.A., Coffin, J.D., Goss, D., Blixt, J.A., Reichert, C.M. and Mercer, J.A. (1999) Absence of the tight junctional protein AF-6 disrupts epithelial cell-cell junctions and cell polarity during mouse development. Curr. Biol. *9*, 880-888.

Zhu, J.J., Qin, Y., Zhao, M., Van Aelst, L. and Malinow, R. (2002) Ras and Rap control AMPA receptor trafficking during synaptic plasticity. Cell *110*, 443-455.

Zondag, G.C., Evers, E.E., ten Klooster, J.P., Janssen, L., van der Kammen, R.A. and Collard, J.G. (2000) Oncogenic Ras downregulates Rac activity, which leads to increased Rho activity and epithelial-mesenchymal transition. J. Cell Biol. *149*, 775-782.

CHAPTER 6

EFFECTORS OF RAS-MEDIATED ONCOGENESIS

EMILY J. CHENETTE[1], GRETCHEN A. REPASKY[2], AND CHANNING J. DER[1,3]
University of North Carolina at Chapel Hill, [1] Curriculum in Genetics and Molecular Biology, [3] Department of Pharmacology, and Lineberger Comprehensive Cancer Center, NC 27599-7295. [2] The Colorado College, Chemistry Department, Colorado Springs, CO 80903

Abstract: Ras proteins activate cytoplasmic signaling cascades that mediate responses in growth, cellular differentiation, and survival. Therefore, it is not surprising that mutationally activated Ras proteins have been found in many human cancers. Determining the effector protein signaling pathways through which Ras causes cellular transformation is important for creating targeted therapeutics that will specifically block the oncogenic effects of activated Ras. In 1993, Raf serine/threonine kinases were identified as key downstream effectors of Ras signaling and transformation. While Raf remains the best characterized Ras effector, the rapid expansion of the Ras effector pool has demonstrated that Ras transforming activity is also mediated by Raf-independent effector signaling pathways. These include phosphatidylinositol 3-kinase and phospholipase regulators of phospholipid metabolism, and guanine nucleotide exchange factors and activators of Ras-related proteins. Further complexity arose when a new and seemingly incongruous group of pro-apoptotic Ras effectors with tumor suppressor function was identified. This chapter will summarize recent findings of mutational activation of B-Raf in human cancers and examine the importance of non-Raf effectors in Ras-mediated signaling and transformation

Keywords: Ras, effector, oncogenesis

1. INTRODUCTION

Since the initial discovery of oncogenic Ras over two decades ago, mutant forms of various *ras* genes (*H-*, *K-*, and *N-Ras*) have been found in a diverse spectrum of human neoplasms (Malumbres and Barbacid, 2003). These oncogenes encode four chronically active Ras proteins (H-Ras, K-Ras4A, K-Ras4B, and N-Ras) that interact with a variety of effector molecules to modulate intracellular signaling pathways with cellular consequences on growth, division, and apoptosis.

C. Der (ed.), RAS Family GTPases, 121–142.

Ras proteins are small GTP binding and hydrolyzing proteins (GTPases) that exist in two distinct structural and functional confirmations: GTP-bound and active, and GDP-bound and inactive (Vetter and Wittinghofer, 2001). The cycling of Ras between the GTP-bound and GDP-bound states is governed by two classes of regulatory proteins: guanine nucleotide exchange factors (GEFs) (Quilliam et al., 2002), which exchange a molecule of GDP for GTP, thereby activating the protein; and GTPase activating proteins (GAPs) (Bernards and Settleman, 2004), which stimulate the slow intrinsic hydrolysis rate of Ras, leading to a GDP-bound and inactive protein. Mutationally activated *ras genes* encode mutated, oncogenic Ras proteins (with point mutations at residues 12, 13 or 61) that are impaired in intrinsic- and GAP-mediated hydrolysis and persist in a chronically active state.

Ras protein conformation is altered upon GDP/GTP cycling in two regions, switch I (residues 30 to 38) and II (residues 59 to 76), of the protein tertiary structure (Vetter and Wittinghofer, 2001). The switch I region, together with flanking residues (residues 25-45), corresponds to the effector domain of Ras. When Ras is bound

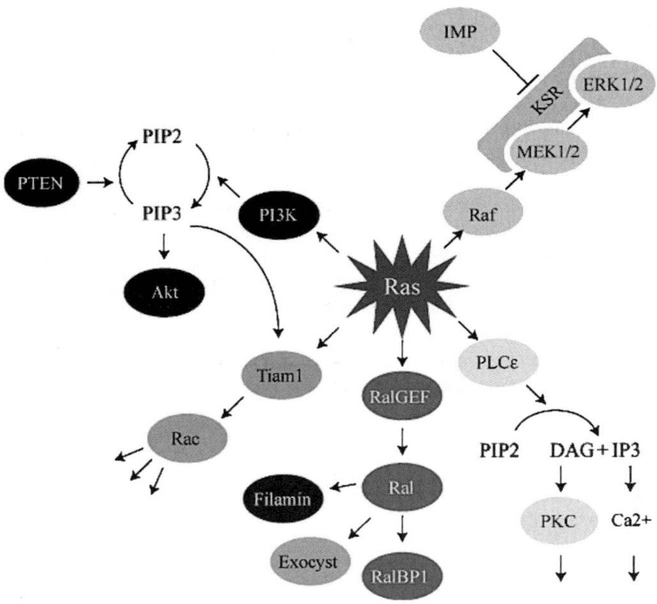

Figure 1. Ras stimulates multiple signaling pathways involved in oncogenesis. Extracellular signals received through membrane-bound receptors stimulate Ras to adopt a GTP-bound and active conformation. In this active state, Ras can signal downstream through its effectors to affect changes in gene transcription, ultimately leading to proliferation and increased survival. A partial list of growth-promoting Ras effectors is illustrated. IMP is a negative regulator of growth, as it interferes with assembly of the KSR MEK-ERK scaffold. PTEN is a tumor suppressor protein, as it counteracts the activity of PI3K. It is tempting to think of cytoplasmic signaling cascades as linear events. However, real signaling networks are much more complex, as Ras effectors commonly signal to more than one downstream target. There is also evidence for extensive crosstalk between different effector pathways (e.g. PI3K and Tiam1)

to GTP, this domain exists as an accessible loop on the surface of the protein and permits high affinity binding to downstream effector proteins that mediate Ras function (Herrmann, 2003).

Ras is localized to the plasma membrane and serves as a critical point of signal convergence, as it transmits signals received from upstream cell surface receptors to downstream cytoplasmic effector pathways. A wide variety of effector proteins and intracellular signaling pathways mediate Ras function, and these effectors control both growth inhibitory and stimulatory intracellular signaling pathways (Repasky et al., 2004). It is clear that Ras cellular function involves activation of multiple effector signaling pathways. In particular, Ras utilizes multiple, functionally distinct classes of effectors to mediate the complex phenotypic changes of the malignant cancer cell (Figure 1 and Table 1).

A Ras effector is defined as a protein that (i) exhibits strong preferential binding to the GTP-bound form of Ras and (ii) is impaired in binding by mutations in the core effector domain of Ras. These biochemical properties do not fully validate the effector status of a protein, as the interaction of endogenous Ras with the effector also needs to be demonstrated. The function of the effector protein should be modulated by interaction with Ras, for example by changes in effector subcellular location (recruitment), intrinsic catalytic activity (allosteric regulation), or interaction with other signaling components (complex formation). The effector also must be involved in Ras signaling. Experimental approaches that have been useful to delineate this last issue include analyses with Ras effector domain mutants,

Table 1. Summary of RAS Effector Function

Effector	Biochemical Function	Target/Substrate	Biological Outcome
c-Raf-1 A-Raf B-Raf	Serine/Threonine kinase	MEK1 and MEK2 serine/threonine kinases	Cellular proliferation Protection from apoptosis Cell migration
p110α p110β p110δ p110γ	Phosphoinositide 3-kinase	Phosphatidylinositol (4,5) bisphosphate	Protection from apoptosis Cellular proliferation Cellular differentiation
RalGDS RGL RGL2/Rlf Rgl3	Guanine nucleotide exchange factor	RalA and/or RalB small GTPases	Endocytosis Exocytosis Tumor formation Actin organization
Tiam1	Guanine nucleotide exchange factor	Rac small GTPase	Actin organization Cell cycle progression
PLCε	Lipase Guanine nucleotide exchange factor	Phosphatidylinositol (4,5) bisphosphate Rap small GTPase	Cellular proliferation Cell adhesion
RASSF1 RASSF2 RASSF4 RASSF5/NORE1	Adaptor	MST1 serine/threonine kinase	Induction of apoptosis

pharmacologic inhibitors of effector signaling, and the use of cells rendered deficient in effector expression, for example by use of small interfering RNA (siRNA).

The majority of Ras effectors contain a well-defined Ras binding domain (Kiel et al., 2005) (Figure 2). To date, at least three distinct ~100 amino acid sequences have been identified as such: the Ras binding domain (RBD) of Raf or Tiam1 (T-lymphoma invasion and metastasis), the RBDs found in class I phosphatidylinositol 3-kinases (PI3K-RBD), and the Ras association (RA) domains found in the majority of Ras effectors (Ponting and Benjamin, 1996). Although they lack primary sequence homology, all three domains exhibit a βαβαβ tertiary structure known as an ubiquitin superfold, which is found in many proteins of different cellular functions (Herrmann, 2003; Wohlgemuth et al., 2005). This common topology accounts for the ability of Ras to interact with effectors with dissimilar RBDs. However, not all proteins that contain RA domains serve as Ras effectors (Kalhammer et al., 1997), and some bind instead to Ras-related proteins (Liao et al., 1999), although conflicting observations have been made about their specificity (Rodriguez-Viciana et al., 2004; Wohlgemuth et al., 2005). In addition, the recently described IMP (Impedes Mitogenic signal Propagation) Ras effector does not exhibit sequence homology with known Ras-interaction sequences (Matheny et al., 2004).

The first Ras effectors to be identified were the Raf serine/threonine kinases (Chong et al., 2003; Mercer and Pritchard, 2003). While Raf-1 was the focus of early research efforts, other effectors were soon discovered. In 1991, Lapetina and colleagues categorized phosphatidylinositol 3-kinase (PI3K) as a Ras effector, and, in 1994, Downward and colleagues identified that the catalytic subunit of PI3K that interacts with Ras (Rodriguez-Viciana et al., 1994; Sjolander et al., 1991; Spaargaren and Bischoff, 1994; Spaargaren et al., 1994). Soon after, GEFs for the Ral family of small GTPases were identified as another family of Ras effectors (Hofer et al., 1994; Kikuchi et al., 1994; Peterson et al., 1996; Shao and Andres, 2000; Spaargaren and Bischoff, 1994; Spaargaren et al., 1994), and shown to contribute, albeit modestly, to Ras transformation of rodent fibroblasts (Urano et al., 1996; White et al., 1996). Since then the list has grown to include more than ten distinct functional classes of effectors (Repasky et al., 2004). A number of these classes contain multiple functionally related isoforms (e.g. A-Raf, B-Raf, and c-Raf-1) (Table 2). Most Ras effectors are expressed in a wide array of tissue and cell types, which gives Ras the potential to utilize a complex array of effectors in any one cell. Finally, it should also be emphasized that the four Ras isoforms exhibit quantitative and qualitative differences in their ability to bind to and activate a particular effector, that some Ras effectors also serve as effectors for other Ras family proteins, and that not all isoforms within a class of effectors have been verified as *bona fide* Ras effectors (Rodriguez-Viciana et al., 2004).

The process of classifying a Ras binding protein as a physiologically relevant Ras effector is a complex one. One complicating factor is the widespread use of ectopic overexpression of activated Ras to determine if effector function is modulated by Ras activation, which raises the concern that Ras overexpression may produce cellular effects not seen at physiological levels. This problem is highlighted in recent comparisons of the consequences of ectopic expression of activated Ras versus

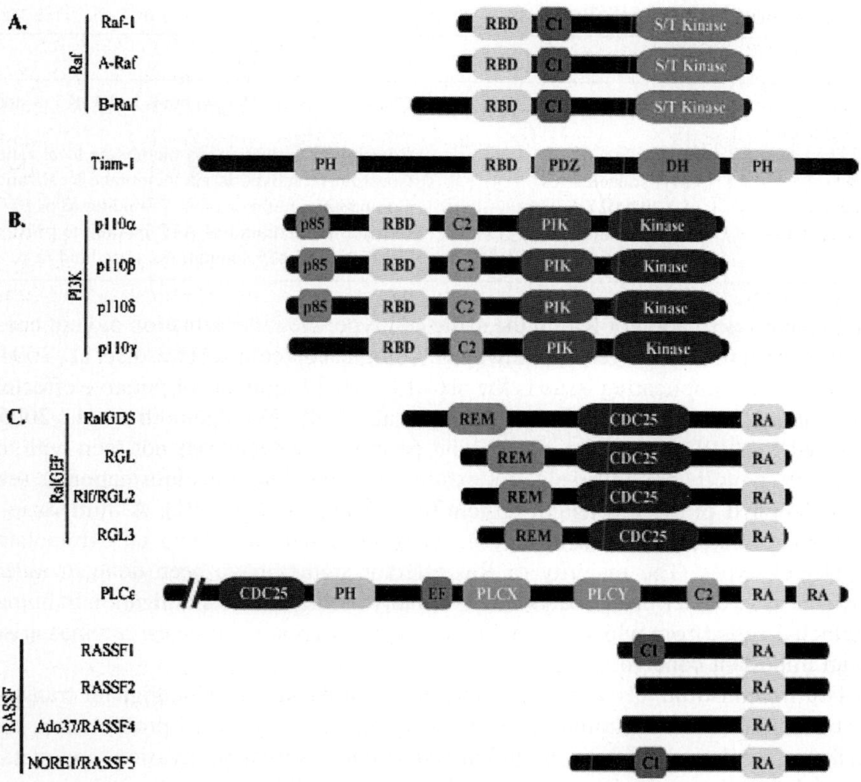

Figure 2. Ras effectors contain Ras association (RA) or Ras binding domains (RBD). Domain architecture of Ras effector proteins was generated by using the SMART database: http://smart.embl-heidelberg.de/. (A) Ras effectors with Raf-like and PI3K Ras binding domains (RBDs). Proteins that contain the Raf-like RBDs include the three Raf serine/threonine kinases Raf-1 (NP 002871), A-Raf (AAH07514), and B-Raf (NP 004324) and the Tiam1 guanine nucleotide exchange factor (NP 003244). Additional domains present in all three Raf proteins are protein kinase C conserved region 1 (C1) cysteine-rich domain and serine/threonine protein kinase domain (S/T Kinase). Additional domains of human Tiam1 (NP 003242) include a pleckstrin homology (PH) domain, a PDZ (present in PSD-95, Dlg, and ZO-1/2) domain, and a Dbl homology (DH) RhoGEF catalytic domain. (B) Proteins that contain the PI3K RBD include p110α (NP 006209), p110β (NP 006210), p110γ (NP 002640), p110δ (NP 005017) catalytic subunits. Additional domains are the p85 PI3K binding domain (p85), a protein kinase C conserved region 2 (C2), a PIK domain that is conserved in all PI3Ks, and a lipid kinase domain (Kinase). (C) Ras effectors with Ras association (RalGDS/AF-6) domains. Four RalGEFs (RalGDS (NP 006257), RGL (NP 055964), Rlf/RGL2 (O15211), and RGL3 (XP 290867)) contain a Ras exchanger motif (REM) and a CDC25 homology GEF domain. PLCε (NP 006217) also contains a CDC25 homology domain, followed by a pleckstrin homology (PH) domain, EF-hand calcium-bindings motifs, the phospholipase C X and Y catalytic boxes (PLCX and PLCY), and a C2 domain. RASSF1A (NP 009113), RASSF1B (NP 733830), and RASSF1C (NP 733831) are the major RASSF1 transcripts expressed in normal tissue. NORE1Aα/RASSF5A (418 aa; NP 872604) shares a similar domain architecture with RASSF1A, with an amino-terminal C1 domain and a carboxyl-terminal sequences, with a truncation within the RA domain. NORE1B/RASSF5C (265 aa; NP 872606) lacks the C1 domain and is transcribed from a different promoter. Unlike RASSF1 and RASSF5, RASSF2 (326 aa; NP 739580) and RASSF4/Ado37 (321 aa; NP 114412) do not contain amino-terminal C1 domains

Table 2a. Tools to study P13K Signaling

	Reagent	Mechanism of action
Positive Regulators	H-Ras(G12V/Y40C)	GTPase-deficient H-Ras; binds to P13K but not Raf or RalGEF
	P110 CAAX	Constitutive activation via membrane localization
	Myrisolated AKT	Constitutive activation via membrane localization
Negative Regulators	LY294002	Competitive inhibitor of ATP binding to p110
	Wortmannin	Competitive inhibitor of ATP binding to p110
	Dominant negative P13K	Truncation of p85 subunit, does not bind to p110

endogenous activation of Ras in the same cell type; the latter situation did not cause activation of the same effector pathway or biological outcome (Tuveson et al., 2004).

A second complicating issue is the use of isolated fragments of putative effectors to evaluate interactions with Ras (Kiel et al., 2005; Wohlgemuth et al., 2005). Isolated RA/RBD sequences may exhibit permissive interactions not seen with the authentic protein, or conversely, not exhibit the same domain conformation as seen in the context of the full-length protein (e.g., Tiam1 and RasIP1). A third issue is that Ras effector utilization in one cell type may not be able to be extrapolated to all cell types. The majority of Ras effector studies have been done in rodent fibroblasts; whether these analyses will reliably predict effector utilization in human epithelial cells (from which the majority of *ras* mutation-positive carcinomas arise) is an important concern.

Fourth, constitutively active Ras induces multiple facets of malignant transformation, both *in vitro* (morphologic transformation, uncontrolled proliferation, and anchorage-independent growth) and *in vivo* (tumor formation, invasion, and metastasis). It is not surprising that a particular effector will be crucial for inducing one, but not another, of these phenotypes in different transformation assays. Therefore, the established role of a specific effector pathway in transformation has so far depended on the assays utilized. Finally, the determination of whether a particular effector is necessary and/or sufficient to promote a transformed phenotype creates some confusion regarding effector function. While activation of an effector may be necessary for Ras-mediated tumor induction or maintenance, it may not be sufficient to mimic an aspect of Ras-mediated transformation. Below, we summarize the current evidence that supports the contribution of specific effectors in promoting the consequences of aberrant Ras signaling in human oncogenesis.

2. RAF IS A KEY EFFECTOR OF RAS-MEDIATED ONCOGENESIS

The Raf serine/threonine kinases (Table 2) were first identified in 1993 as effectors of Ras signaling in mammalian cells (Moodie et al., 1993; Vojtek et al., 1993; Warne et al., 1993; Zhang et al., 1993) as well as in flies and worms (Han et al., 1993; Karim et al., 1996). The three Raf isoforms share significant sequence and functional similarities, yet they exhibit distinct tissue distributions as well as divergent roles in development (Chong et al., 2003; Mercer and Pritchard, 2003). The structural and signaling properties of the Raf kinases are discussed extensively in a different chapter in this book. Raf phosphorylates and activates the MEK1 and MEK2 threonine/tyrosine kinases, which phosphorylate and activate the ERK1 and ERK2 mitogen-activated protein kinases (MAPKs). ERKs phosphorylate and activate Ets family transcription factors and other substrates.

For many years after its initial discovery, Raf was implicated as the most crucial effector in Ras-mediated growth transformation. Various genetic and pharmacologic reagents have been utilized to assess the role of the Raf-MEK-ERK cascade in Ras-mediated oncogenesis (Table 2a). Expression of activated Raf or its only known substrates (MEK1 and MEK2) alone causes morphological transformation of mouse fibroblast NIH 3T3 cells (Cowley et al., 1994; Khosravi-Far et al., 1995; Kolch et al., 1991; Leevers et al., 1994; Mansour et al., 1994; Qiu et al., 1995a; Qiu et al., 1995b; Schaap et al., 1993; Westwick et al., 1994). Three other pieces of data point to an important role for Raf in Ras-mediated transformation of NIH 3T3 cells. First, c-Raf-1 mutants that are constitutively localized to the plasma membrane are able to cause transformation in rodent fibroblasts (Leevers et al., 1994; Stokoe et al., 1994). Second, blocking MEK1 and MEK2 by the use of dominant negatives or pharmacological inhibitors blocks the ability of activated Ras to transform cells. Third, the mammalian Raf-MEK-ERK cascade is conserved downstream of Ras in *C. elegans* and *Drosophila*, and promotes Ras-mediated vulva or eye development, respectively.

Raf signaling is also implicated in human carcinogenesis, and there is extensive evidence that suggests that Ras and Raf are functionally equivalent in the development of at least some human tumors. While over 30 unique missense mutations of B-Raf have been detected in human cancers, more than 90% of Raf mutations in human cancers correspond to a V600E (formerly called V559E) substitution, which increases kinase activity of Raf and facilitates growth transformation of rodent fibroblasts and melanocytes (Wellbrock et al., 2004a; Wellbrock et al., 2004b). Raf and Ras mutations are detected in the same types of human cancers, but in essentially non-overlapping frequency. For example, B-*raf* mutations occur in 70% of melanomas, and N-*ras* mutations occur in 25% of melanomas, but tumors rarely contain mutant N-Ras in combination with mutant B-Raf (Davies et al., 2002). Similar findings have been described for other cancers including colorectal carcinomas (Rajagopalan et al., 2002), papillary thyroid carcinomas (Cohen et al., 2003; Kimura et al., 2003), and serous ovarian carcinomas (Sieben et al., 2004; Singer et al., 2003). The mutual exclusivity of *ras* and *raf* mutations in human cancers suggests that they share redundant roles in oncogenesis, and that the key role of oncogenic Ras is activation of Raf.

3. OTHER EFFECTORS OF RAS-MEDIATED ONCOGENES

While Raf is clearly an important effector of Ras function, there is also substantial evidence that Ras utilizes many non-Raf effectors, with some also contributing to Ras-mediated oncogenesis. For example, mutational activation of *ras*, but not

B-*raf*, is frequently seen in some cancers (e.g., pancreatic carcinomas), suggesting that activation of Raf alone is not functionally equivalent to Ras activation in some tissues. Furthermore, the loss of the *PTEN* tumor suppressor, which results in the activation of a distinct Ras effector pathway (PI3K-Akt), was detected in melanomas that harbor B-*raf* mutations, suggesting that perturbations in other Ras effector pathways may be required to affect transformation (Tsao et al., 2004).

Experimental studies also question the ability of activated Raf to mimic Ras activation. For example, expression of activated Raf is not sufficient to induce morphological transformation in rat intestinal epithelial cells (RIE-1), or human breast epithelial (MCF-10A) or embryonic kidney (HEK) cells (Hamad et al., 2002; Oldham et al., 1996; Schulze et al., 2001). As summarized below, there is now considerable and growing evidence for the importance of non-Raf effectors in Ras-mediated signaling and oncogenesis. Even though mutant B-*raf* seems to play a causative role in many human cancers, and expression of activated Raf phenocopies Ras-mediated growth transformation in mouse fibroblasts, other effector pathways are clearly critical for the function of Ras in human and rodent epithelial cells.

In this section, we summarize the evidence that other effectors may contribute to the function of Ras as an oncogene. After Raf, the next-best characterized effectors of Ras signaling are PI3Ks and RalGEFs. Additional effectors such as Tiam1 and PLCε also contribute to the growth promoting potential of oncogenic Ras, whereas some effectors may serve instead to promote growth inhibitory functions of Ras.

3.1 Phosphatidylinositol 3-Kinase

PI3Ks are a conserved family of proteins, first identified as lipid kinases that phosphorylate the 3' position of the inositol ring of membrane phosphatidylinositol lipids (phosphatidylinositol (PI), phosphatidylinositol 4-phosphate (PIP), phosphatidylinositol 4,5-bisphosphate (PIP2)) and subsequently determined to harbor protein kinase activity (Krasilnikov, 2000). While there are several classes of PI3K enzymes (IA/B, II, III, and IV), only the class I PI3Ks contain functional Ras binding domains. PI3Ks are composed of a catalytic (p110) and a regulatory (p85) subunit (Vanhaesebroeck et al., 1997). Several isoforms of each subunit exist, and multiple isoforms of the p110 subunit (α, β, γ, δ) have been shown to interact with GTP-bound Ras (Rodriguez-Viciana et al., 1997) (Figure 2).

A major function of these lipid kinases is the phosphorylation of PIP2 to produce phosphatidylinositol 3,4,5-triphosphate (PIP3) (Corvera and Czech, 1998), and hence PI3Ks are important in regulating the activation and localization of some Rho family GEFs, where the binding of membrane-associated PIP3 binding to their PH domains facilitates membrane association. Some key experimental tools used to evaluate the role of PI3K in Ras function are described in Table 2b.

PI3K signaling contributes to proliferation, morphology, and survival of transformed cells (Luo et al., 2003). The PI3K pathway is implicated in Ras transformation of rodent fibroblasts, as PIP3 levels are elevated in Ras-transformed cells, and expression of a dominant negative p85 regulatory subunit can block Ras-mediated

Table 2b. Tools to study Raf Signaling

	Reagent	Mechanism of action
Positive Regulators	H-Ras(G12V/T35S)	GTPase-deficient H-Ras; binds to P13K but not P13K or RalGEF
	Raf CAAX	Constitutive activation via membrane localization
	Raf 22W	N-terminal truncation mutant of Raf
Negative	B-Raf V600E	Constitutively active phosphomimetic mutant
Regulators	MEK (S218D/S222D)	Constitutively active phosphomimetic mutant
	PD98059	Non-competitive inhibitor of MEK1
	UO126	Non-competitive inhibitor of MEK1 and MEK2
	MEK(K97A)	Kinase-dead mutant
	MKP-1	MAPK protein phosphatase

transformation of NIH 3T3 cells (Rodriguez-Viciana et al., 1997; Zhang et al., 1996). However, PI3K is not required for Ras-mediated transformation of RIE-1 cells, reflecting the cell type variation in the role of this effector in oncogenic Ras function (McFall et al., 2001; Oldham et al., 1998).

Several signaling molecules have been suggested to function downstream of PI3K and promote its cellular functions of proliferation and survival. One of the best-characterized propagators of the anti-apoptotic signal from PI3K is the Akt/PKB family of serine/threonine kinases (Luo et al., 2003). Oncogenic Ras-mediated inhibition of suspension-induced apoptosis in MDCK canine kidney epithelial cells has been attributed to PI3K-dependent activation of Akt (Khwaja et al., 1997), although other effector pathways are required in other cell types (Eckert et al., 2004; McFall et al., 2001). Akt can phosphorylate and alter the activity of a variety of downstream targets, including the NF-κB and forkhead transcription factors, suggesting that changes in gene expression may be an important outcome of the PI3K/Akt effector pathway (Mitsiades et al., 2004). In addition, Akt promotes the anti-apoptotic activity of Bcl-2, and blocks the pro-apoptotic activity of caspases and GSK3. PI3K may also modulate cell proliferation and morphology by controlling the activation of the Rho family GTPases Rac and Cdc42 (Han et al., 1998; Welch et al., 2002).

The importance of PI3K in Ras-mediated oncogenesis is further supported by other evidence that has implicated aberrant PI3K activation in human cancer development. In particular, the PTEN tumor suppressor and lipid phosphatase is a negative regulator of the PI3K pathway, and the loss of *PTEN* (either through epigenetic silencing or mutation) is a frequent event in cancers (Parsons, 2004). In addition, mutation of the gene encoding the p110α catalytic subunit of the class IA PI3K (*PI3KCA*) has been described in lung, ovarian, stomach, colorectal, brain, and breast cancers (Bachman et al., 2004; Broderick et al., 2004; Samuels et al., 2004; Shayesteh et al., 1999). The majority of mutations fall in either the helical domain (47%) or the kinase domain (33%) of PI3KCA (Samuels et al., 2004). Vogt and colleagues recently determined that PI3KCA mutants with substitution mutations at three "hot-spot" sites (E542K and E545K in the helical domain, or H1047R in the kinase domain) are able to potently transform chicken embryo fibroblasts, with concomitant upregulation of phospholipid kinase and Akt signaling activity (Kang et al., 2005). Whether *PI3KCA* mutants occur in overlapping frequency with *ras* mutations will be interesting to determine.

3.2 Ral Guanine Nucleotide Exchange Factors (RalGEFs)

The Ral guanine nucleotide exchange factors (RalGEFs) consist of four distinct members (RalGDS, RGL, RGL2/Rlf, RGL3) and link Ras proteins to activation of the RalA and RalB small GTPases (Feig, 2003). RalGEFs that lack RBDs, and hence, link Ral GTPases to other mechanism of activation, have also been described (de Bruyn et al., 2000; Quilliam et al., 2002). Ral GTPases are members of the Ras branch of the Ras superfamily (Wennerberg et al., 2005). To date, there have been no reports of Ral or RalGEF mutations in human cancers, but this may also reflect the possibility that no efforts have been made to search for such mutations. Currently, the evidence for the RalGEF-Ral pathway in Ras-mediated oncogenesis has been derived from cell culture and mouse model studies (Table 2C).

Initial reports on Ral determined that, unlike Raf, constitutively activated mutants of Ral were unable to cause transformation of NIH 3T3 mouse fibroblasts. Expression of activated RalGEF alone failed to induce morphological or growth transformation of 208F rodent fibroblasts, or RIE-1 or ROSE rodent epithelial cell lines (McFall et al., 2001; Ulku and Der, 2003; Ulku et al., 2003). However, constitutively active RalA enhanced Ras-mediated focus formation in mouse fibroblasts, and several reports have demonstrated that Ras activation of RalA dramatically increases metastatic potential of rodent fibroblasts (Schulze et al., 2001; Tchevkina et al., 2005). Moreover, expression of RalGDS cooperated with activated Raf to induce synergistic focus formation, and dominant negative Ral blocked Ras-mediated focus formation (Urano et al., 1996; White et al., 1996). Co-expression of the Ras-binding domains from RGL and Rlf likewise inhibited Ras transforming activity (Okazaki et al., 1996; Peterson et al., 1996).

While Ral does not appear to play a major role in Ras-mediated transformation of rodent cells in culture, the substantial evidence that oncogenesis is not facilitated by identical mechanisms in mice and humans (Rangarajan and Weinberg, 2003) prompted Counter and colleagues to compare Ras effector utilization in oncogenesis in human and mouse cells (Hamad et al., 2002). In these studies, Hamad et al. employed constitutively active H-Ras(G12V) with a substitution mutation in the effector domain (H-Ras[12V/37G]) (Hamad et al., 2002). H-Ras(12V/37G) is impaired in its ability to bind to and activate PI3K and Raf, yet retains the ability to activate Ral. In these studies, H-Ras(12V/37G) was not able to elicit potent transformation of rodent fibroblasts, but, surprisingly, was able to cause transformation of primary human fibroblasts, human embryonic kidney (HEK) cells, and human astrocytes, although to a lesser extent that H-Ras(12V). Furthermore, expression of activated Rlf, but not Raf or PI3K, was able to elicit transformation in HEK cells. However, co-expression of H-Ras(12V/37G) and activated Raf or PI3K caused synergistic transformation of rodent cells, while pharmacological inhibition of MEK or PI3K effectively blocked Ras-mediated transformation of HEK cells (unpublished observations), indicating that while the RalGEF pathway is important for Ras signaling in HEK cells, other effector pathways still contribute to Ras-mediated transformation. Finally, recent studies found that activated RalA also promoted HEK growth transformation, and suppression of RalA expression impaired the growth of Ras-transformed HEK cells (Lim, 2005). These data verify that the RalGEF pathway is sufficient to cause transformation of some human cells in culture, but again underscore the necessity of multiple effector pathways in Ras-mediated oncogenesis.

While the data presented by Counter and colleagues present a compelling argument for species- and cell-type differences in effector utilization in Ras-mediated transformation, recent data suggests that the differences in effector utilization lie not only in cells and species, but also in cell culture and *in vivo* animal models. As mentioned earlier, RalGDS was not sufficient to promote Ras-mediated transformation of rodent cells in culture. Recently, Marshall and colleagues documented a role for RalGDS in promoting Ras-induced skin tumors in mice. Using mice that lacked both *RalGDS* alleles, they found that the RalGDS-deficient mice, while developmentally normal, were resistant to skin tumor formation caused by carcinogen-induced mutational activation of H-Ras (Gonzalez-Garcia et al., 2005). Mouse embryo fibroblasts (MEFs) isolated from the RalGDS knockout mice and immortalized and transformed by expression of SV40 large T and activated H-Ras(12V) formed 50% fewer foci in transformation assays when compared to wild-type MEFs. Re-introduction of RalGDS partially rescued the ability of the knockout MEFs to form foci in culture. The reduction in tumor formation was thought to be the result of decreased JNK/SAPK signaling with concomitant upregulation of apoptosis (Gonzalez-Garcia et al., 2005). These data also present a possible role for RalGDS not only in tumor formation, but also in promoting survival of tumor cells in mice. Whether RalGDS or other RalGEFs are also important for tumorigenesis in humans is an interesting question, and may have important implications for the development of targeted anti-cancer therapies.

The mechanism(s) by which Ras-mediated activation of RalGEFs contribute(s) to cell proliferation and transforming activity is not entirely clear. A possible mechanism involves the ability of active, GTP-bound RalA to interact with a Ral effector protein, RalBP1, which has GAP activity toward Rac1 and Cdc42 (Cantor et al., 1995; Jullien-Flores et al., 1995). Thus, the RalGEF effector pathway may modulate the activity of transcription factors and cytoskeletal proteins that are regulated by Rho GTPases. Additionally, Ral GTPase activation of phospholipase D, which cleaves membrane phospholipids, may contribute to cell growth control, via RalGDS (Lucas et al., 2002; Voss et al., 1999). A number of other Ral effectors have been identified (Feig, 2003) and future studies will be needed to determine whether these effectors, or effectors that remain to be discovered, will be key mediators of Ral GTPase transformation.

Lastly, as mentioned above, while RalA and RalB share over 85% sequence identity, they appear to have divergent roles in promoting human tumor cell growth. By using interfering RNA to block expression of either RalA or RalB, White and colleagues showed that RalA was required for proliferation of human tumor cells, while RalB was essential for survival of tumor cells (Chien and White, 2003). Work done by Feig and colleagues also documented differences in RalA and RalB function (Shipitsin and Feig, 2004). Finally, recent studies determined that activated RalA but not RalB promoted HEK growth transformation, and interfering RNA suppression of RalA but not RalB impaired the growth of Ras-transformed HEK cells, as well as for Ras mutation positive pancreatic and other human tumor cell lines (Lim, 2005). Due to their high sequence identity, the distinct roles of RalA and RalB in tumorigenesis may hinge on subtle differences in subcellular localization or regulation of exocyst function (Shipitsin and Feig, 2004).

3.3 Phospholipase C Epsilon (PLCε)

Phospholipase C epsilon (PLCε) was identified in 2001 based on its homology to other known PLC isozymes (Kelley et al., 2001; Lopez et al., 2001; Song et al., 2001). Stimulation of PLCε activates it to cleave PIP2 to generate 1,4,5-triphosphate (IP3) and diacylglycerol (DAG). IP3 stimulates the release of stored Ca^{2+} reserves, while DAG activates protein kinase C. H-Ras and Ras-related proteins bind to and activate PLCε via tandem RA domains (Kelley et al., 2001; Song et al., 2001). Interestingly, PLCε also contains a CDC25 homology domain. CDC25 domains are found in all Ras GEFs and promote the enzymatic activity of the GEF. Current evidence indicates that the CDC25 homology domain of PLCε functions as a GEF for the Ras-related proteins, Rap1A and Rap2, although there are some reports that it functions as a Ras GEF (Rhee, 2001; Song et al., 2001).

A recent report from Kataoka and colleagues documents a concrete role for PLCε in Ras-mediated oncogenesis. As was recently reported for RalGDS (Gonzalez-Garcia et al., 2005), Kataoka et al. found that mice that lacked PLCε expression were resistant to Ras-induced skin tumors (Bai et al., 2004). Tumors that did form in *PLCε*$^{-/-}$ mice did not progress to malignancy. Gene dosage affected tumor formation, as mice that were *PLCε*$^{+/-}$ heterozygotes formed an intermediate number of tumors when compared to their wild-type or PLCε$^{-/-}$ littermates. While the exact mechanism of PLCε activation in H-Ras-mediated oncogenesis remains to be elucidated, the authors speculated that PLCε might have a role in prostaglandin signaling, as prostaglandins are synthesized from DAG, a product of PLCε-mediated cleavage of PIP2. Other studies have shown that prostaglandins are involved in the promotion, as well as progression, of skin tumors (Muller-Decker et al., 2002).

A second recent study characterized the expression of two splice variants of PLCε, designated PLCε1a and PLCε1b (Sorli et al., 2005). Both splice variants interact with GTP-bound H-Ras, and PLC activity is increased upon exogenous expression of constitutively active H-Ras. However, PLCε1a and PLCε1b gene expression appears to be downregulated in a variety of human colon and rectal tumor samples, as well as colorectal cancer cell lines, when compared to normal tissue (Sorli et al., 2005). Transient ectopic expression of PLCε in a PLCε-deficient, Ras mutation positive, colorectal carcinoma cell line caused growth inhibition. In contrast to the mouse studies, these observations suggest that PLCε activity may not facilitate Ras-mediated growth transformation. Thus, the importance of PLCε in Ras-induced human oncogenesis may be complex and distinct for different types of neoplasms.

3.4 Tiam1

Tiam1 was first identified in a screen for genes that conferred invasive ability to previously non-invasive lymphoma cells (Habets et al., 1994) and subsequently shown to function as a Rac-specific GEF (Crompton et al., 2000; Michiels et al., 1995). More recently, Tiam1 has been demonstrated to be a *bona fide* Ras effector and facilitate Ras-mediated activation of Rac (Lambert et al., 2002a; Lambert et al., 2002b). Therefore, Tiam1 presents a clear link to involvement of the Rho family GTPases in Ras signaling. This represents but one of many incidences of crosstalk between Ras and Rho GTPases (Bar-Sagi and Hall, 2000). Reagents that have been developed to study the role of Tiam1 in Ras-mediated transformation are summarized in Table 2D.

A contributory role for Tiam1 in Ras-mediated oncogenesis has been demonstrated and is consistent with previous studies that found that Ras transformation of NIH 3T3 mouse fibroblasts was impaired by interfering with Rac activation (Khosravi-Far and Der, 1995; Khosravi-Far et al., 1995; Qiu et al., 1995a; Qiu et al., 1995b). Mice lacking both alleles of *Tiam1* were largely resistant to skin tumor formation induced by carcinogen-induced activated H-Ras (Malliri et al., 2002). Additionally, Tiam1-deficient primary embryonic fibroblasts were also resistant to Ras-induced focus formation and Rac activation. However, while the tumors that formed in *Tiam1*$^{-/-}$ mice were impaired in growth rate, they showed a surprisingly increased tendency towards malignant progression. Therefore, Tiam1 may serve in two capacities in tumorigenesis: it may first promote Ras-mediated tumor formation, but later antagonize tumor progression.

Further evidence that Tiam1 activation promotes oncogenesis is provided by the identification of Tiam1 gain-of-function missense mutations in human renal cell carcinomas and cell lines (Engers et al., 2000). In addition, Tiam1 expression was upregulated in metastatic human colon cancer cell lines when compared to cells derived from primary tumors (Liu et al., 2005), and migratory cells selected from human colorectal cell lines expressed high levels of Tiam1 and were highly metastatic when injected into nude mice (Minard et al., 2005). These findings support a role for Tiam1 in human metastasis and oncogenesis, but at this point it is not known if *Tiam1* mutations will be found in other human cancers, or if loss of Tiam1 function will impair the growth of Ras-derived tumors. Studies such as these will be important to further validate Tiam1 as an important effector in Ras-mediated oncogenesis.

4. RAS EFFECTORS AS TUMOR SUPPRESSORS

Recent studies have indicated that effectors of Ras activation include not only protein products of oncogenes, but also protein products of tumor suppressor genes. The *RASSF*(1-6) gene family encodes proteins that contain Ras association (RA) domains, but function to inhibit growth. Of the six family members, a role as a Ras effector has been described for Nore1 (also called RASSF5), RASSF1, RASSF2 (Vos et al., 2003a; Vos et al., 2003b), and RASSF4 (Rodriguez-Viciana et al., 2004). These novel effectors may account for the pro-apoptotic and growth inhibitory activity of Ras observed in some research reports (Cox and Der, 2003; Feig and Buchsbaum, 2002).

4.1 Nore1

Nore1 can exist in two alternatively spliced forms, the mRNA of which may be down regulated in some forms of lung carcinoma (Tommasi et al., 2002). Although one study found no methylation of the *Nore1* promoter and only a handful of cases of somatic point mutations responsible for gene silencing (Tommasi et al., 2002), more recent direct experimental evidence of Nore1 shows loss of expression due to partial promoter methylation in various primary tumors, such as renal cell carcinomas and various human lung tumor cell lines. These data, together with the growth inhibitory activity seen with ectopic Nore1 expression, strongly suggest that Nore1 functions as a tumor suppressor, similar to the structurally and functionally related protein, RASSF1 whose loss has been associated with promotion of oncogenesis (Chen et al., 2003; Hesson et al., 2003; Vos et al., 2003a; Vos et al., 2003b).

Nore1 is a proapoptotic, 413 amino acid, non-catalytic protein with putative diacylglyercol- and Src homology 3 (SH3) domain-binding sites in the amino terminus, a zinc finger domain in the central region, and an RA domain in the carboxyl terminus, that interacts specifically with active (GTP-bound) Ras and Ras-related proteins R-Ras and M-Ras (Ortiz-Vega et al., 2002; Vavvas et al., 1998). Interaction of Nore1 with GTP-bound Ras has been observed both *in vitro* and *in vivo*, in response to growth factor or serum stimulation of cells and activation of Ras (Vavvas et al., 1998).

Nore1 mediates its pro-apoptotic cellular function by association with a protein kinase, MST1, which when overexpressed induces apoptosis in several cell types. Specifically, plasma membrane bound Ras-GTP interacts with and recruits a complex of Nore1/MST1, forming a ternary complex with pro-apoptotic function (Khokhlatchev et al., 2002). Expression of Nore1 causes growth inhibition, an effect that is enhanced by co-expression of H-Ras and antagonized by co-expression of dominant inhibitory H-Ras (Vos et al., 2003a; Vos et al., 2003b). However, the importance of the interaction between Nore1 and Ras is unclear, as it has been demonstrated that Nore1-mediated growth inhibition of human tumor cells did not require interaction with Ras (Aoyama et al., 2004). Whether Nore1 interacts with endogenous Ras to promote apoptosis, or whether its growth-inhibitory functions are completely independent of Ras signaling remains to be determined.

4.2 RASSF(1-6)

Ras association domain family 1, or RASSF1, was identified as a putative tumor suppressor gene in a portion of chromosome 3 frequently lost in many types of cancer (Dammann et al., 2000; Dammann et al., 2003a; Dammann et al., 2003b; Lerman and Minna, 2000), and hypermethylation and silencing or the RASSF1 promoter occurs frequently in pancreatic, colorectal, ovarian, prostate, and renal cell carcinomas, as well as malignant melanomas (Dammann et al., 2003a; Dammann et al., 2003b; Dammann et al., 2005; Pfeifer et al., 2002; Spugnardi et al., 2003; van Engeland et al., 2002; Yoon et al., 2001). Clark and colleagues have demonstrated that RASSF1C possesses a pro-apoptotic function that is dependent on Ras signaling (Cox and Der, 2003; Vos et al., 2000). This pro-apoptotic effect is most likely mediated by binding to MST1, as MST1 activity is associated with the onset of apoptosis (Feig and Buchsbaum, 2002). However, the establishment of the RASSF1 family as *bona fide* Ras effectors has been difficult. Several conflicting reports about the activity of the RASSF family have been published recently. While RASSF1C appears to be a viable Ras effector in that it is able to interact with GTP-bound Ras, it is the RASSF1A splice variant that is silenced in human tumors. Overexpression of RASSF1A arrests the cell cycle at the G1/S transition by blocking the accumulation of cyclin D1, but exogenous expression of activated H-Ras cannot bypass the arrest (Shivakumar et al., 2002). Furthermore, one recent study failed to find interaction between RASSF1A or RASSF1C and Ras or Ras-related proteins.

Additional members of the RASSF family also have been implicated as Ras effectors. In addition to the two splice variants of RASSF1 (RASSF1A and 1C), other homologues of RASSF1 have been described. RASSF2 was identified as a K-Ras specific effector with pro-apoptotic and growth-inhibitory properties; it is down-regulated in human lung cancer cell lines (Vos et al., 2003a; Vos et al., 2003b). RASSF4 (also called Ado37) is also down-regulated via promoter methylation in tumor cells, and may also induce Ras-dependent apoptosis (Chow et al., 2004; Eckfeld et al., 2004). Gene knockout analyses in mice, or expression of siRNA in cell culture, will help determine if loss of *RASSF* contributes to Ras-induced tumor formation. These lines of investigation are critical to assess the status of Nore1 and the RASSF proteins as tumor suppressors, and evaluate their roles in Ras signaling.

5. CONCLUSION

The Raf-MEK-ERK MAPK protein kinase cascade remains the best-characterized and validated Ras effector signaling pathway. Our extensive understanding of Raf function in signal transduction, coupled with the successful development of inhibitors of various protein kinases for cancer treatment (Sawyers, 2003), has prompted considerable effort to develop inhibitors of the Raf protein kinase cascade for in the treatment of human cancers (Sebolt-Leopold and Herrera, 2004). However, extensive studies using model cell culture systems and mouse models support critical roles for other effectors in Ras-mediated oncogenesis (Repasky et al., 2004).

These include the PI3K, Tiam1, RalGEF, and PLCε effectors, and future studies are likely to implicate additional effectors in oncogenic Ras function as well as elucidate new roles for existing Ras effectors. While the presence of B-*Raf* mutations in some cancers argues that Raf activation is critical for oncogenesis of some cancers (Garnett and Marais, 2004), the lack of such mutations in other cancers, along with cursory evidence of mutations in other Ras effectors in human cancers, argue that the Raf pathway may not be the dominant pathway of Ras signaling in Ras mutation positive cancers. Will these less well-characterized effectors be better candidates for the development of anti-cancer drugs? The answer to this intriguing question must await further elucidation of the importance of each effector in human oncogenesis.

ACKNOWLEDGEMENTS

We apologize for not being able to cite original work of many colleagues due to space constraints. We thank Misha Rand for assistance in manuscript preparation. Our studies are supported by grants from the National Institutes of Health (CA42978, CA63071, CA69577, and GM65533) to C. J. D.

REFERENCES

Aoyama, Y., Avruch, J. and Zhang, X.F. (2004) Nore1 inhibits tumor cell growth independent of Ras or the MST1/2 kinases. Oncogene *23*, 3426-3433.

Bachman, K.E., Argani, P., Samuels, Y., Silliman, N., Ptak, J., Szabo, S., Konishi, H., Karakas, B., Blair, B.G., Lin, C., Peters, B.A., Velculescu, V.E. and Park, B.H. (2004) The PIK3CA gene is mutated with high frequency in human breast cancers. Cancer Biol. Ther. *3*, 772-775.

Bai, Y., Edamatsu, H., Maeda, S., Saito, H., Suzuki, N., Satoh, T. and Kataoka, T. (2004) Crucial role of phospholipase Cepsilon in chemical carcinogen-induced skin tumor development. Cancer Res. *64*, 8808-8810.

Bar-Sagi, D. and Hall, A. (2000) Ras and Rho GTPases: a family reunion. Cell *103*, 227-238.

Bernards, A. and Settleman, J. (2004) GAP control: regulating the regulators of small GTPases. Trends Cell Biol. *14*, 377-385.

Broderick, D.K., Di, C., Parrett, T.J., Samuels, Y.R., Cummins, J.M., McLendon, R.E., Fults, D.W., Velculescu, V.E., Bigner, D.D. and Yan, H. (2004) Mutations of PIK3CA in anaplastic oligodendrogliomas, high-grade astrocytomas, and medulloblastomas. Cancer Res. *64*, 5048-5050.

Cantor, S.B., Urano, T. and Feig, L.A. (1995) Identification and characterization of Ral-binding protein 1, a potential downstream target of Ral GTPases. Mol. Cell. Biol. *15*, 4578-4584.

Chen, J., Lui, W.O., Vos, M.D., Clark, G.J., Takahashi, M., Schoumans, J., Khoo, S.K., Petillo, D., Lavery, T., Sugimura, J., Astuti, D., Zhang, C., Kagawa, S., Maher, E.R., Larsson, C., Alberts, A.S., Kanayama, H.O. and Teh, B.T. (2003) The t(1;3) breakpoint-spanning genes LSAMP and NORE1 are involved in clear cell renal cell carcinomas. Cancer Cell *4*, 405-413.

Chien, Y. and White, M.A. (2003) RAL GTPases are linchpin modulators of human tumour-cell proliferation and survival. EMBO Rep. *4*, 800-806.

Chong, H., Vikis, H.G. and Guan, K.L. (2003) Mechanisms of regulating the Raf kinase family. Cell. Signal. *15*, 463-469.

Chow, L.S., Lo, K.W., Kwong, J., Wong, A.Y. and Huang, D.P. (2004) Aberrant methylation of RASSF4/AD037 in nasopharyngeal carcinoma. Oncol. Rep. *12*, 781-787.

Cohen, Y., Xing, M., Mambo, E., Guo, Z., Wu, G., Trink, B., Beller, U., Westra, W.H., Ladenson, P.W. and Sidransky, D. (2003) BRAF mutation in papillary thyroid carcinoma. J. Natl. Cancer Inst. *95*, 625-627.

Corvera, S. and Czech, M.P. (1998) Direct targets of phosphoinositide 3-kinase products in membrane traffic and signal transduction. Trends Cell Biol. *8*, 442-446.

Cowley, S., Paterson, H., Kemp, P. and Marshall, C.J. (1994) Activation of MAP kinase kinase is necessary and sufficient for PC12 differentiation and for transformation of NIH 3T3 cells. Cell *77*, 841-852.

Cox, A.D. and Der, C.J. (2003) The dark side of Ras: regulation of apoptosis. Oncogene *22*, 8999-9006.

Crompton, A.M., Foley, L.H., Wood, A., Roscoe, W., Stokoe, D., McCormick, F., Symons, M. and Bollag, G. (2000) Regulation of Tiam1 nucleotide exchange activity by pleckstrin domain binding ligands. J. Biol. Chem. *275*, 25751-25759.

Dammann, R., Li, C., Yoon, J.H., Chin, P.L., Bates, S. and Pfeifer, G.P. (2000) Epigenetic inactivation of a RAS association domain family protein from the lung tumour suppressor locus 3p21.3. Nat. Genet. *25*, 315-319.

Dammann, R., Schagdarsurengin, U., Liu, L., Otto, N., Gimm, O., Dralle, H., Boehm, B.O., Pfeifer, G.P. and Hoang-Vu, C. (2003a) Frequent RASSF1A promoter hypermethylation and K-ras mutations in pancreatic carcinoma. Oncogene *22*, 3806-3812.

Dammann, R., Schagdarsurengin, U., Strunnikova, M., Rastetter, M., Seidel, C., Liu, L., Tommasi, S. and Pfeifer, G.P. (2003b) Epigenetic inactivation of the Ras-association domain family 1 (RASSF1A) gene and its function in human carcinogenesis. Histol. Histopathol. *18*, 665-677.

Dammann, R., Schagdarsurengin, U., Seidel, C., Strunnikova, M., Rastetter, M., Baier, K. and Pfeifer, G.P. (2005) The tumor suppressor RASSF1A in human carcinogenesis: an update. Histol. Histopathol. *20*, 645-663.

Davies, H., Bignell, G.R., Cox, C., Stephens, P., Edkins, S., Clegg, S., Teague, J., Woffendin, H., Garnett, M.J., Bottomley, W., Davis, N., Dicks, E., Ewing, R., Floyd, Y., Gray, K., Hall, S., Hawes, R., Hughes, J., Kosmidou, V., Menzies, A., Mould, C., Parker, A., Stevens, C., Watt, S., Hooper, S., Wilson, R., Jayatilake, H., Gusterson, B.A., Cooper, C., Shipley, J., Hargrave, D., Pritchard-Jones, K., Maitland, N., Chenevix-Trench, G., Riggins, G.J., Bigner, D.D., Palmieri, G., Cossu, A., Flanagan, A., Nicholson, A., Ho, J.W., Leung, S.Y., Yuen, S.T., Weber, B.L., Seigler, H.F., Darrow, T.L., Paterson, H., Marais, R., Marshall, C.J., Wooster, R., Stratton, M.R. and Futreal, P.A. (2002) Mutations of the BRAF gene in human cancer. Nature *417*, 949-954.

de Bruyn, K.M., de Rooij, J., Wolthuis, R.M., Rehmann, H., Wesenbeek, J., Cool, R.H., Wittinghofer, A.H. and Bos, J.L. (2000) RalGEF2, a pleckstrin homology domain containing guanine nucleotide exchange factor for Ral. J. Biol. Chem. *275*, 29761-29766.

Eckert, L.B., Repasky, G.A., Ulku, A.S., McFall, A., Zhou, H., Sartor, C.I. and Der, C.J. (2004) Involvement of Ras activation in human breast cancer cell signaling, invasion, and anoikis. Cancer Res. *64*, 4585-4592.

Eckfeld, K., Hesson, L., Vos, M.D., Bieche, I., Latif, F. and Clark, G.J. (2004) RASSF4/AD037 is a potential ras effector/tumor suppressor of the RASSF family. Cancer Res. *64*, 8688-8693.

Engers, R., Zwaka, T.P., Gohr, L., Weber, A., Gerharz, C.D. and Gabbert, H.E. (2000) Tiam1 mutations in human renal-cell carcinomas. Int. J. Cancer *88*, 369-376.

Feig, L.A. (2003) Ral-GTPases: approaching their 15 minutes of fame. Trends Cell Biol. *13*, 419-425.

Feig, L.A. and Buchsbaum, R.J. (2002) Cell signaling: life or death decisions of ras proteins. Curr. Biol. *12*, R259-R261.

Garnett, M.J. and Marais, R. (2004) Guilty as charged: B-RAF is a human oncogene. Cancer Cell *6*, 313-319.

Gonzalez-Garcia, A., Pritchard, C.A., Paterson, H.F., Mavria, G., Stamp, G. and Marshall, C.J. (2005) RalGDS is required for tumor formation in a model of skin carcinogenesis. Cancer Cell *7*, 219-226.

Habets, G.G., Scholtes, E.H., Zuydgeest, D., van der Kammen, R.A., Stam, J.C., Berns, A. and Collard, J.G. (1994) Identification of an invasion-inducing gene, Tiam-1, that encodes a protein with homology to GDP-GTP exchangers for Rho-like proteins. Cell *77*, 537-549.

Hamad, N.M., Elconin, J.H., Karnoub, A.E., Bai, W., Rich, J.N., Abraham, R.T., Der, C.J. and Counter, C.M. (2002) Distinct requirements for Ras oncogenesis in human versus mouse cells. Genes Dev. *16*, 2045-2057.

Han, J., Luby-Phelps, K., Das, B., Shu, X., Xia, Y., Mosteller, R.D., Krishna, U.M., Falck, J.R., White, M.A. and Broek, D. (1998) Role of substrates and products of PI 3-kinase in regulating activation of Rac-related guanosine triphosphatases by Vav. Science *279*, 558-560.

Han, M., Golden, A., Han, Y. and Sternberg, P.W. (1993) C. elegans lin-45 raf gene participates in let-60 ras-stimulated vulval differentiation. Nature *363*, 133-140.

Herrmann, C. (2003) Ras–effector interactions: after one decade. Curr. Opin. Struct. Biol. *13*, 122-129.

Hesson, L., Dallol, A., Minna, J.D., Maher, E.R. and Latif, F. (2003) NORE1A, a homologue of RASSF1A tumour suppressor gene is inactivated in human cancers. Oncogene *22*, 947-954.

Hofer, F., Fields, S., Schneider, C. and Martin, G.S. (1994) Activated Ras interacts with the Ral guanine nucleotide dissociation stimulator. Proc. Natl. Acad. Sci. U.S.A. *91*, 11089-11093.

Jullien-Flores, V., Dorseuil, O., Romero, F., Letourneur, F., Saragosti, S., Berger, R., Tavitian, A., Gacon, G. and Camonis, J.H. (1995) Bridging Ral GTPase to Rho pathways. RLIP76, a Ral effector with CDC42/Rac GTPase-activating protein activity. J. Biol. Chem. *270*, 22473-22477.

Kalhammer, G., Bahler, M., Schmitz, F., Jockel, J. and Block, C. (1997) Ras-binding domains: predicting function versus folding. FEBS Lett. *414*, 599-602.

Kang, S., Bader, A.G. and Vogt, P.K. (2005) Phosphatidylinositol 3-kinase mutations identified in human cancer are oncogenic. Proc. Natl. Acad. Sci. U.S.A. *102*, 802-807.

Karim, F.D., Chang, H.C., Therrien, M., Wassarman, D.A., Laverty, T. and Rubin, G.M. (1996) A screen for genes that function downstream of Ras1 during Drosophila eye development. Genetics *143*, 315-329.

Kelley, G.G., Reks, S.E., Ondrako, J.M. and Smrcka, A.V. (2001) Phospholipase C (epsilon): a novel Ras effector. EMBO J. *20*, 743-754.

Khokhlatchev, A., Rabizadeh, S., Xavier, R., Nedwidek, M., Chen, T., Zhang, X.F., Seed, B. and Avruch, J. (2002) Identification of a novel Ras-regulated proapoptotic pathway. Curr. Biol. *12*, 253-265.

Khosravi-Far, R. and Der, C.J. (1995) Prenylation analysis of bacterially expressed and insect cell-expressed Ras and Ras-related proteins. Methods Enzymol. *255*, 46-60.

Khosravi-Far, R., Solski, P.A., Clark, G.J., Kinch, M.S. and Der, C.J. (1995) Activation of Rac1, RhoA, and mitogen-activated protein kinases is required for Ras transformation. Mol. Cell. Biol. *15*, 6443-6453.

Khwaja, A., Rodriguez-Viciana, P., Wennstrom, S., Warne, P.H. and Downward, J. (1997) Matrix adhesion and Ras transformation both activate a phosphoinositide 3-OH kinase and protein kinase B/Akt cellular survival pathway. EMBO J. *16*, 2783-2793.

Kiel, C., Wohlgemuth, S., Rousseau, F., Schymkowitz, J., Ferkinghoff-Borg, J., Wittinghofer, F. and Serrano, L. (2005) Recognizing and defining true Ras binding domains II: in silico prediction based on homology modelling and energy calculations. J. Mol. Biol. *348*, 759-775.

Kikuchi, A., Demo, S.D., Ye, Z.H., Chen, Y.W. and Williams, L.T. (1994) ralGDS family members interact with the effector loop of ras p21. Mol. Cell. Biol. *14*, 7483-7491.

Kimura, E.T., Nikiforova, M.N., Zhu, Z., Knauf, J.A., Nikiforov, Y.E. and Fagin, J.A. (2003) High prevalence of BRAF mutations in thyroid cancer: genetic evidence for constitutive activation of the RET/PTC-RAS-BRAF signaling pathway in papillary thyroid carcinoma. Cancer Res. *63*, 1454-1457.

Kolch, W., Heidecker, G., Lloyd, P. and Rapp, U.R. (1991) Raf-1 protein kinase is required for growth of induced NIH/3T3 cells. Nature *349*, 426-428.

Krasilnikov, M.A. (2000) Phosphatidylinositol-3 kinase dependent pathways: the role in control of cell growth, survival, and malignant transformation. Biochemistry (Mosc.) *65*, 59-67.

Lambert, J.M., Karnoub, A.E., Graves, L.M., Campbell, S.L. and Der, C.J. (2002a) Role of MLK3-mediated activation of p70 S6 kinase in Rac1 transformation. J. Biol. Chem. *277*, 4770-4777.

Lambert, J.M., Lambert, Q.T., Reuther, G.W., Malliri, A., Siderovski, D.P., Sondek, J., Collard, J.G. and Der, C.J. (2002b) Tiam1 mediates Ras activation of Rac by a PI(3)K-independent mechanism. Nat. Cell Biol. *4*, 621-625.

Leevers, S.J., Paterson, H.F. and Marshall, C.J. (1994) Requirement for Ras in Raf activation is overcome by targeting Raf to the plasma membrane. Nature *369*, 411-414.

Lerman, M.I. and Minna, J.D. (2000) The 630-kb lung cancer homozygous deletion region on human chromosome 3p21.3: identification and evaluation of the resident candidate tumor suppressor genes. The International Lung Cancer Chromosome 3p21.3 Tumor Suppressor Gene Consortium. Cancer Res. *60*, 6116-6133.

Liao, Y., Kariya, K., Hu, C.D., Shibatohge, M., Goshima, M., Okada, T., Watari, Y., Gao, X., Jin, T.G., Yamawaki-Kataoka, Y. and Kataoka, T. (1999) RA-GEF, a novel Rap1A guanine nucleotide exchange factor containing a Ras/Rap1A-associating domain, is conserved between nematode and humans. J. Biol. Chem. *274*, 37815-37820.

Lim, K.-H., Baines, A.T., Fiordalisi, J.J., Shipitsin, M., Feig, L.A., Cox, A.D., Der, C.J. and Counter, C.M. (2005) Activation of RalA is critical for Ras-induced tumorigenesis of human cells. Cancer Cell *7*, 533-545.

Liu, L., Wu, D.H. and Ding, Y.Q. (2005) Tiam1 gene expression and its significance in colorectal carcinoma. World J. Gastroenterol. *11*, 705-707.

Lopez, I., Mak, E.C., Ding, J., Hamm, H.E. and Lomasney, J.W. (2001) A novel bifunctional phospholipase c that is regulated by Galpha 12 and stimulates the Ras/mitogen-activated protein kinase pathway. J. Biol. Chem. *276*, 2758-2765.

Lucas, L., Penalva, V., Ramirez de Molina, A., Del Peso, L. and Lacal, J.C. (2002) Modulation of phospholipase D by Ras proteins mediated by its effectors Ral-GDS, PI3K and Raf-1. Int. J. Oncol. *21*, 477-485.

Luo, J., Manning, B.D. and Cantley, L.C. (2003) Targeting the PI3K-Akt pathway in human cancer: rationale and promise. Cancer Cell *4*, 257-262.

Malliri, A., van der Kammen, R.A., Clark, K., van der Valk, M., Michiels, F. and Collard, J.G. (2002) Mice deficient in the Rac activator Tiam1 are resistant to Ras-induced skin tumours. Nature *417*, 867-871.

Malumbres, M. and Barbacid, M. (2003) RAS oncogenes: the first 30 years. Nat. Rev. Cancer *3*, 459-465.

Mansour, S.J., Matten, W.T., Hermann, A.S., Candia, J.M., Rong, S., Fukasawa, K., Vande Woude, G.F. and Ahn, N.G. (1994) Transformation of mammalian cells by constitutively active MAP kinase kinase. Science *265*, 966-970.

Matheny, S.A., Chen, C., Kortum, R.L., Razidlo, G.L., Lewis, R.E. and White, M.A. (2004) Ras regulates assembly of mitogenic signalling complexes through the effector protein IMP. Nature *427*, 256-260.

McFall, A., Ulku, A., Lambert, Q.T., Kusa, A., Rogers-Graham, K. and Der, C.J. (2001) Oncogenic Ras blocks anoikis by activation of a novel effector pathway independent of phosphatidylinositol 3-kinase. Mol. Cell. Biol. *21*, 5488-5499.

Mercer, K.E. and Pritchard, C.A. (2003) Raf proteins and cancer: B-Raf is identified as a mutational target. Biochim. Biophys. Acta *1653*, 25-40.

Michiels, F., Habets, G.G., Stam, J.C., van der Kammen, R.A. and Collard, J.G. (1995) A role for Rac in Tiam1-induced membrane ruffling and invasion. Nature *375*, 338-340.

Minard, M.E., Herynk, M.H., Collard, J.G. and Gallick, G.E. (2005) The guanine nucleotide exchange factor Tiam1 increases colon carcinoma growth at metastatic sites in an orthotopic nude mouse model, Oncogene *24*, 2568-2573.

Mitsiades, C.S., Mitsiades, N. and Koutsilieris, M. (2004) The Akt pathway: molecular targets for anti-cancer drug development. Curr. Cancer Drug Targets *4*, 235-256.

Moodie, S.A., Willumsen, B.M., Weber, M.J. and Wolfman, A. (1993) Complexes of Ras. GTP with Raf-1 and mitogen-activated protein kinase kinase. Science *260*, 1658-1661.

Muller-Decker, K., Neufang, G., Berger, I., Neumann, M., Marks, F. and Furstenberger, G. (2002) Transgenic cyclooxygenase-2 overexpression sensitizes mouse skin for carcinogenesis. Proc. Natl. Acad. Sci. U.S.A. *99*, 12483-12488.

Okazaki, M., Kishida, S., Murai, H., Hinoi, T. and Kikuchi, A. (1996) Ras-interacting domain of Ral GDP dissociation stimulator like (RGL) reverses v-Ras-induced transformation and Raf-1 activation in NIH3T3 cells. Cancer Res. *56*, 2387-2392.

Oldham, S.M., Clark, G.J., Gangarosa, L.M., Coffey, R.J. Jr. and Der, C.J. (1996) Activation of the Raf-1/MAP kinase cascade is not sufficient for Ras transformation of RIE-1 epithelial cells. Proc. Natl. Acad. Sci. U.S.A. *93*, 6924-6928.

Oldham, S.M., Cox, A.D., Reynolds, E.R., Sizemore, N.S., Coffey, R.J. Jr. and Der, C.J. (1998) Ras, but not Src, transformation of RIE-1 epithelial cells is dependent on activation of the mitogen-activated protein kinase cascade. Oncogene *16*, 2565-2573.

Ortiz-Vega, S., Khokhlatchev, A., Nedwidek, M., Zhang, X.F., Dammann, R., Pfeifer, G.P. and Avruch, J. (2002) The putative tumor suppressor RASSF1A homodimerizes and heterodimerizes with the Ras-GTP binding protein Nore1. Oncogene *21*, 1381-1390.

Parsons, R. (2004) Human cancer, PTEN and the PI-3 kinase pathway. Semin. Cell Dev. Biol. *15*, 171-176.

Peterson, S.N., Trabalzini, L., Brtva, T.R., Fischer, T., Altschuler, D.L., Martelli, P., Lapetina, E.G., Der, C.J. and White, G.C. 2nd (1996) Identification of a novel RalGDS-related protein as a candidate effector for Ras and Rap1. J. Biol. Chem. *271*, 29903-29908.

Pfeifer, G.P., Yoon, J.H., Liu, L., Tommasi, S., Wilczynski, S.P. and Dammann, R. (2002) Methylation of the RASSF1A gene in human cancers. Biol. Chem. *383*, 907-914.

Ponting, C.P. and Benjamin, D.R. (1996) A novel family of Ras-binding domains. Trends Biochem. Sci. *21*, 422-425.

Qiu, R.G., Chen, J., Kirn, D., McCormick, F. and Symons, M. (1995a) An essential role for Rac in Ras transformation. Nature *374*, 457-459.

Qiu, R.G., Chen, J., McCormick, F. and Symons, M. (1995b) A role for Rho in Ras transformation. Proc. Natl. Acad. Sci. U.S.A. *92*, 11781-11785.

Quilliam, L.A., Rebhun, J.F. and Castro, A.F. (2002) A growing family of guanine nucleotide exchange factors is responsible for activation of Ras-family GTPases. Prog. Nucleic Acid Res. Mol. Biol. *71*, 391-444.

Rajagopalan, H., Bardelli, A., Lengauer, C., Kinzler, K.W., Vogelstein, B. and Velculescu, V.E. (2002) Tumorigenesis: RAF/RAS oncogenes and mismatch-repair status. Nature *418*, 934.

Rangarajan, A. and Weinberg, R.A. (2003) Opinion: Comparative biology of mouse versus human cells: modelling human cancer in mice. Nat. Rev. Cancer *3*, 952-959.

Repasky, G.A., Chenette, E.J. and Der, C.J. (2004) Renewing the conspiracy theory debate: does Raf function alone to mediate Ras oncogenesis? Trends Cell Biol. *14*, 639-647.

Rhee, S.G. (2001) Regulation of phosphoinositide-specific phospholipase C. Annu. Rev. Biochem. *70*, 281-312.

Rodriguez-Viciana, P., Warne, P.H., Dhand, R., Vanhaesebroeck, B., Gout, I., Fry, M.J., Waterfield, M.D. and Downward, J. (1994) Phosphatidylinositol-3-OH kinase as a direct target of Ras. Nature *370*, 527-532.

Rodriguez-Viciana, P., Warne, P.H., Khwaja, A., Marte, B.M., Pappin, D., Das, P., Waterfield, M.D., Ridley, A. and Downward, J. (1997) Role of phosphoinositide 3-OH kinase in cell transformation and control of the actin cytoskeleton by Ras. Cell *89*, 457-467.

Rodriguez-Viciana, P., Sabatier, C. and McCormick, F. (2004) Signaling specificity by Ras family GTPases is determined by the full spectrum of effectors they regulate. Mol. Cell. Biol. *24*, 4943-4954.

Samuels, Y., Wang, Z., Bardelli, A., Silliman, N., Ptak, J., Szabo, S., Yan, H., Gazdar, A., Powell, S.M., Riggins, G.J., Willson, J.K., Markowitz, S., Kinzler, K.W., Vogelstein, B. and Velculescu, V.E. (2004) High frequency of mutations of the PIK3CA gene in human cancers. Science *304*, 554.

Sawyers, C.L. (2003) Opportunities and challenges in the development of kinase inhibitor therapy for cancer. Genes Dev. *17*, 2998-3010.

Schaap, D., van der Wal, J., Howe, L.R., Marshall, C.J. and van Blitterswijk, W.J. (1993) A dominant-negative mutant of raf blocks mitogen-activated protein kinase activation by growth factors and oncogenic p21ras. J. Biol. Chem. *268*, 20232-20236.

Schulze, A., Lehmann, K., Jefferies, H.B., McMahon, M. and Downward, J. (2001) Analysis of the transcriptional program induced by Raf in epithelial cells. Genes Dev. *15*, 981-994.

Sebolt-Leopold, J.S. and Herrera, R. (2004) Targeting the mitogen-activated protein kinase cascade to treat cancer. Nat. Rev. Cancer *4*, 937-947.

Shao, H. and Andres, D.A. (2000) A novel RalGEF-like protein, RGL3, as a candidate effector for rit and Ras. J. Biol. Chem. *275*, 26914-26924.

Shayesteh, L., Lu, Y., Kuo, W.L., Baldocchi, R., Godfrey, T., Collins, C., Pinkel, D., Powell, B., Mills, G.B. and Gray, J.W. (1999) PIK3CA is implicated as an oncogene in ovarian cancer. Nat. Genet. *21*, 99-102.

Shipitsin, M. and Feig, L.A. (2004) RalA but not RalB enhances polarized delivery of membrane proteins to the basolateral surface of epithelial cells. Mol. Cell. Biol. *24*, 5746-5756.

Shivakumar, L., Minna, J., Sakamaki, T., Pestell, R. and White, M.A. (2002) The RASSF1A tumor suppressor blocks cell cycle progression and inhibits cyclin D1 accumulation. Mol. Cell. Biol. *22*, 4309-4318.

Sieben, N.L., Macropoulos, P., Roemen, G.M., Kolkman-Uljee, S.M., Jan Fleuren, G., Houmadi, R., Diss, T., Warren, B., Al Adnani, M., De Goeij, A.P., Krausz, T. and Flanagan, A.M. (2004) In ovarian neoplasms, BRAF, but not KRAS, mutations are restricted to low-grade serous tumours. J. Pathol. *202*, 336-340.

Singer, G., Oldt, R. 3rd, Cohen, Y., Wang, B.G., Sidransky, D., Kurman, R.J. and Shih Ie, M. (2003) Mutations in BRAF and KRAS characterize the development of low-grade ovarian serous carcinoma. J. Natl. Cancer Inst. *95*, 484-486.

Sjolander, A., Yamamoto, K., Huber, B.E. and Lapetina, E.G. (1991) Association of p21ras with phosphatidylinositol 3-kinase. Proc. Natl. Acad. Sci. U.S.A. *88*, 7908-7912.

Song, C., Hu, C.D., Masago, M., Kariyai, K., Yamawaki-Kataoka, Y., Shibatohge, M., Wu, D., Satoh, T. and Kataoka, T. (2001) Regulation of a novel human phospholipase C, PLCepsilon, through membrane targeting by Ras. J. Biol. Chem. *276*, 2752-2757.

Sorli, S.C., Bunney, T.D., Sugden, P.H., Paterson, H.F. and Katan, M. (2005) Signaling properties and expression in normal and tumor tissues of two phospholipase C epsilon splice variants. Oncogene *24*, 90-100.

Spaargaren, M. and Bischoff, J.R. (1994) Identification of the guanine nucleotide dissociation stimulator for Ral as a putative effector molecule of R-ras, H-ras, K-ras, and Rap. Proc. Natl. Acad. Sci. U.S.A. *91*, 12609-12613.

Spaargaren, M., Martin, G.A., McCormick, F., Fernandez-Sarabia, M.J. and Bischoff, J.R. (1994) The Ras-related protein R-ras interacts directly with Raf-1 in a GTP-dependent manner. Biochem. J. *300 (2)*, 303-307.

Spugnardi, M., Tommasi, S., Dammann, R., Pfeifer, G.P. and Hoon, D.S. (2003) Epigenetic inactivation of RAS association domain family protein 1 (RASSF1A) in malignant cutaneous melanoma. Cancer Res. *63*, 1639-1643.

Stokoe, D., Macdonald, S.G., Cadwallader, K., Symons, M. and Hancock, J.F. (1994) Activation of Raf as a result of recruitment to the plasma membrane. Science *264*, 1463-1467.

Tchevkina, E., Agapova, L., Dyakova, N., Martinjuk, A., Komelkov, A. and Tatosyan, A. (2005) The small G-protein RalA stimulates metastasis of transformed cells. Oncogene *24*, 329-335.

Tommasi, S., Dammann, R., Jin, S.G., Zhang, X.F., Avruch, J. and Pfeifer, G.P. (2002) RASSF3 and NORE1: identification and cloning of two human homologues of the putative tumor suppressor gene RASSF1. Oncogene *21*, 2713-2720.

Tsao, H., Goel, V., Wu, H., Yang, G. and Haluska, F.G. (2004) Genetic interaction between NRAS and BRAF mutations and PTEN/MMAC1 inactivation in melanoma. J. Invest. Dermatol. *122*, 337-341.

Tuveson, D.A., Shaw, A.T., Willis, N.A., Silver, D.P., Jackson, E.L., Chang, S., Mercer, K.L., Grochow, R., Hock, H., Crowley, D., Hingorani, S.R., Zaks, T., King, C., Jacobetz, M.A., Wang, L., Bronson, R.T., Orkin, S.H., DePinho, R.A. and Jacks, T. (2004) Endogenous oncogenic K-ras (G12D) stimulates proliferation and widespread neoplastic and developmental defects. Cancer Cell *5*, 375-387.

Ulku, A.S. and Der, C.J. (2003) Ras signaling, deregulation of gene expression and oncogenesis. Cancer Treat. Res. *115*, 189-208.

Ulku, A.S., Schafer, R. and Der, C.J. (2003) Essential role of Raf in Ras transformation and deregulation of matrix metalloproteinase expression in ovarian epithelial cells. Mol. Cancer Res. *1*, 1077-1088.

Urano, T., Emkey, R. and Feig, L.A. (1996) Ral-GTPases mediate a distinct downstream signaling pathway from Ras that facilitates cellular transformation. EMBO J. *15*, 810-816.

van Engeland, M., Roemen, G.M., Brink, M., Pachen, M.M., Weijenberg, M.P., de Bruine, A.P., Arends, J.W., van den Brandt, P.A., de Goeij, A.F. and Herman, J.G. (2002) K-ras mutations and RASSF1A promoter methylation in colorectal cancer. Oncogene *21*, 3792-3795.

Vanhaesebroeck, B., Leevers, S.J., Panayotou, G. and Waterfield, M.D. (1997) Phosphoinositide 3-kinases: a conserved family of signal transducers. Trends Biochem. Sci. *22*, 267-272.

Vavvas, D., Li, X., Avruch, J. and Zhang, X.F. (1998) Identification of Nore1 as a potential Ras effector. J. Biol. Chem. *273*, 5439-5442.

Vetter, I.R. and Wittinghofer, A. (2001) The guanine nucleotide-binding switch in three dimensions. Science *294*, 1299-1304.

Vojtek, A.B., Hollenberg, S.M. and Cooper, J.A. (1993) Mammalian Ras interacts directly with the serine/threonine kinase Raf. Cell *74*, 205-214.

Vos, M.D., Ellis, C.A., Bell, A., Birrer, M.J. and Clark, G.J. (2000) Ras uses the novel tumor suppressor RASSF1 as an effector to mediate apoptosis. J. Biol. Chem. *275*, 35669-35672.

Vos, M.D., Ellis, C.A., Elam, C., Ulku, A.S., Taylor, B.J. and Clark, G.J. (2003a) RASSF2 is a novel K-Ras-specific effector and potential tumor suppressor. J. Biol. Chem. *278*, 28045-28051.

Vos, M.D., Martinez, A., Ellis, C.A., Vallecorsa, T. and Clark, G.J. (2003b) The pro-apoptotic Ras effector Nore1 may serve as a Ras-regulated tumor suppressor in the lung. J. Biol. Chem. *278*, 21938-21943.

Voss, M., Weernink, P.A., Haupenthal, S., Moller, U., Cool, R.H., Bauer, B., Camonis, J.H., Jakobs, K.H. and Schmidt, M. (1999) Phospholipase D stimulation by receptor tyrosine kinases mediated by protein kinase C and a Ras/Ral signaling cascade. J. Biol. Chem. *274*, 34691-34698.

Warne, P.H., Viciana, P.R. and Downward, J. (1993) Direct interaction of Ras and the amino-terminal region of Raf-1 in vitro. Nature *364*, 352-355.

Welch, H.C., Coadwell, W.J., Ellson, C.D., Ferguson, G.J., Andrews, S.R., Erdjument-Bromage, H., Tempst, P., Hawkins, P.T. and Stephens, L.R. (2002) P-Rex1, a PtdIns(3,4,5)P3- and Gbetagamma-regulated guanine-nucleotide exchange factor for Rac. Cell *108*, 809-821.

Wellbrock, C., Karasarides, M. and Marais, R. (2004a) The RAF proteins take centre stage. Nat. Rev. Mol. Cell. Biol. *5*, 875-885.

Wellbrock, C., Ogilvie, L., Hedley, D., Karasarides, M., Martin, J., Niculescu-Duvaz, D., Springer, C.J. and Marais, R. (2004b) V599EB-RAF is an oncogene in melanocytes. Cancer Res. *64*, 2338-2342.

Wennerberg, K., Rossman, K.L. and Der, C.J. (2005) The Ras superfamily at a glance. J. Cell Sci. *118*, 843-846.

Westwick, J.K., Cox, A.D., Der, C.J., Cobb, M.H., Hibi, M., Karin, M. and Brenner, D.A. (1994) Oncogenic Ras activates c-Jun via a separate pathway from the activation of extracellular signal-regulated kinases. Proc. Natl. Acad. Sci. U.S.A. *91*, 6030-6034.

White, M.A., Vale, T., Camonis, J.H., Schaefer, E. and Wigler, M.H. (1996) A role for the Ral guanine nucleotide dissociation stimulator in mediating Ras-induced transformation. J. Biol. Chem. *271*, 16439-16442.

Wohlgemuth, S., Kiel, C., Kramer, A., Serrano, L., Wittinghofer, F. and Herrmann, C. (2005) Recognizing and defining true ras binding domains I: biochemical analysis. J. Mol. Biol. *348*, 741-758.

Yoon, J.H., Dammann, R. and Pfeifer, G.P. (2001) Hypermethylation of the CpG island of the RASSF1A gene in ovarian and renal cell carcinomas. Int. J. Cancer *94*, 212-217.

Zhang, Q.X., Davis, I.D. and Baldwin, G.S. (1996) Controlled overexpression of selected domains of the P85 subunit of phosphatidylinositol 3-kinase reverts v-Ha-Ras transformation. Biochim. Biophys. Acta *1312*, 207-214.

Zhang, X.F., Settleman, J., Kyriakis, J.M., Takeuchi-Suzuki, E., Elledge, S.J., Marshall, M.S., Bruder, J.T., Rapp, U.R. and Avruch, J. (1993) Normal and oncogenic p21ras proteins bind to the amino-terminal regulatory domain of c-Raf-1. Nature *364*, 308-313.

CHAPTER 7

RHO PROTEINS IN RAS SIGNALING AND TRANSFORMATION

ANTOINE E. KARNOUB[1,3], EMILY J. CHENETTE[2,3], AND CHANNING J. DER[1,2,3]
[1] *Department of Pharmacology,*
[2] *Curriculum in Genetics and Molecular Biology, and*
[3] *Lineberger Comprehensive Cancer Center, University of North Carolina at Chapel Hill, Chapel Hill, NC 27599*

Abstract: The important contributions of Ras to tumor progression and maintenance are now well established. These functions couple Ras to a variety of downstream activities that are mediated by multiple effector pathways. Among the critical Ras targets are the Rho GTPases, which have emerged as a family of proteins that plays an essential role in normal and neoplastic cellular growth, particularly in Ras-induced transformation. How Rho family GTPases relay Ras signaling to downstream components and what roles they serve in Ras-induced oncogenesis will be the subject of this chapter

Keywords: Ras, Rho, GTPase, transformation, oncogene, effector, signaling

1. INTRODUCTION

The discovery that *ras* genes were mutated in a wide variety of human cancers sparked an explosion in studies investigating the molecular bases for Ras-mediated oncogenesis (Malumbres and Barbacid, 2003). Ras is mutationally activated in 30% of human cancers. In other tumor subsets, Ras is activated as a consequence of over-expression or mutational activation of receptor tyrosine kinases (e.g., ErbB1/HER1) or loss of negative regulator function (e.g., NF1). While cancer is a multi-step genetic process that requires mutational alteration of multiple genes (Hanahan and Weinberg, 2000), evidence derived from extensive cell culture and mouse model experimental systems suggest that Ras mutations are critical in tumor progression and maintenance. Consequently, there has been intense interest in the development of anti-Ras strategies for cancer treatment (Cox and Der, 2002; Downward, 2003).

Ras proteins function as GDP/GTP-regulated signaling nodes that transmit extracellular signals received by receptor tyrosine kinases, tyrosine kinase-linked

C. Der (ed.), RAS Family GTPases, 143–167.

receptors, G protein-coupled receptors, ion channels and integrins to downstream cytoplasmic signaling cascades. This signaling convergence is mediated by two groups of regulatory molecules: guanine nucleotide exchange factors (GEFs) and GTPase activating proteins (GAPs). GEFs are activated by different receptor-mediated signaling events, and promote formation of active GTP-bound Ras which possesses increased affinity for downstream effector proteins. Activation of Ras is rapid and transient, and is normally balanced by the action of GAPs, which, together with the intrinsic GTPase activity of Ras, promote hydrolysis of GTP to GDP and terminate signaling activity. Tumor-associated Ras proteins harbor missense mutations, most commonly at residues Gly 12 or Gln 61 that render the proteins GAP-insensitive and persistently active in a ligand-independent fashion.

The three mammalian Ras proteins (H-Ras, K-Ras, and N-Ras) are the founding members of a large superfamily of small GTPases comprised of at least 154 human members (Colicelli, 2004; Wennerberg and Der, 2004), with orthologs found in *C. elegans*, *Drosophila*, yeast, and plants (Takai et al., 2001; Valster et al., 2000). Based on sequence and functional similarities, the Ras superfamily has been classified into at least five subfamilies: Ras, Rho, Rab, Arf and Ran. The involvement of many members of this family, in particular members of the Ras and Rho branches, in diverse aspects of normal and neoplastic cellular physiology has made the Ras family of GTPases one of the most intensely researched protein families in cancer cell biology.

Recently, several research efforts have evaluated the role of other members of the Ras superfamily (in particular members of the Rho branch) as effectors of Ras-mediated oncogenesis. The rational behind the involvement of Rho family members in Ras signaling is two-fold: first, Ras-transformed cells exhibit deregulation in processes that control cellular morphology as well as cellular growth and proliferation (Coleman et al., 2004; Pruitt et al., 2002). These cells are typically characterized by an aberrant cell cycle control, increased proliferation, suppressed apoptotic responses, and increased motility (Shields et al., 2000), which are classical phenotypes of Rho activation. Ras also endows cells with an enhanced tolerance for matrix deprivation, a property that facilitates enhanced invasion and metastasis. In doing so, Ras utilizes a variety of inter-related and often synergistic effector-mediated pathways of which Rho GTPases have emerged to be one of the most critical components (Bar-Sagi and Hall, 2000). The various signal transduction pathways relating Ras to its various effectors have been reviewed elsewhere (Repasky et al., 2004). In this chapter, we will focus on the involvement of Rho family GTPases in Ras signaling and transformation and discuss their contributions to Ras-induced oncogenesis.

2. RAS REGULATION AND SIGNALING

As described above, GEFs and GAPs regulate the cycling of Ras between the GDP-bound and the GTP-bound states. GAPs such as p120GAP or neurofibromin (NF1) enhance the intrinsic GTPase activity and hence negatively regulate Ras

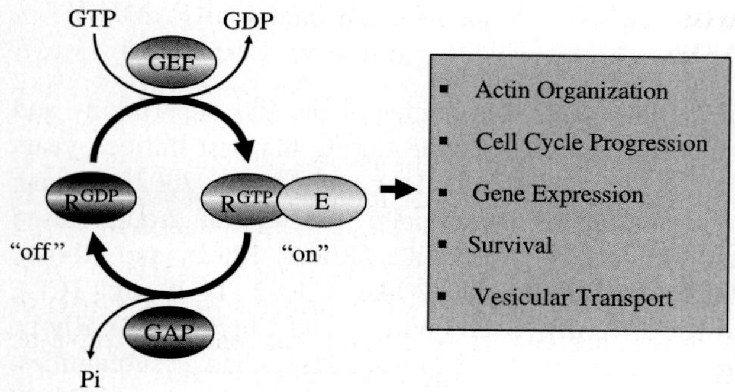

Figure 1. Ras and Rho GTPases function as GDP/GTP-related binary switches. Guanine nucleotide exchange factors stimulate GDP/GTP exchange and formation of the activated GTP-bound GTPase (R^{GTP}), whereas GTPase activating proteins accelerate the intrinsic GTP hydrolysis activity and formation of the inactive GDP-bound GTPase (R^{GDP}). The GTP-bound GTPase preferentially binds to downstream effectors (E) that regulate cytoplasmic signaling networks

protein function (Bernards and Settleman, 2004). Conversely, CDC25 homology domain-containing GEFs (also known as GTP-releasing proteins or GRPs) such as RasGRF, Sos, and RasGRP1-4 catalyze nucleotide ejection and therefore facilitate GTP binding and protein activation (Quilliam et al., 2002) (Fig. 1). The protein conformations of the two nucleotide-bound states of Ras differ in two regions known as switch I (residues 30-38) and switch II (59-67), which comprise the core effector domain of the protein. The effector domain of Ras-GTP forms an accessible loop which facilitates effector binding and signal propagation (Herrmann, 2003; Vetter and Wittinghofer, 2001).

It is now well established that Ras functions are mediated through the concerted actions of multiple effectors (Repasky et al., 2004). Although functionally distinct, these effectors share common structural features that facilitate preferential association with Ras-GTP. These include the Ras binding domains (RBDs) that are characteristic of the Raf and class IA p110 catalytic subunits of phosphatidylinositol 3-kinases (PI3Ks), and the Ras-association (RA) domains found on many other classes of functionally and catalytically diverse proteins. While these domains exhibit very little sequence identity, they share a common tertiary structure and possess ubiquitin fold-like protein conformations (Herrmann, 2003). Recent reviews have described the involvement of effectors in facilitating Ras activation of cytoplasmic signaling networks (Cox and Der, 2003; Feig and Buchsbaum, 2002; Feig, 2003; Repasky et al., 2004). The involvement of canonical Ras effectors in oncogenesis is covered in detail in another chapter in this book. In this chapter, we emphasize the effector pathways that link Ras with Rho GTPases.

3. RHO GTPASES ARE MEMBERS OF THE RAS SUPERFAMILY OF SMALL GTPASES

Rho GTPases comprise one of the major subfamilies of the Ras superfamily and have been the subject of intense research attention during the past thirteen years (Ridley, 2001; Wennerberg and Der, 2004; Wherlock and Mellor, 2002). Human Rho GTPases (22 members) are further sub-divided into several groups based on sequence and functional similarities: RhoA-like (RhoA, RhoB, and RhoC), Rac1-like (Rac1, Rac2, Rac3, and RhoG), Cdc42-like (Cdc42, TC10 and TCL), Rnd-like (Rnd1, Rnd2, and Rnd3/RhoE), Wrch (Wrch-1 and Chp/Wrch-2), RhoD, TTF/RhoH, Rif, Miro (Miro-1 and -2) and RhoBTB (RhoBTB1 and 2) subfamilies. The most intensely-studied and best-described members are Rac1, RhoA, and Cdc42 (Hall, 1998; Van Aelst and D'Souza-Schorey, 1997; Zohn et al., 1998b), but the functions of other members have recently attracted attention (Wennerberg and Der, 2004).

Like Ras, Rho GTPases also cycle between the active GTP-bound and the inactive GDP-bound states, a process that is tightly regulated by Rho-specific GEFs and GAPs (Moon and Zheng, 2003; Wennerberg et al., 2005) (Fig. 1). Members of the Dbl family of Rho GEFs (e.g., Dbl, Vav, Ect2, Lsc, Lfc, Lbc) were first identified as transforming proteins and comprise the largest family of Rho GEFs. Other non-Dbl Rho GEFs include Dock180 and Zizimin1 (Karnoub et al., 2004). The Rho GAPs DLC-1 and DLC-2 have been the focus of many recent research efforts due to their putative tumor suppressor functions and involvement in carcinogenesis (Plaumann et al., 2003; Yuan et al., 1998; Yuan et al., 2003; Yuan et al., 2004). The large number of Rho GEFs and GAPs (>80 human members for each) acting on the smaller number of Rho GTPases (22 human members) reflects the diversity of signaling networks that involve Rho GTPases. Finally, Rho GTPases are controlled by a third group of regulators: the Rho guanine nucleotide dissociation inhibitors (GDIs) (Dermardirossian and Bokoch, 2005). RhoGDIs bind to the prenylated carboxyl terminus of most Rho family proteins, sequestering them in the cytoplasm and preventing their activation.

Like Ras, Rho GTPases also function as signaling nodes, as a diverse array of extracellular stimuli converge on Rho GTPases to modulate their activation and stimulate signal propagation through downstream effector cascades. GEFs are the prime mediators of Rho GTPase stimulation, but signal-mediated regulation of GAPs and GDIs also contributes to regulation of Rho GTPases. Once activated, Rho GTPases interact with a diverse roster of downstream effectors with a multitude of functions, including protein kinases (Bishop and Hall, 2000; Karnoub et al., 2004).

Rho GTPases regulate a plethora of cellular activities (Etienne-Manneville and Hall, 2002) (Fig. 1), with perhaps the best-characterized function being their regulation of actin cytoskeletal organization (Aspenstrom et al., 2004). Each subfamily of Rho proteins is thought to have its own distinct effects on cytoskeletal architecture. Accordingly, while RhoA and its related proteins RhoB and RhoC promote cell adhesion and retraction through actin stress fiber formation and focal adhesion assembly, Rac1 and its related proteins Rac2 and Rac3 promote cell movement via lamellipodia formation. Similarly, Cdc42, Wrch-1 and related proteins (TC10, TLC, Chp) promote actin microspikes and filopodia induction, which are structures involved in sensing the extracellular milieu. In contrast, Rnd3/RhoE and related proteins (Rnd1 and Rnd2) antagonize RhoA function, causing the disruption of actin stress fibers, reduced cell adhesion and cell rounding. The ability of Rho GTPases to regulate actin organization reflects their key involvement in regulation of cell-cell and cell-matrix interactions, cell morphology, cell motility, and cell polarity.

A second major function of Rho GTPases involves regulation of transcription factor activity and induction of gene expression (Benitah et al., 2004). Transcription factors activated by Rho GTPases include serum response factor, transcription factors downstream of the JNK and p38 stress-activated mitogen-activated protein kinases (Jun, ATF-2, Chop, MF2A), STATs and NF-κB. A third important function of Rho GTPases is their role in progression through the G1 phase of the cell cycle (Pruitt and Der, 2001). This is mediated in part by the upregulation of cyclin D1 gene expression, the downregulation of $p27^{KIP1}$ expression and promotion of $p27^{KIP1}$ protein degradation. When taken together with their critical role in actin organization, it is not surprising that aberrant Rho GTPase function has been linked to multiple facets of oncogenesis.

4. RHO GTPASES AND CELLULAR TRANSFORMATION

The presence of mutationally activated Ras genes as well as of their effectors in human cancers (e.g., B-Raf, PI3KCA/p110α) (Samuels et al., 2004; Wellbrock et al., 2004) has provided direct and compelling evidence for the involvement of Ras activation in human oncogenesis. Although there is considerable experimental evidence that the aberrant activation of Rho GTPases can also promote oncogenesis (Boettner and Van Aelst, 2002; Ridley, 2004; Sahai and Marshall, 2002), a causal involvement for Rho GTPases in cancers has been more elusive as mutationally activated forms of Rho GTPases have not been identified in tumor specimen. Instead, what has emerged is that Rho GTPases are deregulated by indirect mechanisms that involve persistent activation by upstream stimuli, by altered gene and protein expression, and by alterations in the activities of select regulators of Rho GTPase GDP/GTP cycling and membrane association.

The first evidence that aberrant Rho GTPase activation can promote cellular transformation comes from the discovery in 1991 that the Dbl transforming protein is a GEF for Rho GTPases (Hart et al., 1991). Hence, the transforming activity of Dbl and other related Dbl oncogenes was ascribed to persistent activation of their Rho GTPase substrates. Subsequent studies with lab-generated, GTPase-deficient mutants of human Rho GTPases (analogous to the tumor-associated Ras mutants) also found a key role for Rho GTPases in regulating several aspects of the transformed phenotype (Khosravi-Far et al., 1995; Lebowitz et al., 1997; Murphy et al., 1999; Prendergast et al., 1995; Qiu et al., 1995a; Qiu et al., 1995b; Qiu et al., 1997; Roux et al., 1997). More importantly, these studies found that Rho GTPase function was essential for Ras-mediated transformation. In the sections below, we summarize the contributions of Rho GTPases to Ras-mediated oncogenesis.

5. MECHANISMS LINKING RAS WITH RHO PROTEINS

The signal transduction pathways linking Ras to various Rho proteins are still incompletely defined, but experimental evidence has delineated several effector pathways that couple Ras signaling with Rho GTPase activation. First, Ras may directly link to Rac1 via phosphatidylinositol 3-kinase (PI3K), a canonical Ras effector (Fig. 2). The binding of Ras to the p110 subunit of PI3K may potentially affect the association of p85 with Rac1, hence regulating its function (Hu et al., 1995; Rodriguez-Viciana et al., 1997). Ras may also indirectly modulate Rac function through PI3K activation. Ras-mediated PI3K activation increases cellular levels of phosphatidylinositol 3,4,5-triphosphate (PIP3), which is critical for membrane localization of GEFs and hence activation of small GTPases. Indeed, PIP3 has been shown to bind the PH domain of the Rac-specific GEFs Tiam1 (Fleming et al., 2000; Sander et al., 1998), Vav (Han et al., 1998), Sos (Nimnual et al., 1998) and P-Rex (Donald et al., 2004; Welch et al., 2002), and modulate their DH activity towards Rac1.

Secondly, Ras binds the RBD of Tiam1 and regulates its activity towards Rac1 (Lambert et al., 2002) (Fig. 2). Third, Ras binding to the Cdc25 Ras GEF domain of Son of sevenless (Sos) may impact the activity of its functional DH-PH unit towards Rac1 (Nimnual et al., 1998), suggesting that Sos-dependent Ras activation may potentially lead to Sos-dependent Rac1 activation. An alternative mechanism consists of Grb2/Sos1 activation leading to the formation of a complex (Sos1-Eps8-E3b1-Abi-1) that is able to bind and possibly modulate Rac1 activity (Scita et al., 1999; Scita et al., 2001). This complex can also recruit the p85 subunit of PI3K, therefore exacerbating Rac activation (Innocenti et al., 2002).

Other potential linkages between Ras and Rho GTPases have been described but not validated. Ras activation of Ral GEFs leads to the activation of the RalA and RalB small GTPases, which in turn bind the Rho GAP-containing protein RalBP1 (also called RLIP), and consequently may modulate its activity towards Rac1 (Cantor et al., 1995; Feig et al., 1996; Jullien-Flores et al., 1995; Park and Weinberg, 1995) (Fig. 2). However, whether Ras regulates the Rho GAP function of RalBP1 remains to be clarified. Another connection involves p120 RasGAP association with p190 RhoGAP (Settleman et al., 1992), but the functional significance of this interaction also remains poorly understood (Fig. 2).

The observations described above suggest a model where Ras activation of Rho GTPases occurs via a Raf-independent pathway(s). Ras-mediated activation of Rho in turn promotes various aspects of Ras transformation in both fibroblasts and epithelial cells. However, recent observations provide evidence that Ras activation of Rho proteins also occurs via MAPK-dependent mechanisms. For example, the Raf-MAPK pathway has been shown to mediate Ras relocalization of p190

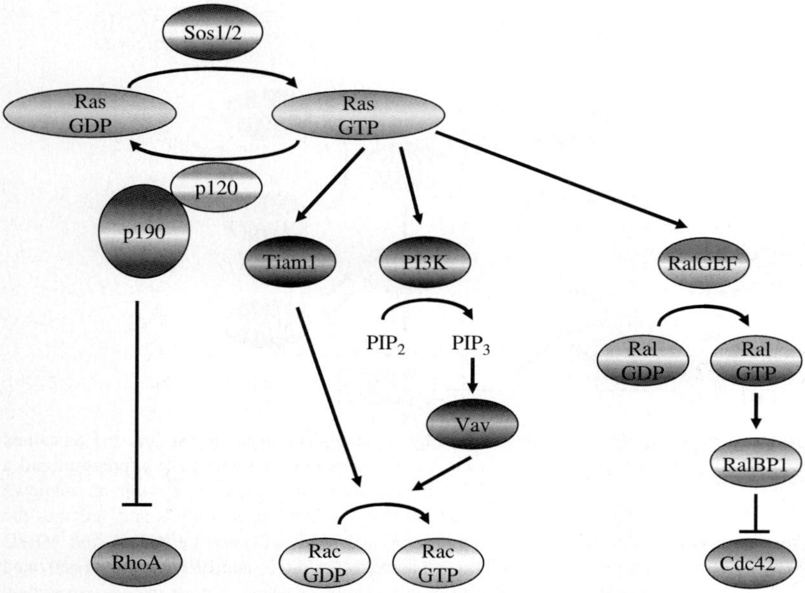

Figure 2. Ras Effector signaling pathways that regulate Rho GTPase activity. Tiam1, the p110 catalytic subunits of PI3K, and Ral guanine nucleotide exchange factors (RalGEfs) directly bind preferentially to activated Ras-GTP. This interaction promotes effector activation, leading to direct or indirect regulation of Rho GTPase function. The RasGEFs, Sos1/2 (and RasGRF1/2) are bifunctional GEFs that activate Ras and Rac. The p120 RasGAP complexes with the p190 RhoGAP, although the functional consequences and significance of this interaction remain poorly understood

RhoGAP to cytoskeletal fractions in NIH 3T3 cells (Chen et al., 2003). This translocation results in inhibition of p190 RhoGAP and activation of endogenous RhoA. Furthermore, Collard and colleagues found that persistent activation of H-Ras or Raf in MDCK cells led to downregulation of Rac, which in turn caused upregulation of Rho activity during the epithelial-mesenchymal transition (Zondag et al., 2000) (Fig. 3). In addition, the finding that ERK regulates the activity of Rac and RhoA in cellular motility (Vial et al., 2003) suggests that the ERK MAPK pathway plays more important roles in Ras-Rho cross-talk than previously anticipated.

Another linkage between Ras and RhoA may be mediated through RhoE/Rnd3 (Fig. 3). Rnd3 inhibits RhoA function and promotes the loss of actin stress fibers causing cell rounding (Guasch et al., 1998; Nobes et al., 1998). Two mechanisms have been described to explain how Rnd3 blocks RhoA function. First, Rnd3 can complex with and inhibit the function of the ROCKI serine/threonine kinase (Riento et al., 2003), a key effector of RhoA-mediated stress fiber formation. Second, Rnd3 can utilize p190 RhoGAP as an effector, causing downregulation of RhoA-GTP binding (Wennerberg et al., 2003). Unlike RhoA or the majority of other Rho GTPases, Rnd3 is persistently GTP-bound and active (Foster et al., 1996). So instead, Rnd3 expression can be upregulated by Ras activation via the

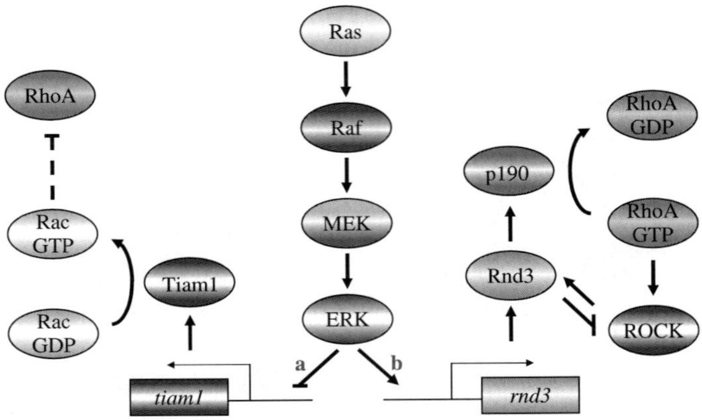

Figure 3. Ras regulation of Rho GTPase function by regulation of gene expression. (a) Sustained Ras activation of the Raf-MEK-ERK cascade causes downregulation of tiam1 gene expression and a reduction in Rac-GTP formation. Since Rac activity downregulates RhoA activity through an unknown mechanism, the reduction in Tiam1 expression leads to RhoA activation. Ras binds and activates the Raf serine/threonine kinase, and activated Raf phosphorylates and activates the MEK1 and MEK2 dual specificity kinases, which then phosphorylate and activate the ERK1 and ERK2 mitogen-activated protein kinases. ERKs phosphorylate Ets family transcription factors, causing changes in gene expression. (b) Ras stimulates rnd3 gene expression. Ras stimulates ERK-dependent upregulation of rnd3 gene expression. Rnd3 antagonizes RhoA function by two distinct mechanisms. First, Rnd3 interacts with p190 RhoGAP as an effector and stimulates p190 RhoGAP function, causing a downregulation of RhoA-GTP. Second, Rnd3 interacts with and inactivates ROCKI function. ROCKI is a key effector that mediates RhoA induction of actin stress fiber formation and RhoA transforming and invasion activity. ROCKI also phosphorylates Rnd3 and enhances Rnd3 protein stability

Raf-MEK-ERK cascade in MDCK cells (Hansen et al., 2000) as well as a variety of other rodent and human epithelial cell types (Singh and Der, unpublished). While some evidence suggests that Rnd3 antagonizes Ras transformation (Villalonga et al., 2004), the upregulation of Rnd3 in Ras-transformed cells suggests that Rnd3 may promote Ras transformation.

6. RHO GTPASES IN RAS-INDUCED TRANSFORMATION

Earlier observations that transient expression of activated Ras caused increased membrane ruffling (Bar-Sagi and Feramisco, 1986), and that activated Ras stimulated Rac-dependent membrane ruffling (Ridley et al., 1992) suggested that Rho GTPases were activated by Ras signaling. These observations, when coupled with the altered morphology and disrupted actin organization of Ras-transformed cells, prompted studies to evaluate a possible contribution of Rho GTPases in Ras transformation (Khosravi-Far et al., 1995; Lebowitz et al., 1997; Murphy et al., 1999; Prendergast et al., 1995; Qiu et al., 1995a; Qiu et al., 1995b; Qiu et al., 1997; Roux et al., 1997). First, co-expression of dominant negative versions of Rac1, RhoA, RhoB, RhoG, Cdc42, and TC10 (variants which impair Rho GTPase activation by forming non-productive complexes with their GEF activators (Feig, 1999)) blocked Ras-induced transformation. Second, activated RhoA, Rac1, or Cdc42 alone promoted anchorage-independent growth and tumor formation in nude mice (Khosravi-Far et al., 1995; Lin et al., 1997; Qiu et al., 1995a; Qiu et al., 1995b) albeit through distinct mechanisms. Third, co-expression of activated versions of Rho GTPases together with activated components of the Raf-MEK-ERK cascade showed cooperative transforming activity (Cobellis et al., 1998; Du et al., 2004; Fischer et al., 1998). Taken together, these observations support a model where Ras activation of Rho GTPases represents a critical step in Ras-mediated transformation.

An important question that arose from the aforementioned studies using dominant negative Rho proteins is whether Rho GTPase activity is elevated persistently in Ras-transformed cells. In light of the existence of multiple GEFs for each Rho GTPase, as well as GEFs that activate multiple Rho proteins (e.g., Vav2), the ability of a specific dominant negative Rho protein to block Ras transformation is difficult to determine unequivocally (Karnoub et al., 2004). Indeed, nonspecific activities of these dominant negatives have been described (Wennerberg et al., 2002). The subsequent development of "pull down" biochemical assays to directly evaluate Rho GTPase activity provided a means to critically reevaluate this issue.

Pull down assays, first developed for Ras (Taylor and Shalloway, 1996), where active GTP-bound GTPases are precipitated using the isolated GTP-dependent binding domains of Rho GTPase effectors (e.g., from the PAK serine/threonine kinase for Rac and Cdc42 activation and from Rhotekin for RhoA activation) have enabled the measurement of the activities of endogenous Rho proteins in response to a variety of stimuli. Using this method, Ras-transformed NIH 3T3 cells were shown to exhibit elevated levels of Cdc42-GTP, and stimulated receptor tyrosine kinases that activate Ras were also found to activated Cdc42 (Nur et al., 1999). Furthermore, oncogenic Ras was shown to stimulate an increase in the levels of Rac1-GTP in COS-1 cells (Walsh and Bar-Sagi, 2001) and Ras-transformed NIH 3T3 cells exhibited persistent elevated levels of Rac-GTP (Lambert et al., 2002). Similarly, Ras-transformed NIH 3T3 cells exhibited an increase in RhoA-GTP levels (Chen et al., 2003).

Although these observations proposed that Rho GTPase activation is a consequence of Ras-mediated transformation, other evidence suggests that this role may differ depending on cell type as well as with the varying functions of Ras in oncogenesis. For example, the loss of actin stress fibers in Ras-transformed cells would be more consistent with RhoA inactivation rather than activation. Consistent with this possibility, Kaibuchi and colleagues found that ectopic expression of activated RhoA or ROCK reversed the transformed morphology of Ras-transformed Rat1 rat fibroblasts (Izawa et al., 1998). In MDCK canine kidney epithelial cells, where transient Ras activation was associated with Rac activation, persistent Ras activation led to a downregulation of Rac activity (Zondag et al., 2000). This decrease in activity was associated with diminished Tiam1 mRNA expression, and downregulation of Rac activity in turn promoted an upregulation of RhoA activity by an unknown mechanism. These findings also correlated with an inhibition of invasion (Hordijk et al., 1997), strengthening the notion that while transient Ras activation causes Rac activation, sustained Ras activation does not, perhaps due to compensatory mechanisms.

A well-established mouse model for Ras-induced oncogenesis is the 7,12-dimethylbenzanthracene (DMBA) and the 12-O-tetradecanoylphorbol-13-acetate (TPA) dual topical chemical treatment, which effectively induces activating H-Ras point mutations in mouse skin cells. When this model was applied to mice deficient in Tiam1, a Rac-specific GEF, a significant reduction in tumor formation was seen (Malliri et al., 2002). Furthermore, the few tumors that did arise grew much slower when compared to tumors in wild type mice, and primary embryonic fibroblasts derived from *tiam1*$^{-/-}$ mice were resistant to Ras(12V)-induced focus formation. However, a greater proportion of tumors in *tiam1*$^{-/-}$ mice progressed to malignancy, suggesting that Tiam1 function was antagonistic to malignant growth. Thus, Rac activation may support Ras-induced tumor formation and growth, but may be antagostic to Ras-induced malignant progression.

A role for other Rho family members, including RhoA and RhoB, in Ras-mediated transformation has been identified, although there is some disagreement about its precise nature. First, it was found that dominant negative RhoB blocked Ras-mediated transformation, whereas activated RhoB cooperated with activated Raf to promote synergistic growth transformation of NIH 3T3 fibroblasts (Prendergast et al., 1995). These results are similar to those found with RhoA and suggest that RhoB promotes Ras transformation. In contrast, RhoB-deficient mice displayed higher numbers of DMBA/TPA-induced skin epithelial tumors when compared to wild type controls (Liu et al., 2001a). In addition, loss of RhoB desensitized Ras-transformed cells to DNA damaging insults suggesting that RhoB may normally function as a negative regulator of Ras oncogenesis (Liu et al., 2001a). Thus, Rho GTPases may play either positive or negative roles in Ras transformation, possibly depending on signal strength and cellular context. Furthermore, while RhoA and RhoB are highly related isoforms, there is increasing evidence that they possess distinct roles in oncogenesis, suggesting that sequence similarity of Rho GTPases does not always correspond with function similarity (Wheeler and Ridley, 2004).

Pharmacologic approaches also implicated Rho GTPase activation in Ras transformation. For example, Treisman and colleagues showed that Y-27632 (a small molecule inhibitor of ROCK serine/threonine kinases) blocked Ras-induced focus formation in NIH 3T3 cells (Sahai et al., 2001). Along the same lines, treatment with the recently identified inhibitor of Rac activation, NSC23766, impaired the growth of Ras-transformed NIH 3T3 cells in soft agar. Similarly, the pyrazoloquinoline derivative SCH 51344 inhibited Ras- and Rac-induced membrane ruffling, as well as the anchorage-independent growth of Ras- or Rac-transformed rat fibroblasts (Walsh et al., 1997). Peptide inhibitors such as a cell-permeable inhibitor of the ACK tyrosine kinase, an effector of Cdc42, also blocked the anchorage-independent growth of Ras-transformed NIH 3T3 cells (Nur et al., 1999). In addition, interfering RNA suppression of ACK-1 expression inhibited the growth of H-Ras-transformed NIH 3T3 cells, indicating that this kinase is a key mediator of Ras and Cdc42-mediated transformation (Nur et al., 2005) and may represent a key target for anti-Ras therapy. An additional such target is the geranylgeranyltransferase I enzyme or GGTase I. GGTase I catalyzes the posttranslational addition of a geranylgeranyl isoprenoid lipid to RhoA, Rac1, Cdc42 and other Rho GTPases, a modification that is critical for Rho GTPase membrane association and transformation (Allal et al., 2000; Murphy et al., 1999; Solski et al., 2002). Fittingly, treatment with GGTI-2154 impaired H-Ras-induced mammary tumor formation in mice, potentially through inhibition of Rac1 and Rac3 (Joyce and Cox, 2003). Taken together, these results support the possibility that pharmacologic inhibition of Rho GTPase signaling may be an effective approach for blocking Ras transformation. In the following sections, we summarize the contributions of Rho GTPases to specific aspects of Ras function.

7. RHO PROTEINS IN RAS REGULATION OF THE CELL CYCLE

Tumorigenic cells acquire the ability of autonomous cellular proliferation which allows escape from the growth constraints imposed by environmental cues in the extracellular matrix (Assoian, 1997). This entails deregulation of the cell cycle machinery and the bypass of several checkpoints which favor cell cycle progression over growth arrest (Pardee, 1989). Consequently, Ras-induced transformation facilitates deregulation of the cell cycle through multiple effector pathways that appear to be cell type-dependent (Hitomi and Stacey, 2001; Pruitt and Der, 2001).

A preponderance of evidence underscores the pivotal role of Rho proteins in controlling cell cycle progression (e.g. (Coleman et al., 2004; Gjoerup et al., 1998; Joyce et al., 1999; Olson et al., 1995; Pruitt and Der, 2001; Welsh, 2004)). In 1993, Narumiya and colleagues presented the first evidence implicating Rho GTPases in cell cycle regulation, where inhibition of RhoA by C3 exoenzyme blocked serum-induced DNA synthesis in rodent fibroblasts (Yamamoto et al., 1993). Although it was later shown that activation of endogenous RhoA is not sufficient to drive a G1 to S transition, microinjection of activated versions of RhoA did enable cell cycle transition and DNA synthesis (Olson et al., 1995).

The mechanistic details of the contribution(s) of Rho proteins to Ras control of the cell cycle have not been fully delineated. What has been established is that the Ras and Rho pathways intersect at three main cell cycle regulatory proteins, cyclin D1, $p21^{cip1}$, and $p27^{kip1}$. Ras upregulation of cyclin D1 expression occurs in both epithelial and fibroblast cells and is thought to be biphasic. The first peak occurs shortly after cells enter the cell cycle from the resting G0 state and is Raf-dependent, while the second peak which coincides with mid-G1 involves PI3K/Akt (Gille and Downward, 1999). Interestingly, activated Cdc42 and Rac1 also stimulate cyclin D1 expression (Gille and Downward, 1999; Westwick et al., 1997); (Welsh and Assoian, 2000). For Rac1, cyclin D1 expression occurs through NF-κB activation in NIH 3T3 cells (Joyce et al., 1999), and through activation of reactive oxygen species (ROS) and ERK in airway smooth muscle cells (Page et al., 1999). It is therefore reasonable to propose that the Ras-PI3K pathway might also feed into a Rac-ROS-NF-κB pathway, culminating in upregulation of cyclin D1 during late G1.

In contrast to Rac1 and Cdc42, RhoA does not upregulate cyclin D1 expression (Gjoerup et al., 1998), although cells treated with C3 toxin exhibit reduced cyclin D1 levels (Welsh et al., 2001). Instead, RhoA inhibits the cyclin-dependent kinase inhibitor $p21^{cip1}$, promoting inactivation of the Rb tumor suppressor, which then allows for G1 progression in response to Ras signaling (Olson et al., 1998). This appears to be a key contribution of RhoA to Ras transformation since loss of $p21^{cip1}$ function is critical for Ras-mediated transformation. Similarly, RhoA also regulates the activity of $p27^{kip1}$ either by protein degradation (Weber et al., 1997), or through enhancement of the activities of cyclinE/CDK2 (Hu et al., 1999).

8. RHO PROTEINS IN RAS-INDUCED APOPTOSIS

Ras governs pathways that promote either cell survival or apoptosis depending on the cell type and signaling context (Cox and Der, 2003; Downward, 1998b). Several reports have indicated that sustained Ras expression is required for maintenance of the transformed phenotype, and is necessary to circumvent apoptosis. For example, attenuation of the H-Ras transgene in melanoma xenografts (Chin et al., 1999) or K-Ras in a lung cancer model (Fisher et al., 2001) induced an apoptotic response in the cancer cells leading to tumor regression. In contrast, active Ras sensitizes NIH 3T3 cells to apoptosis induced by serum deprivation, TNF-α, or lovastatin treatments (Chang et al., 1998; Chang et al., 1999; Liu et al., 2001b), and sensitizes bladder cancer cells to 5-fluorouracil-induced cell death (Tseng et al., 2003).

To date, two effector-mediated pathways have been shown to mediate the pro-survival signals of Ras proteins, and both implicate the Rho GTPase Rac. The first involves PI3K, a critical survival effector downstream of Ras. PI3K stimulation leads to Rac activation (presumably via phosphoinositide-mediated stimulation of a Rac-specific GEF) which in turn leads to ROS-dependent NF-κB activation (Irani et al., 1997; Joneson and Bar-Sagi, 1999; Mayo and Baldwin, 2000; Sulciner et al., 1996). PI3K also activates Akt, which in turn phosphorylates and inactivates Bad, relieving its inhibition of Bcl-2 and Bcl-XL (Downward, 1998a; Downward, 1998b). However, it is not clear at this point whether Rac (or any other Rho GTPase) is involved in Akt-mediated pro-survival signaling downstream of Ras. The second is a PI3K-independent pathway that has been identified recently and involves Ras interaction with the Rac-specific GEF Tiam1 (Lambert et al., 2002). As mentioned above, Tiam1 knock-out mice are resistant to skin tumors due to increased levels of Ras-induced apoptosis (Malliri et al., 2002), suggesting that Tiam1 might constitute a critical anti-apoptotic signal downstream of Ras.

Whether Rac activation of NF-κB mediates Ras anti-apoptotic effects in all systems remains to be fully evaluated. A study by Dajee and colleagues suggests that NF-κB inhibition may promote, and not prevent, squamous cell carcinoma (Dajee et al., 2003). Furthermore, evidence that JNK/c-Jun signaling (also activated by Rac1) is pro-apoptotic downstream of Ras (Verheij et al., 1996) suggests that Rac1 could also be involved in Ras-mediated apoptosis.

Additional Ras effectors such as the RASSF family have been shown to promote Ras-induced apoptotic signals in certain cell types (Repasky et al., 2004) but whether Rho GTPases play any role in these pathways is presently unknown. It is tempting to speculate that the recent characterization of the RhoA effector CNK1 as a binding partner for RASSF1 could form a novel connection for the Ras and Rho pathways in apoptotic regulation (Rabizadeh et al., 2004).

9. RHO PROTEINS IN RAS-INDUCED ANGIOGENESIS

The expansion of a tumor mass beyond 1-2 mm in diameter depends on its ability to develop extensive blood vasculature. This process is often referred to as the "angiogenic switch" where pro-angiogenic factors overcome anti-angiogenic stimuli, leading to the development of new blood vessels permitting further tumor growth (Hanahan and Folkman, 1996). The ability of Ras-initiated cells to induce angiogenic signals has been well established. For example, oncogenic Ras has been shown to induce vascular endothelial growth factor (VEGF) (Rak et al., 1995) which serves as a chemo-attractant for endothelial cells to be recruited into the tumor stroma. Ras also exerts pro-angiogenic responses by down-regulating anti-angiogenic factors such as thrombospondin-1 (Tsp-1; (Rak et al., 2000)). Secreted Tsp-1 inhibits the activity of the matrix metalloproteinase MMP-9, an enzyme that releases VEGF tethered in the ECM (Good et al., 1990; Ribatti et al., 1998; Rodriguez-Manzaneque et al., 2001). Using the Ras effector domain mutants, Watnick and coworkers showed that inhibition of Tsp-1 expression involves PI3K but not Raf or RalGDS (Watnick et al., 2003). Furthermore, Tsp-1 repression was relieved by dominant negative RhoA, and by pharmacologic inhibition of ROCK by Y27632. Expectedly, activated RhoA (and RhoC) also mimicked the actions of Ras on Tsp-1, suggesting that Tsp-1 suppression by Ras occurs through a PI3K-RhoGEF-RhoA-ROCK pathway. Rho proteins may not be involved in all of the Ras-induced angiogenic signals. For instance, dominant negative N-Ras and PI3K, but not Rho, Rac, or Cdc42, could block fibroblast-mediated microvascular network formation when co-mingled with human vascular endothelial cells (HMVEC) (Liu et al., 2003), suggesting that N-Ras may function independently from Rho proteins in angiogenesis.

10. RHO PROTEINS IN RAS-INDUCED INVASION AND METASTASIS

Tumor invasion is a complex process requiring the coordinated actions of multiple signaling cascades which enable cancer cells to venture into adjacent stroma. To invade, cells have to break free from cell-cell contacts, degrade the extracellular matrix (ECM), and overcome detachment-induced apoptosis or anoikis (Frisch and Ruoslahti, 1997).

The ability of Ras to promote invasive and metastatic growth was demonstrated as early as 1985 (Bondy et al., 1985; Muschel et al., 1985) and occurs in many different rodent and human model systems (Gingras et al., 1990; Keely et al., 1999; Pozzatti et al., 1986). Ras promotes invasion and metastasis using multiple signaling pathways that coordinate various aspects of the malignant phenotype. For example, while the Raf/MAPK pathway is important for metastasis of NIH 3T3 cells in nude mice (Webb et al., 1998), PI3K plays a key role in preventing anoikis (Khwaja et al., 1997; McFall et al., 2001). The importance of a third Ras effector pathway RalGDS in metastasis has been highlighted in a series of studies (e.g., (Gildea et al., 2002)). In particular, work by Kelly and colleagues showed that Ral activation is sufficient to induce lung metastases, a process that also requires basal ERK activity (Ward et al., 2001).

Rho GTPases have been tightly linked to cellular motility (Etienne-Manneville and Hall, 2002; Sahai and Marshall, 2002; Schmitz et al., 2000) and as such, regulate several aspects of invasion. The accepted model applied to both fibroblasts and epithelial cells is that localized ruffling at the leading edge of a moving cell involves Rac and Cdc42, and is followed by Rho-dependent contraction of the uropod or lagging tail (Lauffenburger and Horwitz, 1996). In addition, Rho GTPases regulate cell-cell and cell-substratum contacts (Van Aelst and Symons, 2002). In this context, it was found that constitutive activation of Cdc42 protected MDCK cells from anoikis through a mechanism involving PI3K but not ERK, JNK, or p38 MAPK pathways (Cheng et al., 2004). Furthermore, the activities of Rho GTPases are required for the propagation of integrin-induced signals for invasion (e.g., (Shaw et al., 1997)). Similar roles for other members of the Rho family such as RhoD (Tsubakimoto et al., 1999) and Rnd3/RhoE (Guasch et al., 1998) have also been proposed.

In light of these functions, it is not surprising that Rho proteins are implicated in several aspects of Ras-induced invasion, particularly in the epithelial-to-mesenchymal transition (EMT) and in extracellular matrix degradation. EMT involves the regulation of a critical step in metastatic progression of epithelial cells (Thiery, 2003). This process involves the downregulation of E-cadherin, which engenders loss of the epithelial morphology concomitant with acquisition of mesenchymal traits such as expression of vimentin and fibronectin, which promote cell motility and migration. Ras activation induces E-cadherin loss and EMT in MDCK cells, a process that depends on Rac1 inhibition by RhoA (Zondag et al., 2000).

Ras-induced invasion also involves regulation of the activities of a variety of enzymes that degrade the ECM, mainly urokinase plasminogen activator (uPa) and matrix metalloproteases (MMPs). MMPs have been strongly implicated in invasion (Westermarck and Kahari, 1999), and Ras up-regulation of MMPs correlates with the metastatic phenotype (Ballin et al., 1988; Bernhard et al., 1994). Ras control of MMP activation appears to involve Ral- and PI3K-dependent pathways, both causing AP1- and NF-κB-dependent transcriptional activation of MMP gene expression (Okan et al., 2001) accompanied by a repression of TIMP function (Yang et al., 2001). Although Ras-induced MMP activation has not been directly linked to Rho protein function, Ras-induced uPa expression leads to increased activity of MMP2 and MMP9 (Aguirre-Ghiso et al., 1999) and Ras-induced up-regulation of uPa and the uPa receptor were shown to depend on the activity of Rac1 (Muller et al., 2000). RhoA activation can promote MMP-9 and the invasive properties of human microvascular endothelial cells (Abecassis et al., 2003), and MMP-9 is the perhaps the MMP most commonly upregulated upon Ras activation (Ulku and Der, 2003).

11. RHO GTPASES AND THE TRANSFORMING ACTIONS OF OTHER ONCOGENES

In addition to Ras, Rho GTPases also contribute to the transforming functions of other oncogenes. Dbl family proteins comprise the largest class of GEFs for Rho GTPases, and many were identified initially as transforming proteins (Rossman et al., 2005). Therefore, it is not surprising that their transforming activities are dependent on Rho GTPase function. For example, Vav is an activator of multiple Rho GTPases, and inhibition of RhoA, Rac1, or Cdc42 impaired Vav transforming activity (Palmby et al., 2004). Tyrosine kinase oncogenes (e.g., Abl, Met, BCR-Abl, Fps/Fes, ErbB1/HER1) also require Rho GTPase function for transformation (Boerner et al., 2001; Li and Smithgall, 1998; Renshaw et al., 1996; Rodrigues et al., 1997; Skorski et al., 1998). A DNA tumor virus oncogene, the polyomavirus middle-T antigen, causes transformation that is dependent on Rac activity (Urich et al., 1997). Transforming G-protein-coupled receptors (e.g., Mas, G2A, Par-1, KSHV vGPCR) and G alpha subunits ($G\alpha_{12}$ and $G\alpha_{13}$) are activators of Rho GTPases and cause transformation primarily through Rho GTPase activation (Martin et al., 2001; Montaner et al., 2004; Zohn et al., 1998a; Zohn et al., 2000).

Conversely, tumor suppressor function may also be linked to regulation of Rho GTPase function. The function of two tumor suppressor genes involved in the development of neurofibromatosis have been linked to Rho GTPases. The gene involved in the development of neurofibromatosis type 1 (NF1) encodes a Ras GAP, and NF1 deficiency caused an increase in Rac activation that cooperated with Ras activation to promote mast cell proliferation (Ingram et al., 2001). Merlin is encoded by the gene involved in neurofibromatosis type 2 (NF2). Merlin overexpression inhibited Rac-induced signaling, and loss of Merlin was associated with Rac activation (Shaw et al., 2001). Similarly the adenomatous polyposis coli (APC) tumor suppressor, mutated in sporadic and familial colorectal cancers, can activate Asef, a Rac-specific GEF (Kawasaki et al., 2000; Kawasaki et al., 2003). Finally, the DLC-1 tumor suppressor functions as a Rho GAP and DLC-1 expression is extinguished in liver, breast, lung and other cancers (Plaumann et al., 2003; Yuan et al., 1998; Yuan et al., 2003; Yuan et al., 2004). Presumably, loss of DLC-1 function results in hyperactivation of Rho GTPases.

12. CONCLUSIONS AND FUTURE DIRECTIONS

Our knowledge of the molecular details of the Ras-Rho connections in cellular transformation is mounting at a fast pace, but many questions remain unanswered. What is increasingly evident is that Rho proteins are required for a variety of normal as well as oncogenic activities of Ras, and that Ras control of Rho GTPase activation and function involves multiple Ras effectors. The attention has focused thus far on PI3K-, RalGDS-, or Tiam1-initiated pathways, but may also occur via additional routes (e.g., Af6, Rin1, RASSF1, or PLC-ε).

Interestingly, this long-held view of Rho GTPases acting downstream of Ras has been recently altered by evidence showing that Rac1 (and also activated Vav) induced Ras activation by acting on RasGRP (Caloca et al., 2003). Although this may not be generalized to many different cell types, it does create a new paradigm placing Rho GTPases upstream of Ras in certain cellular contexts. Alternatively, Rho proteins could affect Ras function through modulating the activities of its effectors. For example, activated RhoA has been shown to bind to the human homologue of the *Drosophila* connector enhancer of ksr (hCNK1), which associates with two Ras effectors: RalGDS (Jaffe et al., 2004) and RASSF1 (Rabizadeh et al., 2004). RhoA binding therefore potentially modulates their cellular activities. In addition, the roles Rho GTPases play in parallel to Ras signaling should be highlighted, as these signaling pathways could converge, for example, on downstream events such as control of gene expression and regulation of the cell cycle (Pruitt and Der, 2001; Teramoto et al., 2003).

In light of their critical role in Ras-induced signaling and oncogenesis, Rho GTPases may represent attractive targets for anti-tumor therapeutics. Pharmacologic blockage of Ras (such as with FTIs) are already in clinical trials (Sebti and Der, 2003), and there is promise that blockage of Rho GTPases (such as with GGTIs; (Sebti and Hamilton, 2000)) could exacerbate the therapeutic outcome (Joyce and Cox, 2003).

ACKNOWLEDGEMENTS

We thank Misha Rand for assistance in manuscript preparation. We apologize for not being able to cite the work of many colleagues because of space constraints. Our studies were supported by grants from the NIH CA63071 and CA92240 (to CJD), and from the Susan G. Komen Breast Cancer Foundation (to AEK).

REFERENCES

Abecassis, I., Olofsson, B., Schmid, M., Zalcman, G. and Karniguian, A. (2003) RhoA induces MMP-9 expression at CD44 lamellipodial focal complexes and promotes HMEC-1 cell invasion. Exp. Cell Res. *291*, 363-376.

Aguirre-Ghiso, J.A., Frankel, P., Farias, E.F., Lu, Z., Jiang, H., Olsen, A., Feig, L.A., de Kier Joffe, E.B. and Foster, D.A. (1999) RalA requirement for v-Src- and v-Ras-induced tumorigenicity and overproduction of urokinase-type plasminogen activator: involvement of metalloproteases. Oncogene *18*, 4718-4725.

Allal, C., Favre, G., Couderc, B., Salicio, S., Sixou, S., Hamilton, A.D., Sebti, S.M., Lajoie-Mazenc, I. and Pradines, A. (2000) RhoA prenylation is required for promotion of cell growth and transformation and cytoskeleton organization but not for induction of serum response element transcription. J. Biol. Chem. *275*, 31001-31008.

Aspenstrom, P., Fransson, A. and Saras, J. (2004) Rho GTPases have diverse effects on the organization of the actin filament system. Biochem. J. *377*, 327-337.

Assoian, R.K. (1997) Anchorage-dependent cell cycle progression. J. Cell Biol. *136*, 1-4.

Ballin, M., Gomez, D.E., Sinha, C.C. and Thorgeirsson, U.P. (1988) Ras oncogene mediated induction of a 92 kDa metalloproteinase; strong correlation with the malignant phenotype. Biochem. Biophys. Res. Commun. *154*, 832-838.

Bar-Sagi, D. and Feramisco, J.R. (1986) Induction of membrane ruffling and fluid-phase pinocytosis in quiescent fibroblasts by ras proteins. Science *233*, 1061-1068.

Bar-Sagi, D. and Hall, A. (2000) Ras and Rho GTPases: a family reunion. Cell *103*, 227-238.

Benitah, S.A., Valeron, P.F., van Aelst, L., Marshall, C.J. and Lacal, J.C. (2004) Rho GTPases in human cancer: an unresolved link to upstream and downstream transcriptional regulation. Biochim. Biophys. Acta *1705*, 121-132.

Bernards, A. and Settleman, J. (2004) GAP control: regulating the regulators of small GTPases. Trends Cell Biol. *14*, 377-385.

Bernhard, E.J., Gruber, S.B. and Muschel, R.J. (1994) Direct evidence linking expression of matrix metalloproteinase 9 (92-kDa gelatinase/collagenase) to the metastatic phenotype in transformed rat embryo cells. Proc. Natl. Acad. Sci. U.S.A. *91*, 4293-4297.

Bishop, A.L. and Hall, A. (2000) Rho GTPases and their effector proteins. Biochem. J. *348* (Pt2), 241-255.

Boerner, J.L., Danielsen, A., McManus, M.J. and Maihle, N.J. (2001) Activation of Rho is required for ligand-independent oncogenic signaling by a mutant epidermal growth factor receptor. J. Biol. Chem. *276*, 3691-3695.

Boettner, B. and Van Aelst, L. (2002) The role of Rho GTPases in disease development. Gene *286*, 155-174.

Bondy, G.P., Wilson, S. and Chambers, A.F. (1985) Experimental metastatic ability of H-ras-transformed NIH3T3 cells. Cancer Res. *45*, 6005-6009.

Caloca, M.J., Zugaza, J.L., Matallanas, D., Crespo, P. and Bustelo, X.R. (2003) Vav mediates Ras stimulation by direct activation of the GDP/GTP exchange factor Ras GRP1. EMBO J. *22*, 3326-3336.

Cantor, S.B., Urano, T. and Feig, L.A. (1995) Identification and characterization of Ral-binding protein 1, a potential downstream target of Ral GTPases. Mol. Cell. Biol. *15*, 4578-4584.

Chang, M.Y., Jan, M.S., Won, S.J. and Liu, H.S. (1998) Ha-rasVal12 oncogene increases susceptibility of NIH/3T3 cells to lovastatin. Biochem. Biophys. Res. Commun. *248*, 62-68.

Chang, M.Y., Won, S.J., Yang, B.C., Jan, M.S. and Liu, H.S. (1999) Selective activation of Ha-ras (val12) oncogene increases susceptibility of NIH/3T3 cells to TNF-alpha. Exp. Cell Res. *248*, 589-598.

Chen, J.C., Zhuang, S., Nguyen, T.H., Boss, G.R. and Pilz, R.B. (2003) Oncogenic Ras leads to Rho activation by activating the mitogen-activated protein kinase pathway and decreasing Rho-GTPase-activating protein activity. J. Biol. Chem. *278*, 2807-2818.

Cheng, T.L., Symons, M. and Jou, T.S. (2004) Regulation of anoikis by Cdc42 and Rac1. Exp. Cell Res. *295*, 497-511.

Chin, L., Tam, A., Pomerantz, J., Wong, M., Holash, J., Bardeesy, N., Shen, Q., O'Hagan, R., Pantginis, J., Zhou, H., Horner, J.W. 2nd, Cordon-Cardo, C., Yancopoulos, G.D. and DePinho, R.A. (1999) Essential role for oncogenic Ras in tumour maintenance. Nature *400*, 468-472.

Cobellis, G., Missero, C. and Di Lauro, R. (1998) Concomitant activation of MEK-1 and Rac-1 increases the proliferative potential of thyroid epithelial cells, without affecting their differentiation. Oncogene *17*, 2047-2057.

Coleman, M.L., Marshall, C.J. and Olson, M.F. (2004) RAS and RHO GTPases in G1-phase cell-cycle regulation. Nat. Rev. Mol. Cell Biol. *5*, 355-366.

Colicelli, J. (2004) Human RAS superfamily proteins and related GTPases, Sci. STKE *2004*, RE13.

Cox, A.D. and Der, C.J. (2002) Ras family signaling: therapeutic targeting. Cancer Biol. Ther. *1*, 599-606.

Cox, A.D. and Der, C.J. (2003) The dark side of Ras: regulation of apoptosis. Oncogene *22*, 8999-9006.

Dajee, M., Lazarov, M., Zhang, J.Y., Cai, T., Green, C.L., Russell, A.J., Marinkovich, M.P., Tao, S., Lin, Q., Kubo, Y. and Khavari, P.A. (2003) NF-kappaB blockade and oncogenic Ras trigger invasive human epidermal neoplasia. Nature *421*, 639-643.

Dermardirossian, C. and Bokoch, G.M. (2005) GDIs: central regulatory molecules in Rho GTPase activation, Trends Cell Biol. *15*, 356-363.

Donald, S., Hill, K., Lecureuil, C., Barnouin, R., Krugmann, S., John Coadwell, W., Andrews, S.R., Walker, S.A., Hawkins, P.T., Stephens, L.R. and Welch, H.C. (2004) P-Rex2, a new guanine-nucleotide exchange factor for Rac. FEBS Lett. *572*, 172-176.

Downward, J. (1998a) Mechanisms and consequences of activation of protein kinase B/Akt. Curr. Opin. Cell Biol. *10*, 262-267.

Downward, J. (1998b) Ras signalling and apoptosis. Curr. Opin. Genet. Dev. *8*, 49-54.

Downward, J. (2003) Targeting RAS signalling pathways in cancer therapy. Nat. Rev. Cancer *3*, 11-22.

Du, J., Jiang, B., Coffey, R.J. and Barnard, J. (2004) Raf and RhoA cooperate to transform intestinal epithelial cells and induce growth resistance to transforming growth factor beta. Mol. Cancer Res. *2*, 233-241.

Etienne-Manneville, S. and Hall, A. (2002) Rho GTPases in cell biology. Nature *420*, 629-635.

Feig, L.A. (1999) Tools of the trade: use of dominant-inhibitory mutants of Ras-Family GtPases. Nat. Cell Biol. *1*, E25-E27.

Feig, L.A. (2003) Ral-GTPases: approaching their 15 minutes of fame. Trends Cell Biol. *13*, 419-425.

Feig, L.A., Urano, T. and Cantor, S. (1996) Evidence for a Ras/Ral signaling cascade. Trends Biochem. Sci. *21*, 438-441.

Feig, L.A. and Buchsbaum, R.J. (2002) Cell signaling: life or death decisions of ras proteins. Curr. Biol. *12*, R259-R261.

Fisher, G.H., Wellen, S.L., Klimstra, D., Lenczowski, J.M., Tichelaar, J.W., Lizak, M.J., Whitsett, J.A., Koretsky, A. and Varmus, H.E. (2001) Induction and apoptotic regression of lung adenocarcinomas by regulation of a K-Ras transgene in the presence and absence of tumor suppressor genes. Genes Dev. *15*, 3249-3262.

Fischer, R.S., Zheng, Y. and Quinlan, M.P. (1998) Rac1 and extracellularly regulated kinase activation are sufficient for E1A-dependent cooperative transformation of primary epithelial cells, but progression can only be modulated by E1A or Rac1. Cell Growth Differ. *9*, 209-221.

Fleming, I.N., Gray, A. and Downes, C.P. (2000) Regulation of the Rac1-specific exchange factor Tiam1 involves both phosphoinositide 3-kinase-dependent and -independent components. Biochem. J. *351*, 173-182.

Foster, R., Hu, K.Q., Lu, Y., Nolan, K.M., Thissen, J. and Settleman, J. (1996) Identification of a novel human Rho protein with unusual properties: GTPase deficiency and in vivo farnesylation. Mol. Cell. Biol. *16*, 2689-2699.

Frisch, S.M. and Ruoslahti, E. (1997) Integrins and anoikis. Curr. Opin. Cell Biol. *9*, 701-706.

Gildea, J.J., Harding, M.A., Seraj, M.J., Gulding, K.M. and Theodorescu, D. (2002) The role of Ral A in epidermal growth factor receptor-regulated cell motility. Cancer Res. *62*, 982-985.

Gille, H. and Downward, J. (1999) Multiple ras effector pathways contribute to G (1) cell cycle progression. J. Biol. Chem. *274*, 22033-22040.

Gingras, M.C., Jarolim, L., Finch, J., Bowden, G.T., Wright, J.A. and Greenberg, A.H. (1990) Transient alterations in the expression of protease and extracellular matrix genes during metastatic lung colonization by H-ras-transformed 10T1/2 fibroblasts. Cancer Res. *50*, 4061-4066.

Gjoerup, O., Lukas, J., Bartek, J. and Willumsen, B.M. (1998) Rac and Cdc42 are potent stimulators of E2F-dependent transcription capable of promoting retinoblastoma susceptibility gene product hyperphosphorylation. J. Biol. Chem. *273*, 18812-18818.

Good, D.J., Polverini, P.J., Rastinejad, F., Le Beau, M.M., Lemons, R.S., Frazier, W.A. and Bouck, N.P. (1990) A tumor suppressor-dependent inhibitor of angiogenesis is immunologically and functionally indistinguishable from a fragment of thrombospondin. Proc. Natl. Acad. Sci. U.S.A. *87*, 6624-6628.

Guasch, R.M., Scambler, P., Jones, G.E. and Ridley, A.J. (1998) RhoE regulates actin cytoskeleton organization and cell migration. Mol. Cell. Biol. *18*, 4761-4771.

Hall, A. (1998) Rho GTPases and the actin cytoskeleton. Science *279*, 509-514.

Han, J., Luby-Phelps, K., Das, B., Shu, X., Xia, Y., Mosteller, R.D., Krishna, U.M., Falck, J.R., White, M.A. and Broek, D. (1998) Role of substrates and products of PI 3-kinase in regulating activation of Rac-related guanosine triphosphatases by Vav. Science *279*, 558-560.

Hanahan, D. and Folkman, J. (1996) Patterns and emerging mechanisms of the angiogenic switch during tumorigenesis. Cell *86*, 353-364.

Hanahan, D. and Weinberg, R.A. (2000) The hallmarks of cancer. Cell *100*, 57-70.

Hansen, S.H., Zegers, M.M., Woodrow, M., Rodriguez-Viciana, P., Chardin, P., Mostov, K.E. and McMahon, M. (2000) Induced expression of Rnd3 is associated with transformation of polarized epithelial cells by the Raf-MEK-extracellular signal-regulated kinase pathway. Mol. Cell. Biol. *20*, 9364-9375.

Hart, M.J., Eva, A., Evans, T., Aaronson, S.A. and Cerione, R.A. (1991) Catalysis of guanine nucleotide exchange on the CDC42Hs protein by the dbl oncogene product. Nature *354*, 311-314.

Herrmann, C. (2003) Ras-effector interactions: after one decade. Curr. Opin. Struct. Biol. *13*, 122-129.

Hitomi, M. and Stacey, D.W. (2001) Ras-dependent cell cycle commitment during G2 phase. FEBS Lett. *490*, 123-131.

Hordijk, P.L., ten Klooster, J.P., van der Kammen, R.A., Michiels, F., Oomen, L.C. and Collard, J.G. (1997) Inhibition of invasion of epithelial cells by Tiam1-Rac signaling. Science *278*, 1464-1466.

Hu, Q., Klippel, A., Muslin, A.J., Fantl, W.J. and Williams, L.T. (1995) Ras-dependent induction of cellular responses by constitutively active phosphatidylinositol-3 kinase. Science *268*, 100-102.

Hu, W., Bellone, C.J. and Baldassare, J.J. (1999) RhoA stimulates p27 (Kip) degradation through its regulation of cyclin E/CDK2 activity. J. Biol. Chem. *274*, 3396-3401.

Ingram, D.A., Hiatt, K., King, A.J., Fisher, L., Shivakumar, R., Derstine, C., Wenning, M.J., Diaz, B., Travers, J.B., Hood, A., Marshall, M., Williams, D.A. and Clapp, D.W. (2001) Hyperactivation of p21(ras) and the hematopoietic-specific Rho GTPase, Rac2, cooperate to alter the proliferation of neurofibromin-deficient mast cells in vivo and in vitro. J. Exp. Med. *194*, 57-69.

Innocenti, M., Tenca, P., Frittoli, E., Faretta, M., Tocchetti, A., Di Fiore, P.P. and Scita, G. (2002) Mechanisms through which Sos-1 coordinates the activation of Ras and Rac. J. Cell Biol. *156*, 125-136.

Irani, K., Xia, Y., Zweier, J.L., Sollott, S.J., Der, C.J., Fearon, E.R., Sundaresan, M., Finkel, T. and Goldschmidt-Clermont, P.J. (1997) Mitogenic signaling mediated by oxidants in Ras-transformed fibroblasts. Science *275*, 1649-1652.

Izawa, I., Amano, M., Chihara, K., Yamamoto, T. and Kaibuchi, K. (1998) Possible involvement of the inactivation of the Rho-Rho-kinase pathway in oncogenic Ras-induced transformation. Oncogene *17*, 2863-2871.

Jaffe, A.B., Aspenstrom, P. and Hall, A. (2004) Human CNK1 acts as a scaffold protein, linking Rho and Ras signal transduction pathways. Mol. Cell. Biol. *24*, 1736-1746.

Joneson, T. and Bar-Sagi, D. (1999) Suppression of Ras-induced apoptosis by the Rac GTPase. Mol. Cell. Biol. *19*, 5892-5901.

Joyce, D., Bouzahzah, B., Fu, M., Albanese, C., D'Amico, M., Steer, J., Klein, J.U., Lee, R.J., Segall, J.E., Westwick, J.K., Der, C.J. and Pestell, R.G. (1999) Integration of Rac-dependent regulation of cyclin D1 transcription through a nuclear factor-kappaB-dependent pathway. J. Biol. Chem. *274*, 25245-25249.

Joyce, P.L. and Cox, A.D. (2003) Rac1 and Rac3 are targets for geranylgeranyltransferase I inhibitor-mediated inhibition of signaling, transformation, and membrane ruffling. Cancer Res. *63*, 7959-7967.

Jullien-Flores, V., Dorseuil, O., Romero, F., Letourneur, F., Saragosti, S., Berger, R., Tavitian, A., Gacon, G. and Camonis, J.H. (1995) Bridging Ral GTPase to Rho pathways. RLIP76, a Ral effector with CDC42/Rac GTPase-activating protein activity. J. Biol. Chem. *270*, 22473-22477.

Karnoub, A.E., Symons, M., Campbell, S.L. and Der, C.J. (2004) Molecular basis for Rho GTPase signaling specificity. Breast Cancer Res. Treat. *84*, 61-71.

Kawasaki, Y., Senda, T., Ishidate, T., Koyama, R., Morishita, T., Iwayama, Y., Higuchi, O. and Akiyama, T. (2000) Asef, a link between the tumor suppressor APC and G-protein signaling. Science *289*, 1194-1197.

Kawasaki, Y., Sato, R. and Akiyama, T. (2003) Mutated APC and Asef are involved in the migration of colorectal tumour cells. Nat. Cell Biol. *5*, 211-215.

Keely, P.J., Rusyn, E.V., Cox, A.D. and Parise, L.V. (1999) R-Ras signals through specific integrin alpha cytoplasmic domains to promote migration and invasion of breast epithelial cells. J. Cell Biol. *145*, 1077-1088.

Khosravi-Far, R., Solski, P.A., Clark, G.J., Kinch, M.S. and Der, C.J. (1995) Activation of Rac1, RhoA, and mitogen-activated protein kinases is required for Ras transformation. Mol. Cell. Biol. *15*, 6443-6453.

Khwaja, A., Rodriguez-Viciana, P., Wennstrom, S., Warne, P.H. and Downward, J. (1997) Matrix adhesion and Ras transformation both activate a phosphoinositide 3-OH kinase and protein kinase B/Akt cellular survival pathway. EMBO J. *16*, 2783-2793.

Lambert, J.M., Lambert, Q.T., Reuther, G.W., Malliri, A., Siderovski, D.P., Sondek, J., Collard, J.G. and Der, C.J. (2002) Tiam1 mediates Ras activation of Rac by a PI(3)K-independent mechanism. Nat. Cell Biol. *4*, 621-625.

Lauffenburger, D.A. and Horwitz, A.F. (1996) Cell migration: a physically integrated molecular process. Cell *84*, 359-369.

Lebowitz, P.F., Du, W. and Prendergast, G.C. (1997) Prenylation of RhoB is required for its cell transforming function but not its ability to activate serum response element-dependent transcription. J. Biol. Chem. *272*, 16093-16095.

Li, J. and Smithgall, T.E. (1998) Fibroblast transformation by Fps/Fes tyrosine kinases requires Ras, Rac, and Cdc42 and induces extracellular signal-regulated and c-Jun N-terminal kinase activation. J. Biol. Chem. *273*, 13828-13834.

Lin, R., Bagrodia, S., Cerione, R. and Manor, D. (1997) A novel Cdc42Hs mutant induces cellular transformation. Curr. Biol. *7*, 794-797.

Liu, A.X., Rane, N., Liu, J.P. and Prendergast, G.C. (2001a) RhoB is dispensable for mouse development, but it modifies susceptibility to tumor formation as well as cell adhesion and growth factor signaling in transformed cells. Mol. Cell. Biol. *21*, 6906-6912.

Liu, R., Li, B. and Qiu, M. (2001b) Elevated superoxide production by active H-ras enhances human lung WI-38VA-13 cell proliferation, migration and resistance to TNF-alpha. Oncogene *20*, 1486-1496.

Liu, Z.J., Snyder, R., Soma, A., Shirakawa, T., Ziober, B.L., Fairman, R.M., Herlyn, M. and Velazquez, O.C. (2003) VEGF-A and alphaVbeta3 integrin synergistically rescue angiogenesis via N-Ras and PI3-K signaling in human microvascular endothelial cells. FASEB J. *17*, 1931-1933.

Malliri, A., van der Kammen, R.A., Clark, K., van der Valk, M., Michiels, F. and Collard, J.G. (2002) Mice deficient in the Rac activator Tiam1 are resistant to Ras-induced skin tumours. Nature *417*, 867-871.

Malumbres, M. and Barbacid, M. (2003) RAS oncogenes: the first 30 years. Nat. Rev. Cancer *3*, 459-465.

Martin, C.B., Mahon, G.M., Klinger, M.B., Kay, R.J., Symons, M., Der, C.J. and Whitehead, I.P. (2001) The thrombin receptor, PAR-1, causes transformation by activation of Rho-mediated signaling pathways. Oncogene *20*, 1953-1963.

Mayo, M.W. and Baldwin, A.S. (2000) The transcription factor NF-kappaB: control of oncogenesis and cancer therapy resistance. Biochim. Biophys. Acta *1470*, M55-M62.

McFall, A., Ulku, A., Lambert, Q.T., Kusa, A., Rogers-Graham, K. and Der, C.J. (2001) Oncogenic Ras blocks anoikis by activation of a novel effector pathway independent of phosphatidylinositol 3-kinase. Mol. Cell. Biol. *21*, 5488-5499.

Montaner, S., Sodhi, A., Servitja, J.M., Ramsdell, A.K., Barac, A., Sawai, E.T. and Gutkind, J.S. (2004) The small GTPase Rac1 links the Kaposi sarcoma-associated herpesvirus vGPCR to cytokine secretion and paracrine neoplasia. Blood *104*, 2903-2911.

Moon, S.Y. and Zheng, Y. (2003) Rho GTPase-activating proteins in cell regulation. Trends Cell Biol. *13*, 13-22.

Muller, S.M., Okan, E. and Jones, P. (2000) Regulation of urokinase receptor transcription by Ras- and Rho-family GTPases. Biochem. Biophys. Res. Commun. *270*, 892-898.

Murphy, G.A., Solski, P.A., Jillian, S.A., Perez de la Ossa, P., D'Eustachio, P., Der, C.J. and Rush, M.G. (1999) Cellular functions of TC10, a Rho family GTPase: regulation of morphology, signal transduction and cell growth. Oncogene *18*, 3831-3845.

Muschel, R.J., Williams, J.E., Lowy, D.R. and Liotta, L.A. (1985) Harvey ras induction of metastatic potential depends upon oncogene activation and the type of recipient cell. Am. J. Pathol. *121*, 1-8.

Nimnual, A.S., Yatsula, B.A. and Bar-Sagi, D. (1998) Coupling of Ras and Rac guanosine triphosphatases through the Ras exchanger Sos. Science *279*, 560-563.

Nobes, C.D., Lauritzen, I., Mattei, M.G., Paris, S., Hall, A. and Chardin, P. (1998) A new member of the Rho family, Rnd1, promotes disassembly of actin filament structures and loss of cell adhesion. J. Cell Biol. *141*, 187-197.

Nur, E.K.A., Zhang, A., Keenan, S.M., Wang, X.I., Seraj, J., Satoh, T., Meiners, S. and Welsh, W.J. (2005) Requirement of activated Cdc42-associated kinase for survival of v-Ras-transformed mammalian cells. Mol. Cancer Res. *3*, 297-305.

Nur, E.K.M.S., Kamal, J.M., Qureshi, M.M. and Maruta, H. (1999) The CDC42-specific inhibitor derived from ACK-1 blocks v-Ha-Ras-induced transformation. Oncogene *18*, 7787-7793.

Okan, E., Drewett, V., Shaw, P.E. and Jones, P. (2001) The small-GTPase RalA activates transcription of the urokinase plasminogen activator receptor (uPAR) gene via an AP1-dependent mechanism. Oncogene *20*, 1816-1824.

Olson, M.F., Ashworth, A. and Hall, A. (1995) An essential role for Rho, Rac, and Cdc42 GTPases in cell cycle progression through G1. Science *269*, 1270-1272.

Olson, M.F., Paterson, H.F. and Marshall, C.J. (1998) Signals from Ras and Rho GTPases interact to regulate expression of p21Waf1/Cip1. Nature *394*, 295-299.

Page, K., Li, J., Hodge, J.A., Liu, P.T., Vanden Hoek, T.L., Becker, L.B., Pestell, R.G., Rosner, M.R. and Hershenson, M.B. (1999) Characterization of a Rac1 signaling pathway to cyclin D(1) expression in airway smooth muscle cells. J. Biol. Chem. *274*, 22065-22071.

Palmby, T.R., Abe, K., Karnoub, A.E. and Der, C.J. (2004) Vav transformation requires activation of multiple GTPases and regulation of gene expression. Mol. Cancer Res. *2*, 702-711.

Pardee, A.B. (1989) G1 events and regulation of cell proliferation. Science *246*, 603-608.

Park, S.H. and Weinberg, R.A. (1995) A putative effector of Ral has homology to Rho/Rac GTPase activating proteins. Oncogene *11*, 2349-2355.

Plaumann, M., Seitz, S., Frege, R., Estevez-Schwarz, L. and Scherneck, S. (2003) Analysis of DLC-1 expression in human breast cancer. J. Cancer Res. Clin. Oncol. *129*, 349-354.

Pozzatti, R., Muschel, R., Williams, J., Padmanabhan, R., Howard, B., Liotta, L. and Khoury, G. (1986) Primary rat embryo cells transformed by one or two oncogenes show different metastatic potentials. Science *232*, 223-227.

Prendergast, G.C., Khosravi-Far, R., Solski, P.A., Kurzawa, H., Lebowitz, P.F. and Der, C.J. (1995) Critical role of Rho in cell transformation by oncogenic Ras. Oncogene *10*, 2289-2296.

Pruitt, K. and Der, C.J. (2001) Ras and Rho regulation of the cell cycle and oncogenesis. Cancer Lett. *171*, 1-10.

Pruitt, K., Pruitt, W.M., Bilter, G.K., Westwick, J.K. and Der, C.J. (2002) Raf-independent deregulation of p38 and JNK mitogen-activated protein kinases are critical for Ras transformation. J. Biol. Chem. *277*, 31808-31817.

Qiu, R.G., Chen, J., Kirn, D., McCormick, F. and Symons, M. (1995a) An essential role for Rac in Ras transformation. Nature *374*, 457-459.

Qiu, R.G., Chen, J., McCormick, F. and Symons, M. (1995b) A role for Rho in Ras transformation. Proc. Natl. Acad. Sci. U.S.A. *92*, 11781-11785.

Qiu, R.G., Abo, A., McCormick, F. and Symons, M. (1997) Cdc42 regulates anchorage-independent growth and is necessary for Ras transformation. Mol. Cell. Biol. *17*, 3449-3458.

Quilliam, L.A., Rebhun, J.F. and Castro, A.F. (2002) A growing family of guanine nucleotide exchange factors is responsible for activation of Ras-family GTPases. Prog. Nucleic Acid Res. Mol. Biol. *71*, 391-444.

Rabizadeh, S., Xavier, R.J., Ishiguro, K., Bernabeortiz, J., Lopez-Ilasaca, M., Khokhlatchev, A., Mollahan, P., Pfeifer, G.P., Avruch, J. and Seed, B. (2004) The scaffold protein CNK1 interacts with the tumor suppressor RASSF1A and augments RASSF1A-induced cell death. J. Biol. Chem. *279*, 29247-29254.

Rak, J., Mitsuhashi, Y., Bayko, L., Filmus, J., Shirasawa, S., Sasazuki, T. and Kerbel, R.S. (1995) Mutant ras oncogenes upregulate VEGF/VPF expression: implications for induction and inhibition of tumor angiogenesis. Cancer Res. *55*, 4575-4580.

Rak, J., Yu, J.L., Klement, G. and Kerbel, R.S. (2000) Oncogenes and angiogenesis: signaling three-dimensional tumor growth. J. Invest. Dermatol. Symp. Proc. *5*, 24-33.

Renshaw, M.W., Lea-Chou, E. and Wang, J.Y. (1996) Rac is required for v-Abl tyrosine kinase to activate mitogenesis. Curr. Biol. *6*, 76-83.

Repasky, G.A., Chenette, E.J. and Der, C.J. (2004) Renewing the conspiracy theory debate: does Raf function alone to mediate Ras oncogenesis? Trends Cell Biol. *14*, 639-647.

Ribatti, D., Alessandri, G., Vacca, A., Iurlaro, M. and Ponzoni, M. (1998) Human neuroblastoma cells produce extracellular matrix-degrading enzymes, induce endothelial cell proliferation and are angiogenic in vivo. Int. J. Cancer *77*, 449-454.

Ridley, A.J. (2001) Rho family proteins: coordinating cell responses. Trends Cell Biol. *11*, 471-477.

Ridley, A.J. (2004) Rho proteins and cancer. Breast Cancer Res. Treat. *84*, 13-19.

Ridley, A.J., Paterson, H.F., Johnston, C.L., Diekmann, D. and Hall, A. (1992) The small GTP-binding protein rac regulates growth factor-induced membrane ruffling. Cell *70*, 401-410.

Riento, K., Guasch, R.M., Garg, R., Jin, B. and Ridley, A.J. (2003) RhoE binds to ROCK I and inhibits downstream signaling. Mol. Cell. Biol. *23*, 4219-4229.

Rodrigues, G.A., Park, M. and Schlessinger, J. (1997) Activation of the JNK pathway is essential for transformation by the Met oncogene. EMBO J. *16*, 2634-2645.

Rodriguez-Manzaneque, J.C., Lane, T.F., Ortega, M.A., Hynes, R.O., Lawler, J. and Iruela-Arispe, M.L. (2001) Thrombospondin-1 suppresses spontaneous tumor growth and inhibits activation of matrix metalloproteinase-9 and mobilization of vascular endothelial growth factor. Proc. Natl. Acad. Sci. U.S.A. *98*, 12485-12490.

Rodriguez-Viciana, P., Warne, P.H., Khwaja, A., Marte, B.M., Pappin, D., Das, P., Waterfield, M.D., Ridley, A. and Downward, J. (1997) Role of phosphoinositide 3-OH kinase in cell transformation and control of the actin cytoskeleton by Ras. Cell *89*, 457-467.

Rossman, K.L., Der, C.J. and Sondek, J. (2005) GEF means go: turning on RHO GTPases with guanine nucleotide-exchange factors. Nat. Rev. Mol. Cell Biol. *6*, 167-180.

Roux, P., Gauthier-Rouviere, C., Doucet-Brutin, S. and Fort, P. (1997) The small GTPases Cdc42Hs, Rac1 and RhoG delineate Raf-independent pathways that cooperate to transform NIH3T3 Cells. Curr. Biol. *7*, 629-637.

Sahai, E., Olson, M.F. and Marshall, C.J. (2001) Cross-talk between Ras and Rho signalling pathways in transformation favours proliferation and increased motility. EMBO J. *20*, 755-766.

Sahai, E. and Marshall, C.J. (2002) RHO-GTPases and cancer. Nat. Rev. Cancer *2*, 133-142.

Samuels, Y., Wang, Z., Bardelli, A., Silliman, N., Ptak, J., Szabo, S., Yan, H., Gazdar, A., Powell, S.M., Riggins, G.J., Willson, J.K., Markowitz, S., Kinzler, K.W., Vogelstein, B. and Velculescu, V.E. (2004) High frequency of mutations of the PIK3CA gene in human cancers. Science *304*, 554.

Sander, E.E., van Delft, S., ten Klooster, J.P., Reid, T., van der Kammen, R.A., Michiels, F. and Collard, J.G. (1998) Matrix-dependent Tiam1/Rac signaling in epithelial cells promotes either cell-cell adhesion or cell migration and is regulated by phosphatidylinositol 3-kinase. J. Cell Biol. *143*, 1385-1398.

Schmitz, A.A., Govek, E.E., Bottner, B. and Van Aelst, L. (2000) Rho GTPases: signaling, migration, and invasion. Exp. Cell Res. *261*, 1-12.

Scita, G., Nordstrom, J., Carbone, R., Tenca, P., Giardina, G., Gutkind, S., Bjarnegard, M., Betsholtz, C. and Di Fiore, P.P. (1999) EPS8 and E3B1 transduce signals from Ras to Rac. Nature *401*, 290-293.

Scita, G., Tenca, P., Areces, L.B., Tocchetti, A., Frittoli, E., Giardina, G., Ponzanelli, I., Sini, P., Innocenti, M. and Di Fiore, P.P. (2001) An effector region in Eps8 is responsible for the activation of the Rac-specific GEF activity of Sos-1 and for the proper localization of the Rac-based actin-polymerizing machine. J. Cell Biol. *154*, 1031-1044.

Sebti, S.M. and Hamilton, A.D. (2000) Farnesyltransferase and geranylgeranyltransferase I inhibitors and cancer therapy: lessons from mechanism and bench-to-bedside translational studies. Oncogene *19*, 6584-6593.

Sebti, S.M. and Der, C.J. (2003) Opinion: Searching for the elusive targets of farnesyltransferase inhibitors. Nat. Rev. Cancer *3*, 945-951.

Settleman, J., Narasimhan, V., Foster, L.C. and Weinberg, R.A. (1992) Molecular cloning of cDNAs encoding the GAP-associated protein p190: implications for a signaling pathway from ras to the nucleus. Cell *69*, 539-549.

Shaw, L.M., Rabinovitz, I., Wang, H.H., Toker, A. and Mercurio, A.M. (1997) Activation of phosphoinositide 3-OH kinase by the alpha6beta4 integrin promotes carcinoma invasion. Cell *91*, 949-960.

Shaw, R.J., Paez, J.G., Curto, M., Yaktine, A., Pruitt, W.M., Saotome, I., O'Bryan, J.P., Gupta, V., Ratner, N., Der, C.J., Jacks, T. and McClatchey, A.I. (2001) The Nf2 tumor suppressor, merlin, functions in Rac-dependent signaling. Dev. Cell *1*, 63-72.

Shields, J.M., Pruitt, K., McFall, A., Shaub, A. and Der, C.J. (2000) Understanding Ras: 'it ain't over'til it's over'. Trends Cell Biol. *10*, 147-154.

Skorski, T., Wlodarski, P., Daheron, L., Salomoni, P., Nieborowska-Skorska, M., Majewski, M., Wasik, M. and Calabretta, B. (1998) BCR/ABL-mediated leukemogenesis requires the activity of the small GTP-binding protein Rac. Proc. Natl. Acad. Sci. U.S.A. *95*, 11858-11862.

Solski, P.A., Helms, W., Keely, P.J., Su, L. and Der, C.J. (2002) RhoA biological activity is dependent on prenylation but independent of specific isoprenoid modification. Cell Growth Differ. *13*, 363-373.

Sulciner, D.J., Irani, K., Yu, Z.X., Ferrans, V.J., Goldschmidt-Clermont, P. and Finkel, T. (1996) rac1 regulates a cytokine-stimulated, redox-dependent pathway necessary for NF-kappaB activation. Mol. Cell. Biol. *16*, 7115-7121.

Takai, Y., Sasaki, T. and Matozaki, T. (2001) Small GTP-binding proteins. Physiol. Rev. *81*, 153-208.

Taylor, S.J. and Shalloway, D. (1996) Cell cycle-dependent activation of Ras. Curr. Biol. *6*, 1621-1627.

Teramoto, H., Malek, R.L., Behbahani, B., Castellone, M.D., Lee, N.H. and Gutkind, J.S. (2003) Identification of H-Ras, RhoA, Rac1 and Cdc42 responsive genes. Oncogene *22*, 2689-2697.

Thiery, J.P. (2003) Epithelial-mesenchymal transitions in development and pathologies. Curr. Opin. Cell Biol. *15*, 740-746.

Tseng, Y.S., Tzeng, C.C., Chiu, A.W., Lin, C.H., Won, S.J., Wu, I.C. and Liu, H.S. (2003) Ha-ras overexpression mediated cell apoptosis in the presence of 5-fluorouracil. Exp. Cell Res. *288*, 403-414.

Tsubakimoto, K., Matsumoto, K., Abe, H., Ishii, J., Amano, M., Kaibuchi, K. and Endo, T. (1999) Small GTPase RhoD suppresses cell migration and cytokinesis. Oncogene *18*, 2431-2440.

Ulku, A.S. and Der, C.J. (2003) Ras signaling, deregulation of gene expression and oncogenesis. Cancer Treat Res. *115*, 189-208.

Urich, M., Senften, M., Shaw, P.E. and Ballmer-Hofer, K. (1997) A role for the small GTPase Rac in polyomavirus middle-T antigen-mediated activation of the serum response element and in cell transformation. Oncogene *14*, 1235-1241.

Valster, A.H., Hepler, P.K. and Chernoff, J. (2000) Plant GTPases: the Rhos in bloom. Trends Cell Biol. *10*, 141-146.

Van Aelst, L. and D'Souza-Schorey, C. (1997) Rho GTPases and signaling networks. Genes Dev. *11*, 2295-2322.

Van Aelst, L. and Symons, M. (2002) Role of Rho family GTPases in epithelial morphogenesis. Genes Dev. *16*, 1032-1054.

Verheij, M., Bose, R., Lin, X.H., Yao, B., Jarvis, W.D., Grant, S., Birrer, M.J., Szabo, E., Zon, L.I., Kyriakis, J.M., Haimovitz-Friedman, A., Fuks, Z. and Kolesnick, R.N. (1996) Requirement for ceramide-initiated SAPK/JNK signalling in stress-induced apoptosis. Nature *380*, 75-79.

Vetter, I.R. and Wittinghofer, A. (2001) The guanine nucleotide-binding switch in three dimensions. Science *294*, 1299-1304.

Vial, E., Sahai, E. and Marshall, C.J. (2003) ERK-MAPK signaling coordinately regulates activity of Rac1 and RhoA for tumor cell motility. Cancer Cell *4*, 67-79.

Villalonga, P., Guasch, R.M., Riento, K. and Ridley, A.J. (2004) RhoE inhibits cell cycle progression and Ras-induced transformation. Mol. Cell. Biol. *24*, 7829-7840.

Walsh, A.B., Dhanasekaran, M., Bar-Sagi, D. and Kumar, C.C. (1997) SCH 51344-induced reversal of RAS-transformation is accompanied by the specific inhibition of the RAS and RAC-dependent cell morphology pathway. Oncogene *15*, 2553-2560.

Walsh, A.B. and Bar-Sagi, D. (2001) Differential activation of the Rac pathway by Ha-Ras and K-Ras. J. Biol. Chem. *276*, 15609-15615.

Ward, Y., Wang, W., Woodhouse, E., Linnoila, I., Liotta, L. and Kelly, K. (2001) Signal pathways which promote invasion and metastasis: critical and distinct contributions of extracellular signal-regulated kinase and Ral-specific guanine exchange factor pathways. Mol. Cell. Biol. *21*, 5958-5969.

Watnick, R.S., Cheng, Y.N., Rangarajan, A., Ince, T.A. and Weinberg, R.A. (2003) Ras modulates Myc activity to repress thrombospondin-1 expression and increase tumor angiogenesis. Cancer Cell *3*, 219-231.

Webb, C.P., Van Aelst, L., Wigler, M.H. and Woude, G.F. (1998) Signaling pathways in Ras-mediated tumorigenicity and metastasis. Proc. Natl. Acad. Sci. U.S.A. *95*, 8773-8778.

Weber, J.D., Hu, W., Jefcoat, S.C. Jr., Raben, D.M. and Baldassare, J.J. (1997) Ras-stimulated extracellular signal-related kinase 1 and RhoA activities coordinate platelet-derived growth factor-induced G1 progression through the independent regulation of cyclin D1 and p27. J. Biol. Chem. *272*, 32966-32971.

Welch, H.C., Coadwell, W.J., Ellson, C.D., Ferguson, G.J., Andrews, S.R., Erdjument-Bromage, H., Tempst, P., Hawkins, P.T. and Stephens, L.R. (2002) P-Rex1, a PtdIns(3,4,5)P3- and Gbetagamma-regulated guanine-nucleotide exchange factor for Rac. Cell *108*, 809-821.

Wellbrock, C., Karasarides, M. and Marais, R. (2004) The RAF proteins take centre stage. Nat. Rev. Mol. Cell. Biol. *5*, 875-885.

Welsh, C.F. (2004) Rho GTPases as key transducers of proliferative signals in g1 cell cycle regulation. Breast Cancer Res. Treat. *84*, 33-42.

Welsh, C.F. and Assoian, R.K. (2000) A growing role for Rho family GTPases as intermediaries in growth factor- and adhesion-dependent cell cycle progression. Biochim. Biophys. Acta *1471*, M21-M29.

Welsh, C.F., Roovers, K., Villanueva, J., Liu, Y., Schwartz, M.A. and Assoian, R.K. (2001) Timing of cyclin D1 expression within G1 phase is controlled by Rho. Nat. Cell Biol. *3*, 950-957.

Wennerberg, K., Ellerbroek, S.M., Liu, R.Y., Karnoub, A.E., Burridge, K. and Der, C.J. (2002) RhoG signals in parallel with Rac1 and Cdc42. J. Biol. Chem. *277*, 47810-47817.

Wennerberg, K., Forget, M.A., Ellerbroek, S.M., Arthur, W.T., Burridge, K., Settleman, J., Der, C.J. and Hansen, S.H. (2003) Rnd proteins function as RhoA antagonists by activating p190 RhoGAP. Curr. Biol. *13*, 1106-1115.

Wennerberg, K. and Der, C.J. (2004) Rho-family GTPases: it's not only Rac and Rho (and I like it). J. Cell Sci. *117*, 1301-1312.

Wennerberg, K., Rossman, K.L. and Der, C.J. (2005) The Ras superfamily at a glance. J. Cell Sci. *118*, 843-846.

Westermarck, J. and Kahari, V.M. (1999) Regulation of matrix metalloproteinase expression in tumor invasion. FASEB J. *13*, 781-792.

Westwick, J.K., Lambert, Q.T., Clark, G.J., Symons, M., Van Aelst, L., Pestell, R.G. and Der, C.J. (1997) Rac regulation of transformation, gene expression, and actin organization by multiple, PAK-independent pathways. Mol. Cell. Biol. *17*, 1324-1335.

Wheeler, A.P. and Ridley, A.J. (2004) Why three Rho proteins? RhoA, RhoB, RhoC, and cell motility. Exp. Cell Res. *301*, 43-49.

Wherlock, M. and Mellor, H. (2002) The Rho GTPase family: a Racs to Wrchs story. J. Cell Sci. *115*, 239-240.

Yamamoto, M., Marui, N., Sakai, T., Morii, N., Kozaki, S., Ikai, K., Imamura, S. and Narumiya, S. (1993) ADP-ribosylation of the rhoA gene product by botulinum C3 exoenzyme causes Swiss 3T3 cells to accumulate in the G1 phase of the cell cycle. Oncogene *8*, 1449-1455.

Yang, J.Q., Zhao, W., Duan, H., Robbins, M.E., Buettner, G.R., Oberley, L.W. and Domann, F.E. (2001) v-Ha-RaS oncogene upregulates the 92-kDa type IV collagenase (MMP-9) gene by increasing cellular superoxide production and activating NF-kappaB. Free Radic. Biol. Med. *31*, 520-529.

Yuan, B.Z., Miller, M.J., Keck, C.L., Zimonjic, D.B., Thorgeirsson, S.S. and Popescu, N.C. (1998) Cloning, characterization, and chromosomal localization of a gene frequently deleted in human liver cancer (DLC-1) homologous to rat RhoGAP. Cancer Res. *58*, 2196-2199.

Yuan, B.Z., Zhou, X., Durkin, M.E., Zimonjic, D.B., Gumundsdottir, K., Eyfjord, J.E., Thorgeirsson, S.S. and Popescu, N.C. (2003) DLC-1 gene inhibits human breast cancer cell growth and in vivo tumorigenicity. Oncogene *22*, 445-450.

Yuan, B.Z., Jefferson, A.M., Baldwin, K.T., Thorgeirsson, S.S., Popescu, N.C. and Reynolds, S.H. (2004) DLC-1 operates as a tumor suppressor gene in human non-small cell lung carcinomas. Oncogene *23*, 1405-1411.

Zohn, I.E., Symons, M., Chrzanowska-Wodnicka, M., Westwick, J.K. and Der, C.J. (1998a) Mas oncogene signaling and transformation require the small GTP-binding protein Rac. Mol. Cell. Biol. *18*, 1225-1235.

Zohn, I.E., Klinger, M., Karp, X., Kirk, H., Symons, M., Chrzanowska-Wodnicka, M., Der, C.J. and Kay, R.J. (2000) G2A is an oncogenic G protein-coupled receptor. Oncogene *19*, 3866-3877.

Zohn, I.M., Campbell, S.L., Khosravi-Far, R., Rossman, K.L. and Der, C.J. (1998b) Rho family proteins and Ras transformation: the RHOad less traveled gets congested. Oncogene *17*, 1415-1438.

Zondag, G.C., Evers, E.E., ten Klooster, J.P., Janssen, L., van der Kammen, R.A. and Collard, J.G. (2000) Oncogenic Ras downregulates Rac activity, which leads to increased Rho activity and epithelial-mesenchymal transition. J. Cell Biol. *149*, 775-782.

CHAPTER 8

GLOBAL EFFECTS OF RAS SIGNALING ON THE GENETIC PROGRAM IN MAMMALIAN CELLS

REINHOLD SCHÄFER, OLEG I. TCHERNITSA, AND CHRISTINE SERS

Laboratories of Molecular Tumor Pathology and of Functional Genomics,Charité, Universitätsmedizin Berlin, Schumannstr. 20/21, D-10117 Berlin, Germany. Phone: +49 30 450 536 072; Fax: +49 30 450 536 909; e-mail: reinhold.schaefer@charite.de

Abstract: While target genes deregulated by oncogenic Ras signaling were identified in a low-throughput manner in the past, recent expression profiling technologies have permitted transcriptome-wide screening approaches. The number of known deregulated genes has increased substantially, as has the proportion of candidate genes involved in various aspects of Ras-induced transformation. This chapter summarizes the results of gene expression profiling studies based on microarrays and advanced cDNA subtraction procedures

Keywords: expression profiling, transcriptome, Ras targets, transcriptional stimulation, transcriptional repression, pathway interference

Abbreviations: SSH: suppressive subtraction hybridization; RDA: representational difference analysis; MEF: mouse embryonic fibroblasts; ROSE: rat ovarian surface epithelial cells; HOSE: Human ovarian surface epithelial cells; NSCLC: non-small cell lung carcinoma; MMTV: mouse mammary tumor virus; EGF: epidermal growth factor; VEGF: vascular endothelial growth factor; EST: expressed sequence tag; TGF: transforming growth factor; HMG protein: high mobility group protein; PHRP: parathyroid hormone-related protein; MEC: mammary epithelial cells; hTERT: human telomerase reverse transcriptase; EMT: epithelial-mesenchymal transition; siRNA: small interfering RNA; shRNA: short hairpin RNA; RNAi: RNA interference

1. INTRODUCTION

Ras signaling impinges on cytoplasmic and nuclear targets via its numerous downstream effectors and pathways. Ras signaling pathways regulate the organization of the actin cytoskeleton, cell cycle progression and gene expression [for review see (Campbell et al., 1998; Shields et al., 2000)]. Ras signaling activity is linked to the activation of the transcription factors Ets-1, Ets-2, Elk1, NFκB, SRF, c-Fos,

C. Der (ed.), RAS Family GTPases, 169–198.

c-Jun, c-Myc and E2F. Their decisive role in the transformation process has been assessed in studies based on specific antisense sequences targeting these factors, on dominant negative mutants and gene-deficient cells [for review see (Murphy et al., 2002)]. In NIH/3T3 cells, oncogenic transformation mediated by Ras signaling can be blocked, if c-Myc is depleted by incorporation of antisense sequences. Similarly, dominant negative mutants of c-Fos and c-Jun which normally dimerize to form the functionally active transcriptional activator (AP-1) complexes, of Ets-1, Ets-2 can inhibit Ras transformation. Mouse embryo fibroblasts deficient in c-Jun are refractory toward Ras-mediated transformation. The inhibition of NFκB induced apoptosis in Ras-expressing cells, thereby preventing transformed phenotypes. The role of transcription factors downstream of the Ras signaling cascades is not restricted to cultured cells, in which the conversion to transformed phenotypes as well as the reversion to the normal phenotype can be easily assessed. For example, mice deficient in c-Fos carrying the mutated H-Ras transgene were prone to hyperkeratosis and papilloma formation following treatment of their skin with tumor promoters. However, these mice did not develop progressively growing tumors, in contrast to Ras-transgenic mice carrying wild-type c-Fos.

In view of the large number of transcription factors stimulated by Ras signaling activity, a highly complex set of target genes responding to oncogenic Ras-mediated stimuli has to be expected. Putative Ras-responsive genes may be predicted using in silico analysis of common cis-regulatory elements in their promoters. For example, transcription factor binding sites upstream of genes regulated by phosphatidylinositol 3-kinase and Mek/Erk signaling pathways were recently identified in this way and scrutinized for their functional involvement (Tullai et al., 2004). Although algorithms for predicting cis-regulatory elements in evolutionarily conserved non-coding sequences on a genome-wide scale are currently being developed (Down and Hubbard, 2004), the predictions tend to be hampered by the high incidence of false-positives (Kielbasa et al., 2001). At the single gene level, however, transcription factor binding sites have been identified in "classical" Ras-responsive genes and their functional relevance determined experimentally. The conventional approach for identifying targets of Ras signaling is based on studies aiming to find out if and how Ras-mediated signal activation impinges on genes whose products were known to contribute to cellular transformation. These genes can transform nontumorigenic recipient cells of rodent origin following forced expression. Alternatively, their experimental ablation can reduce or abolish transformed phenotypes in Ras-transformed cells. Well-studied examples for up-regulated genes are TGF-α, cyclin D1, cyclooxygenase-2 (Cox-2), matrix metalloproteases (MMPs) and vascular endothelial growth factor (VEGF) [reviewed in (Murphy et al., 2002)]. Down-regulated genes comprise tropomyosins, collagens, fibronectin, lysyl oxidase, plasminogen activator inhibitor I, α-actin, gelsolin, vinculin, cytochrome b, cytochrome c oxidase subunit II as well as NADH dehydrogenases 1 and 4 [reviewed in (Schäfer, 1994; Murphy et al., 2002). In phenotypic revertants derived from Ras-transformed cells, the expression of the down-regulated genes was at least partially restored. This suggested that the reversible loss of expression indeed contributed to transformed phenotypes (Schäfer, 1994).

2. HIGHLY PARALLEL ANALYSIS OF GENE EXPRESSION FOR THE IDENTIFICATION OF RAS-RESPONSIVE GENES

On a genome-wide scale, the identification of Ras-responsive genes become feasible with the advent of differential display RT-PCR (Liang and Pardee, 1992; Liang et al., 1992; Jo et al., 2001) and improved techniques for cDNA subtraction (suppressive subtraction hybridization, SSH) (Diatchenko et al., 1996) and representational difference analysis (RDA) (Lisitsyn and Wigler, 1993; Hubank and Schatz, 1994). Typically, cDNA subtraction library screening involves the 50- to 100-fold enrichment of differentially expressed cDNA sequences from mixtures of cDNAs isolated from a normal precursor and a Ras-transformed derivative (Hajnal et al., 1994). The refined subtraction methods permit an approx. 1000-fold enrichment of preferentially expressed sequences in a given biological sample such as a cell line or tumor tissue and are thus superior to the classical cDNA subtraction techniques (Diatchenko et al., 1996; Zuber et al., 2000). The cDNA subtraction approaches provide a largely unbiased recovery of differentially expressed sequences, perhaps only limited by the efficiencies of integrated PCR steps and the ligation of tester cDNA to adapter sequences. Microarray technology (Chung et al., 2002; Churchill, 2002; Petricoin III et al., 2002) relies on pre-selected sets of genes represented by partial cDNA sequences (Habets et al., 2001) or oligonucleotides (Schulze et al., 2001). Microarrays typically represent partial transcriptomes comprising annotated genes and expressed sequence tags available at the time of production. Only recently, whole genome arrays have become a reality, however, the fraction of sequences without known function is still significant. The microarray approach does not require extensive sequencing work like RDA and SSH, however, a thorough statistical evaluation for interpreting the results of a microarray experiment is essential and alternate microarray platforms may yield different results (Tan et al., 2003). Discrepancies may be explained by different procedures for normalization, scoring of perfect sequence matches, different sequence annotation and by divergent technical peculiarities of the microarray experiment. Notably, because of the massively parallel type of analysis, results from microarrays are only partially validated by independent methods such as northern blotting, RT-PCR or western blotting. Usually, some 10-20 genes are analyzed in this way and the results are extrapolated to the entire data set. Nevertheless, microarray analysis has become a premier technology for identifying molecular signatures related to tumor classification and even clinical parameters such as prognosis, not recognizable by alternative diagnostic techniques [for example see (Golub et al., 1999; van de Vijver et al., 2002)]. Accordingly, many groups have published microarray data related to Ras-induced transformation and tumorigenesis (see below).

A number of Ras-responsive genes have been recovered by differential display or cDNA subtraction library screening and characterized in some detail. Liang and colleagues identified 12 genes that were differentially expressed in rat embryo fibroblasts transformed by oncogenic Ras and a dominant negative form of the tumor suppressor p53. The α-chemokine Mob-1 was found up-regulated (Liang et al., 1994) and rCop-1, a member the family of cysteine-rich growth regulators (CCN) down-regulated in transformed cells. Ectopic expression of rCop-1 results in cell death in transformed cells (Zhang et al., 1998). McMahon and colleagues used the differential display technique to identify heparin binding (HB)-EGF as a factor stimulated by Raf-signaling in Ras-transformed cells (Mccarthy et al., 1995). Another member of the EGF family of growth factors, epiregulin, was identified by Baba et al. as a gene up-regulated by mutated K-ras in HCT116 human colon carcinoma cells. Forced expression in a HCT 116 variant that lacks the transforming K-ras allele enhanced tumor formation in nude mice (Baba et al., 2000). Furthermore, we have identified a number of genes down-regulated in Ras-transformed rat 208F fibroblasts and reexpressed in phenotypic revertants derived from them (Hajnal et al., 1993a; Hajnal et al., 1993b). One of them, designated H-rev107 (Hajnal et al., 1994), qualifies as a class II tumor suppressor in that it is frequently down-regulated in cancer cells (Sers et al., 2002) and forced expression in Ras-transformed cells can suppress transformed phenotypes in vitro and in vivo (Sers et al., 1997).

To contrast expression profiles of normal and Ras-transformed cells on a transcriptome-wide scale several groups have used established cell culture models of cellular transformation, albeit under different experimental conditions. Some authors have compared actively proliferating cells in order to avoid transcriptional alterations due to growth factor stimulation that may mimic true oncogene-induced changes (Zuber et al., 2000). Others have taken advantage of inducible expression systems (Schulze et al., 2001) to minimize the identification of secondary effects which might occur as a consequence of forced expression rather than representing the primary transcriptional response. However, it is difficult to distinguish between these two aspects, since Ras proteins by definition do not directly impinge on transcription. Both fibroblasts and epithelial cells of either human (Schulze et al., 2001; Gadal et al., 2003; Liu et al., 2004; Grill et al., 2004), mouse (Desai et al., 2002; Jechlinger et al., 2003; Teramoto et al., 2003; Liu et al., 2004; Brem et al., 2001; Vasseur et al., 2003; Rajasekhar et al., 2003; Huang et al., 2003) or rat origin (Ordway et al., 2004; Zuber et al., 2000; Tchernitsa et al., 2004; Yoon et al., 2002) have been used for profiling Ras-induced changes. In summary, it is difficult to compare the results of independent studies, because of the use of different technical platforms and validation experiments. Rather than discussing and comparing single target genes identified in individual studies, this review focuses mainly on the magnitude of transcriptional alterations, the major conclusions with respect to functional implications drawn by the authors and on the strategies for dealing with complex data sets in order to assess biological significance. Ideally, the results of all studies would be stored in a common database with individual expression profiles and experimental parameters etc. presented in a uniform format as to enable easy comparisons. Such a database would need to be constantly controlled and updated, because sequence information, gene annotations, proven functions of gene products will vary considerably. The complexity and heterogeneity of data sets, however, preclude manual entries to such a database. Computational methods are being developed in various places and may be available in the not too distant future.

3. COMPARING NORMAL, PRE-NEOPLASTIC CELLS WITH RAS-TRANSFORMED DERIVATIVES

3.1 Fibroblasts

A summary of the different approaches for identifying Ras-responsive genes is presented in table 1. To obtain a general catalogue of transcriptional alterations related to permanent Ras oncogene expression, we chose the preneoplastic rat 208F fibroblast line (Quade, 1979) and the malignant, HRAS-transformed derivative FE-8 (Griegel et al., 1986). These cell lines exhibit a near-diploid karyotype without gross numerical or structural chromosomal aberrations. Thus, the cell system appears to be an excellent model for transcriptome profiling. The oncogene is permanently expressed in FE-8 cells. Therefore, *de novo* chromosomal abnormalities which frequently accumulate in cells expressing inducible Ras do not mask the transcriptional changes governed by oncogenic signaling (Denko et al., 1994; Denko et al., 1995). 208F cells exhibit a low incidence of spontaneous transformation, while FE-8 cells form rapidly progressing tumors in nude mice or new-born rats (Sers et al., 1997). To contrast differential gene expression profiles in these two cell lines, we have used subtractive suppression hybridization (SSH), a PCR-based cDNA subtraction technique (Diatchenko et al., 1996). This approach permitted unbiased recovery of sequences and did not rely on pre-selected sets of genes present on available microarrays. The SSH method also allows an equal representation of high and low abundance gene sequences. Differentially expressed sequences were recovered from forward and reverse subtraction using tester cDNA derived from 208F cells and driver cDNA from FE-8 cells, respectively. The expression pattern was verified by conventional and reverse northern analysis.

Sequence and expression analysis of more than 1,200 subtracted cDNA fragments revealed transcriptional stimulation or repression of 104 ESTs, 45 novel sequences and 244 known genes in HRAS-transformed cells compared with normal cells (Zuber et al., 2000). In the meantime, the proportion of annotated genes and ESTs has changed due to the sequencing of human, man and rat genomes. The study identified known Ras targets such as the metastasis glycoprotein CD44, the transcription factor Fra-1, the α-chemokine Mob-1, metalloproteinases Mmp-1 and Mmp-3 and myosin regulatory light chain. We also recovered sequences known to be transcriptionally repressed in Ras-transformed cells. These genes included α-actin, collagen α-1, entactin/nidogen, fibronectin, the TGF-β-stimulated gene TSC-36,

Table 1. Summary of global expression studies

Cell types (normal precursors)	Species	Oncogene(s) mediating transformed phenotypes	Mode of introduction into cells	Technology used for expression profiling	No. of deregulated genes/ESTs	References
Fibroblasts						
208F fibroblast line	rat	HRAS (G12V) KRAS (C12V) NRAS (G12D)	Stable transfection	SSH	393	(Zuber et al., 2000)
NIH/3T3 Preneoplastic cell line	mouse	H-ras (V12) RhoA Rac1 Cdc42	Stable transfection	cDNA, 19,117 unique genes	62 (1,184)[1]	(Teramoto et al., 2003)
embryonic fibroblasts	mouse	Ras(V12)/E1A	Retroviral infection	MGu74Av2 Genechip (Affymetrix)	815[2]	(Vasseur et al., 2003)
Quiescent embryonic fibroblasts	mouse	H-ras(L61) v-Myc E2F1/2/3	Adenoviral infection	Mu11K (Affymetrix)	50[3] 250[4]	(Huang et al., 2003)
208F fibroblast line	rat	Ras(V12) v-fos Dnmt1	Stable transfection	Rat Genome U34A (Affymetrix)	65[5] (500–>650)	(Ordway et al., 2004)
Epithelial cells						
MCF-10A mammary epithelial line	human	Raf	Retroviral infection, conditional expression	6,000 chip (Affymetrix)	120	(Schulze et al., 2001)
Tumors (epithelial)	mouse	MMTV-c-Myc MMTV-neu MMTV-Ha-ras MMTV-PyMT WAP-SV40T	Transgenes targeted to mammary gland	GEM 18.7K (Incyte) NCI 2.7K	387 (930)[1]	(Desai et al., 2002)
RK3E, Kidney epithelial cells, stably transfected with E1A	rat	H-ras (V12)	transfection	Rat UniGene clones (4,608)	124	(Yoon et al., 2002)
EpH4 Mammary epithelial line	mouse	MEK1 const. active	transfection	1.2 K mouse cDNA array (Clontech)	19	(Pinkas and Leder, 2002)
EpH4 Mammary epithelial line	mouse	H-ras (V12) H-ras effector domain mutants (S35, C40)	Stable transfection	Mu11K (Affymetrix)	294	(Jechlinger et al., 2003)
MCF-12A Mammary epithelial cell line	human	Mek1 const. Active mutant	Adenoviral infection (24h)	U95A Genechip (Affymetrix)	735	(Grill et al., 2004)

HOSE642-1 Ovarian surface epithelial line	human	Raf	Retroviral infection, conditional expression	cDNA array Sanger human 10K 1.2.1	135	(Schulze et al., 2004)
IOSE-80 IOSE-29 Ovarian surface epithelial cells transfected with SV40 large T and small t antigens	human	(hTERT) Ha-ras (V12) Ki-ras (V12)	Retroviral infection	U133 (Affymetrix)	32[6]	(Liu et al., 2004)
ROSE199 Ovarian surface epithelials cell line	rat	KRAS (C12V)	Stable transfection	SSH	367	(Tchernitsa et al., 2004)
Other cell types						
PB-3c Mast cell line	mouse	v-H-ras	Stable transfection	Mu11KsubA and B (Affymetrix)	75[7] (400)	(Brem et al., 2001)
MCF-7 Carcinoma cell line	human	Ha-ras (V12)	Stable transfection	CDNA subtraction VGID™	166	(Gadal et al., 2003)
Primary glial cell cultures expressing ALV receptor	mouse	K-ras (G12D) Akt const. active	Retroviral infection	U74Av2	66 (705)[8]	(Rajasekhar et al., 2003)

[1] Total number of deregulated genes in all cell pairs tested shown in brackets;
[2] alterations induced by cooperating Ras and E1A oncogenes;
[3] number of genes that provide Ras discrimination;
[4] number of genes used for Myc discrimination;
[5] intersect of genes differentially expressed in cells transfected with any of three genes as indicated, total number of genes identified in all cell lines in brackets;
[6] number of genes dregulated in 3/6 infected cell lines;
[7] intersect of 3 cell lines, total number of genes in brackets;
[8] mRNA alterations, changes in polysomal RNA in brackets.

lysyl oxidase, smooth muscle myosin light chain and NADH dehydrogenase. Other genes known to be deregulated in conjunction with oncogenic Ras signaling were not recovered such as glucose transporter, the proto-oncogene Myc, PDGF receptor and Cox-2. This may be due to an intrinsic bias of the cDNA subtraction procedure, insufficient sequencing of subtracted sequences or be explained by the cell types and experimental conditions used.

HRAS-transformed cells expressed significantly elevated levels of target genes capable of mediating invasion and metastasis. These encode the laminin receptor, Mmp-1 (collagenase), Mmp-3 (stromelysin-1), Mmp-10 (stromelysin-2) and CD44. While the level of target up-regulation exceeded a 30-fold increase for many genes, there was also a close link between Ras signaling and target gene down-regulation. Many genes appeared to be down-regulated to mRNA levels not detectable on northern blots. Factors responsible for anti-proliferative, anti-invasive and anti-angiogenic functions were repressed, including the genes encoding syndecan-2, tissue inhibitor of metalloproteases-2 (Timp-2), lysyl oxidase, thrombospondin-1, protein kinase A II, the myristoylated alanine-rich C-kinase substrate (Marcks) and Gas-1, growth-arrest specific protein. The changes reflect the capacity of Ras-expressing cells for acquiring as well as maintaining the malignant state.

To determine to what extent the Raf/Mek/Erk signaling cascade downstream of Ras affects transcriptional patterns and phenotypes, we treated FE-8 cells with the Mek inhibitor PD98059 (Dudley et al., 1995). On inhibitor treatment for 48 h, FE-8 cells showed a more normal morphology and a reduced capacity of anchorage independent growth. Levels of $p21^{ras}$ remained unaltered, but the levels of phospho-p44/42 MAPK were reduced. We found 61 Ras-responsive genes (36 down-regulated, 25 up-regulated) to be sensitive to blocking Mek, while 116 targets analyzed were unaffected. This suggested that non-Erk signaling pathway significantly impinge on transcriptional control in Ras-transformed cells. Since individual Ras isoforms are preferentially mutated in different types of cancer (Bos, 1989), we sought to determine how KRAS and NRAS expression would affect the target genes identified in our initial screening. We generated transformants after transfection of KRAS (C12V) and NRAS (G12D) into normal 208F cells, respectively. Approx. 90% of 237 sequences sensitive to HRAS-mediated transformation showed a similar expression pattern in cells transformed by the two other Ras isoforms. The transcript levels related to 26 targets exhibited distinct differences indicating that isoform-specific effects on transcription do exist [cf. (Malumbres, 1998)].

In addition to Ras, other members of the superfamily of small GTP-binding proteins such as RhoA, Rac1 and Cdc42 can play a role in signaling to the nucleus and cell growth control (Bar-Sagi and Hall, 2000). To understand the contribution of genes regulated by Ras and Rho GTPases to their complex biological effects, Teramoto et al. investigated the global gene expression patterns induced by activated forms of H-Ras, RhoA, Rac1 and Cdc42 in NIH/3T3 cells by interrogating cDNA microarrays representing 19,117 unique elements (Teramoto et al., 2003). The authors identified 1,184 genes up- or down-regulated by at least twofold. Hierarchical cluster analysis revealed the existence of common and unique patterns of gene regulation, related to H-Ras (V12), RhoA, Rac1 and Cdc42 activation. For example, H-Ras (V12) up-regulated osteopontin and Akt 1, and H-Ras and RhoA stimulated cyclin G1, cyclin-dependent kinase 8, cyclin A2 and HMGI-C, while Rac1 and Cdc42 up-regulated extracellular matrix and cell adhesion proteins such as α-actinin 4, procollagen type I and V and neuropilin. Furthermore, H-Ras (V12) down-regulated 52 genes by >8-fold, while RhoA, Rac1 and Cdc42 down-regulated three genes only. Overall, 7.3% of all genes examined were affected by expression of GTPases.

Vasseur and colleagues established a catalog of mouse genes whose expression is altered in mouse embryonic fibroblasts (MEFs) transduced with the pBabe-ras^{V12}/E1A retroviral vector directing expression of mutated Ras and E1A (Vasseur et al., 2003). Mutated Ras and E1A cooperating oncogenes transform primary rodent fibroblasts (Land et al., 1983) and abandon Ras-induced senescence (Serrano et al., 1997). Among the approximately 12,000 genes and ESTs analyzed, 815 showed altered expression in ras^{V12}/E1A-transformed fibroblasts, compared to control fibroblasts. Among the genes with known function, 202 were up-regulated and 410 were down-regulated. About one half of genes encoding transcription factors, signaling proteins, membrane proteins, channels or apoptosis-related proteins was up-regulated, whereas the other half was down-regulated. Interestingly, genes encoding structural proteins, secretory proteins, receptors, extracellular matrix components, and cytosolic proteins were preferentially down-regulated. Genes encoding DNA-associated proteins (involved in DNA replication and reparation) and cell growth-related proteins were up-regulated. The authors conclude that this gene expression pattern may explain, at least in part, the behavior of transformed cells in that down-regulation of structural proteins, extracellular matrix components, secretory proteins and receptors is consistent with the morphological transformation and re-organization of cytoskeletal architecture. Similarly, they suggested that up-regulation of cell growth-related proteins and DNA-associated proteins is consistent with the accelerated growth of transformed cells. Unexpectedly, proteases and inhibitors of proteases as well as all eight angiogenic factors present on the array were down-regulated in transformed fibroblasts despite their general up-regulation in cancers. The authors speculate that, in human cancers, proteases, protease inhibitors and angiogenic factors could be regulated through a mechanism disconnected from Ras activation. However, other groups have shown that proteases capable of mediating invasive properties and pro-angiogenic factors are up-regulated in rodent and human cells expressing Ras or downstream effectors (Zuber et al., 2000; Breier et al., 2002; White et al., 1997; Grill et al., 2004). Overall, the high number of deregulated genes identified by Vasseur et al. may reflect the combined actions of mutant Ras and the E1A oncogene which mediate both immortalization and neoplastic transformation. Cultured MEFs have been shown to resist oncogene-mediated transformation (Land et al., 1983). Oncogene "resistance" is associated with complex transcriptional alterations as shown by comparing the transcriptomes of fibroblasts refractory toward transformation by Ras with that of cells sensitive to oncogenesis (Tchernitsa et al., 1999).

In the publications discussed so far, catalogues of Ras-responsive, co-regulated genes were reported and their functional significance at least partially investigated (Zuber et al., 2000; Teramoto et al., 2003; Vasseur et al., 2003). Huang, Nevins and colleagues have studied the impact of oncogene expression on the transcriptome by a bioinformatic approach (Huang et al., 2003). They applied singular value decomposition methods based on Bayesian regression models developed for predicting relative probabilities of clinical outcomes (West et al., 2001) to cell culture models harboring oncogenes. Using complex data sets provided by microarray analysis, linear combinations of individual gene expression values were established that together constitute “metagenes”. Metagenes were then used to assess the relative probabilities of clinical prognosis for novel, unknown samples (West et al., 2001). In their recent paper, Huang et al. describe metagenes that have the capacity to classify and predict cellular phenotypes resulting from the deregulation of oncogenic pathways related to Ras, Myc and members of the E2F transcription factor family (Huang et al., 2003). Recombinant adenoviruses controlling expression of Ras, Myc or E2F proteins were used to infect quiescent mouse embryo fibroblasts. RNA was prepared 18 h after infection. In total, 55 experiments were performed for profiling expression patterns in these cells and their controls. Metagens were presented that discriminate between samples of cells infected with each different recombinant vector including Ras-, Myc- and individual E2Fs. The predictive ability of metagenes was then tested in a series of mammary tumors in transgenic mice with mouse mammary tumor virus enhancer-driven Ras-, Myc- or ErbB2 (HER-2/neu) expression. Each of the Ras tumors was correctly predicted and separated from ErbB2 tumors and normal mammary tissue, while Myc tumors were indistinguishable. The authors explained this by the finding that MMTV-Myc tumors in transgenic mice often have sustained mutations in the K-ras gene (D’Cruz et al., 2001), which imposes a Ras-related expression signature on the tumor cell population. In view of the strong anti-proliferative effects of forced Ras expression on embryo fibroblasts in the absence of an immortalizing oncogene (Land et al., 1986; Serrano et al., 1997), the expression profile underlying the metagene with predictive power for Ras-induced tumorigenesis appears to cover-up expression changes related to growth arrest, premature senescence and/or apoptosis. In conclusion, the gene expression phenotypes reported by Huang, Nevins et al. have the potential to characterize the complex genetic alterations that typify the neoplastic state *in vitro* and *in vivo*.

Tom Curran’s group have taken a microarray-based gene expression approach to compare differentially expressed genes in 208F fibroblasts transformed by the oncogenes c-fos, v-fos, ras (V12) or DNA (cytosine 5) methyltransferase 1 (Dnmt1) (Ordway et al., 2004). The authors report a large number of genes scored as increased or decreased by at least a factor of two (ranging between 300 and 450 up-regulated genes and >200 down-regulated genes). The large number of deviations in gene expression profiles is explained by clonal variation among cell lines and secondary changes occurring downstream of the oncogenic process. While the number of affected genes was similar for the cells transformed by v-fos, ras (V12) or Dnmt1, cells transformed by c-fos exhibited similar changes with respect to decreased genes, but less changes in up-regulated genes. Interestingly, the authors scored 14 genes as up-regulated in all four cell lines and 51 commonly down-regulated genes. The cohort of genes differentially expressed in all four transformation systems includes an over-representation of repressed genes, many of which have been functionally implicated in the suppression of transformation or tumorigenesis such as SSeCKS (the ortholog of human gravin) and Lot-1, BMP4 and N-cadherin *(cdh2)*. To exclude alterations due to clonality of cells, authors have analyzed a cell system harboring conditional v-fos. This allowed them to follow the time course of gene down-regulation and to assess the reversibility of alterations. Twenty-one out of 51 genes repressed in all stably transformed cell lines were conditionally repressed in the v-fos-mediated transformation/reversion. The four potential tumor suppressor genes are subject to epigenetic transcriptional repression in transformed cells. This indicates that inappropriate epigenetic transcription regulation may be a common route of Ras- and Fos-induced oncogenesis, and that cell transformation may model aspects of the epigenetic deregulation that often occurs in tumors (Table 1).

3.2 Epithelial Cells

A. Schulze, J. Downward and colleagues have analyzed the transcriptional program induced by Raf in serum-starved human MCF-10A mammary epithelial cells (Schulze et al., 2001). The activation of the Raf/MAP kinase pathway is a critical event in tumorigenesis induced by RAS and other oncogenes. The group used an inducible form of Raf for particularly identifying early transcriptional changes. Out of 6,000 genes represented on microarrays, more than 120 exhibited significant changes in mRNA level. Genes capable of promoting cell proliferation, invasiveness, and angiogenesis featured prominently. Further analysis focussed on one of the most strongly induced genes encoding growth factors of the EGF family, heparin-binding EGF-like growth factor (HB-EGF). Autocrine activation of the EGF receptor was shown to be responsible for the ability of Raf activation to protect MCF-10A cells from anoikis, a form of programmed cell death induced by detachment of cells from the extracellular matrix. The ability to survive under conditions without anchorage to the substratum provided by the environment is a critical component of the transformed phenotype.

Desai et al. have contrasted gene expression profiles of normal mammary glands and of mammary tumors induced by targeted transgenic overexpression of c-Myc, c-Neu, c-Ha-ras, polyoma middle T-antigen (PyMT) or simian virus 40 large T antigen (T-ag) (Desai et al., 2002). Overall, 930 genes were differentially expressed between the tumor models. Regardless of the transgenic tumor, genes for the glycolytic pathway including lactate dehydrogenase were highly induced. Increased expression of translation elongation factors and structural RNA genes reflect the accelerated metabolic rate in tumors as compared to the normal mammary tissue. Furthermore, cell cycle regulators, signaling receptors and effectors, downstream transcription factors as well as protein tyrosine phosphatases were induced in all tumors. Among the commonly repressed genes Desai et al. identified soluble protein tyrosine kinases. The authors pointed out that a proportion of "repressed" genes do not reflect a specific down-regulation in Ras-expressing cells but rather the diminished representation of non-epithelial tissue in the transgene-induced tumors. In addition to these common patterns, the authors identified oncogene-specific signatures. Tumors expressing Neu, Ras or PyMT oncogenes clustered tightly. Specific transcriptional alterations affected GTPase activating proteins (GAPs), E2F and cyclin D1. Most unique changes were found in the T-antigen-related gene cluster (>100 genes affected), while the Myc-cluster exhibited less, yet overlapping alterations. The authors concluded that comparative target gene identification may facilitate the development of lesion-specific therapeutics and pre-clinical testing. It also provides a solid basis for evaluating similarities and discrepancies between human breast cancer and mouse tumor models.

Rat kidney epithelial cells stably transfected with E1A were transformed by introduction of mutated H-ras. Expression profiling of normal versus Ras-transformed cells revealed deregulation of 124 target genes (Yoon et al., 2002). Known functions of Ras-responsive genes include cytoskeletal architecture, cell adhesion, signal transduction, apoptosis, transcriptional control, protein biosynthesis and metabolism. There was only little overlap with the expression profile of stable transfectants expressing the zinc finger transcription factor GLI1, which mediates Sonic hedgehog signaling during development (Platt et al., 1997) and is abundantly expressed in basal cell carcinoma (Dahmane et al., 1997) and other human cancers.

Another large set of genes deregulated in conjunction with Ras oncogene expression was described by the groups of H. Beug and N. Kraut (Jechlinger et al., 2003). Instead of using standard expression profiling procedures, the authors have adapted polysome-bound mRNA expression profiling to high density microarrays interrogating approx. 11,000 transcription units. Polysome-bound mRNA better reflects the protein level as compared to total RNA and permits to identify regulated genes controlled at the level of translation (Zong et al., 1999; Mikulits et al., 2000). In the initial screening, Jechlinger et al. identified 104 transcripts up-regulated at least 4-fold and 190 transcripts down-regulated in Ha-ras transformed EpH4 mouse mammary epithelial cells which have undergone epithelial-mesenchymal transition dependent on TGF-β. Overall, 75% of these genes were regulated exclusively at the transcriptional level, 18% showed exclusive regulation at the level of translation and 7% were regulated at both levels. The expression changes affected genes encoding cell adhesion proteins, extracellular matrix proteins, cell surface proteins, proteases and their inhibitors, signaling molecules, transcriptional regulators and metabolic enzymes. The authors describe an experimental strategy to reduce and prioritize the number of candidate genes involved in the distinct biological properties of Ras-transformed mammary epithelial cells. Based on the normal EpH4 cell line, derivatives were generated by introduction of oncogenes other than Ras (Fos, Bcl-2) and of Ras effector domain mutants that exhibit distinct aspects of epithelial cell plasticity relevant to cell migration, local invasion and metastasis. In addition, the effects of TGF-β and of a dominant-negative TGF-β receptor were assessed in a related metastatic colon carcinoma cell line. Cluster analysis of expression data obtained in appropriate pairs of EpH4 and transfectants revealed that smaller sets of deregulated genes were closely associated with proliferation, polarity, survival, trans-differentiation and invasive behavior.

A more limited set of genes was identified in EpH4 cells expressing constitutively activated MEK1 (Pinkas and Leder, 2002). Nineteen genes were found either up- or down-regulated relative to normal EpH4 cells. Although Mek-expressing cells do not undergo epithelial-mesenchymal transition in culture, Mek activity was sufficient to mediate tumorigenicity, invasiveness and metastasis.

Since MAPK (mitogen-activated protein kinase) pathways activated by Ras expression constitute major regulators of cellular transcriptional programs, Grill et al. analyzed the ERK1,2 (extracellular-signal-regulated kinase 1,2) transcriptome in a non-transformed MEC (mammary epithelial cell) line, MCF-12A (Grill et al., 2004). The cells were infected with a recombinant adenovirus encoding constitutively active MEK1 (MAPK/ERK kinase 1). The infection with the recombinant adenovirus controlling Mek1 expression induced morphological changes and DNA synthesis which were inhibited by the MEK1,2 inhibitor PD184352. Hierarchical clustering of data derived from seven time points over 24 h identified 430 and 305 co-ordinately up-regulated and down-regulated genes, respectively. The authors identified c-Myc binding sites in the promoters of most up-regulated genes. A total of 46 candidate effectors of the Raf/MEK/ERK1,2 pathway in MECs have also been described in the previous study of Raf-1-responsive genes in MCF-10A, a similar nontransformed mammary epithelial cell line (Schulze et al., 2001). The study by Grill and colleagues confirms the coordinate induction of multiple ErbB ligands. In addition, VEGF and PHRP (parathyroid hormone-related protein) were up-regulated by the Raf/Mek/Erk pathway. PHRP mediates humoral hypercalcaemia in progressed cancer patients. Inhibitor studies revealed that PHRP secretion is blocked by Mek inhibitors but not by ErbB inhibitors. The authors conclude that Mek1,2 inhibitors may be of therapeutic importance for treating advanced, PHRP-positive tumors.

Several groups have reported Ras-signaling related gene expression profiles in model systems for ovarian cancer. In their follow-up paper, J. Downward's group have extended the study of Raf-induced transcriptional alterations to human ovarian epithelial cells (HOSE642-1) (Schulze et al., 2004). They showed that the majority of Raf-induced transcriptional changes (135 significant alterations in total) are blocked in the presence of PD 98059, a pharmacological inhibitor of the Raf-substrate Mek. The functional importance of autocrine stimulation of the EGF receptor prompted the authors to assess the contribution of EGF signaling by Raf-induced EGF-related growth factors to the Raf transcriptional response. Treatment of cells with PD 168393, which inhibits EGFR by irreversibly binding to the kinase domain, revealed that about 50% of the transcriptional response to Raf-induction is dependent on EGFR function. EGFR-dependent transcriptional responses may be positive (up-regulation) and negative (down-regulation), while the set of EGFR-independent Raf-responsive genes were exclusively up-regulated.

To generate a model for ovarian cancer driven by oncogenic Ras signaling, Liu et al. have performed a cancer reconstruction experiment as has been described for human mesenchymal and epithelial cells (Hahn et al., 1999). Liu et al. introduced the catalytic subunit of human telomerase reverse transcriptase (hTERT), the SV40 early genomic region, and the oncogenic alleles of human HRAS or KRAS into human ovarian surface epithelial cells (HOSE) (Liu et al., 2004). Disruption of the p53 and Rb pathway by SV40 early genomic region and hTERT immortalized but did not transform HOSE cells. Following introduction of HRAS(V12) or KRAS(V12) into the immortalized cells, however, tumors were formed after subcutaneous injection into immuno-deficient mice. The peritoneal injection of the Ras-transformed HOSE cells produced undifferentiated carcinomas or malignant mixed Mullerian tumors and developed ascites; the tumor cells were focally positive for CA125 and mesothelin. Since disruptions of the p53, retinoblastoma (Rb), and RAS signaling pathways and activation of hTERT are common steps in the formation of human ovarian cancer, the tumorigenic HOSE cells provide a well-defined model, in which the global effects of oncogenic Ras expression can be assessed. Microarray analysis of transformed cells using Affymetrix U133A chips (probing $>27,000$ genes) revealed elevated expression of several cytokines, including interleukin-1β, IL-6, and IL-8, that are up-regulated by the NFκB pathway, which is known to contribute to the tumor growth of naturally occurring ovarian cancer cells. Overall, 23 genes were up-regulated and 9 genes were down-regulated 2.5-fold in at least 3 out of six cell lines analyzed. The elevated expression of IL-1β and IL-8 is of functional significance, because incubation with antibodies to either interleukin led to apoptosis in the Ras-transformed cells and ovarian cancer cells, but not in immortalized HOSE cells (Liu et al., 2004).

Following our previous work on normal and Ras-transformed fibroblasts, we also analyzed Ras-responsive expression profiles in an epithelial background. ROSE 199 is a spontaneously immortalized cell line derived from the continuous passage of primary rat ovarian surface epithelial cells. ROSE 199 cells express epithelial and mesenchymal characteristics and do not transform spontaneously in vitro. In dense cultures, the cells form multi-layers resembling histologically serous papillary cystadenomas of borderline malignancy (Adams and Auersperg, 1985). We generated a stable KRAS-transformed derivative of ROSE 199 cells, designated ROSE A2/5. As reported earlier, Ras-transformed ROSE cells represent a late stage in ovarian transformation (Auersperg et al., 1999). We recovered fragments of genes differentially expressed in normal ROSE 199 and KRAS-transformed A2/5 cells from two subtracted SSH libraries. Overall, we identified 192 differentially expressed genes, which could be classified with respect to known functional properties of their products, 140 expressed sequence tags without known function and 35 sequences without match in the public sequence databases (Tchernitsa et al., 2004). Of note, 44 sequences recovered by SSH were below the detection limit of reverse northern analysis. However, we assume that approx. 60% of these low-abundance transcripts are differentially expressed as well, as indicated by the proportion of recovered sequences, of which differential expression was verified by northern analysis. Stable KRAS-expression in the ROSE model of ovarian carcinoma stimulated transcription of genes capable of controlling cell signaling, gene activity, proteolysis, angiogenesis, invasion and metastasis. About 20 % of the transcriptional changes were independently identified in the fibroblast model described earlier (Zuber et al., 2000). The most prominent feature of KRAS-transformed ROSE cells is the down-regulation of a multitude of transcriptional targets, particularly of genes capable of adversely affecting growth factor-induced signal transduction, transformation and tumor progression (Table 2). Complete loss of mRNA expression indicates total functional loss. Interestingly, functional impairment in ovarian cancer has been previously shown for 10 of the down-regulated Ras-responsive genes (see references in table 2). The remaining negative growth regulators have been implicated in other types of cancers. Thus, their role in the neoplastic transformation and progression of the ovarian epithelium remains to be analyzed (Table 2).

Inhibition of Raf/Mek/Erk signaling by exposing A2/5 cells to PD98059 reverted the transcriptional alterations of 98 target genes (54% of all targets with known function) and caused a partial loss of transformed properties. Particularly, epithelial-mesenchymal transition (EMT) was reversed. To assess the role of PI3K-signaling on Ras-dependent transcriptional alterations, we treated A2/5 cells with the inhibitor LY 294002 (Vlahos et al., 1994). We chose an inhibitor concentration which was not cytotoxic to the cells and induced also a 50% reduction in anchorage-independent proliferation, but did not affect cell morphology and EMT. Only 27 up-regulated target genes were sensitive to blocking the PI3K-pathway. Nineteen of them were co-regulated by Mek-signaling as well. Down-regulated genes were not affected by LY 294002. The pathway-blocking experiments suggest that Ras-responsive genes are organized in signal-regulated transcriptional modules (SITMs) (Figure 1). Almost 50% of identified targets are regulated by Mek- and PI3K-independent pathways. Among them are several genes with crucial functions in neoplastic transformation. The inhibitor experiments suggest that they are not directly involved in EMT, but may play a significant role in invasion and metastasis (Figure 1).

3.3 Other Cell Types

DNA microarrays representing 11,000 genes were interrogated by Brem et al. using RNA prepared from normal mouse PB-3c mast cells and three independently derived v-Ha-ras-transformed derivatives (Brem et al., 2001). Stable Ha-ras transfectants induce tumor formation in vivo when implanted into mice. Such tumor cells are characterized by an autocrine interleukin-3 loop. The expression of about 400 genes was modulated in each tumor. A subset of 75 genes was shared and up- or down-regulated in all three lines. The genes present in this limited set possess functions related to tumorigenesis such as cell adhesion, signaling or transcriptional regulation. Apart from a number of ESTs the authors emphasize the down-regulation of four interferon-inducible genes in the tumor lines. Microarray analysis recapitulated some of previous findings by differential display PCR (Buess et al., 1999). The authors extrapolated their data to the complete mouse genome

Table 2. Parallel down-regulation of negative growth-regulators in KRAS- transformed rat ovarian surface epithelial cells

Down-regulated gene	Expression pattern in tumorigenic cells	Function related to neoplastic and malignant transformation	References
Genes known to be involved in ovarian cancer			
TGF-β II receptor	Decreased in progressed ovarian carcinoma	Impairing receptor function reduces Smad binding activity and growth inhibition by TGF-ß in carcinomas	(Evangelou et al., 2000; Paterson et al., 2001)
E-cadherin	Decreased in EMT and metastatic ovarian carcinoma	Establishes cell-cell adhesion; suppresses invasion	(Auersperg et al., 1999; Boyer et al., 2000)
P-cadherin	Co-expressed with E-cadherin in most epithelia	Exhibits redundant function like E-cadherin	(Lewis et al., 1994)
TIMP2	Down-regulated in epithelial cells overexpressing c-Ha-ras and c-erbB-2 oncogenes and in mucinous ovarian neoplasms of the borderline category	Reduces tumor growth and invasion	(Giunciuglio et al., 1995; Imren et al., 1996; Furuya et al., 2000)
Tropomyosins (multiple genes recovered)	Repressed in many transformed cells including ovarian carcinoma	Improve microfilament architecture and establish normal growth control in transformed cells	(Gimona et al., 1996; Alaiya et al., 1997)
Connexin 43	Down-regulated in transformed cells and adenocarcinomas in parallel to loss of gap junctional intercellular communication (GJIC)	Controls GJIC in normal ovarian epithelial cells	(Hanna et al., 1999; Umhauer et al., 2000)
Lot-1	Down-regulated by EGFR ligands	Encodes a nuclear zinc finger protein, candidate tumor suppressor	(Abdollahi et al., 1997; Abdollahi et al., 1999)
WT1	Repression by epigenetic mechanisms in papillary serous carcinoma of the peritoneum and in advanced ovarian carcinoma	Wilms tumor suppressor, suppresses RAS-induced transformation	(Luo et al., 1995; Schorge et al., 2000)
Sparc/osteonectin		Extracellular protein with tumor-suppressing activity in ovarian epithelial cells	(Mok et al., 1996)

thrombospondin-1	Down-regulated in Ras-transformed cells	Inhibits angiogenesis	(Zabrenetzky et al., 1994; Alvarez et al., 2001)
IP6K2 (Inositol hexakis-phosphate kinase 2)		Interferon-regulated gene, mediates growth suppressive and anti-apoptotic effects in ovarian carcinoma cells	(Morrison et al., 2001)
Candidate genes involved in malignant transformation of ovarian epithelium			
HTS1/ST5		p70 isoform suppresses activation of MAPK/ERK2 in response to EGF stimulation; suppresses transformed phenotype in mammalian cells	(Majidi et al., 2000)
Follistatin-related protein precursor (TSC-36)		Activin-binding protein positively regulated by TGF-ß and negatively controlled by AP-1; causes growth arrest in cancer cell lines and inhibits invasive properties of oncogene-transformed cells	(Johnston et al., 2000; Sumitomo et al., 2000)
Gas-1		Negatively controls serum-dependent transition from G0 to S-phase of the cell cycle	(Delsal et al., 1992; Delsal et al., 1994)
Collagens (multiple genes recovered)	Down-regulated in RAS-transformed cells	Cytoskeletal tumor suppressors	(Slack et al., 1992; Travers et al., 1996; Lombardi et al., 1990)
Alpha-parvin		Focal adhesion protein related to the α-actinin family, may mediate cell-matrix adhesion and act as a cytoskeletal suppressor protein similar to α-actinin	(Olski et al., 2001; Glück et al., 1993; Glück and Benzeev, 1994)
Lysyl oxidase	Down-regulated in tumors and oncogene-transformed cells	Targeting by anti-sense RNA results in transformation, regulates NFκB signaling	(Hajnal et al., 1993b; Giampuzzi et al., 2001; Jeay et al., 2003)
NO3/DAN	Candidate tumor suppressor	Not known	(Ozaki and Sakiyama, 1994)
ARPP-19	Highly homologous to the putative brain tumor suppressor gene C4-2	Not known	(Sehgal et al., 1997a; Sehgal et al., 1997b)

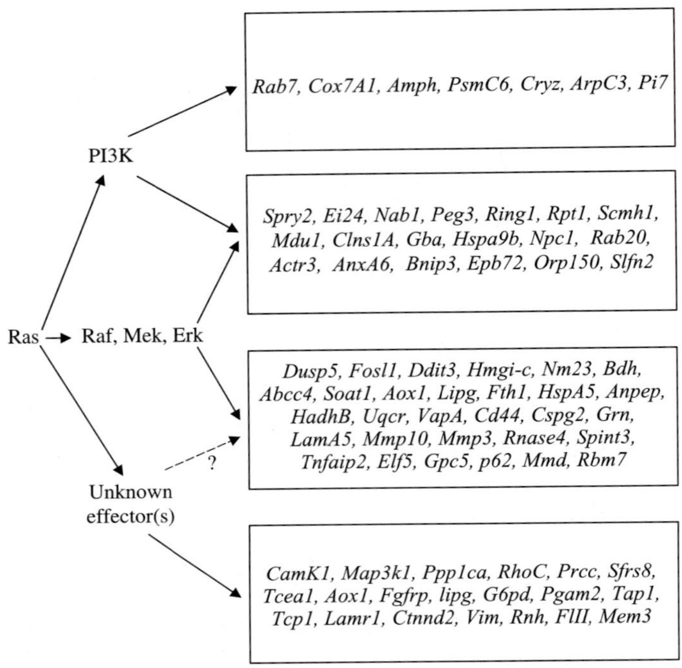

Figure 1A. Organization of Ras-responsive genes in signal-regulated transcriptional modules (SITM). **A: Up-regulated genes:** PI3K-module (top), up-regulation blocked by treatment of KRAS-transformed ROSE cells by LY294002; mixed regulation, (upper middle), up-regulation blocked by LY294002 and PD98059; Erk-module 1, (lower middle), up-regulation blocked by PD98059; Non-Erk/non-PI3K-module 1 (bottom), up-regulation not blocked by either inhibitor.

and estimated that about 500 genes were differentially expressed in tumor cells compared to the precursor cell PB-3c. Based on the sequence information of the entire mouse genome and the predicted number of genes being available now, the overall number of Ras targets would have to be reduced to approx. 220.

Using a subtractive hybridization method designated VGID^{TM}, Gadal and colleagues have contrasted gene expression in the human mammary tumor cell line MCF-7 and a Ha-ras transformed derivative. They identified 166 over- and under-expressed genes which are known to be involved in different aspects of tumorigenic transformation such as signaling pathways, cellular growth, protection against apoptosis, extracellular matrix and cytoskeletal architecture and remodeling (Gadal et al., 2003). The differential expression of recovered genes was partially verified by microarray analysis. The authors used an analytical procedure based upon published data in order to understand the physiological mechanisms potentially affected by the genes confirmed to be differentially expressed. The authors claim that this approach permits an unbiased selection of genes for further, more detailed analysis with regard to the transformation process and also modes for intervention with the process. The model predicted calcium-dependent pathways and protein

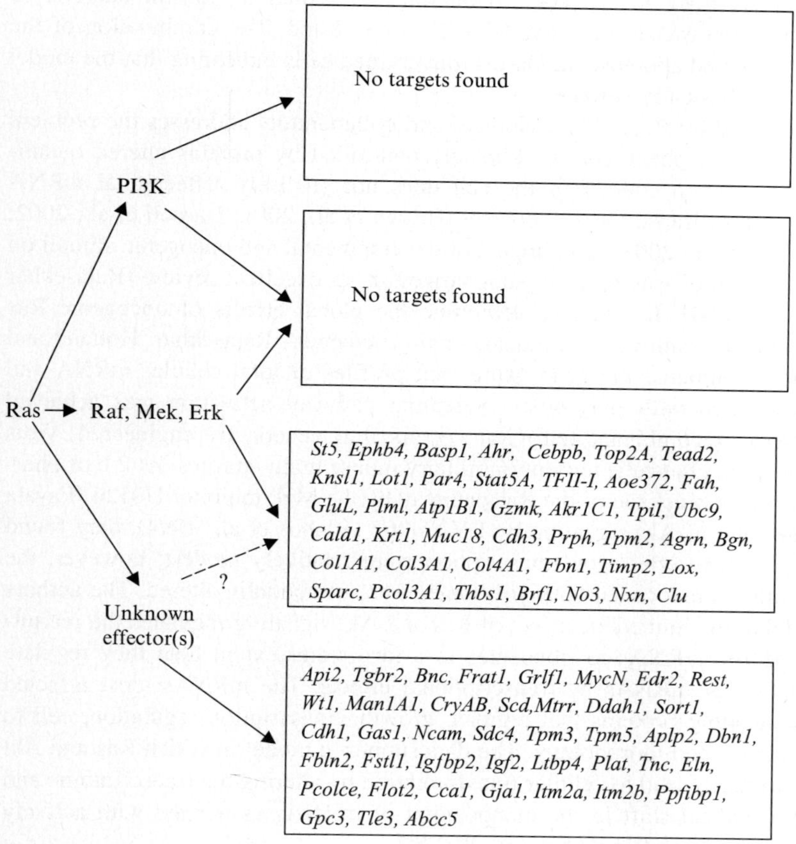

Figure 1B. **B: Down-regulated genes:** Erk-module 2, (lower middle), down-regulation blocked by PD98059; Non-Erk/non-PI3K module (bottom), down-regulation not blocked by inhibitors. Gene names according to Genbank symbols, nomenclature was adjusted for the rat, although sequence comparisons of subtracted cDNAs frequently matched with corresponding human gene sequences. For the following genes Genbank symbols were not found, although recovered sequences could be aligned with database entries: ARPP-19 cAMP-regulated phosphoprotein (Erk-module 1), MUK2 (Non-Erk/non-PI3K-module 2), PKC-zeta-interacting protein (Non-Erk/non-PI3K-module 1), Coxsackie and adenovirus receptor homologue (Non-Erk/non-PI3K-module 2), glutamine synthetase (glnA) similar to GLUL (Erk-module 2), UDP-galactose transporter related isozyme 1 (mixed up-regulation), PS-PLA1 (serine-phospholipid-selective phospholipase A) (Erk-module 1), epithelial cell transmembrane proteinantigen precursor RTI40, human DNA sequence, clone 1049G16 on chromosome 20q12-13.2, interferon-ß induced mRNA, megakaryocyte potentiating factor (all non-Erk/non-PI3K module 2), mitsugumin 23 (Erk-module 2), paladin, T16 (non-Erk/non-PI3K module 2), TADA1 protein (Erk-module 2), sequence similar to XLC12 X. laevis mRNA (non-Erk/non-PI3K module 2), endoplasmic reticulum ATPase (PI3K-module), neuritin, S-100 related sequence, tescalcin (all non-Erk/non-PI3K module 1)

kinase A-dependent mechanisms as points for intervention. Therefore, the effects of the Ca^{2+} ionophore AG23187 and dibutyryl-cAMP which prevents a decrease in endogenous cAMP levels while maintaining PKA activity, on the survival of MCF-ras and non-transformed MCF-7 cells were tested. The combination of the two drugs mediated apoptosis in Ha-ras transformed cells indicating that the model prediction was basically correct.

Work reported by Rajasekhar, Holland and collaborators addresses the problem that the neoplastic phenotype is ultimately controlled by proteins altered quantitatively. The protein content of the cell does not precisely reflect total mRNA levels due to post-transcriptional control (Brown et al., 2001; Darnell et al., 2002; Pradet-Balade et al., 2001). The impact of developmental and oncogenic stimuli on translational control was recently summarized in an excellent review (Rajasekhar and Holland, 2004). In order to determine the global effects of oncogenic Ras and Akt signaling pathways on translational efficiencies, Rajasekhar, Holland and collaborators compared the gene expression profiles of total cellular mRNA and mRNA associated with polysomes. Signaling pathway activation was achieved by introducing recombinant retroviral vectors into genetically engineered, virus receptor-positive glial cells present in primary mouse brain cultures. At 2 h of pharmacologic signal blockade of the Ras pathway by the Mek inhibitor U0126 (Favata et al., 1998) and of Akt signaling by LY294002 (Vlahos et al., 1994), they found that the immediate effects on transcription were relatively modest; however, the profile of mRNA associated with polysomes was substantially altered. The authors concluded that the immediate effect of Ras and Akt signaling regulates the recruitment of specific mRNAs to ribosomes to a far greater extent than they regulate the production of mRNAs by transcriptional effects. The mRNAs most affected are those encoding proteins that regulate growth, transcription regulation, cell to cell interactions, and morphology. The data support a model in which Ras and Akt signaling primarily lead to cellular transformation by altering the transcriptome and producing a radical shift in the composition of mRNAs associated with actively translating polysomes (Rajasekhar et al., 2003).

4. SCREENING FOR RAS TARGETS IN TUMORIGENIC CELLS WITHOUT NORMAL COUNTERPARTS OR PRECURSORS

Several groups have used established tumorigenic cell lines harboring Ras mutations for identifying gene expression profiles which are partially related to oncogenic Ras signaling and to interactions with other signaling pathways, e.g. TGF-β signaling. The impact of Ras was measured in an indirect way by introducing a dominant negative HRAS(s71N) mutant into the pancreatic cancer cell line PANC-1 (Fensterer et al., 2004). The authors interrogated a customized array containing 1,264 cDNA clones, previously identified as being differentially expressed in pancreatic cancer, oncogenes and factors responding to serum and/or EGF stimulation, phosphatases, cell cycle-associated genes and others. A total of 109 genes responded to the block to Ras-expression. SW480 colorectal cancer cells were transfected with rationally selected K-ras antisense nucleotides. Expression profiling using 4,132 cDNA clones spotted onto nylon membranes detected altered expression of 32 genes in SW480 cells, when K-ras expression was impaired (Ross et al., 2001). Four pancreatic cell lines were infected with an adenovirus vector expressing an antisense K-ras RNA (Ohnami et al., 2003). Changes of gene expression were analyzed by oligonucleotide-based microarrays containing 12,626 genes. Among the genes showing more than 2-fold differences in the expression levels between control- and antisense-transduced cells, 7 genes were up-regulated and 4 genes down-regulated. Gene expression profiles of metastatic and non-metastatic medulloblastoma were compared by MacDonald et al. (MacDonald et al., 2001). Treatment of cells with U0126 (Favata et al., 1998) identified downstream targets of Ras/MAPK signaling particularly in metastatic medulloblastoma. Expression profiling of CC531 colon carcinoma cells harboring a K-ras mutation identified 7 differentially expressed genes known as targets of the Ras and/or β-catenin pathways (Germann et al., 2003). A list of gene differentially expressed in the NSCLC cell line A 549 transiently transfected with the tumor suppressor gene RASSF1A significantly overlaps with genes found to be de-regulated by Ras signaling (Agathanggelou et al., 2003). Interestingly, some genes found to be down-regulated by Ras (e.g. the class II tumor suppressor Sparc/osteonectin) were up-regulated by RASSF1A. This supports the notion that Ras-induced transcriptional alterations can be reversed by enhanced activity of a potential upstream regulator.

5. OUTLOOK

Gene expression profiling primarily provides correlative information. So far, only a minority of genes identified in the screening experiments has been analyzed for functional involvement in the process of Ras-mediated transformation. Microarray analysis and subtracted cDNA library screening have identified a number of known targets with an established role in the transformation process and thus recapitulated previous work. However, the vast number of genes deregulated by oncogenic signaling reemphasizes the complexity of cancer phenotypes. The new information on hundreds of genes deregulated in conjunction with oncogenic Ras signaling indicates that yet unknown factors regulating or executing this process exist and may be expressed in a cell-type specific manner. Some of these factors may act redundantly, others may be components of hierarchically organized pathways and networks. Can any up-regulated target theoretically been considered a candidate effector which stimulates or even executes Ras-related phenotypic changes? Moreover, can any down-regulated target been regarded a candidate involved in negative growth control, because its expression is reduced or abolished during the transformation process? Is the systematic deregulation of transcription essential for the cancer phenotype or are we mainly looking at secondary events? To address these questions, functional gene assays are indispensable.

In view of the large lists of transcriptional targets obtained in fibroblasts, epithelial cells and other systems, the conclusion emerges that up to 5% or more of transcripts expressed in cells are deregulated as a consequence of oncogenic Ras signaling. Interestingly, the quantity of changes had been predicted long before whole genome sequences and new technologies for gene expression profiling became a reality (Groudine and Weintraub, 1980; Augenlicht et al., 1987). Before we can draw far-reaching conclusions on the systems biology (Ideker et al., 2001) of a cancer cell expressing Ras as one of the most frequent oncogenes, we are in need of effective methods for assessing the function of deregulated genes in the cell systems that gave rise to their identification. RNA interference has been heralded as a revolutionary approach for studying gene function (Paddison and Hannon, 2002; Tijsterman et al., 2002). Targeting siRNAs to up-regulated genes followed by phenotypic changes potentially can provide information whether the target is essential or dispensable for a given aspect of Ras-transformation. For example, we and others have tested the contribution of Fra-1 to neoplastic transformation. The transcriptional regulator Fra-1 is frequently up-regulated in tumors (Risse et al., 1998; Zajchowski et al., 2001) as well as in Ras-transformed fibroblasts (Zuber et al., 2000) and epithelial cells (Tchernitsa et al., 2004). The Fra-1 protein shares a DNA-binding domain with other members of the Fos-family, but lacks a transcriptional activator domain. One of the consequences of Fra-1 protein accumulation in cancer cells may be the reduction of Fos/Jun (AP-1) transcriptional activator complexes and accumulation of inactive Fra-1/c-Jun heterodimers. This can attenuate Fos-regulated gene expression (Kessler et al., 1999). Fra-1 is expressed in an Erk-dependent manner. Recent work done by Chris Marshall's group, using an siRNA-based strategy, addressed the function of Fra-1 in BE human colon carcinoma cells carrying K-Ras (G13D) and B-Raf (G463V) mutations. Silencing of Fra-1 expression resulted in the loss of cell polarization, motility and invasiveness *in vitro*. The proposed model for Fra-1 function involves the down-regulation of a RhoA-ROCK pathway and low β1-integrin activity, resulting in a decrease of stress fibers and destabilization of focal contacts. In addition, depression of RhoA activity appeared to be necessary to permit a second Erk-dependent signaling event via uPAR, the receptor for urokinase plasminogen activator, for activating Rac and thereby promoting motility (Vial et al., 2003). We have analyzed the effect of Fra-1 ablation on KRAS-transformed ROSE cells. Cellular proliferation is reduced by only 50%, indicating that high Fra-1 expression contributes to but is not essential for growth (Tchernitsa et al., 2004). However, the main function of Fra-1 may be independent of proliferation control. These results show that functional experiments based on gene silencing require well-defined phenotypic read-outs reaching beyond simple proliferation assays in order to elucidate the precise function of a given target. Even if current technical obstacles such as the efficient delivery of siRNA (Elbashir et al., 2001; Brummelkamp et al., 2002) or shRNA (Paddison et al., 2002), off-target RNAi effects (Jackson et al., 2003) and non-specific anti-proliferative effects via interferon signaling (Sledz et al., 2003) can be overcome, functional experiments using hundreds of potential Ras-responsive targets are difficult and time-consuming to perform.

Studying the function of down-regulated Ras targets may be even more challenging. Applying the strategy of RNAi, such experiments involve the silencing of targets in normal precursor cells. However, can we expect that ablation of a single gene will result in a measurable phenotypic change shifting the recipient cells to more transformed characteristics? Fortunately, there is an increasing number of gene silencing studies aiming at defining the role of down-regulated genes or tumor suppressor genes in the transformation process. Down-regulation of lysyl oxidase, a frequent Ras target, by antisense expression resulted in the re-transformation of normal cells (Contente et al., 1990). Suppression of p53 expression through RNAi in MEFs led to rapid cell cycle re-entry and immortalization (Dirac and Bernards, 2003). The familial cylindromatosis tumour suppressor gene (*CYLD*) was identified in a screen based on RNA interference vectors designed to suppress de-ubiquitinating enzymes and to study their role in cancer-relevant pathways (Brummelkamp et al., 2003). For large-scale loss-of-function screens in mammalian cells, bar-coded shRNA expression libraries have been constructed. The libraries can be used to rapidly identify individual siRNA vectors associated with a specific phenotype (Berns et al., 2004; Paddison et al., 2004). Alternatively, the classical approach for studying genes down-regulated by Ras is to restore their function by stable or conditional transfection. A variety of genes capable of antagonizing Ras-transformation have been identified in this manner (Schäfer, 1994; Sers et al., 1997; Luo et al., 1995).

Much of our current understanding of Ras-mediated signaling comes from a combination of biochemical experiments conducted in mammalian cells and genetic screening performed in *Drosophila* and *Caenorhabditis elegans* [for review see (Wassarman et al., 1995; Sternberg and Han, 1998)]. In the not too distant future, these model organisms may provide further deep insights into the relationships between Ras signaling and gene control. Gene expression profiles related to Ras signal transduction have been described in over-proliferating *Drosophila* hemocytes (Asha et al., 2003), transgenic nematodes expressing let-60/Ras (G12V) (Romagnolo et al., 2002) and a *Saccharomyces cerevisiae* mutant with a constitutively activated Ras/cAMP pathway (Jones et al., 2003). Regardless of the biological system studied, integrating the information from gene lists obtained by expression profiling and from functional screens based on RNA interference or alternate technologies will be a major challenge and source of scientific excitement for many years to come.

ACKNOWLEDGEMENTS

The authors' work is supported by Deutsche Forschungsgemeinschaft (SFB 506, 618), Deutsche Krebshilfe (Dr. Mildred-Scheel-Stiftung), Stifterverband für die Deutsche Wissenschaft and Berliner Krebsgesellschaft (Ernst-v.-Leyden Fellowship Program).

REFERENCES

Abdollahi, A., Godwin, A.K., Miller, P.D., Getts, L.A., Schultz, D.C., Taguchi, T., Testa, J.R. and Hamilton, T.C. (1997) Identification of a gene containing zinc-finger motifs based on lost expression in malignantly transformed rat ovarian surface epithelial cells. Cancer Res. *57*, 2029-2034.

Abdollahi, A., Bao, R. and Hamilton, T.C. (1999) LOT1 is a growth suppressor gene down-regulated by the epidermal growth factor receptor ligands and encodes a nuclear zinc-finger protein. Oncogene *18*, 6477-6487.

Adams, A.T. and Auersperg, N. (1985) A cell line, ROSE 199, derived from normal rat ovarian surface epithelium. Exp. Cell Biol. *53*, 181-188.

Agathanggelou, A., Bieche, I., Ahmed-Choudhury, J., Nicke, B., Dammann, R., Baksh, S., Gao, B., Minna, J.D., Downward, J., Maher, E.R. and Latif, F. (2003) Identification of novel gene expression targets for the Ras association domain family 1 (RASSF1A) tumor suppressor gene in non-small cell lung cancer and neuroblastoma. Cancer Res. *63*, 5344-5351.

Alaiya, A.A., Franzen, B., Fujioka, K., Moberger, B., Schedvins, K., Silfversvard, C., Linder, S. and Auer, G. (1997) Phenotypic analysis of ovarian carcinoma: polypeptide expression in benign, borderline and malignant tumors. Int. J. Cancer *73*, 678-683.

Alvarez, A.A., Axelrod, J.R., Whitaker, R.S., Isner, P.D., Bentley, R.C., Dodge, R.K. and Rodriguez, G.C. (2001) Thrombospondin-1 expression in epithelial ovarian carcinoma: association with p53 status, tumor angiogenesis, and survival in platinum-treated patients. Gynecol. Oncol. *82*, 273-278.

Asha, H., Nagy, I., Kovacs, G., Stetson, D., Ando, I. and Dearolf, C.R. (2003) Analysis of Ras-induced overproliferation in Drosophila hemocytes. Genetics *163*, 203-215.

Auersperg, N., Pan, J., Grove, B.D., Peterson, T., Fisher, J., Maines, B.S., Somasiri, A. and Roskelley, C.D. (1999) E-cadherin induces mesenchymal-to-epithelial transition in human ovarian surface epithelium. Proc. Natl. Acad. Sci. U.S.A. *96*, 6249-6254.

Augenlicht, L.H., Wahrman, M.Z., Halsey, H., Anderson, L., Taylor, J. and Lipkin, M. (1987) Expression of cloned sequences in biopsies of human colonic tissue and in colonic carcinoma cells induced to differentiate in vitro. Cancer Res. *47*, 6017-6021.

Baba, I., Shirasawa, S., Iwamoto, R., Okumura, K., Tsunoda, T., Nishioka, M., Fukuyama, K., Yamamoto, K., Mekada, E. and Sasazuki, T. (2000) Involvement of deregulated epiregulin expression in tumorigenesis in vivo through activated Ki-Ras signaling pathway in human colon cancer cells. Cancer Res. *60*, 6886-6889.

Bar-Sagi, D. and Hall, A. (2000) Ras and Rho GTPases: a family reunion. Cell *103*, 227-238.

Berns, K., Hijmans, E.M., Mullenders, J., Brummelkamp, T.R., Velds, A., Heimerikx, M., Kerkhoven, R.M., Madiredjo, M., Nijkamp, W., Weigelt, B., Agami, R., Ge, W., Cavet, G., Linsley, P.S., Beijersbergen, R.L. and Bernards, R. (2004) A large-scale RNAi screen in human cells identifies new components of the p53 pathway. Nature *428*, 431-437.

Bos, J.L. (1989) Ras oncogenes in human cancer: a review. Cancer Res. *49*, 4682-4689.

Boyer, B., Valles, A.M. and Edme, N. (2000) Induction and regulation of epithelial-mesenchymal transitions. Biochem. Pharmacol. *60*, 1091-1099.

Breier, G., Blum, S., Peli, J., Groot, M., Wild, C., Risau, W. and Reichmann, E. (2002) Transforming growth factor-beta and Ras regulate the VEGF/VEGF-receptor system during tumor angiogenesis. Int. J. Cancer *97*, 142-148.

Brem, R., Certa, U., Neeb, M., Nair, A.P. and Moroni, C. (2001) Global analysis of differential gene expression after transformation with the v-H-ras oncogene in a murine tumor model. Oncogene *20*, 2854-2858.

Brown, V., Jin, P., Ceman, S., Darnell, J.C., O'Donnell, W.T., Tenenbaum, S.A., Jin, X., Feng, Y., Wilkinson, K.D., Keene, J.D., Darnell, R.B. and Warren, S.T. (2001) Microarray identification of FMRP-associated brain mRNAs and altered mRNA translational profiles in fragile X syndrome. Cell *107*, 477-487.

Brummelkamp, T.R., Bernards, R. and Agami, R. (2002) A system for stable expression of short interfering RNAs in mammalian cells. Science *296*, 550-553.

Brummelkamp, T.R., Nijman, S.M., Dirac, A.M. and Bernards, R. (2003) Loss of the cylindromatosis tumour suppressor inhibits apoptosis by activating NF-kappaB. Nature *424*, 797-801.

Buess, M., Engler, O., Hirsch, H.H. and Moroni, C. (1999) Search for oncogenic regulators in an autocrine tumor model using differential display PCR: identification of novel candidate genes including the calcium channel mtrp6. Oncogene *18*, 1487-1494.

Campbell, S.L., Khosravi-Far, R., Rossman, K.L., Clark, G.J. and Der, C.J. (1998) Increasing complexity of Ras signaling. Oncogene 17, 1395-1413.

Chung, C.H., Bernard, P.S. and Perou, C.M. (2002) Molecular portraits and the family tree of cancer. Nat. Genet. *32* (Suppl.), 533-540.

Churchill, G.A. (2002) Fundamentals of experimental design for cDNA microarrays. Nat. Genet. *32* (Suppl.), 490-495.

Contente, S., Kenyon, K., Rimoldi, D. and Friedman, R.M. (1990) Expression of gene rrg is associated with reversion of NIH 3T3 transformed by LTR-c-Ha-ras. Science *249*, 796-798.

D'Cruz, C.M., Gunther, E.J., Boxer, R.B., Hartman, J.L., Sintasath, L., Moody, S.E., Cox, J.D., Ha, S.I., Belka, G.K., Golant, A., Cardiff, R.D. and Chodosh, L.A. (2001) c-MYC induces mammary tumorigenesis by means of a preferred pathway involving spontaneous Kras2 mutations. Nat. Med. *7*, 235-239.

Dahmane, N., Lee, J., Robins, P., Heller, P. and Altaba, A. (1997) Activation of the transcription factor Gli1 and the Sonic hedgehog signalling pathway in skin tumours. Nature *389*, 876-881.

Darnell, R.B. (2002) RNA logic in time and space. Cell *110*, 545-550.

Delsal, G., Ruaro, M.E., Philipson, L. and Schneider, C. (1992) The Growth Arrest-Specific Gene, gas1, Is Involved in Growth Suppression. Cell *70*, 595-607.

Delsal, G., Collavin, L., Ruaro, M.E., Edomi, P., Saccone, S., Dellavalle, G. and Schneider, C. (1994) Structure, Function, and Chromosome Mapping of the Growth-Suppressing Human Homologue of the Murine Gas1 Gene. Proc. Natl. Acad. Sci. U.S.A. *91*, 1848-1852.

Denko, N., Stringer, J., Wani, M. and Stambrook, P. (1995) Mitotic and post mitotic consequences of genomic instability induced by oncogenic Ha-ras. Somat. Cell Mol. Genet. *21*, 241-253.

Denko, N.C., Giaccia, A.J., Stringer, J.R. and Stambrook, P.J. (1994) The human Ha-ras oncogene induces genomic instability in murine fibroblasts within one cell cycle. Proc. Natl. Acad. Sci. U.S.A. *91*, 5124-5128.

Desai, K.V., Xiao, N., Wang, W., Gangi, L., Greene, J., Powell, J.I., Dickson, R., Furth, P., Hunter, K., Kucherlapati, R., Simon, R., Liu, E.T. and Green, J.E. (2002) Initiating oncogenic event determines gene-expression patterns of human breast cancer models. Proc. Natl. Acad. Sci. U.S.A. *99*, 6967-6972.

Diatchenko, L., Lau, Y.-F., Campbell, A.P., Chenchik, A., Moqadam, F., Huang, B., Lukyanov, S.A., Lukyanov, K.A., Gurskaya, N.D., Sverdlov, E.D. and Siebert, P.D. (1996) Suppression subtractive hybridization: a method for generating differentially regulated or tissue-specific cDNA probes and libraries. Proc. Natl. Acad. Sci. U.S.A. *93*, 6025-6030.

Dirac, A.M. and Bernards, R. (2003) Reversal of senescence in mouse fibroblasts through lentiviral suppression of p53. J. Biol. Chem. *278*, 11731-11734.

Down, T.A. and Hubbard, T.J. (2004) What can we learn from noncoding regions of similarity between genomes? BMC. Bioinformatics *5*, 131.

Dudley, D.T., Pang, L., Decker, S.J., Bridges, A.J. and Saltiel, A.R. (1995) A synthetic inhibitor of the mitogen-activated protein kinase cascade. Proc. Natl. Acad. Sci. U.S.A. *92*, 7686-7689.

Elbashir, S.M., Harborth, J., Lendeckel, W., Yalcin, A., Weber, K. and Tuschl, T. (2001) Duplexes of 21-nucleotide RNAs mediate RNA interference in cultured mammalian cells. Nature *411*, 494-498.

Evangelou, A., Jindal, S.K., Brown, T.J. and Letarte, M. (2000) Down-regulation of transforming growth factor beta receptors by androgen in ovarian cancer cells. Cancer Res. *60*, 929-935.

Favata, M.F., Horiuchi, K.Y., Manos, E.J., Daulerio, A.J., Stradley, D.A., Feeser, W.S., Van, D.D., Pitts, W.J., Earl, R.A., Hobbs, F., Copeland, R.A., Magolda, R.L., Scherle, P.A. and Trzaskos, J.M. (1998) Identification of a novel inhibitor of mitogen-activated protein kinase kinase. J. Biol. Chem. *273*, 18623-18632.

Fensterer, H., Giehl, K., Buchholz, M., Ellenrieder, V., Buck, A., Kestler, H.A., Adler, G., Gierschik, P. and Gress, T.M. (2004) Expression profiling of the influence of RAS mutants on the TGFB1-induced phenotype of the pancreatic cancer cell line PANC-1. Genes Chromosomes Cancer *39*, 224-235.

Furuya, M., Ishikura, H., Kawarada, Y., Ogawa, Y., Sakuragi, N., Fujimoto, S. and Yoshiki, T. (2000) Expression of matrix metalloproteinases and related tissue inhibitors in the cyst fluids of ovarian mucinous neoplasms. Gynecol. Oncol. *78*, 106-112.

Gadal, F., Bozic, C., Pillot-Brochet, C., Malinge, S., Wagner, S., Le Cam, A., Buffat, L., Crepin, M. and Iris, F. (2003) Integrated transcriptome analysis of the cellular mechanisms associated with

Ha-ras-dependent malignant transformation of the human breast epithelial MCF7 cell line. Nucleic Acids Res. *31*, 5789-5804.

Germann, A., Dihlmann, S., Hergenhahn, M., Doeberitz, M.K. and Koesters, R. (2003) Expression profiling of CC531 colon carcinoma cells reveals similar regulation of beta-catenin target genes by both butyrate and aspirin. Int. J. Cancer *20* (106), 187-197.

Giampuzzi, M., Botti, G., Cilli, M., Gusmano, R., Borel, A., Sommer, P. and Di Donato, A. (2001) Down-regulation of lysyl oxidase-induced tumorigenic transformation in NRK-49F cells characterized by constitutive activation of ras proto-oncogene. J. Biol. Chem. *276*, 29226-29232.

Gimona, M., Kazzaz, J.A. and Helfman, D.M. (1996) Forced expression of tropomyosin 2 or 3 in v-Ki-ras-transformed fibroblasts results in distinct phenotypic effects. Proc. Natl. Acad. Sci. U.S.A. *93*, 9618-9623.

Giunciuglio, D., Culty, M., Fassina, G., Masiello, L., Melchiori, A., Paglialunga, G., Arand, G., Ciardiello, F., Basolo, F. and Thompson, E.W. (1995) Invasive phenotype of MCF10A cells overexpressing c-Ha-ras and c-erbB-2 oncogenes. Int. J. Cancer *63*, 815-822.

Glück, U., Kwiatkowski, D.J. and Ben-Ze'ev, A. (1993) Suppresion of tumorigenicity in simian virus 40-transformed 3T3 cells transfected with a-actinin cDNA. Proc. Natl. Acad. Sci. U.S.A. *90*, 383-387.

Glück, U. and Benzeev, A. (1994) Modulation of alpha-actinin levels affects cell motility and confers tumorigenicity on 3T3 cells. J. Cell Sci. *107*, 1773-1782.

Golub, T.R., Slonim, D.K., Tamayo, P., Huard, C., Gaasenbeek, M., Mesirov, J.P., Coller, H., Loh, M.L., Downing, J.R., Caligiuri, M.A., Bloomfield, C.D. and Lander, E.S. (1999) Molecular classification of cancer: class discovery and class prediction by gene expression monitoring. Science *286*, 531-537.

Griegel, S., Traub, O., Willecke, K. and Schäfer, R. (1986) Suppression and re-expression of transformed phenotype in hybrids of Ha-ras1 transformed Rat-1 cells and early passage rat embryo fibroblasts. Int. J. Canc. *38*, 697-705.

Grill, C., Gheyas, F., Dayananth, P., Jin, W., Ding, W., Qiu, P., Wang, L., Doll, R.J. and English, J.M. (2004) Analysis of the ERK1,2 transcriptome in mammary epithelial cells. Biochem. J. *381*, 635-644.

Groudine, M. and Weintraub, H. (1980) Activation of cellular genes by avian RNA tumor viruses. Proc. Natl. Acad. Sci. U.S.A. *77*, 5351-5354.

Habets, G.G., Knepper, M., Sumortin, J., Choi, Y.J., Sasazuki, T., Shirasawa, S. and Bollag, G. (2001) cDNA array analyses of K-ras-induced gene transcription. Methods Enzymol. *332*, 245-260.

Hahn, W.C., Counter, C.M., Lundberg, A.S., Beijersbergen, R.L., Brooks, M.W. and Weinberg, R.A. (1999) Creation of human tumour cells with defined genetic elements. Nature *400*, 464-468.

Hajnal, A., Klemenz, R. and Schäfer, R. (1993a) Suppression of ras-mediated transformation. Differential expression of genes encoding extracellular matrix proteins in normal, transformed and revertant cells. Adv. Enzyme Regul. *33*, 267-280.

Hajnal, A., Klemenz, R. and Schäfer, R. (1993b) Upregulation of lysyl oxidase in spontaneous revertants of H-ras transformed rat fibroblasts. Canc. Res. *53*, 4670-4675.

Hajnal, A., Klemenz, R. and Schäfer, R. (1994) Subtraction cloning of H-Rev107, a gene specifically expressed in H-ras resistant fibroblasts. Oncogene *9*, 479-490.

Hanna, E.A., Umhauer, S., Roshong, S.L., Piechocki, M.P., Fernstrom, M.J., Fanning, J.D. and Ruch, R.J. (1999) Gap junctional intercellular communication and connexin43 expression in human ovarian surface epithelial cells and ovarian carcinomas in vivo and in vitro. Carcinogenesis *20*, 1369-1373.

Huang, E., Ishida, S., Pittman, J., Dressman, H., Bild, A., Kloos, M., D'Amico, M., Pestell, R.G., West, M. and Nevins, J.R. (2003) Gene expression phenotypic models that predict the activity of oncogenic pathways. Nat. Genet. *34*, 226-230.

Hubank, M. and Schatz, D.G. (1994) Identifying differences in mRNA expression by representational difference analysis of cDNA. Nucleic Acids Res. *22*, 5640-5648.

Ideker, T., Galitski, T. and Hood, L. (2001) A new approach to decoding life: systems biology. Annu. Rev. Genomics Hum. Genet. *2*, 343-372.

Imren, S., Kohn, D.B., Shimada, H., Blavier, L. and DeClerck, Y.A. (1996) Overexpression of tissue inhibitor of metalloproteinases-2 retroviral-mediated gene transfer in vivo inhibits tumor growth and invasion. Cancer Res. *56*, 2891-2895.

Jackson, A.L., Bartz, S.R., Schelter, J., Kobayashi, S.V., Burchard, J., Mao, M., Li, B., Cavet, G. and Linsley, P.S. (2003) Expression profiling reveals off-target gene regulation by RNAi. Nat. Biotechnol. *21*, 635-637.

Jeay, S., Pianetti, S., Kagan, H.M. and Sonenshein, G.E. (2003) Lysyl oxidase inhibits ras-mediated transformation by preventing activation of NF-kappa B. Mol. Cell. Biol. *23*, 2251-2263.

Jechlinger, M., Grunert, S., Tamir, I.H., Janda, E., Ludemann, S., Waerner, T., Seither, P., Weith, A., Beug, H. and Kraut, N. (2003) Expression profiling of epithelial plasticity in tumor progression. Oncogene *22*, 7155-7169.

Jo, H., Cho, Y.J., Zhang, H. and Liang, P. (2001) Differential display analysis of gene expression altered by ras oncogene. Methods Enzymol. *332*, 233-244.

Johnston, I.M., Spence, H.J., Winnie, J.N., McGarry, L., Vass, J.K., Meagher, L., Stapleton, G. and Ozanne, B.W. (2000) Regulation of a multigenic invasion programme by the transcription factor, AP-1: re-expression of a down-regulated gene, TSC-36, inhibits invasion. Oncogene *19*, 5348-5358.

Jones, D.L., Petty, J., Hoyle, D.C., Hayes, A., Ragni, E., Popolo, L., Oliver, S.G. and Stateva, L.I. (2003) Transcriptome profiling of a Saccharomyces cerevisiae mutant with a constitutively activated Ras/cAMP pathway. Physiol. Genomics *16*, 107-118.

Kessler, R., Zacharova, A.A., Laursen, N.B., Kalousek, M. and Klemenz, R. (1999) Attenuated expression of the serum responsive T1 gene in ras transformed fibroblasts due to the inhibition of c-fos gene activity. Oncogene *18*, 1733-1744.

Kielbasa, S.M., Korbel, J.O., Beule, D., Schuchhardt, J. and Herzel, H. (2001) Combining frequency and positional information to predict transcription factor binding sites. Bioinformatics *17*, 1019-1026.

Land, H., Parada, L.F. and Weinberg, R.A. (1983) Tumorigenic conversion of primary embryo fibroblasts requires at least two cooperating oncogenes. Nature *304*, 596-602.

Land, H., Chen, A.C., Morgenstern, J.P., Parada, L.F. and Weinberg, R.A. (1986) Behavior of myc and ras oncogenes in transformation of rat embryo fibroblasts. Mol. Cell. Biol. *6*, 1917-1925.

Lewis, J.E., Jensen, P.J. and Wheelock, M.J. (1994) Cadherin function is required for human keratinocytes to assemble desmosomes and stratify in response to calcium. J. Invest. Dermatol. *102*, 870-877.

Liang, P. and Pardee, A.B. (1992) Differential display of eukaryotic messenger RNA by means of the polymerase chain reaction. Science *257*, 967-971.

Liang, P., Averboukh, L., Keyomarsi, K., Sager, R. and Pardee, A.B. (1992) Differential display and cloning of messenger RNAs from human breast cancer versus mammary epithelial cells. Cancer Res. *52*, 6966-6968.

Liang, P., Averboukh, L., Zhu, W.M. and Pardee, A.B. (1994) Ras activation of genes: Mob-1 as a model. Proc. Natl. Acad. Sci. U.S.A. *91*, 12515-12519.

Lisitsyn, N. and Wigler, M. (1993) Cloning the Differences Between 2 Complex Genomes. Science *259*, 946-951.

Liu, J., Yang, G., Thompson-Lanza, J.A., Glassman, A., Hayes, K., Patterson, A., Marquez, R.T. and Auersperg, N., Yu, Y., Hahn, W.C., Mills, G.B. and Bast, R.C. Jr. (2004) A genetically defined model for human ovarian cancer. Cancer Res. *64*, 1655-1663.

Lombardi, L., Ballinari, D., Bongarzone, I., Migliari, M., Mondellini, P., Traversari, C. and Modina, S. (1990) Ultrastructural cytoskeleton alterations and modification of actin expression in the NIH/3T3 cell line after transformation with Ha-ras-activated oncogene. Cell Motil. Cytoskel. *15*, 220-229.

Luo, X.N., Reddy, J.C., Yeyati, P.L., Idris, A.H., Hosono, S., Haber, D.A., Licht, J.D. and Atweh, G.F. (1995) The tumor suppressor gene WT1 inhibits ras-mediated transformation. Oncogene *11*, 743-750.

MacDonald, T.J., Brown, K.M., LaFleur, B., Peterson, K., Lawlor, C., Chen, Y., Packer, R.J., Cogen, P. and Stephan, D.A. (2001) Expression profiling of medulloblastoma: PDGFRA and the RAS/MAPK pathway as therapeutic targets for metastatic disease. Nat. Genet. *29*, 143-152.

Majidi, M., Gutkind, J.S. and Lichy, J.H. (2000) Deletion of the COOH terminus converts the ST5 p70 protein from an inhibitor of RAS signaling to an activator with transforming activity in NIH-3T3 cells. J. Biol. Chem. *275*, 6560-6565.

Malumbres, M. and Pellicer, A. (1998) Ras pathways to cell cycle control and cell transformation. Front. Biosci. *3*, 887-912.

Mccarthy, S.A., Samuels, M.L., Pritchard, C.A., Abraham, J.A. and McMahon, M. (1995) Rapid induction of heparin-binding epidermal growth factor/diphtheria toxin receptor expression by Raf and Ras oncogenes. Genes Dev. *9*, 1953-1964.

Mikulits, W., Pradet-Balade, B., Habermann, B., Beug, H., Garcia-Sanz, J.A. and Mullner, E.W. (2000) Isolation of translationally controlled mRNAs by differential screening. FASEB J. *14*, 1641-1652.

Mok, S.C., Chan, W.Y., Wong, K., Muta, M.G. and Berkowitz, R.S. (1996) SPARC, an extracellular matrix protein with tumor-suppressing activity in human ovarian epithelial cells. Oncogene *12*, 1895-1901.

Morrison, B.H., Bauer, J.A., Kalvakolanu, D.V. and Lindner, D.J. (2001) Inositol hexakisphosphate kinase 2 mediates growth suppressive and apoptotic effects of interferon-beta in ovarian carcinoma cells. J. Biol. Chem. *276*, 24965-24970.

Murphy, G.A. and Der, C.J. (2002) Ras-mediated deregulation of gene expression and contribution to oncogenesis. In: Oncogene-Directed Therapies. J.W. Rak, ed. (Totowa NJ: Humana Press), pp. 77-100.

Ohnami, S., Aoki, K., Yoshida, K., Ohnami, S., Hatanaka, K., Suzuki, K., Sasaki, H. and Yoshida, T. (2003) Expression profiles of pancreatic cancer cell lines infected with antisense K-ras-expressing adenoviral vector. Biochem. Biophys. Res. Commun. *309*, 798-803.

Olski, T.M., Noegel, A.A. and Korenbaum, E. (2001) Parvin, a 42 kDa focal adhesion protein, related to the alpha-actinin superfamily. J. Cell Sci. *114*, 525-538.

Ordway, J.M., Williams, K. and Curran, T. (2004) Transcription repression in oncogenic transformation: common targets of epigenetic repression in cells transformed by Fos, Ras or Dnmt1. Oncogene *23*, 3737-3748.

Ozaki, T. and Sakiyama, S. (1994) Tumor-Suppressive Activity of N03 Gene Product in v-src- Transformed Rat 3Y1 Fibroblasts. Cancer Res. *54*, 646-648.

Paddison, P.J. and Hannon, G.J. (2002) RNA interference: the new somatic cell genetics? Cancer Cell *2*, 17-23.

Paddison, P.J., Caudy, A.A., Bernstein, E., Hannon, G.J. and Conklin, D.S. (2002) Short hairpin RNAs (shRNAs) induce sequence-specific silencing in mammalian cells. Genes Dev. *16*, 948-958.

Paddison, P.J., Silva, J.M., Conklin, D.S., Schlabach, M., Li, M., Aruleba, S., Balija, V., O'Shaughnessy, A., Gnoj, L., Scobie, K., Chang, K., Westbrook, T., Cleary, M., Sachidanandam, R., McCombie, W.R., Elledge, S.J. and Hannon, G.J. (2004) A resource for large-scale RNA-interference-based screens in mammals. Nature *428*, 427-431.

Paterson, I.C., Matthews, J.B., Huntley, S., Robinson, C.M., Fahey, M., Parkinson, E.K. and Prime, S.S. (2001) Decreased expression of TGF-beta cell surface receptors during progression of human oral squamous cell carcinoma. J. Pathol. *193*, 458-467.

Petricoin, E.F. III, Hackett, J.L., Lesko, L.J., Puri, R.K., Gutman, S.I., Chumakov, K., Woodcock, J., Feigal, D.W. Jr., Zoon, K.C. and Sistare, F.D. (2002) Medical applications of microarray technologies: a regulatory science perspective. Nat. Genet. *32* (Suppl.), 474-479.

Pinkas, J. and Leder, P. (2002) MEK1 signaling mediates transformation and metastasis of EpH4 mammary epithelial cells independent of an epithelial to mesenchymal transition. Cancer Res. *62*, 4781-4790.

Platt, K.A., Michaud, J. and Joyner, A.L. (1997) Expression of the mouse Gli and Ptc genes is adjacent to embryonic sources of hedgehog signals suggesting a conservation of pathways between flies and mice. Mech. Dev. *62*, 121-135.

Pradet-Balade, B., Boulme, F., Beug, H., Mullner, E.W. and Garcia-Sanz, J.A. (2001) Translation control: bridging the gap between genomics and proteomics? Trends Biochem. Sci. *26*, 225-229.

Quade, K. (1979) Transformation of mammalian cells by avian myelocytomatosis virus and avian erythroblastosis virus. Virology *98*, 461-465.

Rajasekhar, V.K., Viale, A., Socci, N.D., Wiedmann, M., Hu, X. and Holland, E.C. (2003) Oncogenic Ras and Akt signaling contribute to glioblastoma formation by differential recruitment of existing mRNAs to polysomes. Mol. Cell *12*, 889-901.

Rajasekhar, V.K. and Holland, E.C. (2004) Postgenomic global analysis of translational control induced by oncogenic signaling. Oncogene *19*, 3248-3264.

Risse, H.G., Adamkiewicz, J., Wimmel, A. and Schuermann, M. (1998) Transition from SCLC to NSCLC phenotype is accompanied by an increased TRE-binding activity and recruitment of specific AP-1 proteins. Oncogene *16*, 3057-3068.

Romagnolo, B., Jiang, M., Kiraly, M., Breton, C., Begley, R., Wang, J., Lund, J. and Kim, S.K. (2002) Downstream targets of let-60 Ras in Caenorhabditis elegans. Dev. Biol. *247*, 127-136.

Ross, P.J., George, M., Cunningham, D., DiStefano, F., Andreyev, H.J., Workman, P. and Clarke, P.A. (2001) Inhibition of Kirsten-ras expression in human colorectal cancer using rationally selected Kirsten-ras antisense oligonucleotides. Mol. Cancer Ther. *1*, 29-41.

Schäfer, R. (1994) Suppression of ras oncogene-mediated transformation. Rev. Physiol. Biochem. Pharmacol. *124*, 29-92.

Schorge, J.O., Miller, Y.B., Qi, L.J., Muto, M.G., Welch, W.R., Berkowitz, R.S. and Mok, S.C. (2000) Genetic alterations of the WT1 gene in papillary serous carcinoma of the peritoneum. Gynecol. Oncol. *76*, 369-372.

Schulze, A., Lehmann, K., Jefferies, H.B.J., McMahon, M. and Downward, J. (2001) Analysis of the transcriptional program induced by Raf in epithelial cells. Genes Dev. *15*, 981-994.

Schulze, A., Nicke, B., Warne, P.H., Tomlinson, S. and Downward, J. (2004) The transcriptional response to Raf activation is almost completely dependent on MEK activity and shows a major autocrine component. Mol. Biol. Cell. *15*, 673-685.

Sehgal, A., Keener, C., Boynton, A.L., Young, R.F., Vermeulen, S.S., Yonemura, K.S., Kohler, E.P., Aldape, H.C., Simrell, C.R. and Murphy, G.P. (1997a) Characterization of C4-2 as a tumor-suppressor gene in human brain tumors. J. Surg. Oncol. *64*, 102-108.

Sehgal, A., Ricks, S., Keener, C., Boynton, A.L., Young, R.F., Vermeulen, S.S., Yonemura, K.S., Kohler, E.P., Aldape, H.C., Simrell, C.R. and Murphy, G.P. (1997b) Cloning, sequence, and developmental expression analysis of C4-2, a potential brain tumor-suppressor gene. J. Surg. Oncol. *65*, 249-257.

Serrano, M., Lin, A.W., McCurrach, M.E., Beach, D. and Lowe, S.W. (1997) Oncogenic ras provokes premature cell senescence associated with accumulation of p53 and p16 ^{INK4a}. Cell *88*, 593-602.

Sers, C., Emenegger, U., Husmann, K., Bucher, K., Andres, A.-C. and Schäfer, R. (1997) Growth-inhibitory activity and downregulation of the class II tumor-suppressor gene H-rev107 in tumor cell lines and experimental tumors. J. Cell. Biol. *136*, 935-944.

Sers, C., Husmann, K., Nazarenko, I., Reich, S., Wiechen, K., Zhumabayeva, B., Adhikari, P., Schroder, K., Gontarewicz, A. and Schäfer, R. (2002) The class II tumour suppressor gene H-REV107-1 is a target of interferon-regulatory factor-1 and is involved in IFNgamma-induced cell death in human ovarian carcinoma cells. Oncogene *21*, 2829-2839.

Shields, J.M., Pruitt, K., McFall, A., Shaub, A. and Der, C.J. (2000) Understanding Ras: 'it ain't over'til it's over'. Trends Cell Biol. *10*, 147-154.

Slack, J.L., Parker, M.I., Robinson, V.R. and Bornstein, P. (1992) Regulation of Collagen-I Gene Expression by ras. Mol. Cell Biol. *12*, 4714-4723.

Sledz, C.A., Holko, M., de Veer, M.J., Silverman, R.H. and Williams, B.R. (2003) Activation of the interferon system by short-interfering RNAs. Nat. Cell Biol. *5*, 834-839.

Sternberg, P.W. and Han, M. (1998) Genetics of RAS signaling in C. elegans. Trends Genet. *14*, 466-472.

Sumitomo, K., Kurisaki, A., Yamakawa, N., Tsuchida, K., Shimizu, E., Sone, S. and Sugino, H. (2000) Expression of a TGF-beta1 inducible gene, TSC-36, causes growth inhibition in human lung cancer cell lines. Cancer Lett. *155*, 37-46.

Tan, P.K., Downey, T.J., Spitznagel, E.L. Jr, Xu, P., Fu, D., Dimitrov, D.S., Lempicki, R.A., Raaka, B.M. and Cam, M.C. (2003) Evaluation of gene expression measurements from commercial microarray platforms. Nucleic Acids Res. *31*, 5676-5684.

Tchernitsa, O.I., Zuber, J., Sers, C., Brinckmann, R., Britsch, S.K., Adams, V. and Schäfer, R. (1999) Gene expression profiling of fibroblasts resistant toward oncogene-mediated transformation reveals preferential transcription of negative growth regulators [In Process Citation]. Oncogene *18*, 5448-5454.

Tchernitsa, O.I., Sers, C., Zuber, J., Hinzmann, B., Grips, M., Schramme, A., Lund, P., Schwendel, A., Rosenthal, A. and Schäfer, R. (2004) Transcriptional basis of KRAS oncogene-mediated cellular transformation in ovarian epithelial cells. Oncogene. *23*, 4536-4555.

Teramoto, H., Malek, R.L., Behbahani, B., Castellone, M.D., Lee, N.H. and Gutkind, J.S. (2003) Identification of H-Ras, RhoA, Rac1 and Cdc42 responsive genes. Oncogene *22*, 2689-2697.

Tijsterman, M., Ketting, R.F. and Plasterk, R.H. (2002) The genetics of RNA silencing. Annu. Rev. Genet. *36*, 489-519.

Travers, H., French, N.S. and Norton, J.D. (1996) Suppression of tumorigenicity in Ras-transformed fibroblasts by alpha 2(I) collagen. Cell Growth Differ. *7*, 1353-1360.

Tullai, J.W., Schaffer, M.E., Mullenbrock, S., Kasif, S. and Cooper, G.M. (2004) Identification of transcription factor binding sites upstream of human genes regulated by the phosphatidylinositol 3-kinase and MEK/ERK signaling pathways. J. Biol. Chem. *279*, 20167-20177.

Umhauer, S., Ruch, R.J. and Fanning, J. (2000) Gap junctional intercellular communication and connexin 43 expression in ovarian carcinoma. Am. J. Obstet. Gynecol. *182*, 999-1000.

van de Vijver, M.J., He, Y.D., van't Veer, L.J., Dai, H., Hart, A.A., Voskuil, D.W., Schreiber, G.J., Peterse, J.L., Roberts, C., Marton, M.J., Parrish, M., Atsma, D., Witteveen, A., Glas, A., Delahaye, L., van d. V., Bartelink, H., Rodenhuis, S., Rutgers, E.T., Friend, S.H. and Bernards, R. (2002) A gene-expression signature as a predictor of survival in breast cancer. N. Engl. J. Med. *19* (347), 1999-2009.

Vasseur, S., Malicet, C., Calvo, E.L., Labrie, C., Berthezene, P., Dagorn, J.C. and Iovanna, J.L. (2003) Gene expression profiling by DNA microarray analysis in mouse embryonic fibroblasts transformed by rasV12 mutated protein and the E1A oncogene. Mol. Cancer *19* (2), 19.

Vial, E., Sahai, E. and Marshall, C.J. (2003) ERK-MAPK signaling coordinately regulates activity of Rac1 and RhoA for tumor cell motility. Cancer Cell *4*, 67-79.

Vlahos, C.J., Matter, W.F., Hui, K.Y. and Brown, R.F. (1994) A specific inhibitor of phosphatidylinositol 3-kinase, 2-(4-morpholinyl)-8-phenyl-4H-1-benzopyran-4-one (LY294002). J. Biol. Chem. *269*, 5241-5248.

Wassarman, D.A., Therrien, M. and Rubin, G.M. (1995) The Ras signaling pathway in Drosophila. Curr. Opin. Genet. Dev. *5*, 44-50.

West, M., Blanchette, C., Dressman, H., Huang, E., Ishida, S., Spang, R., Zuzan, H., Olson, J.A. Jr, Marks, J.R. and Nevins, J.R. (2001) Predicting the clinical status of human breast cancer by using gene expression profiles. Proc. Natl. Acad. Sci. U.S.A. *98*, 11462-11467.

White, F.C., Benehacene, A., Scheele, J.S. and Kamps, M. (1997) VEGF mRNA is stabilized by ras and tyrosine kinase oncogenes, as well as by UV radiation – evidence for divergent stabilization pathways. Growth Factors *14*, 199-212.

Yoon, J.W., Kita, Y., Frank, D.J., Majewski, R.R., Konicek, B.A., Nobrega, M.A., Jacob, H., Walterhouse, D. and Iannaccone, P. (2002) Gene expression profiling leads to identification of GLI1-binding elements in target genes and a role for multiple downstream pathways in GLI1-induced cell transformation. J. Biol. Chem. *277*, 5548-5555.

Zabrenetzky, V., Harris, C.C., Steeg, P.S. and Roberts, D.D. (1994) Expression of the extracellular matrix molecule thrombospondin inversely correlates with malignant progression in melanoma, lung and breast carcinoma cell lines. Int. J. Cancer *59*, 191-195.

Zajchowski, D.A., Bartholdi, M.F., Gong, Y., Webster, L., Liu, H.L., Munishkin, A., Beauheim, C., Harvey, S., Ethier, S.P. and Johnson, P.H. (2001) Identification of gene expression profiles that predict the aggressive behavior of breast cancer cells. Cancer Res. *61*, 5168-5178.

Zhang, R., Averboukh, L., Zhu, W., Zhang, H., Jo, H., Dempsey, P.J., Coffey, R.J., Pardee, A.B. and Liang, P. (1998) Identification of rCop-1, a new member of the CCN protein family, as a negative regulator for cell transformation. Mol. Cell. Biol. *18*, 6131-6141.

Zong, Q., Schummer, M., Hood, L. and Morris, D.R. (1999) Messenger RNA translation state: the second dimension of high-throughput expression screening. Proc. Natl. Acad. Sci. U.S.A. *96*, 10632-10636.

Zuber, J., Tchernitsa, O.I., Hinzmann, B., Schmitz, A.-C., Grips, M., Hellriegel, M., Sers, C., Rosenthal, A. and Schäfer, R. (2000) A genome-wide survey of Ras transformation targets. Nature Genet. *24*, 144-152.

CHAPTER 9

RAS SIGNALING IN *C. ELEGANS*

A Genetic Overview

DENNIS J. EASTBURN AND MIN HAN

Howard Hughes Medical Institute, Department of Molecular Cellular and Developmental Biology, University of Colorado at Boulder, USA

Abstract: *C. elegans* is one of several model genetic organisms that have contributed substantially to our understanding of the role *ras* plays in signal transduction and development. Although *ras* has many developmental roles in *C. elegans*, it was the unbiased genetic identification and analysis of genes affecting vulval development that first provided a picture of *ras* function in a universal RTK-RAS-MAPK pathway. However, beyond the discovery of this seemingly simple linear pathway, *C. elegans* has furthered our appreciation for the complexities inherent in the regulation of RAS signaling. In this chapter, we summarize our current understanding of RAS signaling in *C. elegans*

Keywords: let-60, vulval induction, EGF signaling, MAP Kinase

1. INTRODUCTION

Genetic analysis in model organisms has made possible the identification of developmental roles for a number of signaling pathways as well as the elucidation of their molecular components and their ordered arrangement. There are few better examples of the value in a genetic approach to studying signaling pathways than the study of the RAS-MAPK pathway in yeast, *Drosophila* and the nematode *Caenorhabditis elegans*. Here, we will review the unique contributions that *C. elegans* has provided to our understanding of RAS signaling.

C. elegans is an ideal genetic system for the study of cell signaling during development. The invariant cell lineage, ability to observe individual cells and the relative ease of genetic and molecular analysis have aided investigators working with this organism. These features have been used to address fundamental questions surrounding cell signaling and pattern formation. One important achievement in *C. elegans* was the discovery of a developmental role for the proto-oncogene *ras*.

C. Der (ed.), RAS Family GTPases, 199–225.

In *C. elegans*, RAS is encoded by the *let-60* locus and functions in a variety of developmental and cellular processes. This chapter will focus on the role of *let-60 ras* in the proper specification of the vulval cell fate since much of our knowledge of RAS signaling in *C. elegans* has come through the genetic dissection of vulval development. By studying RAS signaling in such a developmental context, we have the opportunity not only to better understand the complex web of regulatory events governing RAS signaling, but also to view how an RTK-RAS-MAPK pathway is integrated with other cell signaling events to properly specify cell fate.

2. IDENTIFICATION OF RAS IN *C. ELEGANS*

The *C. elegans* vulva is derived from three of six initially equivalent vulval precursor cells (VPCs), P3.p-P8.p, which reside in an anterior to posterior row along the ventral midline of the hermaphrodite (Sulston and Horvitz, 1977). In wild-type animals, P5.p, P6.p and P7.p are induced by an epidermal growth factor (EGF)-like signal from the anchor cell (AC) in the overlying somatic gonad to adopt the vulval cell fate (Hill and Sternberg, 1992; 1993) (Figure 1A). P6.p adopts the primary (1°) vulval cell fate and divides a total of three times to contribute 8 cells to the vulval lineage. Both P5.p and P7.p adopt the secondary (2°) vulval cell fate and give rise to seven daughter cells each. Together, these 22 cells form the complete

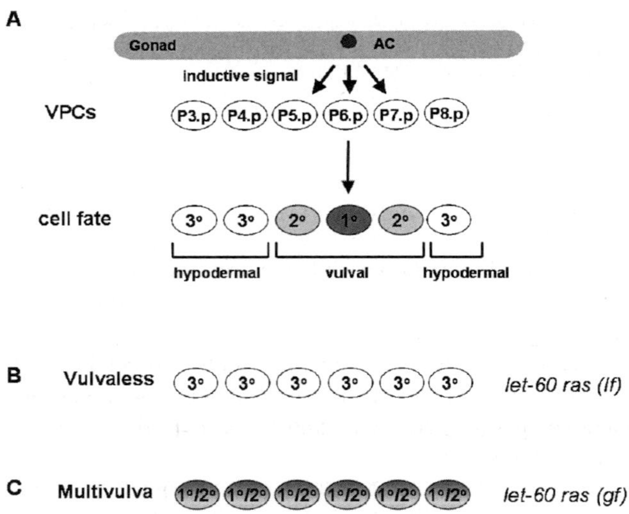

Figure 1. In wild-type hermaphrodites, an inductive signal produced by the anchor cell (AC) induces three of six equivalent vulval precursor cells (VPCs) to invariantly adopt either the 1° or 2° vulval cell fate (A). In the absence of an inductive signal or when RAS pathway activity has been compromised, the VPCs adopt the nonvulval 3° fate and fuse with the hypodermis creating a Vulvaless (Vul) phenotype (B). Conversely, hyperactivation of the RAS pathway results in a Multivulva (Muv) animal where ectopic VPCs are induced to adopt a vulval cell fate (C)

vulval lineage. The 22 vulval cells then undergo cell migrations, fusion events and morphogenesis to form the mature egg laying and reproductive vulval structure. The three un-induced VPCs (P3.p, P4.p and P8.p) divide once, adopt the nonvulval or tertiary (3°) fate and fuse to the overlying hyp7 syncytial hypodermal cell.

Although these fate decisions are invariant in wild-type animals, all six VPCs are equivalent and can adopt any of the three fates in various genetically mutant backgrounds (Kimble, 1981; Sternberg and Horvitz, 1986). Mutants that lack an inductive signal from the anchor cell or are incapable of responding to it, have VPCs that adopt the 3° fate and fuse to the hypodermis. These animals display a Vulvaless (Vul) phenotype and are clearly distinguished by an egg laying defective (Egl) phenotype (Figure 1B). Conversely, mutants with excessive signaling from the anchor cell or that have VPCs induced in a signal-independent manner are Multivulva (Muv) (Figure 1C). These animals are easily identified by ectopic pseudovulvae present on the ventral surface of the hermaphrodite. Thus, *C. elegans* researchers have an easily detectable readout for screening for mutants defective in vulval cell fate decisions (reviewed by Sternberg and Horvitz, 1991).

C. elegans RAS is encoded by the *let-60* gene (Han and Sternberg, 1990). *let-60*, as its name indicates, was first identified through studies of larval lethal mutations on chromosome IV (Rogalski et al., 1982; Clark et al., 1988). A series of genetic observations subsequently led to the realization that *let*-60 was a key component in vulval cell fate specification. Maternal rescue of strong loss-of-function (lf) mutations in *let-60* results in a Vul phenotype (Han et al., 1990). Additionally, a number of dominant negative (dn) and partial lf alleles of let-60 were isolated as dominant suppressors of the Muv phenotype of a *lin-15 (lf)* (a negative regulator of vulval induction) allele (Beitel et al., 1990; Han et al., 1990; Han and Sternberg, 1991). All of these dn alleles cause a Vul phenotype and most of them are recessively lethal. Gain-of-function (gf) Muv alleles of *let-60* (initially referred to as *lin-34*) were also isolated from wild-type strains and as suppressors of the Vul phenotype caused by mutation in several previously identified positive regulators of vulval induction (Ferguson et al., 1985; Beitel et al., 1990; Han et al., 1990). Thus, *let-60* appears to function as a genetic switch whose activity controls the vulval cell fate decision.

The mapping and molecular cloning of the *let-60* locus led to the discovery that it encodes the *C. elegans* homolog of Ras proteins (Han and Sternberg, 1990). *C. elegans let-60 ras* is 184 amino acids in length and contains all of the functional domains and motifs of mammalian Ras proteins including the C-terminal CAAX sequence that has been shown to be the site for post-translational farnesylation and membrane attachment. As in other systems, targeting LET-60 RAS to the membrane appears to play an important role in its activity since farnesyltransferase inhibitors reduce VPC induction in *C. elegans* (Hara and Han, 1995).

The conserved features between *C. elegans* LET-60 RAS and Ras proteins in other systems imply that many of the observations from studying *let-60 ras* are relevant to those systems as well. In support of this, mutations identified in the *let-60 ras* gene produce analogous defects when the corresponding residues are mutated in vertebrate *ras* genes. Five independently isolated gf *let-60 ras* alleles all result in the same G13E mutation (Beitel et al., 1990). In mammalian systems such a change has been found in Ras oncoproteins and was shown to reduce Ras GTPase activity and to produce a constitutively active form of the protein (Bos, 1989). Interestingly, no G12V gf mutation, which is the most common type of mutation in Ras oncoproteins, has been isolated in *let-60 ras*. This could be due to a stronger mutational effect resulting from a lower level of GTPase activity associated with the RAS (G12V) mutant protein as compared to that associated with RAS (G13E). A novel temperature sensitive gf allele of *let-60 ras* caused by a leucine to phenylalanine change at amino acid 19, conserved in all Ras proteins, has also been identified (Eisenmann and Kim, 1997). When introduced into human H-Ras, this mutation conferred a temperature-dependent effect on the GTPase activity of this protein. Similarly, lf mutations in *let-60 ras* have been shown to be relevant to mammalian Ras function as well (Howe and Marshall, 1993). However, despite these similarities, it is worth noting that there are three mammalian *ras* genes and *let-60 ras* is most similar in sequence to K-*ras*. Thus, some aspects of *let-60 ras* function might not be applicable to all mammalian *ras* genes. This could be especially true with respect to effectors of *let-60 ras*.

3. LET-60 RAS FUNCTIONS IN A CANONICAL RTK-RAS-MAPK PATHWAY IN THE VPCs

Genetic screens, particularly suppressor screens, have been very successful in identifying genes that appear to act in a common signal transduction pathway with *let-60 ras* (reviewed by Sternberg and Han, 1998; Kornfeld, 1997). Specifically, mutations in *let-23*, *sem-5*, *let-341/sos-1*, *let-60* and *lin-45* were identified as suppressors of the Muv phenotype of *lin-15 (lf)* alleles, while mutations in *lin-45*, *mek-2* and *mpk-1* were isolated as suppressors of the Muv phenotype of activated *let-60 ras* alleles. A number of these key genes required for correctly specifying vulval cell fates were cloned in the early 1990's and their epistatic relationship to *let-60 ras* and one another was determined. Through these efforts, it became clear that RAS was but one component of a canonical signal transduction pathway acting from the plasma membrane to the nucleus (Figure 2). Genetic analysis in yeast and flies as well as biochemical data from mammalian systems has demonstrated that this pathway is highly conserved across species and has many functions. Some components of this pathway are in fact so conserved that substitution of the *C. elegans* gene for the mammalian gene confers near wild-type activity to the pathway. This core RTK-RAS-MAPK pathway is now quite familiar to biologists and has been reviewed extensively so here we provide only a brief description.

let-23 encodes the only known homolog of the epidermal growth factor receptor (EGFR) in *C. elegans* (Aroian et al., 1990). This receptor tyrosine kinase (RTK) is activated by the EGF-like signal produced by the anchor cell (AC), LIN-3 (Hill and Sternberg, 1992). Loss of either of these two components results in a Vul phenotype (Ferguson et al., 1987; Aroian and Sternberg, 1991; Hill and Sternberg, 1992).

Furthermore, LET-23 RTK is asymmetrically localized to the basolateral region of the plasma membrane where it is more easily able to respond to the LIN-3 inductive signal (Simske et al., 1996). Three genes, *lin-2*, *lin-7* and *lin-10*, were identified by isolating mutations that cause a Vul phenotype (Ferguson et al., 1985, 1987). Subsequently it was determined that these genes encode PDZ domain containing proteins that form a complex which binds to the intracellular portion of LET-23 RTK and mediates its proper basolateral localization (Hoskins et al., 1996; Simske et al., 1996; Kaech et al., 1998; Whitfield et al., 1999).

sem-5 and *let-341/sos-1* are two genes thought to be responsible for transducing the signal from activated LET-23 RTK to LET-60 RAS. *sem-5* encodes a Grb2 like protein with SH2 and SH3 domains (Clark et al., 1992). *let-341/sos-1* encodes the *C. elegans* homolog of Sos (son of sevenless) (Chang et al., 2000). Like Sos in other systems, SOS-1 contains the CDC25 and DH domains responsible for guanine nucleotide exchange factor activity upon Ras family GTPases. The SEM-5 adapter protein is thought to bind to activated LET-23 RTK and thereby recruit SOS-1 to the plasma membrane where it can activate LET-60 RAS. One interesting aspect to SOS-1 function in the vulva is that it does not appear to be capable of transducing the totality of the signal provided by LIN-3. Thus, it is possible there are other factors acting at this level that have not yet been identified.

LET-60 RAS can be thought of as a molecular switch. Once LET-60 is induced to switch to its GTP bound conformation, it triggers the activation of a kinase cascade involving *lin-45* (RAF), *mek-2* (MAP kinase kinase) and *mpk-1* (MAP kinase) (Han et al., 1993; Lackner et al., 1994; Wu and Han, 1994; Kornfeld et al., 1995; Wu et al., 1995; Church et al., 1995; Hsu et al., 2002). Hypomorphic mutations in any of these three essential kinases can suppress an activated *let-60 ras* mutation or cause a highly penetrant Vul phenotype on their own. Upon activation, members of the MAP kinase family can translocate to the nucleus where they regulate changes in gene expression through the phosphorylation of a variety of transcription factors.

4. MODIFIERS OF LET-60 RAS SIGNALING

The canonical RTK-RAS-MAPK pathway described above defines a direct and linear route from the plasma membrane to the nucleus of the cell comprised of genes essential for vulval development. In reality, this pathway must be integrated into a complex cellular environment and is subject to regulation and modification at every step. However, in *C. elegans*, mutation of a factor that positively or negatively modifies the RTK-RAS-MAPK pathway often has no overt effect on vulval cell fate specification in an otherwise wild-type animal. Only through the use of sensitive suppressor and enhancer screens using alleles of genes such as *let-60 ras* and *let-23 RTK*, have researchers been able to identify numerous components that modify the level of RAS signaling (Figure 2). (Note: *lip-1* and *par-1* are negative modifiers of RAS signaling but will be discussed elsewhere in this chapter.)

4.1 Negatively Acting Modifiers

Inappropriate RTK-RAS-MAPK pathway signaling can have disastrous consequences for development or lead to the onset of human cancers. Thus there are likely numerous and redundant mechanisms to negatively regulate and prevent ligand-independent activation of this pathway. A number of genes that negatively modify RAS signaling in *C. elegans* have been identified and will be discussed in this section. Although all of these genes are negative regulators of RAS signaling, loss of any one of these genes is not sufficient to bypass the requirement for canonical RAS pathway components. Double mutants of these modifiers show differing levels of RTK-RAS-MAPK pathway hyper-activation, demonstrating a degree of redundancy in the negative regulation of RAS signaling.

The gene *sli-1* was implicated as a negative regulator of vulval cell fate specification through a screen for suppressors of a hypomorphic allele of *let-23* (Jongeward et al., 1995). Loss-of-function mutations in *sli-1* also suppress many of the defects in *sem-5 (lf)* animals but can not significantly suppress *let-60 (lf)* mutations, suggesting that SLI-1 is functioning at or near the level of the LET-23 RTK mediated step in the pathway. *sli-1* is highly similar to the mammalian proto-oncogene *c-cbl* and it was through *sli-1* analysis in *C. elegans* that the first link between *c-cbl* and RTKs was established (Yoon et al., 1995). SLI-1 and human c-Cbl share several functional domains including, a divergent SH2 domain and a C-terminal RING finger domain implicated in ubiquitination of active RTKs (reviewed in Thien and Langdon, 2001). It has been proposed from work on mammalian c-Cbl that ubiquitination of active EGFRs can direct their endocytic sorting and degradation (Levkowitz et al., 1998). Probably in a manner similar to c-Cbl, SLI-1 binds to an inhibitory tyrosine in activated LET-23 RTK and directs receptor degradation via ubiquitination. However, ubiquitination of LET-23 RTK alone is not sufficient to explain the mechanism for SLI-1 action since the RING finger domain of SLI-1 appears to be partially dispensable for its function (Yoon et al., 2000).

UNC-101 is a homolog of the medium chain of the *trans*-Golgi clathrin-associated protein complex, AP47 (Lee et al., 1994). Loss-of-function mutations in this gene result in a similar phenotype to *sli-1* alleles in that they place UNC-101 action at the level of LET-23 RTK and upstream of LET-60 RAS. Subsequently it was realized that *unc-101* is functioning redundantly in vulval cell fate specification with *apm*-1, another homolog of AP47 (Shim et al., 2000). Although a clear mechanism for UNC-101 and APM-1 in VPC induction does not exist, these proteins might attenuate RTK-RAS-MAPK signaling by functioning to control the intracellular trafficking of LET-23 RTK. Alternatively, APM-1 and UNC-101 could indirectly control the degradation of LET-23 RTK by regulating the sorting of a factor necessary for this process.

Mutations in *ark-1* were found by screening for synthetic enhancers of an allele of *sli-1* (Hopper et al., 2000). *ark-1* mutations cause overinduction of the VPCs and embryonic lethality when combined with *sli-1*, or *unc-101* mutations. These phenotypes are thought to result from hyperactivation of LET-60 RAS. Like SLI-1 and UNC-101, ARK-1 also acts at or near the level of LET-23 RTK and is dependent on SEM-5 activity. ARK-1 is a predicted tyrosine kinase and has a similar domain structure to the Ack subfamily of kinases. A C-terminal proline rich region of ARK-1 was found to interact with SEM-5 in a yeast two-hybrid assay. An untested model has been proposed for ARK-1 that has SEM-5 recruiting ARK-1 to the LET-23 RTK signaling complex where it can negatively regulate pathway activity redundantly with other factors like SLI-1, UNC-101, APM-1 and GAP-1.

Acting in opposition to RAS guanine nucleotide exchange factors are a group of proteins (RasGAPs) that help "switch" RAS to an inactive state by stimulating its intrinsic GTPase activity. *gap-1* encodes a RasGAP and mutations in this gene have been isolated that suppress both the Vul and larval lethal phenotypes of RAS pathway mutants (Hajnal et al., 1997). Loss of RasGAP homologs in *Drosophila*, yeast and mammals cause constitutive RAS pathway activation. However, lf mutations in *gap*-1 alone are unable to cause ectopic VPC inductions characteristic of *let-60 ras* activation. This may be due to the fact that GAP-1 plays only a limited role in the negative regulation of LET-60 RAS in *C. elegans* or, there is another unidentified RasGAP that functions redundantly with GAP-1 in the VPCs.

In addition to its well defined positive role in VPC induction, LET-23 RTK itself appears to cell non-autonomously negatively regulate RAS signaling. A Hyperinduced (Hin) phenotype is seen in weak loss of function *let-23* alleles (particularly those that affect LIN-3 EGF binding or receptor localization), alleles of *lin-2*, *lin-7* and *lin-10* and double mutants with the negative modifiers of LET-23 RTK signaling discussed above (Aroian and Sternberg, 1991; Jongeward et al., 1995; Hajnal et al., 1997; Hopper et al., 2000). This phenotype is different from the Muv phenotype in several ways, including the requirement for the LIN-3 signal from the AC. A model where LET-23 RTK, expressed in P6.p (the VPC closest to the AC) and to a lesser extent in P5.p and P7.p, binds to and titrates the diffusible LIN-3 signal from the extracellular environment thus preventing it from reaching and activating LET-23 RTK in the distal VPCs has been proposed to account for this Hin phenotype (Hajnal et al., 1997). Ligand sequestration might be a general mechanism for the localization and attenuation of signals activating RTK linked RAS pathways.

sur-5 was defined through mutations that suppress the Vul phenotype caused by the dominant negative (K16N) *let-60 ras* allele (Gu et al., 1998). *sur-5* encodes a novel protein with two AMP binding motifs and one ATP/GTP binding motif. A potential human homolog of *sur-5* has been identified that is 35% identical to the worm protein. Unlike the previous negative modifiers discussed in this section, SUR-5 appears to function specifically at the level of LET-60 RAS. Interestingly, loss of *sur*-5 fails to suppress the Vul phenotypes caused by hypomorphic alleles of *let-23* and *sem*-5. This suggests that the mechanism of suppression for *sur-5 (lf)* is specific to dominant negative alleles of *ras*. Additionally, *sur-5 (lf)* alleles can only suppress one class of dn *let-60 ras* mutants (group I alleles with mutations in loop 1 of RAS). Since all dn *let-60 ras* mutants can be suppressed by a gf (G13E) *let-60 ras* mutation in *trans*-heterozygotes, it is likely that the dominant negative RAS mutations function by titrating upstream activators such as guanine nucleotide exchange factors (Han and Sternberg, 1990; Beitel et al., 1990; Han and Sternberg, 1991). This implies that SUR-5 normally functions to inhibit an upstream activator of RAS and that the dn *let-60 ras* alleles that *sur-5 (lf)* is not able to suppress are toxic to another activator of RAS. The idea that there could be multiple activators of LET-60 RAS is supported by the previously mentioned observation that SOS-1 is unable to transduce the totality of the LET-23 RTK signal to LET-60 RAS (Chang et al., 2000).

Lastly, the G-protein-coupled receptor (GPCR) SRA-13 and its Gα subunit GPA-5 were recently reported to negatively modify RTK-RAS-MAPK signaling at the level of or upstream of MPK-1 (Battu et al., 2003). Furthermore, these two proteins function via an unknown mechanism to link environmental conditions such as the availability of food to the RAS pathway and vulval development. At present, it is not clear if SRA-13 and GPA-5 act cell autonomously in the VPCs or non-autonomously to control a secondary signal which globally affects RAS signaling.

4.2 Positively Acting Modifiers

Failure to achieve appropriately high levels of RAS signaling can also have deleterious developmental consequences. Positive modifiers of RAS signaling in *C. elegans* have been identified that converge at multiple points on the pathway to regulate and facilitate the transduction of an extracellular signal to the nucleus. Like the negative modifiers of RAS signaling in *C. elegans*, there is a level of redundancy in the genes classified as positive modifiers and when individually mutated, they display no VPC induction defects. These proteins appear to primarily function parallel to or downstream of LET-60 RAS in the VPCs.

A putative null allele of *ptp-2* was isolated and shown to suppress the Muv phenotype of *let-23 RTK* and *let-60 ras* gf alleles in the vulva, indicating that *ptp-2* normally functions to positively regulate RAS signaling (Gutch et al., 1998). Although PTP-2 is not required for normal VPC induction, it performs an essential role in oogenesis (discussed in section 7). Epistasis analysis with activated LET-60 RAS suggests that PTP-2 is functioning downstream of or parallel to RAS. PTP-2 is similar to SHP (SH2 domain containing protein tyrosine phosphatases) proteins and has the characteristic tandem SH2 domains and C-terminal protein tyrosine phosphatase (PTP) catalytic domain. As has been proposed for SHPs in other systems, it is possible that PTP-2 is directly interacting with activated LET-23 and positively transducing a signal for VPC induction parallel to the RAS-MAPK pathway.

One possible way to regulate the kinetics of RAS-MAPK pathway activation is through the use of scaffolding proteins (Levchenko et al., 2000). Screens for suppressors of the Muv phenotype caused by activated LET-60 RAS in *C. elegans* have identified positive regulators that could perform such a scaffold like function. *sur-8/soc-2* is such a positive regulator, which has been genetically placed between *let-60 ras* and *lin-45 raf* (Sieburth et al., 1998; Selfors et al., 1998; Yoder et al., 2004). *sur-8* encodes a protein that contains 18 tandem leucine-rich repeats (LRRs) which have been implicated in protein-protein interactions. Interestingly, yeast adenylate cyclase, which is an effector of Ras in yeast, also contains a large LRR domain. It was shown that SUR-8 and a functional homolog in humans could physically interact with K-Ras and N-Ras (Sieburth et al., 1998). Additional work demonstrated that human SUR-8 also binds Raf and forms a ternary complex together with Ras (Li et al., 2000). SUR-8 was able to stimulate EGF-induced Raf and ERK activation but was unable to affect activation of ERK by activated Raf or MEK. These findings support the genetic data and suggest a model of SUR-8 acting as a scaffold to facilitate the interaction between RAF and RAS.

Another positive modifier that was identified as a suppressor of gf *let-60 ras* and that is thought to function as a scaffolding protein in the RAS-MAPK pathway is *ksr-1* (Kornfeld et al., 1995; Sundaram and Han, 1995). KSR-1 has a C-terminal kinase domain with some similarity to proteins of the Raf serine/threonine kinase family. There have been conflicting reports on whether mammalian KSR kinase activity exists or is important (reviewed in Morrison, 2001). In *C. elegans* however, a predicted kinase dead *ksr-1* transgene is still able to efficiently compliment *ksr-1* alleles arguing against an essential role for this kinase activity (Stewart et al., 1999). Genetic analysis using a gf *Drosophila raf* transgene in *C. elegans* suggested that *ksr-1* functions upstream of *raf* in VPC induction (Sieburth et al., 1999). However, by using a constitutively active *C. elegans lin-45 (gf) raf* transgene (Chong et al., 2001) in epistasis experiments, *ksr-1* was shown to function genetically downstream of *lin-45 raf* (Yoder et al., 2004). This new genetic data is more consistent with biochemical experiments showing KSR likely serves as a scaffolding protein for RAF and MEK (Roy et al., 2002). Recently a second KSR gene, *ksr-2*, was discovered and shown to function redundantly with *ksr-1* in a number of LET-60 RAS mediated processes including vulval cell fate specification (Ohmachi et al., 2002).

sur-6 mutations were also isolated as suppressors of the Muv phenotype caused by activated LET-60 RAS (Sieburth et al., 1999). SUR-6 is a *C. elegans* homolog of the regulatory B subunit of protein phosphatase 2A (PP2A). It is likely that SUR-6 PP2A-B is a positive regulator of the PP2A catalytic core since mutations in the catalytic subunit of PP2A, *let-92* (PP2A-C), also reduce vulval induction (Sieburth et al., 1999; Kao et al., 2004). Genetic analysis further demonstrated that *sur-6* acts downstream of *lin-45 raf* but upstream of *ksr-1/2* (Yoder et al., 2004; Kao et al., 2004). Targets of SUR-6 PP2A-B and the LET-92 PP2A-C catalytic core could include inhibitory phosphates on KSR-1/2 or LIN-45 RAF; however, the inhibitory phosphorylation sites mutated in the *lin-45 (gf) raf* transgene are excluded as possible sites of PP2A complex action since the Muv phenotype caused by this transgene is dependent on SUR-6 activity.

Two additional genes, *cdf-1* and *sur-7*, were defined by mutations that suppress the vulval defects of activated LET-60 RAS. *cdf-1* and *sur-7* function in a similar genetic manner to each other and to *ksr-1* and *sur-6* (Jakubowski and Kornfeld, 1999; Bruinsma et al., 2002; Yoder et al., 2004). Thus, *cdf-1*, *sur-7*, *ksr-1*/2 and *sur-6* are all likely to comprise a regulatory convergence point at the level of LIN-45

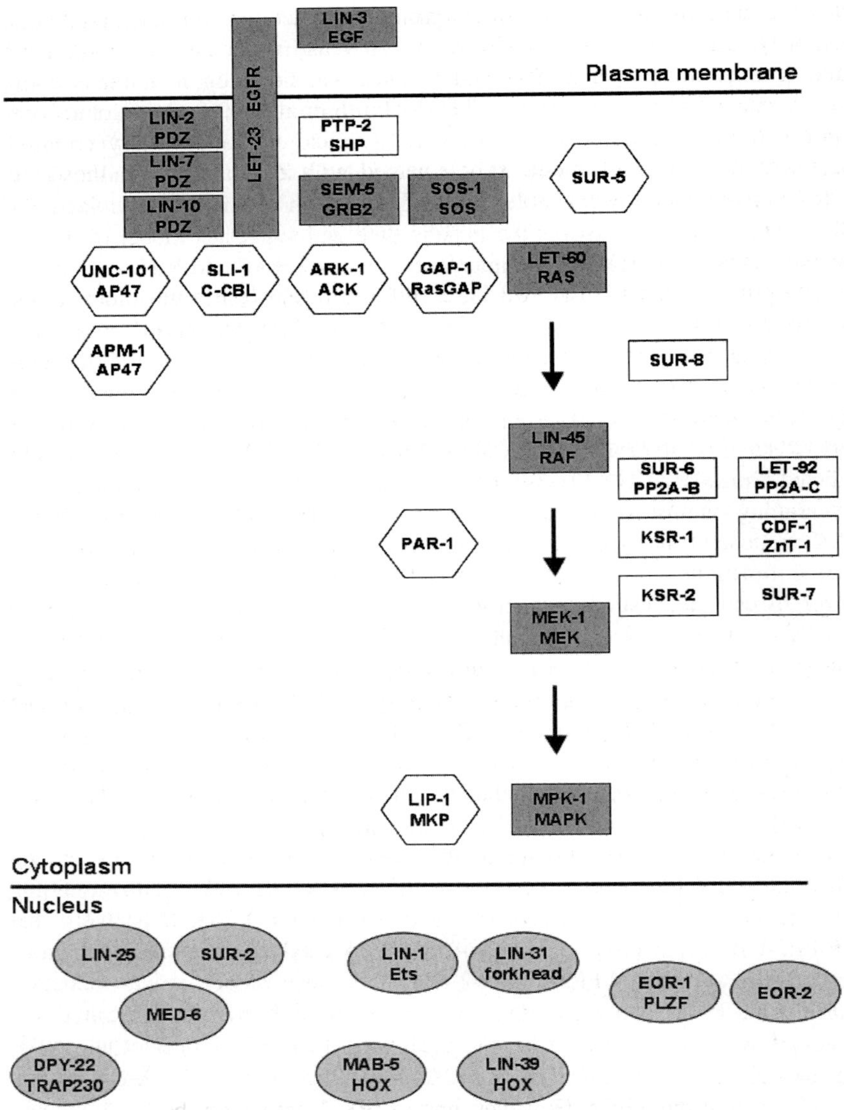

Figure 2. Schematic representation of the RTK-RAS-MAPK signaling pathway in the VPCs. Essential RTK-RAS-MAPK pathway components are represented as shaded squares, positive modifiers are open squares, negative modifiers are hexagons and factors thought to act at the transcriptional level downstream of MPK-1 are ovals

RAF activation of MEK-2. *cdf-1* encodes a member of the cation diffusion facilitator (CDF) family of proteins that is similar to vertebrate ZnT-1 (Jakubowski and Kornfeld, 1999; Bruinsma et al., 2002). *sur-7* encodes a more distantly related CDF family member (Yoder et al., 2004). CDF proteins are thought to lower cytosolic Zn^{2+} concentration and animals carrying *cdf-1* or *sur-7* mutations display increased sensitivity to Zn^{2+}. Growth media supplemented with Zn^{2+} were also shown to suppress the Muv phenotype of activated LET-60 RAS. Therefore, cytosolic Zn^{2+} appears to negatively regulate RAS-MAPK signaling in the VPCs.

The mechanism by which Zn^{2+} regulates RAS-MAPK signaling is likely to involve phosphorylation of KSR-1/2. The addition of Zn^{2+}, but not other heavy metal ions, to mammalian cells lead to an increase in phosphorylated KSR that was not a result of PP2A phosphatase inhibition (Yoder et al., 2004). Mammalian C-TAK1 kinase negatively regulates RAS-MAPK activity by phosphorylating KSR (Muller et al., 2001) and the *C. elegans* homolog of C-TAK1, *par*-1, was subsequently shown to function genetically upstream of *ksr-1* to negatively regulate vulval induction (Yoder et al., 2004; Kao et al., 2004). Thus, PAR-1 or a currently unidentified kinase could be regulating KSR-1/2 via phosphorylation in response to elevated Zn^{2+} levels.

5. FACTORS DOWNSTREAM OF MPK-1

As mentioned earlier, one of the main mechanisms by which the RAS-MAPK pathway is thought to exert its cellular response is by altering the regulation of transcription. It is also at this point that due to branching of the signal and the parallel function of the components involved, epistasis analysis fails to define a clear linear order to the pathway. In the VPCs, two proteins are thought to make up the transcriptional terminus of the RAS pathway, LIN-1 and LIN-31 (Figure 2). Both *lin-1* and *lin-31* were identified as negative regulators of vulval cell fate specification from early screens for Muv animals (Ferguson et al., 1987). LIN-1 is an Ets domain transcription factor family member (Beitel et al., 1995). LIN-31 is an HNF-3/forkhead transcription factor homolog and has an expression pattern that is restricted to the VPCs (Miller et al., 1993; Tan et al., 1998). Consequently, LIN-31 could be a major determinant of the vulval specific response achieved through activation of the rather ubiquitous RAS-MAPK pathway. It is thought that in the absence of RAS signaling, LIN-1 and LIN-31 form a heterodimer that antagonizes vulval induction in the VPCs. However, upon MPK-1 activation, both LIN-1 and LIN-31 are phosphorylated causing disruption of this heterodimer and progression of vulval cell fate specification (Tan et al., 1998). In support of this, gf alleles of *lin-1* were isolated with mutations in a C-terminal region of the protein shown to be critical for phosphorylation by MAP kinase (Jacobs et al., 1998). Although both LIN-1 and LIN-31 have negative roles in RAS-MAPK signaling, closer genetic and phenotypic analysis has elucidated a more complicated picture of their roles in vulval cell fate specification. The Vul phenotype and/or deregulation of vulval cell fates, often seen in many *lin-31 (lf)* animals, led to the proposal that LIN-31 was also acting as a positive regulator in VPC induction (most likely after MPK-1 activation) (Miller et al., 1993; Tan et al., 1998; Miller et al., 2000). A positive role, which may require the function of *sur-2* and *lin-25*, has more recently been proposed for LIN-1 in VPC induction as well (Howard and Sundaram, 2002).

sur-2 and *lin-25* are two genes that positively regulate vulval induction downstream of MPK-1 and have additional roles in RAS signaling outside the VPCs (Singh and Han, 1995; Tuck and Greenwald, 1995; Nilsson et al., 2000). SUR-2 and LIN-25 are predicted to function together in vulval cell fate specification (Nilsson et al., 1998). Loss of either *sur-2* or *lin-25* causes a relatively strong Vul phenotype, suggesting they help transduce a large part of the signal from RAS. *lin-25* encodes a protein with no known homologs. *sur-2* encodes a protein whose homologs function in the Mediator transcriptional coactivator complex. The Mediator complex consists of 20-30 proteins that are thought to transmit transcriptional instructions from gene specific transcription factors to the basal transcriptional machinery (Boube et al., 2002). Human SUR-2 physically interacts with and promotes transcriptional activation by the ETS-family member Elk-1 in response to MAPK phosphorylation but is not required for the transcriptional activity of numerous other activators (Boyer et al., 1999; Stevens et al., 2002). This data, taken together with the limited phenotypes and restricted expression patterns of *C. elegans sur-2*, suggest that SUR-2 is responsible for functionally bridging a subset of activators downstream of RAS-MAPK signaling to the Mediator complex.

Another gene thought to function in a *C. elegans* transcriptional Mediator complex is *dpy-22/sop-1*, the homolog of human TRAP230. DPY-22/SOP-1 was first identified as a specific inhibitor of BAR-1 dependent Wnt signaling (Zhang and Emmons, 2000). Subsequently, alleles of *dpy-22/sop-1* were identified that enhanced VPC induction (Moghal and Sternberg, 2003). Although DPY-22/SOP-1 negatively regulates Wnt signaling, it appears that, in the VPCs, this protein primarily acts to inhibit the RAS pathway. *dpy-22/sop-1* alleles display a wide range of phenotypes but they do not cause significant lethality or appear to affect all cell types equally. Thus, like SUR-2, DPY-22/SOP-1 functions with only a subset of tissue specific transcription factors and signaling events during development.

In contrast to the Mediator subunits discussed above, MED-6 (also known as LET-425) appears to be a subunit of the Mediator complex that is required broadly for transcription (Kwon et al., 1999; Boube et al., 2002). *med-6* is essential in yeast, and is required throughout development in *C. elegans*. Although loss of *med-6* results in embryonic lethality, maternally rescued animals demonstrated that *med-6* has a role in positively regulating RAS signaling and VPC induction (Kwon and Lee, 2001). It is likely MED-6 functions as an essential core component of a universal Mediator complex.

eor-1 and *eor-2* both positively regulate VPC induction and function downstream of MPK-1. These genes were identified in a screen for enhancers of the larval lethal phenotype and vulval defects of a partial lf allele of *lin-45 raf* (Rocheleau et al., 2002). From phenotypic and genetic analysis, it is thought that both *eor-1* and *eor-2* function together at the same point in the RAS pathway (Howard and

Sundaram, 2002). Additionally, these two genes function redundantly with the Mediator genes, *sur-2* and *lin-25*, to achieve wild-type levels of RAS signaling. Unlike *sur-2* and *lin*-25, eor-1 or *eor-2* single mutants do not overtly compromise RAS signaling; therefore, they are not essential components of the RAS pathway transcriptional response. In agreement with a role for *eor-1* and *eor-2* acting downstream of MPK-1, *eor-1* encodes the ortholog of human PLZF, a BTB/zinc-finger transcription factor that is fused to RARα in acute promyelocytic leukemia. *eor-2* encodes a novel protein with no recognizable homologs. Both EOR-1 and EOR-2 are also localized in the nucleus. In addition to their role in the RAS-MAPK pathway, EOR-1 and EOR-2 also positively contribute to a Wnt signaling pathway that promotes vulval cell fate specification. The Wnt pathway and the roles of EOR-1 and EOR-2 with regard to that pathway will be discussed below.

6. SIGNAL SPECIFICITY AND INTEGRATION WITH OTHER PATHWAYS

A canonical RTK-RAS-MAPK pathway is used in multiple processes throughout development yet can achieve unique and tissue specific responses for each of these events (see section 7). One way in which the specificity of RTK-RAS-MAPK signaling can be achieved is through the regulation of signaling kinetics. This regulation can take place at the receptor or any other point in the pathway and is modulated in a cell type specific manner by genes such as the positive or negative modifiers of RAS signaling discussed earlier. Specificity can also be achieved through pathway interaction with unique or cell type specific combinations of transcriptional components. Both the modifiers of RTK-RAS-MAPK signaling in the vulva and the unique transcriptional components, such as LIN-31, are expressed in the VPCs before inductive signaling and help define VPC competence to respond uniquely to that signal. There are a number of spatial and temporal regulatory events governing VPC competence that we will not describe in this chapter (for review see Wang and Sternberg, 2001). Lastly, another way in which cell fate decisions are properly specified in response to a common RTK-RAS-MAPK pathway is through interactions with other signaling pathways. These other pathways or components can be either antagonistic or synergistic. Here we will describe some of the other signaling events in the VPCs and examine how they are integrated with RAS signaling to properly specify vulval cell fates (Figure 3).

6.1 Wnt Signaling and the *lin-39* Hox Gene Promote VPC Fate Specification with LET-60 RAS

lin-39 encodes a homeobox gene that is most similar to the *Drosophila* homeotic proteins Deformed and Sex combs reduced (Wang et al., 1993; Clark et al., 1993). Like other homeotic genes, *lin-39* is thought to convey anteroposterior information to cells in the developing animal. Consequently, *lin-39* is an essential spatial cue in the proper specification of the VPC fate. *lin-39* function is required to keep the VPCs from fusing to the hypodermis and allow the VPCs to remain competent to respond to the LIN-3 inductive signal. In addition to this role, *lin-39* has a later function in positively regulating vulval induction with the RAS pathway (Clandinin et al., 1997; Maloof and Kenyon, 1998). LIN-39 expression levels increase during the period of vulval induction; this increase in expression was shown to be dependent on the RAS pathway (Maloof and Kenyon, 1998). However, a Wnt/*bar-1* signaling pathway was also shown to be essential for the upregulated expression of LIN-39 (Eisenmann et al., 1998). *bar-1* encodes a β-catenin/Armadillo homolog that functions in a conserved Wnt signaling pathway (Eisenmann et al., 1998). Mutation of *bar-1* causes defects in VPC specification and competence as well as VPC induction. These defects were correlated with an inability to upregulate LIN-39 expression in a *bar-1* mutant background since forced expression of LIN-39 was able to partially overcome the defects associated with the loss of *bar-1* (Eisenmann et al., 1998). Therefore, both the RAS pathway and a *bar-1* mediated Wnt pathway converge on the expression of *lin-39* to properly specify the vulval cell fate.

After the initial observation that Wnt signaling affects vulval cell fate specification, a number of other Wnt pathway components were shown to have vulval defects (Eisenmann and Kim, 2000; Hoier et al., 2000; Gleason et al., 2002). Mutations in components of the Wnt pathway do not cause as severe a Vul phenotype as do mutations in members of the canonical RAS-MAPK pathway, suggesting that the Wnt pathway plays a lesser role in VPC induction. Despite not being necessary for vulval cell fate specification, hyper-activation of the Wnt pathway is sufficient to bypass the requirement for RAS signaling in VPC induction (Gleason et al., 2002). A gf mutation in *bar-1* or the loss of *pry-1* (a functional Axin homolog and negative regulator of Wnt signaling) results in an Overinduced phenotype that does not appear to be dependent on the components of the RAS pathway (Korswagen et al., 2002; Gleason et al., 2002). However, a requirement for low or basal levels of RAS signaling on this Overinduced phenotype has not been ruled out.

Mutations in either *eor-1* or *eor-2* are able to suppress the Overinduction phenotype caused by *pry-1 (lf)* implying that they are also mediators of the Wnt pathway and do not exclusively function in RAS signaling (Howard and Sundaram, 2002). It is not likely that the Wnt pathway functions solely through *lin-39* to promote VPC induction since overexpression of LIN-39 during the period of induction could not phenocopy the Overinduction seen for Wnt pathway hyperactivation (Maloof and Kenyon, 1998; Gleason et al., 2002). Accordingly, epistasis experiments suggest that *eor-1* and *eor-2* function parallel to *lin-39* as key components responsible for the integration of signals from both the RAS and Wnt pathways in vulval cell fate specification (Howard and Sundaram, 2002).

Although the six VPCs are thought of as equivalent in their ability to respond to the LIN-3 signal, in fact, this is not entirely true. Another Hox gene, *mab-5*, is expressed in the two most posterior VPCs, P7.p and P8.p, and acts antagonistically to *lin-39* to inhibit the vulval cell fate (Salser et al., 1993; Clandinin et al., 1997). It is likely that *mab*-5 functions at the level of VPC competence rather than directly inhibiting RTK-RAS-MAPK signaling.

6.2 LIN-12/Notch Mediated Lateral Signaling Acts Antagonistically to RAS Pathway Activation

Correct vulval cell fate specification is a more complicated process than simply receiving and responding to the LIN-3 inductive signal. There are two vulval fates, 1° (adopted by P6.p) and 2° (adopted by P5.p and P7.p). The decision of which fate to adopt is controlled by a lateral signaling pathway mediated by LIN-12, a protein similar to the Notch family of receptors (Greenwald et al., 1983; Yochem et al., 1988; Sternberg, 1988; Sternberg and Horvitz, 1989). LIN-12 appears to be both necessary and sufficient for the adoption of the 2° fate (Greenwald et al., 1983). A sequential signaling model in which P6.p receives the most LIN-3 inductive signal from the AC, adopts the 1° fate, and then signals to the adjacent VPCs, P5.p and P7.p, via LIN-12 to adopt the 2° fate was proposed based on genetic mosaic analyses with *let-23 RTK* (Koga and Ohshima, 1995; Simske and Kim, 1995). In accordance with this model, the ligands for LIN-12 are expressed specifically in P6.p and in response to LIN-3 (Chen and Greenwald, 2004). However, the LIN-3 gradient produced by the anchor cell was shown to initially activate RAS-MAPK signaling in P5.p and P7.p and abnormally high levels of LIN-3 can override lateral signaling causing adjacent VPCs to adopt 1° fates (Katz et al., 1995; Yoo et al., 2004). Thus, there are likely to be mechanisms for LIN-12 lateral signaling and the RAS pathway to antagonize each other and reinforce a 1°/2° fate decision that is initially established by a graded distribution of the LIN-3 inductive signal. Recent work has examined the mechanisms of crosstalk between these two pathways.

lip-1 was identified by searching the *C. elegans* genome for homologs of vertebrate MAP kinase phosphatases (MKPs) (Berset et al., 2001). Genetic analysis of *lip-1* suggests that it functions to negatively regulate VPC induction at the level of *mpk-1*. Furthermore, *lip-1* was expressed in 2° but not 1° lineages and this expression was dependent on *lin-12*. Hence, it is likely that LIP-1 is functioning to promote the 2° fate by negatively modifying RAS signaling at the level of MPK-1 in a LIN-12 dependent manner. Loss of *lip-1* alone was not sufficient to disrupt lateral signaling and promote ectopic 1° fates; therefore, it is probably functioning redundantly with other LIN-12 dependent factors to promote the 2° fate in P5.p and P7.p. In support of this idea, a number of additional genes were identified, using a biocomputational approach, as being transcriptional targets of LIN-12 signaling (Yoo et al., 2004). These *lateral signal target* (*lst*) genes were shown to negatively regulate RAS-MAPK activity in the presumptive 2° cells and at least two of them probably participate in degrading LET-23 RTK. These studies demonstrate that a major function of LIN-12/Notch signaling in 2° cell fate determination is the downregulation of RTK-RAS-MAPK pathway.

A converse mechanism to that proposed above appears to be functioning in P6.p to reinforce RAS mediated signaling for the 1° fate. Concomitant with an increase in LET-23 expression, LIN-12 protein levels are downregulated in P6.p which adopts the 1° fate (Simske et al., 1996; Levitan and Greenwald, 1998). This downregulation of expression was shown to be due to a 'downregulation target sequence' (DTS) located in the cytoplasmic domain of the LIN-12 protein (Shaye and Greenwald, 2002). Upon removal of the DTS or in *sur-2* mutants, which are defective in lateral signaling, there is a failure in the downregulation of LIN-12 expression (Singh and Han, 1995; Shaye and Greenwald, 2002). Therefore, the RAS pathway activated in P6.p is likely to target LIN-12 for destruction through the DTS. Sequence conservation of potential DTSs in vertebrate Notch proteins suggests this mechanism of receptor downregulation is conserved.

6.3 Two Redundant Pathways Inhibit Basal Levels of RTK-RAS-MAPK Activity

The synthetic multivulva (synMuv) genes are comprised of two classes, A and B, of functionally redundant genes (Horvitz and Sulston, 1980; Ferguson and Horvitz, 1989). Mutation of either a class A or class B gene alone has no phenotype, but simultaneous mutation of both classes of genes produces a Multivulva phenotype. Thus, these genes function to redundantly antagonize vulval induction. The list of genes identified as an A or B class synMuv gene seems to be constantly expanding; there are currently at least 19 classified members.

Some class B synMuv genes encode homologs of the Rb/E2F transcriptional regulatory complex. These include *lin-35* which encodes a *C. elegans* protein similar to the pRB tumor suppressor as well as *efl-1* E2F and *dpl-1* DP (Lu and Horvitz, 1998; Ceol and Horvitz, 2001). A number of other class B synMuv genes encode members of the NuRD nucleosomal remodeling and deacetylase complex(e.g., Lu and Horvitz, 1998; Solari and Ahringer, 2000; von Zelewsky et al., 2000; Chen and Han, 2001b). Biochemical work in mammalian tissue culture has indicated that these two complexes interact with each other to inhibit transcription. Accordingly, many of the class B synMuv genes could function in a complex acting to repress transcription at promoters regulated by *efl-1* E2F and *dpl-1* DP. However, no such complex has yet been detected in *C. elegans* and furthermore, it appears as though there are some important genetic differences between the Rb/E2F and NuRD complexes during vulval development (Chen and Han, 2001a) and cell proliferation in general (Boxem and van den Heuvel, 2002; Fay et al., 2002). The redundantly functioning synMuv A genes either encode novel proteins or remain uncloned (Clark et al., 1994; Huang et al., 1994).

Although many of the synMuv genes function cell autonomously in the VPCs, genetic mosaic analyses have indicated that several of them, such as *lin-15A*, *lin-37* and *lin-13*, function non-autonomously in the overlying hypodermal cell, hyp7 (Herman and Hedgecock, 1990; Hedgecock and Herman, 1995; Melendez and Greenwald, 2000). This could be evidence for the existence of a signaling pathway between the hyp7 cell and the VPCs.

Epistasis experiments done to examine the genetic relationship between the synMuv genes and the RAS pathway have demonstrated that the synMuv genes are functioning upstream of or parallel to the RTK-RAS-MAPK pathway to inhibit vulval induction. This is based on the observation that loss or a significant reduction in the activity of any of the essential components of the RAS pathway can suppress

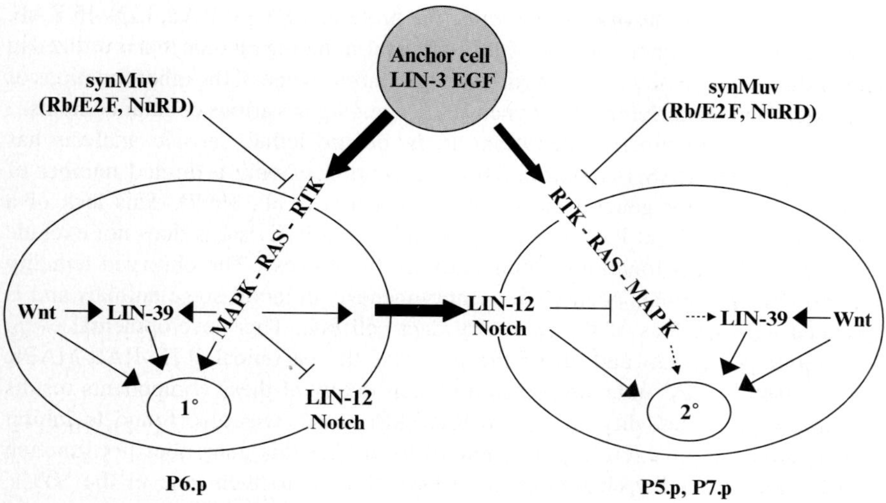

Figure 3. Multiple pathways influence RTK-RAS-MAPK signaling to properly specify vulval cell fates

the synMuv phenotype (e.g., Ferguson et al., 1987; Han et al., 1990; Beitel et al., 1990; Lu and Horvitz, 1998). However, a puzzling result is that loss-of-function mutations in the *lin-3* gene which encodes the EGF-like signal, do not suppress the synMuv phenotype (Ferguson et al., 1987; Lu and Horvitz, 1998). A possible model to explain this is that the synMuv genes act to inhibit basal levels of RAS signaling and promote a "default" hypodermal 3° cell fate in the absence of the LIN-3 inductive signal (Thomas and Horvitz, 1999). One attractive mechanism by which the synMuv genes could promote this default fate is based on the well documented negative role of pRB and pRB-associated components on cell cycle progression in other systems. The synMuv genes could antagonize basal levels of RAS signaling through cell cycle regulation in the VPCs as well (reviewed by Kaelin, 1999). Additionally, the synMuv genes might be acting more directly to promote the hypodermal fate. Recent analysis of several *C. elegans* NuRD components suggests that they antagonize basal levels of RAS pathway activity by inhibiting the expression and activity of LIN-39 thus promoting the hypodermal fate (Chen and Han, 2001a). There are of course still other models for synMuv function, which are not necessarily mutually exclusive (for review Fay and Han, 2000).

7. OTHER DEVELOPMENTAL ROLES FOR LET-60 RAS

In this chapter we have outlined LET-60 RAS function in the specification of the vulval cell fate since the majority of research on RAS in *C. elegans* has been centered on this process. However, LET-60 RAS is broadly expressed and null alleles of *let-60* cause a larval lethal phenotype, suggesting that it is involved in other developmental events (Clark et al., 1988; Han et al., 1990; Beitel et al., 1990; Dent and Han,

1998) . In *C. elegans*, as in other systems, the proteins LET-60 RAS, LIN-45 RAF, MEK-2 and MPK-1 can be thought of as a universal signaling module that is utilized in many different processes. Here we will briefly examine some of the other functions of this module and the differences between RAS signaling in various cellular contexts.

Although loss-of-function mutations in *let-60* are lethal, mosaic analysis has demonstrated that RAS is required for the execution of only a limited number of cell fates and not for general proliferation (Yochem et al., 1997). This lack of a requirement for LET-60 RAS in *C. elegans* mitotic cell divisions does not exclude the possibility that it functions redundantly in this process. The observed lethality in *let-60 ras (lf)* animals occurs after embryogenesis in larval stage animals and is correlated with the loss of the excretory duct cell fate. This developmental event likely requires LET-23 and all of the genes of the canonical RTK-RAS-MAPK pathway used in VPC induction since mutation in any of these components results in a similar larval lethality. A second RasGAP, *gap-2*, was also found to inhibit RAS signaling in the excretory duct fate decision, but this gene does not function in VPC induction (Hayashizaki et al., 1998). Similar to their role in the VPCs, downstream components such as *lin-1*, *eor-1*, *eor-2*, *sur-2* and *lin-25* appear to be important for this fate decision, although to a lesser extent (Nilsson et al., 2000; Howard and Sundaram, 2002).

Mutations in *egl-15*, a homolog of the fibroblast growth factor receptor (FGFR), also cause larval lethality and other phenotypes likely associated with defects in the excretory system (Nelson and Riddle, 1984; Kokel et al., 1998 reviewed in Borland et al., 2001). It is not known precisely in which cells destined to give rise to the excretory system *egl-15* normally functions. In addition to affecting viability through the excretory duct system, *egl-15* and the fibroblast growth factor-like signal encoded by *egl-17* are required to attract the sex myoblasts (SMs) and refine their position within the animal during sexual development (Stern and Horvitz, 1991; DeVore et al., 1995; Burdine et al., 1998). *egl-15* activity in both viability and SM migration was shown to be dependent on *let-60 ras* (Sundaram et al., 1996; Chen et al., 1997). Indeed, many of the same RAS pathway components involved in VPC induction were isolated through suppressor screens of the clear (Clr) phenotype caused by hyperactive *egl-15* activity (for review Borland et al., 2001). Although the FGFR signaling pathway requires SEM-5 and SOS-1, the exact mechanism by which EGL-15 FGFR signals to these two components is not known. In mammals, the FRS2/SNT1 adaptor phosphoprotein is thought to link FGFR signaling to GRB-2 and SOS. Such a protein has not yet been implicated in *C. elegans* FGF signaling. However, the SOC-1 protein may perform an analogous role. Loss-of-function mutations in *soc-1* suppress phenotypes associated with increased EGL-15 activity; therefore, it is thought to positively regulate FGFR signaling (Schutzman et al., 2001). *soc-1* encodes a protein containing a pleckstrin homology domain and is similar to the *Drosophila* DOS and mammalian GAB1 adapter proteins that link RTKs with downstream pathways through the recruitment of SH2 and SH3 domain containing signaling components like SEM-5. Another component thought to function uniquely in *egl-15* signaling is *clr-1*. In contrast to *soc-1*, *clr-1* negatively regulates *egl-15* mediated RAS pathway activity. *clr-1* encodes a protein similar to receptor tyrosine phosphatases (RTPs) (Kokel et al., 1998). Additionally, a CLR-1 membrane proximal phosphatase domain was shown to have phosphatase activity and this domain was required for proper CLR-1 function. The target of CLR-1 phosphatase activity is not known, but phosphorylated tyrosine residues in EGl-15 are good candidates.

let-60 ras has also been implicated in two distinct steps of germ cell development. *let-60 ras*, *mek-2* and *mpk-1* mutations cause a sterile phenotype correlated with a failure of meiotic germ cells to exit the pachytene stage of meiosis prophase I (Church et al., 1995). Mosaic analysis further demonstrated that these three genes likely function within the germline cells to control meiotic progression. The RAS-MAPK pathway is subsequently inactivated after pachytene exit and throughout diakinesis. This inactivation requires the MAP kinase phosphatase LIP-1 and the EFL-1 E2F and DPL-1 DP synMuv components (Page et al., 2001; Hajnal and Berset, 2002). MPK-1 is later reactivated by the presence of the major sperm cytoskeletal protein (MSP), which promotes oocyte meiotic maturation and ovulation (Miller et al., 2001). MSP activates the RAS-MAPK pathway by binding to and inhibiting an Eph receptor (VAB-1) dependent pathway and a CEH-18 dependent pathway (Miller et al., 2003). As stated earlier, *ptp-2* mutant animals have defects in oogenesis (Gutch et al., 1998). These defects do not appear to mimic the block in pachytene exit characteristic of RAS-MAPK pathway mutants. However, the *ptp-2 (lf)* oogenesis defects can be rescued by an activated *let-60 ras* allele, suggesting *ptp-2* acts upstream of, or parallel to, *let-60 ras* in the germline. Thus, *ptp-2* could positively function in MSP mediated oocyte maturation or in an as yet unidentified step in oogenesis.

This review has discussed extensively the RTK-RAS-MAPK pathway activated by the anchor cell signal LIN-3. However, there is an anchor cell signal sent later in vulval development which utilizes the same core LET-23 RTK pathway to promote patterning of the 1° vulval lineage (Wang and Sternberg, 2000). In contrast to VPC induction, *lin-31* and the synMuv genes do not appear to play a role in this signaling event.

The male tail copulatory structure in *C. elegans* is comprised of a number of spicules. These spicules are used both in a sensory capacity and for the physical transfer of sperm into the hermaphrodite. During spicule development, four pairs of precursor cells respond to positional signals to specify their anterior-posterior fate. The anterior fates are specified by the LIN-3 EGF signal (Chamberlin and Sternberg, 1994). Additionally, *let-23*, *sem-5*, *let-60 ras* and *lin-45 raf* are essential for the response to this anterior signal. The synMuv genes of the *lin-15* locus, responsible for repressing basal levels of RAS activity in the VPCs, also function in this fate decision process and help promote the posterior fate.

The two most posterior P cells, P11 and P12, also utilize a RAS signaling pathway to establish a fate decision. Activation of LET-23 RTK and the RAS pathway in response to the LIN-3 signal specifies the unique P12 fate (Jiang and Sternberg, 1998). Much like in vulval development, RAS pathway activation upregulates the expression of a HOX gene, in this case *egl-5*, which is required for proper P12 specification. The Wnt pathway genes *lin-44* (Wnt-like) and *lin-17* (Frizzled receptor-like) are required for this fate decision and could converge on *egl-5* transcription. The transcriptional components *eor-1* and *eor-2* positively regulate P12 specification and function parallel to *egl-5* to integrate both RAS and Wnt signaling (Howard and Sundaram, 2002). Again, the synMuv components that comprise the *lin-15* locus antagonize RAS signaling in this process and likely help specify the P11 fate. We see many of the same signaling features in P12 fate specification that are seen in other developmental events such as spicule fate specification and VPC induction. The repetitive use in *C. elegans* of these signals and their similar modes of convergence suggest that cell fate decisions in other organisms might be regulated in a comparable manner.

A role for RAS signaling in *C. elegans* neuronal olfaction has also been identified. Either inactivation or hyperactivation of the RAS-MAPK pathway in the AWC and AWA olfactory neurons was shown to cause defects in chemotaxis to a set of volatile odorants (Hirotsu et al., 2000). The activation of the RAS-MAPK pathway in these neurons does not appear to involve SEM-5 but instead is dependent on other components required for an odorant response such as the nucleotide-gated channel TAX-2/TAX-4 and the voltage-activated calcium channel subunit UNC-2. Additionally, SRA-13 GPCR and GPA-5 Gα subunit inhibit the odorant induced activation of the RAS pathway in AWC and AWA (Battu et al., 2003). The mechanism by which the RAS-MAPK pathway is activated during olfaction and how it influences this process is not currently known.

Lastly, the genes of the conserved RAS-MAPK signaling module are responsible for the promotion of protein degradation in muscle cells independently of mechanisms that promote protein degradation under conditions of starvation (Szewczyk et al., 2002). Activation of *let-60 ras* or *mpk-1* promoted protein degradation while loss of *lin-45 raf*, *mek-2* and *mpk-1* proved inhibitory. It was also shown that the EGL-15 FGF receptor and SEM-5 are required for activation of the canonical RAS-MAPK pathway in this process (Szewczyk and Jacobson, 2003).

8. CONCLUSIONS AND FUTURE DIRECTIONS

In *C. elegans*, the RAS pathway has been systematically dissected and its components analyzed. The study of VPC fate specification has also given us a complex picture of the regulation of RAS signaling and its integration into a coordinated developmental decision. Additionally, much of the work done in *C. elegans* has been and continues to be translated to other model systems and generally appears to be applicable. Thus, the study of *C. elegans* demonstrates the utility of genetic approaches to study signaling events and the utility of *C. elegans* as a model organism.

The RAS-MAPK signaling pathway is thought to terminate with changes in gene transcription. These changes are likely to determine the specific cellular response to pathway activation. The identity of the transcriptional targets of this pathway remains a major unanswered question and an area of active research. Recently, microarray experiments in *C. elegans* have identified a number of potential RAS pathway targets that await further study (Romagnolo et al., 2002). The completion of the *C. elegans* genome project and the expanding use of RNAi as a tool for reverse genetics are likely to be instrumental to the future analysis of transcriptional targets and other components affecting RAS signaling in *C. elegans*. It is also worth mentioning that forward genetics is unlikely to become obsolete anytime soon. A number of the genes discussed in this chapter are defined by only one or two mutations, suggesting the screens from which they were isolated are not yet saturated. Moreover, the identification of non-null alleles obtained through traditional genetic screens is often the easiest or only way to elucidate the complete function of certain genes. There is no better example of this than the dominant alleles of *let-60 ras* that led to the identification of its role as a molecular switch in vulval cell fate specification.

Note added in proof: A number of significant papers have been published that further refine our knowledge of Ras signaling since the writing of this review. We apologize to those authors for not being able to include their contributions in this review.

ACKNOWLEDGEMENTS

We would like to thank Dan Starr and Yo Suzuki for critical reading of this manuscript. This work was supported by grants from the NIH and the Howard Hughes Medical Institute, with which M.H. is an investigator.

REFERENCES

Aroian, R.V., Koga, M., Mendel, J.E., Ohshima, Y. and Sternberg, P.W. (1990) The let-23 gene necessary for Caenorhabditis elegans vulval induction encodes a tyrosine kinase of the EGF receptor subfamily. Nature *348*, 693-699.

Aroian, R.V. and Sternberg, P.W. (1991) Multiple functions of let-23, a Caenorhabditis elegans receptor tyrosine kinase gene required for vulval induction. Genetics *128*, 251-267.

Battu, G., Hoier, E.F. and Hajnal, A. (2003) The C. elegans G-protein-coupled receptor SRA-13 inhibits RAS/MAPK signalling during olfaction and vulval development. Development *130*, 2567-2577.

Beitel, G.J., Clark, S.G. and Horvitz, H.R. (1990) Caenorhabditis elegans ras gene let-60 acts as a switch in the pathway of vulval induction. Nature *348*, 503-509.

Beitel, G.J., Tuck, S., Greenwald, I. and Horvitz, H.R. (1995) The Caenorhabditis elegans gene lin-1 encodes an ETS-domain protein and defines a branch of the vulval induction pathway. Genes Dev. *9*, 3149-3162.

Berset, T., Hoier, E.F., Battu, G., Canevascini, S. and Hajnal, A. (2001) Notch inhibition of RAS signaling through MAP kinase phosphatase LIP-1 during C. elegans vulval development. Science *291*, 1055-1058.

Borland, C.Z., Schutzman, J.L. and Stern, M.J. (2001) Fibroblast growth factor signaling in Caenorhabditis elegans. Bioessays *23*, 1120-1130.

Bos, J.L. (1989) ras oncogenes in human cancer: a review. Cancer Res. *49*, 4682-4689.

Boube, M., Joulia, L., Cribbs, D.L. and Bourbon, H. (2002) Evidence for a mediator of RNA polymerase II transcriptional regulation conserved from yeast to man. Cell *110*, 143-151.

Boxem, M. and van den Heuvel, S. (2002) C. elegans class B synthetic multivulva genes act in G (1) regulation. Curr. Biol. *12*, 906-911.

Boyer, T.G., Martin, M.E., Lees, E., Ricciardi, R.P. and Berk, A.J. (1999) Mammalian Srb/mediator complex is targeted by adenovirus E1A protein. Nature *399*, 276-279.

Bruinsma, J.J., Jirakulaporn, T., Muslin, A.J. and Kornfeld, K. (2002) Zinc ions and cation diffusion facilitator proteins regulate Ras-mediated signaling. Dev. Cell *2*, 567-578.

Burdine, R.D., Branda, C.S. and Stern, M.J. (1998) EGL-17 (FGF) expression coordinates the attraction of the migrating sex myoblasts with vulval induction in C. Elegans. Development *125*, 1083-1093.

Ceol, C.J. and Horvitz, H.R. (2001) dpl-1 DP and efl-1 E2F act with lin-35 Rb to antagonize Ras signaling in C. elegans vulval development. Mol. Cell *7*, 461-473.

Chamberlin, H.M. and Sternberg, P.W. (1994) The lin-3/let-23 pathway mediates inductive signalling during male spicule development in Caenorhabditis elegans. Development *120*, 2713-2721.

Chang, C., Hopper, N.A. and Sternberg, P.W. (2000) Caenorhabditis elegans SOS-1 is necessary for multiple RAS-mediated developmental signals. EMBO J. *19*, 3283-3294.

Chen, E.B., Branda, C.S. and Stern, M.J. (1997) Genetic enhancers of sem-5 define components of the gonad-independent guidance mechanism controlling sex myoblast migration inCaenorhabditis elegans hermaphrodites. Dev. Biol. *182*, 88-100.

Chen, N. and Greenwald, I. (2004) The lateral signal for LIN-12 Notch in C. elegans vulval development comprises redundant secreted and transmembrane DSL proteins. Dev. Cell *6*, 183-192.

Chen, Z. and Han, M. (2001a) C. elegans Rb, NuRD, and Ras regulate lin-39-mediated cell fusion during vulval fate specification. Curr. Biol. *11*, 1874-1879.

Chen, Z. and Han, M. (2001b) Role of C. elegans lin-40 MTA in vulval fate specification and morphogenesis. Development *128*, 4911-4921.

Chong, H., Lee, J. and Guan, K.L. (2001) Positive and negative regulation of Raf kinase activity and function by phosphorylation. EMBO J. *20*, 3716-3727.

Church, D.L., Guan, K.L. and Lambie, E.J. (1995) Three genes of the MAP kinase cascade, mek-2, mpk-1/sur-1 and let-60 ras, are required for meiotic cell cycle progression in Caenorhabditis elegans. Development *121*, 2525-2535.

Clandinin, T.R., Katz, W.S. and Sternberg, P.W. (1997) Caenorhabditis elegans HOM-C genes regulate the response of vulval precursor cells to inductive signal. Dev. Biol. *182*, 150-161.

Clark, D.V., Rogalski, T.M., Donati, L.M. and Baillie, D.L. (1988) The unc-22 (IV) region of Caenorhabditis elegans: genetic analysis of lethal mutations. Genetics *119*, 345-353.

Clark, S.G., Stern, M.J. and Horvitz, H.R. (1992) C. elegans cell-signalling gene sem-5 encodes a protein with SH2 and SH3 domains. Nature *356*, 340-344.

Clark, S.G., Chisholm, A.D. and Horvitz, H.R. (1993) Control of cell fates in the central body region of C. elegans by the homeobox gene lin-39. Cell *74*, 43-55.

Clark, S.G., Lu, X. and Horvitz, H.R. (1994) The Caenorhabditis elegans locus lin-15, a negative regulator of a tyrosine kinase signaling pathway, encodes two different proteins. Genetics *137*, 987-997.

Dent, J.A. and Han, M. (1998) Post-embryonic expression pattern of C. elegans let-60 ras reporter constructs. Mech. Dev. *72*, 179-182.

DeVore, D.L., Horvitz, H.R. and Stern, M.J. (1995) An FGF receptor signaling pathway is required for the normal cell migrations of the sex myoblasts in C. elegans hermaphrodites. Cell *83*, 611-620.

Eisenmann, D.M. and Kim, S.K. (1997) Mechanism of activation of the Caenorhabditis elegans ras homologue let-60 by a novel, temperature-sensitive, gain-of-function mutation. Genetics *146*, 553-565.

Eisenmann, D.M., Maloof, J.N., Simske, J.S., Kenyon, C. and Kim, S.K. (1998) The beta-catenin homolog BAR-1 and LET-60 Ras coordinately regulate the Hox gene lin-39 during Caenorhabditis elegans vulval development. Development *125*, 3667-3680.

Eisenmann, D.M. and Kim, S.K. (2000) Protruding vulva mutants identify novel loci and Wnt signaling factors that function during Caenorhabditis elegans vulva development. Genetics *156*, 1097-1116.

Fay, D.S. and Han, M. (2000) The synthetic multivulval genes of C. elegans: functional redundancy, Ras-antagonism, and cell fate determination. Genesis *26*, 279-284.

Fay, D.S., Keenan, S. and Han, M. (2002) fzr-1 and lin-35/Rb function redundantly to control cell proliferation in C. elegans as revealed by a nonbiased synthetic screen. Genes Dev. *16*, 503-517.

Ferguson, E.L. and Horvitz, H.R. (1985) Identification and characterization of 22 genes that affect the vulval cell lineages of the nematode Caenorhabditis elegans. Genetics *110*, 17-72.

Ferguson, E.L., Sternberg, P.W. and Horvitz, H.R. (1987) A genetic pathway for the specification of the vulval cell lineages of Caenorhabditis elegans. Nature *326*, 259-267.

Ferguson, E.L. and Horvitz, H.R. (1989) The multivulva phenotype of certain Caenorhabditis elegans mutants results from defects in two functionally redundant pathways. Genetics *123*, 109-121.

Gleason, J.E., Korswagen, H.C. and Eisenmann, D.M. (2002) Activation of Wnt signaling bypasses the requirement for RTK/Ras signaling during C. elegans vulval induction. Genes Dev. *16*, 1281-1290.

Greenwald, I.S., Sternberg, P.W. and Horvitz, H.R. (1983) The lin-12 locus specifies cell fates in Caenorhabditis elegans. Cell *34*, 435-444.

Gu, T., Orita, S. and Han, M. (1998) Caenorhabditis elegans SUR-5, a novel but conserved protein, negatively regulates LET-60 Ras activity during vulval induction. Mol. Cell. Biol. *18*, 4556-4564.

Gutch, M.J., Flint, A.J., Keller, J., Tonks, N.K. and Hengartner, M.O. (1998) The Caenorhabditis elegans SH2 domain-containing protein tyrosine phosphatase PTP-2 participates in signal transduction during oogenesis and vulval development. Genes Dev. *12*, 571-585.

Hajnal, A., Whitfield, C.W. and Kim, S.K. (1997) Inhibition of Caenorhabditis elegans vulval induction by gap-1 and by let-23 receptor tyrosine kinase. Genes Dev. *11*, 2715-2728.

Hajnal, A. and Berset, T. (2002) The C. elegans MAPK phosphatase LIP-1 is required for the G (2)/M meiotic arrest of developing oocytes. EMBO J. *21*, 4317-4326.

Han, M. and Sternberg, P.W. (1990) let-60, a gene that specifies cell fates during C. elegans vulval induction, encodes a ras protein. Cell *63*, 921-931.

Han, M., Aroian, R.V. and Sternberg, P.W. (1990) The let-60 locus controls the switch between vulval and nonvulval cell fates in Caenorhabditis elegans. Genetics *126*, 899-913.

Han, M. and Sternberg, P.W. (1991) Analysis of dominant-negative mutations of the Caenorhabditis elegans let-60 ras gene. Genes Dev. *5*, 2188-2198.

Han, M., Golden, A., Han, Y. and Sternberg, P.W. (1993) C. elegans lin-45 raf gene participates in let-60 ras-stimulated vulval differentiation. Nature *363*, 133-140.

Hara, M. and Han, M. (1995) Ras farnesyltransferase inhibitors suppress the phenotype resulting from an activated ras mutation in Caenorhabditis elegans. Proc. Natl. Acad. Sci. U.S.A. *92*, 3333-3337.

Hayashizaki, S., Iino, Y. and Yamamoto, M. (1998) Characterization of the C. elegans gap-2 gene encoding a novel Ras-GTPase activating protein and its possible role in larval development. Genes Cells *3*, 189-202.

Hedgecock, E.M. and Herman, R.K. (1995) The ncl-1 gene and genetic mosaics of Caenorhabditis elegans. Genetics *141*, 989-1006.

Herman, R.K. and Hedgecock, E.M. (1990) Limitation of the size of the vulval primordium of Caenorhabditis elegans by lin-15 expression in surrounding hypodermis. Nature *348*, 169-171.

Hill, R.J. and Sternberg, P.W. (1992) The gene lin-3 encodes an inductive signal for vulval development in C. elegans. Nature *358*, 470-476.

Hill, R.J. and Sternberg, P.W. (1993) Cell fate patterning during C. elegans vulval development. Development (Suppl.) 9-18.

Hirotsu, T., Saeki, S., Yamamoto, M. and Iino, Y. (2000) The Ras-MAPK pathway is important for olfaction in Caenorhabditis elegans. Nature *404*, 289-293.

Hoier, E.F., Mohler, W.A., Kim, S.K. and Hajnal, A. (2000) The Caenorhabditis elegans APC-related gene apr-1 is required for epithelial cell migration and Hox gene expression. Genes Dev. *14*, 874-886.

Hopper, N.A., Lee, J. and Sternberg, P.W. (2000) ARK-1 inhibits EGFR signaling in C. elegans. Mol. Cell *6*, 65-75.

Horvitz, H.R. and Sulston, J.E. (1980) Isolation and genetic characterization of cell-lineage mutants of the nematode Caenorhabditis elegans. Genetics *96*, 435-454.

Hoskins, R., Hajnal, A.F., Harp, S.A. and Kim, S.K. (1996) The C. elegans vulval induction gene lin-2 encodes a member of the MAGUK family of cell junction proteins. Development *122*, 97-111.

Howard, R.M. and Sundaram, M.V. (2002) C. elegans EOR-1/PLZF and EOR-2 positively regulate Ras and Wnt signaling and function redundantly with LIN-25 and the SUR-2 mediator component. Genes Dev. *16*, 1815-1827.

Howe, L.R. and Marshall, C.J. (1993) Identification of amino acids in p21ras involved in exchange factor interaction. Oncogene *8*, 2583-2590.

Hsu, V., Zobel, C.L., Lambie, E.J., Schedl, T. and Kornfeld, K. (2002) Caenorhabditis elegans lin-45 raf is essential for larval viability, fertility and the induction of vulval cell fates. Genetics *160*, 481-492.

Huang, L.S., Tzou, P. and Sternberg, P.W. (1994) The lin-15 locus encodes two negative regulators of Caenorhabditis elegans vulval development. Mol. Biol. Cell *5*, 395-411.

Jacobs, D., Beitel, G.J., Clark, S.G., Horvitz, H.R. and Kornfeld, K. (1998) Gain-of-function mutations in the Caenorhabditis elegans lin-1 ETS gene identify a C-terminal regulatory domain phosphorylated by ERK MAP kinase. Genetics *149*, 1809-1822.

Jakubowski, J. and Kornfeld, K. (1999) A local, high-density, single-nucleotide polymorphism map used to clone Caenorhabditis elegans cdf-1. Genetics *153*, 743-752.

Jiang, L.I. and Sternberg, P.W. (1998) Interactions of EGF, Wnt and HOM-C genes specify the P12 neuroectoblast fate in C. elegans. Development *125*, 2337-2347.

Jongeward, G.D., Clandinin, T.R. and Sternberg, P.W. (1995) sli-1, a negative regulator of let-23-mediated signaling in C. elegans. Genetics *139*, 1553-1566.

Kaech, S.M., Whitfield, C.W. and Kim, S.K. (1998) The LIN-2/LIN-7/LIN-10 complex mediates basolateral membrane localization of the C. elegans EGF receptor LET-23 in vulval epithelial cells. Cell *94*, 761-771.

Kaelin, W.G. Jr. (1999) Functions of the retinoblastoma protein. Bioessays *21*, 950-958.

Kao, G., Tuck, S., Baillie, D. and Sundaram, M. (2004) C. elegans SUR-6/PR55 cooperates with LET-92/protein phosphatase 2A and promotes Raf activity independently of inhibitory Akt phosphorylation sites. Development *131*, 755-765.

Katz, W.S., Hill, R.J., Clandinin, T.R. and Sternberg, P.W. (1995) Different levels of the C. elegans growth factor LIN-3 promote distinct vulval precursor fates. Cell *82*, 297-307.

Kimble, J. (1981) Alterations in cell lineage following laser ablation of cells in the somatic gonad of Caenorhabditis elegans. Dev. Biol. *87*, 286-300.

Koga, M. and Ohshima, Y. (1995) Mosaic analysis of the let-23 gene function in vulval induction of Caenorhabditis elegans. Development *121*, 2655-2666.

Kokel, M., Borland, C.Z., DeLong, L., Horvitz, H.R. and Stern, M.J. (1998) clr-1 encodes a receptor tyrosine phosphatase that negatively regulates an FGF receptor signaling pathway in Caenorhabditis elegans. Genes Dev. *12*, 1425-1437.

Kornfeld, K. (1997) Vulval development in Caenorhabditis elegans. Trends Genet. *13*, 55-61.

Kornfeld, K., Guan, K.L. and Horvitz, H.R. (1995a) The Caenorhabditis elegans gene mek-2 is required for vulval induction and encodes a protein similar to the protein kinase MEK. Genes Dev. *9*, 756-768.

Kornfeld, K., Hom, D.B. and Horvitz, H.R. (1995b) The ksr-1 gene encodes a novel protein kinase involved in Ras-mediated signaling in C. elegans. Cell *83*, 903-913.

Korswagen, H.C., Coudreuse, D.Y., Betist, M.C., van de Water, S., Zivkovic, D. and Clevers, H.C. (2002) The Axin-like protein PRY-1 is a negative regulator of a canonical Wnt pathway in C. elegans. Genes Dev. *16*, 1291-1302.

Kwon, J.Y., Park, J.M., Gim, B.S., Han, S.J., Lee, J. and Kim, Y.J. (1999) Caenorhabditis elegans mediator complexes are required for developmental-specific transcriptional activation. Proc. Natl. Acad. Sci. U.S.A. *96*, 14990-14995.

Kwon, J.Y. and Lee, J. (2001) Biological significance of a universally conserved transcription mediator in metazoan developmental signaling pathways. Development *128*, 3095-3104.

Lackner, M.R., Kornfeld, K., Miller, L.M., Horvitz, H.R. and Kim, S.K. (1994) A MAP kinase homolog, mpk-1, is involved in ras-mediated induction of vulval cell fates in Caenorhabditis elegans. Genes Dev. *8*, 160-173.

Lee, J., Jongeward, G.D. and Sternberg, P.W. (1994) unc-101, a gene required for many aspects of Caenorhabditis elegans development and behavior, encodes a clathrin-associated protein. Genes Dev. *8*, 60-73.

Levchenko, A., Bruck, J. and Sternberg, P.W. (2000) Scaffold proteins may biphasically affect the levels of mitogen-activated protein kinase signaling and reduce its threshold properties. Proc. Natl. Acad. Sci. U.S.A. *97*, 5818-5823.

Levitan, D. and Greenwald, I. (1998) LIN-12 protein expression and localization during vulval development in C. elegans. Development *125*, 3101-3109.

Levkowitz, G., Waterman, H., Zamir, E., Kam, Z., Oved, S., Langdon, W.Y., Beguinot, L., Geiger, B. and Yarden, Y. (1998) c-Cbl/Sli-1 regulates endocytic sorting and ubiquitination of the epidermal growth factor receptor. Genes Dev. *12*, 3663-3674.

Li, W., Han, M. and Guan, K.L. (2000) The leucine-rich repeat protein SUR-8 enhances MAP kinase activation and forms a complex with Ras and Raf. Genes Dev. *14*, 895-900.

Lu, X. and Horvitz, H.R. (1998) lin-35 and lin-53, two genes that antagonize a C. elegans Ras pathway, encode proteins similar to Rb and its binding protein RbAp48. Cell *95*, 981-991.

Maloof, J.N. and Kenyon, C. (1998) The Hox gene lin-39 is required during C. elegans vulval induction to select the outcome of Ras signaling. Development *125*, 181-190.

Melendez, A. and Greenwald, I. (2000) Caenorhabditis elegans lin-13, a member of the LIN-35 Rb class of genes involved in vulval development, encodes a protein with zinc fingers and an LXCXE motif. Genetics *155*, 1127-1137.

Miller, L.M., Gallegos, M.E., Morisseau, B.A. and Kim, S.K. (1993) lin-31, a Caenorhabditis elegans HNF-3/fork head transcription factor homolog, specifies three alternative cell fates in vulval development. Genes Dev. *7*, 933-947.

Miller, L.M., Hess, H.A., Doroquez, D.B. and Andrews, N.M. (2000) Null mutations in the lin-31 gene indicate two functions during Caenorhabditis elegans vulval development. Genetics *156*, 1595-1602.

Miller, M.A., Nguyen, V.Q., Lee, M.H., Kosinski, M., Schedl, T., Caprioli, R.M. and Greenstein, D. (2001) A sperm cytoskeletal protein that signals oocyte meiotic maturation and ovulation. Science *291*, 2144-2147.

Miller, M.A., Ruest, P.J., Kosinski, M., Hanks, S.K. and Greenstein, D. (2003) An Eph receptor sperm-sensing control mechanism for oocyte meiotic maturation in Caenorhabditis elegans. Genes Dev. *17*, 187-200.

Moghal, N. and Sternberg, P.W. (2003) A component of the transcriptional mediator complex inhibits RAS-dependent vulval fate specification in C. elegans. Development *130*, 57-69.

Morrison, D.K. (2001) KSR: a MAPK scaffold of the Ras pathway? . J. Cell Sci. *114*, 1609-1612.

Muller, J., Ory, S., Copeland, T., Piwnica-Worms, H. and Morrison, D.K. (2001) C-TAK1 regulates Ras signaling by phosphorylating the MAPK Scaffold KSR1. Mol. Cell *8*, 983-993.

Nelson, F.K. and Riddle, D.L. (1984) Functional study of the Caenorhabditis elegans secretory-excretory system using laser microsurgery. J. Exp. Zool. *231*, 45-56.

Nilsson, L., Li, X., Tiensuu, T., Auty, R., Greenwald, I. and Tuck, S. (1998) Caenorhabditis elegans lin-25: cellular focus, protein expression and requirement for sur-2 during induction of vulval fates. Development *125*, 4809-4819.

Nilsson, L., Tiensuu, T. and Tuck, S. (2000) Caenorhabditis elegans lin-25: a study of its role in multiple cell fate specification events involving Ras and the identification and characterization of evolutionarily conserved domains. Genetics *156*, 1083-1096.

Ohmachi, M., Rocheleau, C.E., Church, D., Lambie, E., Schedl, T. and Sundaram, M.V. (2002) C. elegans ksr-1 and ksr-2 have both unique and redundant functions and are required for MPK-1 ERK phosphorylation. Curr. Biol. *12*, 427-433.

Page, B.D., Guedes, S., Waring, D. and Priess, J.R. (2001) The C. elegans E2F- and DP-related proteins are required for embryonic asymmetry and negatively regulate Ras/MAPK signaling. Mol. Cell *7*, 451-460.

Rocheleau, C.E., Howard, R.M., Goldman, A.P., Volk, M.L., Girard, L.J. and Sundaram, M.V. (2002) A lin-45 raf enhancer screen identifies eor-1, eor-2 and unusual alleles of Ras pathway genes in Caenorhabditis elegans. Genetics *161*, 121-131.

Rogalski, T.M., Moerman, D.G. and Baillie, D.L. (1982) Essential genes and deficiencies in the unc-22, IV region of Caenorhabditis elegans. Genetics *102*, 725-736.

Romagnolo, B., Jiang, M., Kiraly, M., Breton, C., Begley, R., Wang, J., Lund, J. and Kim, S.K. (2002) Downstream targets of let-60 Ras in Caenorhabditis elegans. Dev. Biol. *247*, 127-136.

Roy, F., Laberge, G., Douziech, M., Ferland-McCollough, D. and Therrien, M. (2002) KSR is a scaffold required for activation of the ERK/MAPK module. Genes Dev. *16*, 427-438.

Salser, S.J., Loer, C.M. and Kenyon, C. (1993) Multiple HOM-C gene interactions specify cell fates in the nematode central nervous system. Genes Dev. *7*, 1714-1724.

Schutzman, J.L., Borland, C.Z., Newman, J.C., Robinson, M.K., Kokel, M. and Stern, M.J. (2001) The Caenorhabditis elegans EGL-15 signaling pathway implicates a DOS-like multisubstrate adaptor protein in fibroblast growth factor signal transduction. Mol. Cell. Biol. *21*, 8104-8116.

Selfors, L.M., Schutzman, J.L., Borland, C.Z. and Stern, M.J. (1998) soc-2 encodes a leucine-rich repeat protein implicated in fibroblast growth factor receptor signaling. Proc. Natl. Acad. Sci. U.S.A. *95*, 6903-6908.

Shaye, D.D. and Greenwald, I. (2002) Endocytosis-mediated downregulation of LIN-12/Notch upon Ras activation in Caenorhabditis elegans. Nature *420*, 686-690.

Shim, J., Sternberg, P.W. and Lee, J. (2000) Distinct and redundant functions of mu1 medium chains of the AP-1 clathrin-associated protein complex in the nematode Caenorhabditis elegans. Mol. Biol. Cell *11*, 2743-2756.

Sieburth, D.S., Sun, Q. and Han, M. (1998) SUR-8, a conserved Ras-binding protein with leucine-rich repeats, positively regulates Ras-mediated signaling in C. elegans. Cell *94*, 119-130.

Sieburth, D.S., Sundaram, M., Howard, R.M. and Han, M. (1999) A PP2A regulatory subunit positively regulates Ras-mediated signaling during Caenorhabditis elegans vulval induction. Genes Dev. *13*, 2562-2569.

Simske, J.S. and Kim, S.K. (1995) Sequential signalling during Caenorhabditis elegans vulval induction. Nature *375*, 142-146.

Simske, J.S., Kaech, S.M., Harp, S.A. and Kim, S.K. (1996) LET-23 receptor localization by the cell junction protein LIN-7 during C. elegans vulval induction. Cell *85*, 195-204.

Singh, N. and Han, M. (1995) sur-2, a novel gene, functions late in the let-60 ras-mediated signaling pathway during Caenorhabditis elegans vulval induction. Genes Dev. *9*, 2251-2265.

Solari, F. and Ahringer, J. (2000) NURD-complex genes antagonise Ras-induced vulval development in Caenorhabditis elegans. Curr. Biol. *10*, 223-226.

Stern, M.J. and Horvitz, H.R. (1991) A normally attractive cell interaction is repulsive in two C. elegans mesodermal cell migration mutants. Development *113*, 797-803.

Sternberg, P.W. (1988) Lateral inhibition during vulval induction in Caenorhabditis elegans. Nature *335*, 551-554.

Sternberg, P.W. and Horvitz, H.R. (1986) Pattern formation during vulval development in C. elegans. Cell *44*, 761-772.

Sternberg, P.W. and Horvitz, H.R. (1989) The combined action of two intercellular signaling pathways specifies three cell fates during vulval induction in C. elegans. Cell *58*, 679-693.

Sternberg, P.W. and Horvitz, H.R. (1991) Signal transduction during C. elegans vulval induction. Trends Genet. *7*, 366-371.

Sternberg, P.W. and Han, M. (1998) Genetics of RAS signaling in C. elegans. Trends Genet. *14*, 466-472.

Stevens, J.L., Cantin, G.T., Wang, G., Shevchenko, A. and Berk, A.J. (2002) Transcription control by E1A and MAP kinase pathway via Sur2 mediator subunit. Science *296*, 755-758.

Stewart, S., Sundaram, M., Zhang, Y., Lee, J., Han, M. and Guan, K.L. (1999) Kinase suppressor of Ras forms a multiprotein signaling complex and modulates MEK localization. Mol. Cell. Biol. *19*, 5523-5534.

Sulston, J.E. and Horvitz, H.R. (1977) Post-embryonic cell lineages of the nematode, Caenorhabditis elegans. Dev. Biol. *56*, 110-156.

Sundaram, M. and Han, M. (1995) The C. elegans ksr-1 gene encodes a novel Raf-related kinase involved in Ras-mediated signal transduction. Cell *83*, 889-901.

Sundaram, M., Yochem., J. and Han, M. (1996) A Ras-mediated signal transduction pathway is involved in the control of sex myoblast migration in Caenorhabditis elegans. Development *122*, 2823-2833.

Szewczyk, N.J., Peterson, B.K. and Jacobson, L.A. (2002) Activation of Ras and the mitogen-activated protein kinase pathway promotes protein degradation in muscle cells of Caenorhabditis elegans. Mol. Cell. Biol. *22*, 4181-4188.

Szewczyk, N.J. and Jacobson, L.A. (2003) Activated EGL-15 FGF receptor promotes protein degradation in muscles of Caenorhabditis elegans. EMBO J. *22*, 5058-5067.

Tan, P.B., Lackner, M.R. and Kim, S.K. (1998) MAP kinase signaling specificity mediated by the LIN-1 Ets/LIN-31 WH transcription factor complex during C. elegans vulval induction. Cell *93*, 569-580.

Thien, C.B. and Langdon, W.Y. (2001) Cbl: many adaptations to regulate protein tyrosine kinases. Nat. Rev. Mol. Cell. Biol. *2*, 294-307.

Thomas, J.H. and Horvitz, H.R. (1999) The C. elegans gene lin-36 acts cell autonomously in the lin-35 Rb pathway. Development *126*, 3449-3459.

Tuck, S. and Greenwald, I. (1995) lin-25, a gene required for vulval induction in Caenorhabditis elegans. Genes Dev. *9*, 341-357.

von Zelewsky, T., Palladino, F., Brunschwig, K., Tobler, H., Hajnal, A. and Muller, F. (2000) The C. elegans Mi-2 chromatin-remodelling proteins function in vulval cell fate determination. Development *127*, 5277-5284.

Wang, B.B., Muller-Immergluck, M.M., Austin, J., Robinson, N.T., Chisholm, A. and Kenyon, C. (1993) A homeotic gene cluster patterns the anteroposterior body axis of C. elegans. Cell *74*, 29-42.

Wang, M. and Sternberg, P.W.L. (2000) Patterning of the 1P^{oP} vulval lineage by Ras and Wnt pathways. Development *127*, 5047-5058.

Wang, M. and Sternberg, P.W. (2001) Pattern formation during C. elegans vulval induction. Curr. Top. Dev. Biol. *51*, 189-220.

Whitfield, C.W., Benard, C., Barnes, T., Hekimi, S. and Kim, S.K. (1999) Basolateral localization of the Caenorhabditis elegans epidermal growth factor receptor in epithelial cells by the PDZ protein LIN-10. Mol. Biol. Cell *10*, 2087-2100.

Wu, Y. and Han, M. (1994) Suppression of activated Let-60 ras protein defines a role of Caenorhabditis elegans Sur-1 MAP kinase in vulval differentiation. Genes Dev. *8*, 147-159.

Wu, Y., Han, M. and Guan, K.L. (1995) MEK-2, a Caenorhabditis elegans MAP kinase kinase, functions in Ras-mediated vulval induction and other developmental events. Genes Dev. *9*, 742-755.

Yochem, J., Weston, K. and Greenwald, I. (1988) The Caenorhabditis elegans lin-12 gene encodes a transmembrane protein with overall similarity to Drosophila notch. Nature *335*, 547-550.

Yochem, J., Sundaram, M. and Han, M. (1997) Ras is required for a limited number of cell fates and not for general proliferation in Caenorhabditis elegans. Mol. Cell. Biol. *17*, 2716-2722.

Yoder, J.H., Chong, H., Guan, K. and Han, M. (2004) Modulation of KSR activity in Caenorhabditis elegans by Zn ions, PAR-1 kinase and PP2A phosphatase. EMBO J. *23*, 111-119.

Yoo, A.S., Bais, C. and Greenwald, I. (2004) Crosstalk between the EGFR and LIN-12/Notch Patways in C. elegans vulval development Science *303*, 663-666.

Yoon, C.H., Lee, J., Jongeward, G.D. and Sternberg, P.W. (1995) Similarity of sli-1, a regulator of vulval development in C. elegans, to the mammalian proto-oncogene c-cbl. Science *269*, 1102-1105.

Yoon, C.H., Chang, C., Hopper, N.A., Lesa, G.M. and Sternberg, P.W. (2000) Requirements of multiple domains of SLI-1, a Caenorhabditis elegans homologue of c-Cbl, and an inhibitory tyrosine in LET-23 in regulating vulval differentiation. Mol. Biol. Cell *11*, 4019-4031.

Zhang, H. and Emmons, S.W. (2000) A C. elegans mediator protein confers regulatory selectivity on lineage-specific expression of a transcription factor gene. Genes Dev. *14*, 2161-2172.

CHAPTER 10

RAS FAMILY G-PROTEINS IN *SACCHAROMYCES CEREVISIAE* AND *SCHIZOSACCHAROMYCES POMBE*

Ras1/2, Rsr1/Bud1 and Rheb

ANGEL P. TABANCAY, JR., MELISSA J. COMISO, AND FUYUHIKO TAMANOI

Department of Microbiology, Immunology, and Molecular Genetics, 1602 Molecular Sciences Building, University of California, Los Angeles, California 90095-1489

Abstract: The Ras superfamily G proteins are signaling proteins that regulate a variety of physiological events within the cell. The superfamily consists of distinct subfamilies including the Ras, Rho, Rab, ARF, Ran, and Rheb branches. Much of what is known about these subfamilies has come from work done in the simple unicellular eukaryotes, *Saccharomyces cerevisiae* and *Schizosaccharomyces pombe.* Using homologous proteins found in these yeasts, clues as to their physiological function in higher eukaryotic systems have been deduced. In addition, an understanding of how these proteins are regulated has also been obtained from yeast studies. This chapter will focus on the Ras superfamily G-proteins in yeast; more specifically on Ras and Rheb from both yeasts, as well as Rsr1/Bud1 of *S. cerevisiae.* Their physiological functions as well as modes of regulation with be addressed

Keywords: Ras, Bud1, Rheb, GAP, GEF, farnesylation, geranylgeranylation

1. INTRODUCTION

The Ras superfamily of G-proteins are involved in a number of cellular processes (Bourne et al., 1990; Downward, 1990; Bourne et al., 1991; Reuther and Der, 2000). The superfamily consists of various subfamilies; each making up a distinct branch of small guanine triphosphatases (GTPases), including the Ras, Rho, Rab, ARF, Ran, and Rheb subfamilies (Bourne et al., 1990). The Rho family proteins have been shown to regulate actin cytoskeleton organization and exocytosis. The Rab and ARF families play a critical role in vesicular trafficking. The Ran subfamily of proteins controls nuclear transport as well as microtubule spindle formation. This chapter focuses on the Ras, Rsr, and Rheb subfamilies in *Saccharomyces cerevisiae*

C. Der (ed.), RAS Family GTPases, 227–256.

A.

GTPase	*S. cerevisiae*	*S. pombe*
Ras	Cell Cycle Pseudohyphal growth	Mating Morphology
Rsr1/Bud1	Budding	No homologue
Rheb	Nutrient uptake	Cell cycle Cell growth Nutrient uptake

B.

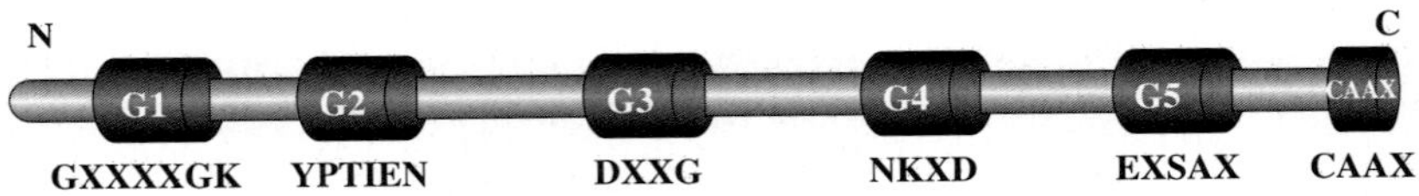

Figure 1. The Ras superfaimly G proteins of *Saccharomyces cerevisiae* and *Schizosaccharomyces pombe*. A) Table shows Ras GTPases in *S. cerevisiae* and *S. pombe* and their physiological functions. B) General structure of Ras GTPases. G1-G5 denote the G boxes characteristic of all small GTPases. These domains are involved in binding and hydrolysis of the guanine nucleotide. G2 is also called the effector domain and interacts with downstream effectors. The C-terminal CAAX box is where post-translational modification occurs by protein farnesyltransferase. The sequences below each domain denote the consensus sequence of each region

and *Schizosaccharomyces pombe*. The Ras family proteins play essential roles in growth and differentiation. The Rsr protein is involved in bud site selection in budding yeast. The Rheb family proteins have been implicated in nutrient uptake and cell cycle regulation. We will discuss various aspects regarding Ras, Rsr, and Rheb proteins in yeast including structural features, biochemical activities, and physiological function [Figure 1A].

An alignment of Ras, Rap, and Rheb proteins in yeast reveals a number of key elements (Bourne et al., 1991). The proteins consist of 5 highly conserved G-boxes (G1-G5) [Figure 1B]. G1, G3, G4, and G5 have been shown to be critical for binding to guanine nucleotides as well as hydrolyzing bound GTP to GDP. The G2 box, which is also known as the effector domain, has been shown to be essential for binding downstream effector molecules. Although this region is highly conserved, the minor differences in amino acid sequence of the effector domains has been shown to alter the specificity of the small GTPase for their downstream effectors. This can explain how this superfamily of structurally similar proteins is involved in a number of functions within the cell. Mutants have been made throughout these conserved regions showing that these regions are important for guanine nucleotide binding, hydrolysis, and effector interaction. Analysis of the C-terminal regions of the small GTPases reveals another conserved region. This is the *CAAX* motif (C is for cysteine, A is any aliphatic amino acid, and X is the C-terminal amino acid), which directs the post-translational addition of a lipid moiety by the enzyme protein farnesyltransferase (FTase) and protein geranylgeranyltransferase type I (GGTase I) (Sattler and Tamanoi, 1996). FTase preferentially modifies proteins with *CAAX* boxes ending in A, Q, S, M, or C, while GGTase I modifies proteins containing a *CAAX* box terminating with L or F. Ras and Rheb proteins have been shown to be farnesylated, while Rap proteins are modified by GGTase I (Sattler and Tamanoi, 1996). Post-translational modification of Ras, Rap, and Rheb have been shown to be essential for the proper function of these proteins in the cell (Sattler and Tamanoi, 1996).The process of farnesylation and geranylgeranylation will also be discussed in this chapter.

Ras, Rsr, and Rheb are small GTPases (Bourne et al., 1990; Downward, 1990; Bourne et al., 1991). As members of this family of GTPases, these proteins exhibit specific intrinsic properties. First, small GTPases have been shown to bind guanosine 5′-triphosphate (GTP) and guanosine 5′-diphosphate (GDP). GTPases cycle between an inactive GDP-bound form and an active GTP-bound form. When GTP is bound, the GTPase contains an intrinsic ability to hydrolyze the γ-phosphate of the bound GTP to form GDP with the loss of inorganic phosphate. In this inactive state, the GDP is able to dissociate from the GTPase, where a new molecule of GTP can be bound, thus converting the GTPase to an active form. These properties are intrinsic abilities of the small GTPases. However, there are regulatory proteins in the cell that serve to enhance these intrinsic properties. GTPase activating proteins (GAPs) bind to the small GTP binding protein and enhance the hydrolysis of the bound GTP to GDP. Guanine nucleotide exchange factors (GEFs) are proteins that can stimulate exchange of GDP, the product of hydrolysis, for a new molecule of GTP. Thus, GAPs are negative regulators, while GEFs serve as activators of small GTPases. Ras and Rsr proteins have specific GAPs and GEFs that regulate their function *in vivo*. Recent evidence from *Drosophila* and mammalian cells has suggested that the tuberous sclerosis tumor suppressor protein, Tsc2, complexed with Tsc1, serves as a GAP for dRheb as well as human Rheb (Castro et al., 2003; Inoki et al., 2003; Tee et al., 2003; Zhang et al., 2003).

The focus of this chapter is on the Ras family G-proteins of two yeasts, *S. cerevisiae* and *S. pombe*. Yeast genetics has provided a powerful tool in elucidating the functions of these small GTPases. A number of clues have been obtained from using yeast as a model system. These clues have been carried over into higher eukaryotic systems to gain insight into the function of small G-proteins in more complex organisms.Due to the smaller number of genes present, the use of yeast systems has allowed exhaustive genetic screens to be carried out as compared to other organisms such as *Drosophila* or human.

2. RAS FAMILY PROTEINS IN *SACCHAROMYCES CEREVISIAE*

2.1 *S. cerevisiae* Ras

2.1.1 Ras function: Adenylate cyclase activation and other functions

S. cerevisiae contains two *RAS* genes, *RAS1* and *RAS2*. Ras1 and Ras2 proteins were identified by their strong homology to the mammalian counterpart, H-ras (DeFeo-Jones et al., 1983; Powers et al., 1984). Ras1p and Ras2p are 36 and 40kDa respectively and are thus larger than mammalian 21kDa Ras proteins, although functionally homologous (Kataoka et al., 1984; Tatchell et al., 1984; DeFeo-Jones et al., 1985). They diverge from mammalian ras homologues at their C-terminal domain of approximately 120 amino acids. In the early 1980s, Ras1p and Ras2p were shown to activate adenylate cyclase (Broek et al., 1985; Toda et al., 1985). Yeast lacking *RAS1* and *RAS2* displayed a similar phenotype to yeast lacking adenylate cyclase (Toda et al., 1985). In addition, yeast containing an activated mutation, $Ras2^{Val19}$, which is analogous to the mutation that activates mammalian *ras* genes, looks very similar to yeast that have elevated levels of adenylate cyclase. The intracellular cyclic AMP (cAMP) levels are significantly higher than that of wild type yeast. Yeast cells expressing activated Ras exhibit phenotypes including heat shock sensitivity, nutrient starvation sensitivity, and decreased glycogen storage. Biochemical evidence also provided clues as to the function of Ras1 and Ras2. Purified Ras1 and Ras2 were able to activate adenylate cyclase in the presence of guanine nucleotides (Broek et al., 1985). The similarity of yeast *RAS* genes to mammalian *ras* genes suggested a conserved function for the protein. This functional conservation was shown by complementation studies in yeast. H-ras was shown to complement the loss of Ras1 and Ras2 proteins in yeast (Kataoka et al., 1984; DeFeo-Jones et al., 1985). It was demonstrated that purified yeast Ras as well as mammalian ras could stimulate the magnesium and GTP-dependent adenylate cyclase activation (Broek et al., 1985). However, *RAS* genes are not involved in adenylate cyclase regulation in mammalian cells (Levitzki, 1996). It has been shown that cells lacking *RAS2* are unable to grow on non-fermentable carbon source (Breviario et al., 1988). However, when *ras2* deficient cells are grown on a fermentable carbon source, cells are now able to grow (Breviario et al., 1986). This is due to the fact that Ras1 is expressed when grown on a fermentable carbon source and is able to complement the lack of Ras2 in the cells.

The activation of adenylate cyclase regulates the cAMP dependent protein kinase pathway. When Ras binds to GTP, thus making it active, it can now activate adenylate cyclase, which is encoded by the gene *CYR1*. Adenylate cyclase catalyzes the formation of cAMP from ATP (Broek et al., 1985). Adenylate cyclase binds to the cyclase associated protein (CAP), forming the adenylate cyclase-CAP complex (Gerst et al., 1991). The association of Cyr1 with CAP is essential for its activation by posttranslationally modified Ras (Gerst et al., 1991). The associated CAP contributes to the formation of a Ras-binding site of Cyr1, which mediates Cyr1 activation, other than the leucine rich repeat domain, the primary Ras-binding site in Cyr1 (Mintzer and Field, 1994). The resulting elevated level of cAMP within the cells leads to activation of the cAMP-dependent protein kinase pathway. The protein kinase A (PKA) consists of a catalytic subunit Tpk and a regulatory subunit Bcy (Toda et al., 1987). Three genes TPK1, 2, and 3 encode the catalytic subunit (Toda et al., 1987). These proteins take the signal from Ras1/2 and activate events leading to the transcription of genes involved in metabolism, proliferation, and stress resistance. A protein called Sok2 was reported to function downstream of the PKA to regulate the expression of genes important in growth and development (Ward and Garrett, 1994) and this regulation has been suggested to be mediated

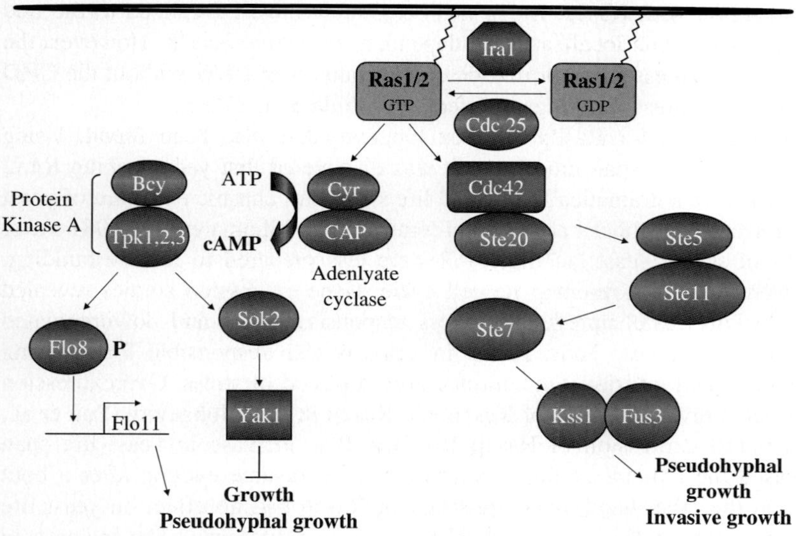

Figure 2. Ras1/Ras2p in *Saccharomyces cerevisiae*. The relevant pathways are highlighted below. Activation of Ras1/2p leads to adenylate cyclase activation, resulting in growth and pseudohyphal growth through the cAMP pathway. Ras1/2-GTP can also signal through Cdc42p and downstream players to result in pseudohyphal growth

by Yak1 (Garrett and Broach, 1989). The Ras-activated PKA pathway can also activate the transcription factor Flo8p which activates the transcription of *FLO11*, a gene involved in pseudohyphal growth (Pan and Heitman, 1999) [Figure 2].

Another role for Ras2 in yeast is in filamentation and invasive growth. Ras2 has been shown to signal through the above mentioned Cyr1p/cAMP/PKA pathway, resulting in expression of filamentation response genes (Mosch et al., 1999). An additional pathway that Ras2 signals through is the Cdc42p/Ste20/MAPK pathway. Ras2 can activate Cdc42p which then signals through a complex consisting of Ste20, Ste11, Ste7 and Ste5 (Mosch et al., 1999). This results in the expression of filamentation response genes by the transcription factors, Ste12 and Tec1.

Ras has been shown to play a role in mitotic exit. A yeast strain lacking both *RAS1* and *RAS2* genes are defective in mitotic exit (Morishita et al., 1995). Therefore, a downstream effector of Ras proteins was believed to function in mitotic exit. This protein was identified as Lte1p (low temperature essential) (Shirayama et al., 1994a). Lte1 has been shown to serve as a GEF for the small GTP-binding protein Tem1p and it is required for the proper mitotic exit at lower temperature (Shirayama et al., 1994b; Bardin et al., 2000). Lte1p was shown to be a multicopy suppressor of the heat shock sensitivity of the *ira1-1* and $ras2^{Val19}$ mutants (Shirayama et al., 1994a) as well as the *ras1 ras2 cyr1* mutant (Morishita et al., 1995). Further work showed a direct interaction between Lte1p and Ras2p-GTP both *in vivo* and *in vitro* (Yoshida et al., 2003). The Cdc25 homology domain (CHD) of Lte1p is essential for the interaction with Ras2p. Lte1p has been shown to be localized to the bud cortex of yeast and this localization is dependent on active Ras2p. However, the CHD of Lte1p is not essential for mitotic exit as mutants of *LTE1* without the CHD are still able to suppress mitotic exit defects (Yoshida et al., 2003).

An additional role for Ras2p in yeast longevity has also been found. Using thermal stress as a life span modulator, it was discovered that yeast lacking *RAS2* and not *RAS1* have a dramatically reduced life span after chronic exposure of yeast cells to recurring heat shocks at sublethal temperatures (Shama et al., 1998). After these bouts of stress, yeast lacking *RAS2* were less prepared to resume budding, and thus were unable to re-enter the cell cycle. Gene expression studies revealed that yeast lacking *RAS2* upregulated stress responsive genes and downregulated growth promoting genes. Thus, Ras2p in yeast is also responsible for ensuring that growth and cell division can continue after a period of stress. Overexpression studies revealed divergent roles of Ras1p and Ras2p in yeast longevity (Sun et al., 1994). Increased expression of Ras2p led to a 30% increase in yeast life span and decreased the amount of time for yeast cells to resume cycling after a bout of stress. On the other hand, overexpression of Ras1p had no effect on yeast life span. Yeast lacking *RAS1*, in fact, displayed a longer life span. The increase in longevity does not seem to be a result of the cyclic AMP/PKA pathway as increased intracellular cyclic AMP levels actually decrease lifespan. Additionally, an effector domain mutant of Ras2p that is unable to activate adenylate cyclase leads to a life span extension similar to that of the wild type Ras2 protein.

Recently, a novel functional link between MAP kinase cascades and the Ras/cAMP pathway that regulates survival has been revealed (Cherkasova et al., 2003). Kss1p and Fus3p, which are key players in the yeast MAPK cascade, have been shown to act upstream of the Ras/cAMP pathway. Previously, Kss1p and Fus3p were known to promote filamentous growth and cell integrity (Lee and Elion, 1999; Morillon et al., 2000) as well as cross-regulate the Ras/cAMP pathway during growth and mating (Arkinstall et al., 1991; Elion et al., 1991; Elion et al., 1991b; Francois et al., 1991; Papasavvas et al., 1992). However, direct evidence for this cross-regulation was lacking. The disruption of *FUS3* results in increased cAMP and poor long-term survival and stress resistance (Cherkasova et al., 2003). In addition, an activated Kss1p also increases cAMP, but a catalytically inactive Fus3p decreases cAMP levels. Because both Fus3p and Kss1p have been reported to interact with and phosphorylate the RasGEF Cdc25 (Schlessinger and Bar-Sagi, 1994; Fan et al., 1998; Thevelein and de Winde, 1999; Cherkasova et al., 2003), it is believed that Fus3p and Kss1p exert their effects through Cdc25. Therefore, the yeast Fus3/Kss1/MAPK pathway is able to crosstalk with the Ras/cAMP pathway.

2.1.2 Ras1p and Ras2p regulators

2.1.2.1 Ras1p and Ras2p GEF: CDC25 Cdc25p is the GEF for both Ras1 and Ras2 proteins. It has a C-terminal region highly homologous (47%) to other members of the Ras GEF family [Figure 3A]. Its GEF activity towards Ras1p and Ras2p has been demonstrated by the stimulation of GDP/GTP exchange

(Broek et al., 1985; Toda et al., 1985; Boy-Marcotte et al., 1993; Crechet et al., 1993; Lai et al., 1993). The *CDC25* gene was discovered through its involvement in the regulation of cAMP and currently appears to be the only GEF regulating the activation of Ras1p and Ras2p (Camonis et al., 1986; Martegani et al., 1986; Broek et al., 1987; Daniel et al., 1987; Marshall et al., 1987; Robinson et al., 1987; Jones et al., 1991). Cdc25p is 1,589 amino acids with a predicted molecular weight of 180kDa (Camonis et al., 1986; Broek et al., 1987).

The C-terminal region of Cdc25p (amino acids1102-1589) is homologous to the catalytic domain of Ras GEFs of all organisms and is exclusively involved in catalyzing GDP/GTP exchange (Crechet et al., 1993). The N-terminal region (amino acids 11-1253) is involved in the regulation of glucose-induced rise in cAMP and is also believed to be involved in the negative feedback, or attenuation, of this pathway

A.

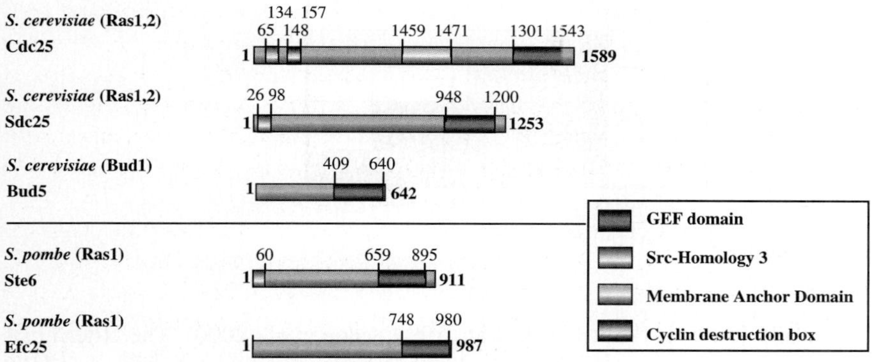

Percent Identity: GEF domain

	Cdc25	Sdc25	Ste6	Efc25	Bud5
Cdc25		50	35	34	26
Sdc25			36	33	22
Ste6				32	29
Efc25					27
Bud5					

http://www.ncbi.nlm.nih.gov/blast/bl2seq/bl2.html

Figure 3. Saccharomyces cerevisiae and *Schizosaccharomyces pombe* GEF and GAP domains and percent identity. (A) Each exchange factor exhibits a core GEF domain at the C-terminal with a percentage identity indicated in the corresponding chart, as determined by the NCBI blast website: http://www.ncbi.nlm.nih.gov/blast/bl2seq/bl2.html. *S. cerevisiae* Cdc25, Sdc25 and *S. pombe* Ste6 contain an SH3 domain at their N-terminal. In addition, *S. cerevisiae* Cdc25 contains a membrane anchor domain and a cyclin destruction box. (B) *Saccharomyces cerevisiae* and *Schizosaccharomyces pombe* GAP domains and percent identity also determined by the NCBI blast website. Each GAP exhibits a conserved GAP related domain with a percentage identity as indicated in the corresponding chart. *S. cerevisiae* Bud2 contains a C2 domain, which characteristically binds Ca^{2+} and phospholipids

B.

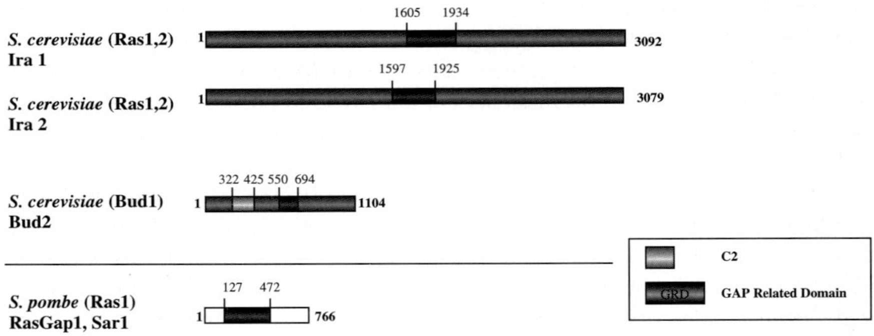

Amino Acid identity between GAP homologues in conserved core region

	Ira1	Ira2	RasGAP/Sar1	Bud2
Ira 1		49	27	20
Ira 2			23	19
RasGAP/Sar1				22
Bud2				

http://www.ncbi.nlm.nih.gov/blast/bl2seq/bl2.html

Figure 3. (Continued)

(Munder and Kuntzel, 1989; Gross et al., 1999; Chen et al., 2000). The N-terminal region contains seven consensus sites for cAMP-dependent protein kinase (cAPK phosphorylation) that is required for its phosphorylation following glucose induction. Cdc25p is rapidly phosphorylated by cAPK following the addition of glucose, which is reported to attenuate the glucose signal. It has been hypothesized that since the phosphorylated residues are close to the SH3 domain, it might weaken the interaction of Cdc25p with adenylate cyclase (Cdc35/Cyr1) (Gross et al., 1999). This N-terminal region, and the C-terminal 37 amino acids are furthermore believed to be involved in yeast response to glucose levels and in sustaining the basal levels of cAMP in un-stimulated cells. Since the direct interaction of Cdc25p with the macromolecular complex Cdc35/Cyr1 cylase has been shown, it is a possibility that the glucose signal is transmitted to this complex (Munder and Kuntzel, 1989; Freeman et al., 1996; Gross et al., 1999). The N-terminal region of Cdc25p also has a hydrophobic membrane localization domain (1459-1471aa), a cyclin destruction box and the previously mentioned SH3 domain (60-130aa) [Figure 3A]. The SH3 domain interacts with adenylate cylase and is believed to facilitate Ras activation of cAMP signaling (Mintzer and Field, 1999). Cdc25p shows membrane localization though fractionation experiments and immunofluorescence microscopy, which is reliant upon amino acids 1441-1552 (Garreau et al., 1996). The regulation of

Cdc25p activity may depend upon protein expression or its change in localization. Cdc25p cellular concentration is very low, due to the fact that it is a very unstable protein with a half-life of 15-20 minutes. This characteristic is dependent upon the presence of the cyclin destruction box that is also found in mitotic cyclins and is involved in degradation. The quick degradation of Cdc25p led to the proposal that Ras and cAMP activity are regulated through this GEF. As discussed earlier, Cherkasova et al. (2003) have demonstrated an interaction between Cdc25p and the MAPKs Kss1p and Fus3p. They show that these kinases act upstream of Ras/cAMP in regulating survival. They further propose that Cdc25p phosphorylation by Kss1p could be the regulation point for cell survival and resistance to stress (Cherkasova et al., 2003).

Cdc25p functions upstream of Ras as its GEF and is required for normal Ras function (Broek et al., 1987; Daniel et al., 1987; Marshall et al., 1987; Robinson et al., 1987). Disruption of *CDC25* is lethal, but could be suppressed by $RAS2^{Val19}$ or exogenous cAMP (Broek et al., 1985; Martegani et al., 1986; Robinson et al., 1987). Cdc25p binds to the nucleotide free form of Ras, suggesting that it stabilizes Ras2p in the nucleotide free form, to allow for GTP to enter into the nucleotide binding site (Jones et al., 1991; Lai et al., 1993; Haney and Broach, 1994). External glucose stimulation of adenylate cyclase results in the upregulation of cAMP, which is required for the metabolic processing of the sugar.

Sdc25p (suppression of Cdc25-5 mutation) was discovered to have GEF activity towards Ras in a disruption screen searching for genes that suppressed a *cdc25* mutation which results in the failure of *S. cerevisiae* to bud (Boy-Marcotte et al., 1993). The C-terminal GEF domain of Sdc25p is 50% identical to Cdc25p, with the highest homology at the C-terminal region. It also contains an SH3 domain, like that of Cdc25p [Figure 3A]. Sdc25p rescues the null mutation of *cdc25* and exhibits GDP-GTP exchange activity on Ras2p (Crechet et al., 1993). In these experiments, only the C-terminal domain of Sdc25p, which is structurally related to Cdc25p, overcame the lack of cAMP production and growth, whereas the intact *SDC25* gene could not. This suggests that the other domains of Sdc25p act to specify its targets, presenting the possibility that it may not be an authentic exchange factor for Ras (Crechet et al., 1993). Sdc25p is also transcriptionally regulated differently than Cdc25p (Boy-Marcotte et al., 1996). Unlike Cdc25p, Sdc25p is not detectable during growth when glucose is the carbon source. It is transcribed when nutrients are depleted at the end of growth, and in cells grown on nonfermentable carbon sources (Boy-Marcotte et al., 1996). Finally, an important difference between *CDC25* and *SDC25* is that *CDC25* was deemed an essential gene in deletion studies, whereas *SDC25* deletion caused no detectable phenotype (Damak et al., 1991).

2.1.2.2 Ras1p and Ras2p GAPs: Ira1p and Ira2p Ira1p was the first GAP found for Ras1 and 2 proteins, followed by Ira2p, both of which serve as upstream negative regulators of the Ras proteins (Tanaka et al., 1989; Tanaka et al., 1990b; Tanaka et al., 1990; Tanaka et al., 1991). Ira1p and Ira2p have close sequence homology to mammalian GAP (Tanaka et al., 1990b; Tanaka et al., 1990). By disrupting the two genes, the lethality of *cdc25* is suppressed, there is an increased level of intracellular cAMP and sporulation defects, and increased sensitivity to heat shock and nitrogen starvation (Tanaka et al., 1990; Tanaka et al., 1990b). These phenotypes are also observed in strains expressing activated Ras2^{Val19}. The two proteins are most highly homologous in the middle and carboxyl terminal regions (52% and 53%, respectively), and are the least homologous at the amino terminal (31%) (Tanaka et al., 1990).

IRA1 was initially isolated as a mutant that was sensitive to nutrient limitation and failed to sporulate (Matsumoto et al., 1985). Although originally believed to encode 2,938 amino acids, 154 extra amino acids were later sequenced at the N-terminus and thus was shown to be 3,092 amino acids in total (Zagulski et al., 1994). Disruption of *IRA1* leads to phenotypes similar to those of the *ras2*^{Val19}mutant, indicating that it is necessary for GTPase activity of Ras2p. Ira1p has potential sites for A-kinase phosphorylation Arg-Arg-X-Ser at serines located at the amino acids 1753 and 3004 [Figure 3B]. These phosphorylation sites are believed to be involved in the feedback control of cAMP formation conducted by A-kinase (PKA) (Tanaka et al., 1989). The region between 1763 and 2145 amino acids is sufficient to suppress the *ira2* mutation. This region is homologous to the Ras GAP domain of p120Gap and NF1. Ira1p and Ira2p interact with Ras at the effector domain. Involvement of the α3 region of Ras2 in the interaction is suggested from the identification of the dominant activating mutants Ras2E99K and E130K (Wilson et al., 1993; Wood et al., 1994). Ira1p has also been shown to interact with adenylate cyclase (Mitts et al., 1991).

IRA2 was later discovered as a gene that suppressed the heat shock sensitivity of *ira1* mutations (Tanaka et al., 1990b). Ira2p was verified as a GAP protein by demonstrating GAP activity similar to that of mammalian GAP proteins (Tanaka et al., 1991). An effector domain Ras2p mutant, P41S, was characterized to be a novel dominantly activated mutant resulting from its failure to respond to Ira2p activation (Tanaka et al., 1992). Ira2p is 3,079 amino acids and also has one A-kinase phosphorylation RRYS site at about the position as one of the Ira1p sites, with the serine located at amino acid 1018 (Tanaka et al., 1990). High levels of Ras1-GTP forms were found in *ira2* mutant cells and in the double mutant (*ira1*, *ira2*), both Ras1p and Ras2p showed increased activity, resulting in the over-activation of the cAMP pathway, thus demonstrating the negative regulatory functions of Ira1p and Ira2p (Tanaka et al., 1990b; Tanaka et al., 1990; Parrini et al., 1995). Ira1p was not found to suppress the heat shock sensitivity phenotype of *ira2*, however Ira2p was found to partially suppress *ira1* mutations, suggesting differences in their functions (Tanaka et al., 1990).

2.1.2.3 Biosynthesis of S. cerevisiae *Ras* Yeast Ras is synthesized in the cytosol as an unmodified precursor protein. The first N-terminal methionine is removed immediately after the synthesis of the protein possibly cotranslationally (Fujiyama and Tamanoi, 1990). A series of C-terminal modification takes place that involves the addition of a farnesyl group, removal of C-terminal three amino acids, carboxymethylation and the addition of palmitic acid (Fujiyama and Tamanoi, 1986; Fujiyama et al., 1987; Fujiyama and Tamanoi, 1990).

Farnesylation is catalyzed by protein farnesyltransferase (FTase) that consists of two subunits, Dpr1/Ram1p and Ram2p (Goodman et al., 1988; Goodman et al., 1990; He et al., 1991). This enzyme is a member of the protein prenyltransferase family that included protein geranylgeranyltransferase type I (GGTase I) and protein geranylgeranyltransferase type II (GGTase II). GGTase I consists of two subunits, Cdc43/Cal1p and Ram2p, and catalyzes the addition of a geranylgeranyl group on proteins such as Rho1p and Cdc42p (Finegold et al., 1991; Ohya et al., 1991). GGTase II consists of three subunits, Bet2p, Mad2p, and Mrs6p (Jiang and Ferro-Novick, 1994), and catalyzes the addition of two geranylgeranyl groups to proteins such as Ypt involved in protein secretion. Mutational analysis of *DPR1/RAM1* and *RAM2* have been carried out that revealed significance of residues involved in substrate recognition (Trueblood et al., 1993; Mitsuzawa et al., 1995; Del Villar et al., 1997; Trueblood et al., 1997).

After farnesylation, three C-terminal amino acids are removed by a protease encoded by *RCE1* (Boyartchuk et al., 1997). This gene was identified from a screen to identify genes affecting a-factor biosynthesis (Boyartchuk et al., 1997). While C-terminal modification of the a-factor is identical to that of Ras, this protein undergoes additional cleavage at its N-terminal region and *STE24* is a gene that catalyzes both these cleavage events (Tam et al., 2001). The exposed C-terminus of Ras is then methylated by the action of carboxymethyltransferase. This enzyme is encoded by *STE14* (Hrycyna and Clarke, 1990). Interestingly, Rce1p and Ste14p are localized in the endoplasmic reticulum suggesting that Ras biosynthesis involves modification at the endoplasmic reticulum (Romano et al., 1998).

Yeast Ras1 and Ras2 proteins are modified by the addition of palmitic acid (Fujiyama and Tamanoi, 1986; Deschenes and Broach, 1987; Fujiyama et al., 1987). A palmitoyltransferase responsible for this modification has recently been identified and has been shown to be encoded by two genes *ERF2* and *ERF4* (Lobo et al., 2002). Earlier results have also shown that the post-translational farnesylation of Ras2p is important for its interaction and activation of downstream effector molecule, adenylate cyclase (Kuroda et al., 1993).

2.2 *S. cerevisiae* Rsr1/Bud1p

2.2.1 *Rsr1/Bud1p is involved in budding*

S. cerevisiae divide by budding and exhibit two different budding patterns depending on their ploidy in a process that is reliant on the GTPase, Rsr1/Bud1p (Chant and Herskowitz, 1991; Cabib et al., 1998). Yeast cells develop polarity during the dividing and mating periods of their life cycle. In their haploid state (*MAT***a** and *MAT*α), *S. cerevisiae* divide by axial budding, where the new bud emerges from the same pole as the birth scar (orderly deposits of cell wall materials left behind from previous cell divisions). In diploids (*MAT* $\mathbf{a}/\alpha$), both mating type information is expressed, causing budding to occur in a bipolar manner. During this type of cell division, the bud emerges from the opposite cell pole of the birth scar (Freifelder, 1960; Hicks and Fink, 1977; Hicks and Strathern, 1977). Physical landmarks, or complexes of proteins, establish where buds are to emerge. This landmark forms a ring around the neck of the budding daughter cell and is retained on both the newly budded cell as well as the mother following the budding process. Haploid protein landmarks include proteins such as Bud3p, Bud4p and Bud10/Axl2p (Chant and Herskowitz, 1991; Fujita et al., 1994; Herskowitz et al., 1995; Halme et al., 1996; Sanders and Herskowitz, 1996). Diploid cortical landmarks include proteins such as Bud8p, Bud9p, and Rax2p (Chen et al., 2000b). Once the budding site is presented, the polarity establishment proteins are recruited for the budding process. Members of the polarity establishment complex include Cdc42p, Cdc24p and Bem1p. These proteins organize the actin filaments into mobile cortical patches at the future bud site and at the tip of the growing bud (Cabib et al., 1998) [Figure 4].

Rsr1/Bud1p is a Ras related GTP-binding protein required for bud site selection in both axial and bipolar budding. Introduction of a yeast genomic library into *cdc24* mutants revealed *RSR1/BUD1* as a multicopy suppressor of the *cdc24* mutation that inhibited cell polarity and budding (Bender and Pringle, 1989). Deletion of *RSR1/BUD1* gene was shown to be non-lethal, however the normal pattern of bud site selection is disrupted. It is believed to activate the downstream effector Cdc24p, a GEF for Cdc42p, in order to organize the cytoskeleton at the bud site (Cabib et al., 1998). Rsr1/Bud1p is not localized to a specific bud site, rather is distributed uniformly throughout the plasma membrane (Zheng et al., 1995). However, specific

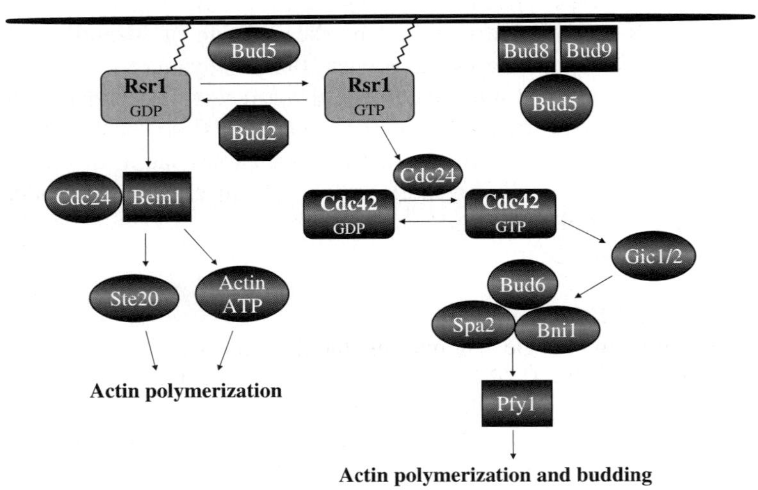

Figure 4. Rsrl/Budl of *Saccharomyces cerevisiae*. GTP-bound Rsrlp can signal through Cdc42p resulting in actin polymerization and the eventual budding of the yeast cell. Interestingly, GDP-bound Rsrlp can activate the Cdc24p/Bemlp complex to cause actin polymerization. Bud5p is the GEF and Bud2p is the GAP for Rsrl/Budlp

Rsr1/Bud1p localization to the proper bud site (axial or distal) is crucial for budding cystoskeleton organization. Therefore, it is thought that the activity of Rsr1p is localized via localization of Bud5p and Bud2p, the GEF and GAP of Rsr1/Bud1p, respectively (Bender and Pringle, 1989; Chant et al., 1991; Chant and Herskowitz, 1991; Park et al., 1993; Zheng et al., 1995).

Rsr1 in its GDP-bound state has also been found to have a cellular function. GDP-bound Rsr1p can bind to Bem1p complexed to Cdc24p (Chant et al., 1991). Upon binding, Bem1p interacts with Ste20p as well as actin-ATP and profilin to stimulate actin polymerization in budding (Elion, 2000).

A recent paper by Kozminski et al. (2003) has shown genetic and physical evidence that Rsr1/Bud1p interacts with a Rho family GTPase, Cdc42p. This provides a novel link between the ability of Rsr1/Bud1p to select for a growth site and the function of Cdc42p to establish cell polarity in yeast. The interaction between Rsr1/Bud1p and Cdc42p can be enhanced by the presence of Cdc24p, which functions as a GEF for Cdc42p (Kozminski et al., 2003).

As of yet, no homologous protein to Rsr1/Bud1p has been found in *S. pombe*. However, the amino acid sequence of Rsr1/Bud1p is most closely related to the Rap proteins (Ruggieri et al., 1992).

2.2.2 *Rsr1/Bud1p regulators*

2.2.2.1 Rsr1/Bud1p GEF: BUD5 Random budding occurs in the absence of either regulatory protein, as well as the landmark proteins Bud3p, Axl2p for haploid cells and Bud8p for diploid cells. It has been proposed that Bud5p and Bud2p, the respective GEF and GAP of Rsr1/Bud1p, bind to specific cell markers in the various cell types (diploid versus haploid) and regulate the activity of Bud1p at these locations (Kang et al., 2001). This is particularly evident in localization experiments of Bud5p and Bud2p, whose localization patterns have been studied in haploid and diploid cells (Arkowitz et al., 2001; Kang et al., 2001; Marston et al., 2001).

Bud5p is homologous to Cdc25p, the GEF for *S. cerevisiae* Ras (Chant and Herskowitz, 1991; Powers et al., 1991). It is 20% homologous to Cdc25p over the C-terminal GEF region [Figure 3A] (Jones et al., 1991; Powers et al., 1991). Mutations in Bud5p result in its mislocalization as well as bipolar budding defects of diploid cells (Kang et al., 2001). A Bud5p-GFP fusion protein localized to a presumptive patch at the periphery of small buds and maintained its localization to form a double ring that encircled the mother-bud neck. Upon cell division, the double ring split, resulting in a single ring marking the division site in both mother and daughter cells. A new Bud5p-GFP patch was then observed at the presumptive bud site, adjacent to the previous bud scar in the mother cell, suggesting that Bud5p is recruited to the new presumptive bud site (Kang et al., 2001; Marston et al., 2001). It has been shown that Bud5p depends on axial and bipolar cortical landmarks for proper localization and activation of Rsr1/Bud1p (Chant et al., 1995; Pringle et al., 1995; Roemer et al., 1996; Sanders and Herskowitz, 1996; Kang et al., 2001). Mutants of *bud5* are believed to be defective in recognizing bipolar landmarks because they fail to localize to the small diploid buds after cells continued to progress through the cell cycle. In haploid cells, mutations in the putative landmark protein, Bud10/Axl2p, had an effect on Bud5p localization, as did the mutations in Bud3p and Bud4p, although to a lesser degree (Kang et al., 2001). Bud5p was also shown to physically interact *in vitro* with Axl2p (Kang et al., 2001). Bud5-GFP localization in *bud8* mutant diploid cells resulted in Bud5p localization proximal to the birth scar, rather than the distal poles (Zahner et al., 1996). This was not so apparent with the *bud9* mutant. Loss of Bud9p did not alter the localization of Bud5p to either distal or proximal poles, suggesting that it is not involved in initial proximal pole targeting (Zahner et al., 1996). This indicates that the Bud5 protein requires the axial landmark Bud10/Axl2p and the distal landmark Bud8p for localization. Bud5p and Axl2p are both equally expressed in all cell types, but how Axl2p would be able to localize Bud5 protein only in haploid cells is not clear, thus other parameters are most likely required. In addition, the loss of Rax2p, another protein required for maintaining bipolar budding pattern, resulted in unbudded cells with Bud5p localized only to the birth pole (Arkowitz et al., 2001; Kang et al., 2001).

2.2.2.2 Rsr1/Bud1 GAP: BUD2 Bud2p was isolated from a complementation screen that caused random budding and has been shown to function as a GTPase-activating protein for Rsr1/Bud1p (Chant et al., 1991; Chant and Herskowitz, 1991; Bender, 1993; Park et al., 1993). Chromosome deletion and overexpression of Bud2p results in random budding, and its deletion does not affect cell growth, indicating that it is mainly utilized for proper bud site selection (Park et al., 1993). Bud2p contains 1,104 amino acids with a M_r of 127kDa [Figure 3B]. It contains a region similar to the mammalian GAP domain. This GAP domain has 21.5% identity to NF1, 19.5% identity to Ira2p over 231 residues, and 20.1% similarity to bovine GAP (Park et al., 1993). Most *bud2* mutants were further found to exhibit non-axial budding patterns, unlike wild-type strains (Chant and Herskowitz, 1991).

Interestingly, it is proposed that the regulatory proteins of Rsr1/Bud1p play a major role in bud site selection for oriented cell division by localizing to the future bud site at the cell membrane. In this way, they promote the repeated activation of Rsr/Bud1p at this specific site, since Rsr1/Bud1p is localized throughout the cell (Michelitch and Chant, 1996). The importance of Rsr1/Bud1p cycling was demonstrated when the overexpression of the wild type protein resulted in better *cdc24* suppression properties than $bud1^{G12V}$ (GTP-bound) or $bud1^{K16N}$ (GDP-bound) mutants (Ruggieri et al., 1992).

2.2.3 Biosynthesis of Rsr1/Bud1p

Rsr1/Bud1p ends with the C-terminal *CAAL* motif. This protein is modified by the addition of a geranylgeranyl group (Park et al., 2002). It has also been shown that mutation of the lysine repeat in the hypervariable region of Rsr1/Bud1p abolishes its plasma membrane localization (Park et al., 2002). In addition, mutation of the CAAX box of Rsr1/Bud1p abolished both plasma membrane and internal membrane association (Park et al., 2002).

2.3 *S. cerevisiae* Rheb

2.3.1 ScRheb protein is involved in nutrient uptake

Ras homologue enriched in brain (Rheb) was first identified in rat as a gene rapidly induced by electroconvulsive seizures and NMDA-mediated synaptic activity (Yamagata et al., 1994). Rheb was shown to both antagonize and synergize with *ras*-induced transformation (Clark et al., 1997; Yee and Worley, 1997; Im et al., 2002). Rheb was expressed ubiquitously with the highest expression occurring in the cardiac and skeletal muscle (Gromov et al., 1995). Rheb is a highly conserved protein. Since the discovery of Rheb in rat, homologues of Rheb have been found in various organisms from yeast to human (Urano et al., 2000). Analysis of the amino acid sequence of the various Rheb homologues reveals a number of unique features of this subfamily of protein (Urano et al., 2000). First, the Rheb family proteins contain the characteristic G-boxes found in other small GTPases. Second, in the G1 box of Rheb proteins, there is a highly conserved arginine residue corresponding to the glycine 12 of ras proteins. Third, the G2 effector domain is highly conserved. Lastly, Rheb proteins contain a C-terminal *CAAX* motif allowing for farnesylation and subsequent membrane localization (Urano et al., 2000; Yang et al., 2001).

Homologues of Rheb were identified in both budding and fission yeast (Urano et al., 2000). *S. cerevisiae* Rheb (ScRheb) plays a role in arginine and lysine uptake (Urano et al., 2000). Disruption of *SCRHEB* gene showed that the gene is non-essential (Urano et al., 2000). Cells lacking ScRheb displayed hypersensitivity to the toxic arginine analogue, canavanine, as well as increased uptake of radioactively labeled arginine (Urano et al., 2000). *SCRHEB* disruptants also displayed hypersensitivity to the toxic analogue of lysine, thialysine, as well as increased uptake of radioactively labeled lysine (Urano et al., 2000). Further studies have shown that this regulation can take place at the level of the amino acid permease, Can1. Cells lacking the Can1 permease did not demonstrate an increased uptake of arginine or lysine. When *SCRHEB* was disrupted in a *can1* mutant background, a differential effect on arginine uptake was no longer observed (Urano et al., 2000).

2.3.2 Biosynthesis of ScRheb

ScRheb protein ends with the *CAAX* motif and is farnesylated (Urano et al., 2000). The modification is critical for the function of ScRheb, as a mutant form of ScRheb that has the *CAAX* motif modified is unable to complement the canavanine sensitivity of the *scrheb* mutant strain (Urano et al., 2000).

3. RAS FAMILY PROTEINS IN *SCHIZOSACCHAROMYCES POMBE*

3.1 *S. Pombe* Ras1

3.1.1 Ras1 is involved in making response and morphological changes

A single homolog of the mammalian *ras* genes was discovered in the fission yeast, *S. pombe*. This gene, $ras1^{+}$, was detected by Southern hybridization (Fukui et al., 1986). Disruption of $ras1^{+}$ gene resulted in an inability for yeast cells to mate, while expression of $ras1^{val17}$, the analogous mutation to a transforming mutant of mammalian *ras*, rescued this phenotype (Fukui and Yamamoto, 1988). An interesting feature of *S. pombe* Ras1 is that it does not activate adenylate cyclase in fission yeast. Ras1 is essential for mating, but not for vegetative growth (Fukui et al., 1986).

A downstream effector of Ras1 is Byr2 (Nadin-Davis and Nasim, 1990; Wang et al., 1991b; Masuda et al., 1995). Byr2 is a protein kinase that has been shown to act downstream of Ras1 in a mating pheromone signal transduction system. Ras1 binds a region in the N-terminal 206 amino acids of Byr2 in a GTP-dependent manner. Genetic studies have shown that Byr1 lies downstream of Byr2 (Xu et al., 1994). Byr1 then activates Spk1 kinase, a structural and functional homolog of the vertebrate MAP kinases (Hughes et al., 1993; Neiman et al., 1993). Spk1 plays a role in conjugation and sporulation [Figure 5].

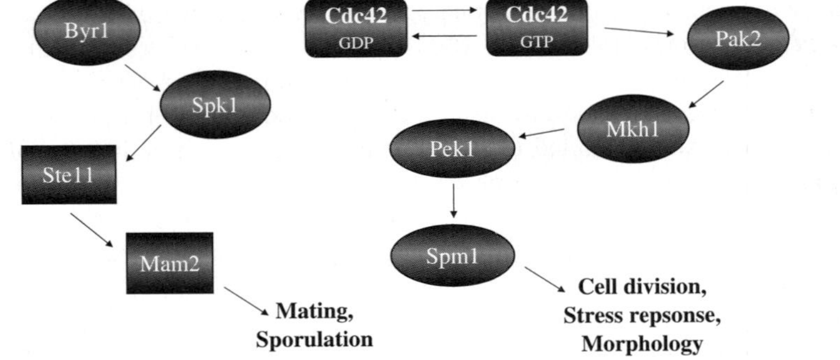

Figure 5. Rasl in *Schizosaccharomyces pombe*. GTP-bound Ras1 can signal to Byr2, resulting in mating and sporulation pathway. In addition, activated Ras1 can interact with GTP-bound Cdc42, thus resulting in cell division, stress response, and morphological changes

S. pombe Ras1 also has role in morphology. Ras1 activates Scd1, a guanine nucleotide exchange factor for Cdc42 (Chang et al., 1994; Hughes, 1995).The activation of Cdc42, which is a Rho-like GTP-binding protein can regulate the dynamics of the actin cytoskeleton (Hughes, 1995). Cdc42 signals to Pak2 which activates Mkh1 followed by Pek1 and Spm1 (Ottilie et al., 1995; Merla and Johnson, 2001). This signal transduction system leads to control of morphology and cell polarity in fission yeast.

An interesting feature of Ras1 signal transduction in *S. pombe* lies in how it determines the activation of the Byr2 versus the Scd1 pathway. As will be discussed below, Ras1 interacts with the two GEFs, Ste6 and Efc25. Ste6 is a GEF for Ras1 that specifically activates Ras1 for the Byr2 pathway, while Efc25 specifically activates Ras1 for the Scd1 pathway. The end result of Ras1 activation is thus dependent on both the level and subcellular localization of GEF expression within the cell. Section 3.1.2 further discusses these exchange factors in depth and their role in Ras1 regulation.

3.1.2 Ras1p Regulators

3.1.2.1 Ras1 GEFs: Ste6, Efc25 Two guanine nucleotide exchange factors (GEF) have been found for Ras1, Ste6 and the more recently identified GEF, Efc25 (Papadaki et al., 2002). Ste6 was first reported to be a homologue of Cdc25p with significant homology to its C-terminal region [Figure 3A] (Hughes et al., 1990). Ste6 is 911 amino acids long and contains an SH3 domain that may be involved in Ras1 activation, as well as a GEF domain (Hughes et al., 1990). Ste6 was found to be essential for mating and act as a GEF for Ras1 that specifically activates Ras1 for the mating response pathway (Hughes et al., 1990). Mating is triggered in haploid cells upon nitrogen starvation so that cells of opposite mating types mate in response to released pheromones in order to form a diploid zygote. This diploid undergoes meiosis and sporulation, and forms four haploid spores. *ste6*$^-$ mutants behave similarly to *ras1*$^-$ mutants in that they are sterile and unable to respond to mating pheromones (Fukui et al., 1986; Nadin-Davis et al., 1986; Leupold et al., 1991). Mutants of *ste6*$^-$ cause a block in sexual differentiation and sterility but do not affect cell morphology (Papadaki et al., 2002). These characteristics are also similar to the phenotype of *byr2*$^-$ mutants (Wang et al., 1991b). Such evidence suggests that Ste6 GEF activity activates Ras1 so that it specifically interacts with the downstream effector Byr2 to trigger the downstream pathway involved in sexual differentiation and sterility.

Ste6 expression is low during vegetative growth and is induced following nitrogen starvation or external signaling from mating pheromones, which both lead to mating and sexual differentiation (Hughes et al., 1994). Mutations in genes encoding pheromone response pathway kinases Byr2, Byr1 and Spk1, resulted in the loss of *ste6*$^+$ RNA induced expression in response to nitrogen starvation (Hughes et al., 1990). Ste6 expression is also increased in *ras1*$^+$ activated mutants and is lowered in *ras1*$^-$ mutants, indicating that Ras1 might act in a positive feedback loop to induce its own activator (Hughes et al., 1990). This positive feedback is believed to be pheromone dependent. Ste11, an HMG-box protein, was shown to positively upregulate *ste6*$^+$ gene expression, acting as a transcriptional activator of Ste6 (Sugimoto et al., 1991; Hughes et al., 1994). Hughes et al. proposed that during nitrogen starvation, Ste11 is activated, transcriptionally upregulates Ste6 by binding upstream of the *ste6*$^+$ promoter. This increase in expression would then result in Ste6 regulated activation of Ras1 and the Byr1-pheromone signaling pathway, which further upregulates Ste6 in a positive feedback loop (Hughes et al., 1994).

Efc25 (exchange factor cdc25-like), was discovered by isolating genes encoding a subunit of the DNA polymerase δ (Tratner et al., 1997). It is 987 amino acids (112kDa) and contains a C-terminal region that is highly homologous with other Ras GEFs including *S. cerevisiae* Cdc25, Scd25 and *S. pombe* Ste6, although it is unable to complement *CDC25* thermosensitivity in *S. cerevisiae* [Figure 3A] (Tratner et al., 1997). Efc25 has a GEF domain, which is essential for its function, but requires the N-terminus for regulation and efficiency (Papadaki et al., 2002). Efc25 has shown involvement in the Scd1 effector pathway, which regulates cellular morphology, chromosome segregation and spindle formation (Tratner et al., 1997; Papadaki et al., 2002).

In comparison to sterile *ste6*$^-$ mutants, *efc25*$^-$ mutants are round but fertile (Tratner et al., 1997; Papadaki et al., 2002). Scd1 is a presumptive GEF for Cdc42. Scd1 inactivation results in changes in cell morphology from elongated to round, defects in chromosome segregation and spindle formation, as well as sterility, although seemingly not resulting from abnormalities in mating pheromone signaling (Chang et al., 1994; Chen et al., 1999; Li et al., 2000).

Unlike Ste6, which is barely detectable during vegetative growth and upregulated expression during mating, Efc25 expression is constitutive and thus appears to be controlled by a different promoter (Papadaki et al., 2002). Interestingly, overexpression of Efc25 results in hyper-elongation of cells, but not in *ras1*$^-$ and *scd1*$^-$ cells (Papadaki et al., 2002). This demonstrates that Efc25 must be activating the Scd1 pathway, rather than the Byr2 pathway, particularly since the cells are sterile. They propose the model suggesting that Ras1 must be preferentially recruited to activate Scd1 during vegetative growth due to the presence of constitutively expressed Efc25, resulting in Ras1 performing more morphogenic functions including controlling polarized cell extension and mitotic fidelity. Efc25 would regulates Ras1 by competitively recruiting it away from Byr2. Alternatively in this model, during sexual differentiation, the GEF Ste6 is upregulated and would thus recruit Ras1 to activate the Byr2 pathway.

It is not known how each GEF regulates Ras1 to interact with specific downstream effectors. It has been proposed that the presence of a given GEF could induce a certain conformational change in Ras1 in order to favor the binding of a specific effector, which could be further aided by scaffolding proteins (Papadaki et al., 2002). Such a conformational change has been known to occur in the mammalian GEF, SOS, which binds to Ras, causing a dramatic conformational change in the Ras Switch I region, which includes the effector binding domain (Hall et al., 2001). Alternatively, it is also proposed that two pathways are spacially segregated within the cell, or that both conformational change and spacial segregation explanations are possible (Papadaki et al., 2002). Strains that are *efc25*¯ and *ste6*¯ behave similarly to *ras1*¯ deleted strains, providing evidence that these are the only GEFs for Ras1, although they do not seem interchangeable in their function (Chang et al., 1994; Papadaki et al., 2002). Rather, the GEFs appear to play an important role in directing Ras1 to activate either Byr2 or Scd1 through their expression, which is regulated by mating signals.

3.1.2.2 Ras1 Gap: Ras Gap/Sar1 *S. pombe* RasGAP/Sar1 was first discovered as a Ras GAP homologue, which complemented loss of *ira* function in *S. cerevisiae* (Imai et al., 1991; Wang et al., 1991). This gene was also identified by a loss of function screen that results in hypersensitivity of *S. pombe* to the mating factor and an inability to perform efficient mating when mutated. These phenotypes are similar to those of activated $Ras1^{Val17}$. This provided evidence that RasGap/Sar1 was a negative regulator of Ras (Imai et al., 1991). RasGAP/Sar1 is 766 amino acids and contains a conserved GAP core sequence, with relatively short sequences flanking the core [Figure 3B] (Imai et al., 1991). RasGAP/Sar1 is homologous to *S. cerevisiae* Ira GAPs, particularly within the GAP regions, although it is much smaller and does not contain the long N- and C-terminal sequences. It is also closely related to the mammalian IQGAP1, where it shares 54% similarity at the GAP related domain (GRD) and 27% aa sequence identity outside of the GRD (Weissbach et al., 1994).

3.1.3 Biosynthesis of Ras1p

Like *S. cerevisiae* Ras1 and Ras2, *S. pombe* Ras1 undergoes C-terminal modification including farnesylation (Yang et al., 2000). The farnesylation is catalyzed by protein farnesyltransferase that consists of Cpp1 and Cwp1 (Danjoh and Fujiyama, 1996; Yang et al., 2000). Disruption of $cpp1^+$ gene leads to a mutant defective in protein farnesyltransferase. The disruptant cells exhibit altered morphology that is suppressed by the expression of Ras1 mutant that is farnesylation-independent (Yang et al., 2000). The *cpp1*¯ disruptant also exhibits canavanine sensitivity and accumulation of G0/G1 phase cells. These phenotypes are suppressed by the expression of Rheb mutant that can bypass farnesylation (Yang et al., 2000).

3.2 *S. pombe* Rheb

3.2.1 S. pombe Rheb is Involved in Cell Cycle

$Sprheb^+$ is an essential gene, as haploid cells with $rheb^+$ disruption could not be obtained (Mach et al., 2000; Yang et al., 2001). Inhibition of SpRheb expression results in the arrest in the G1 phase of the cell cycle (Mach et al., 2000; Yang et al., 2001). The arrested cells have a rounded morphology as compared to the normal rod-like shape of wild type *S. pombe* cells (Mach et al., 2000; Yang et al., 2001). These phenotypes are reminiscent of cells undergoing nitrogen starvation. In fact, *sprheb*¯ cells were found to upregulate the genes, $mei2^+$ and $fnx1^+$ (Mach et al., 2000).

Expression of these genes is also induced during nitrogen starvation. SpRheb also affects arginine uptake, as cells defective in protein farnesyltransferase exhibit increased uptake of arginine and this phenotype is complemented by a modified form of SpRheb that could bypass farnesylation (Yang et al., 2000). Therefore, it seems that through evolution, SpRheb has a role in controlling cell cycle in *S. pombe* in addition to regulating amino acid uptake.

Using the above phenotypes of inhibiting Rheb expression, a screen was devised to identify dominant negative mutants of SpRheb (Tabancay et al., 2003). This screen was necessary, as a mutation analogous to RasS17N did not exhibit any dominant negative phenotypes in Rheb (Tabancay et al., 2003). The assay involved expressing an *sprheb* mutant library constructed by PCR mutagenesis in a wild type strain. Because the inhibition of Rheb results in increased *fnx1*$^{+}$ expression, this wild type strain contains a *fnx1*-lacZ reporter. We screened for growth inhibition and lacZ induction upon overexpression of the mutant *rheb*$^{+}$ library. This screen allowed us to identify a mutant D60V that resulted in dominant negative phenotypes, G1 arrest, rounded cells, and *fnx1*$^{+}$ induction. This D60 residue was, then, mutated to all 20 amino acids, and we found two mutants, D60K and D60I, with a more potent dominant negative effect. The mutants, D60V and D60I, were shown to bind GDP preferentially, while the mutant D60K had lost the ability to bind to both GDP and GTP (Tabancay et al., 2003). The mechanism by which these D60 mutants lead to dominant negative effects appears to be titrating out the guanine nucleotide exchange factor (GEF) for Rheb. Similar D60 mutants in Ras have been shown to bind tightly to exchange factor (Feig, 1999).

Utilizing a system in which the expression of SpRheb can be modulated, it was discovered that human Rheb, but not ScRheb, can complement the loss of *sprheb*$^{+}$ gene (Yang et al., 2001). In addition, farnesylation was shown to be necessary for the proper function of SpRheb as a *CAAX* mutant is unable to complement the loss of SpRheb (Yang et al., 2001). Expression of an alternatively modified version of SpRheb, but not of Ras1, in *S. pombe* cells lacking protein farnesyltransferase was able to suppress the arginine uptake defect, possibly demonstrating that Ras and Rheb do not serve overlapping functions in fission yeast (Yang et al., 2000).

Recent studies with *Drosophila* Rheb (dRheb) have revealed a role for Rheb in cell growth. Overexpression of dRheb in the imaginal disc results in a large head and eye. Similarly, dRheb overexpression in the wing and salivary gland causes an enlargement of these organs (Patel et al., 2003; Saucedo et al., 2003; Stocker et al., 2003). In *Drosophila* tissue culture cells, dRheb inhibition results in cell cycle block at the G1 phase, while dRheb overexpression leads to the increase of S phase cells (Patel et al., 2003). Further analysis in *Drosophila* has placed dRheb in the TOR/S6K pathway, a major signaling pathway that controls nutrient uptake, protein synthesis, and cell cycle (Patel et al., 2003; Saucedo et al., 2003; Stocker et al., 2003). dRheb heterozygous mutants are hypersensitive to rapamycin (Patel et al., 2003). Recent work in mammalian systems has also shown that Rheb is involved in the mTOR/S6K pathway (Castro et al., 2003; Inoki et al., 2003; Tabancay et al., 2003; Tee et al., 2003). Transfection of HEK293 cells with Human Rheb1 and

Rheb2 leads to S6K activation, and this S6K activation is inhibited by rapamycin (Tabancay et al., 2003). In addition, serum- or nutrient-induced activation of S6K can be blocked by the expression of the human Rheb dominant negative mutants (Tabancay et al., 2003). Thus, Rheb has been shown to play a role in the mTOR/S6K pathway in both *Drosophila* and mammalian cells. *S. pombe* contains homologues of mTOR, SpTor1 and SpTor2 (Kawai et al., 2001; Weisman and Choder, 2001; Weisman, 2004). Further work is currently being conducted to see if SpRheb plays a similar role in the TOR pathway in fission yeast.

3.2.2 SpRheb Regulators

Until presently, no GEFs or GAPs have been found as the SpRheb regulators. Recently, the tuberous sclerosis complex, TSC1/2, has been proposed as the GAP for *Drosophila* Rheb and human Rheb (Castro et al., 2003; Garami et al., 2003; Inoki et al., 2003; Zhang et al., 2003). A recent report has described a genetic interaction between a TSC2 homologue in *S. pombe*, *tsc2*$^+$, and *sprheb*$^+$ (Van Slegtenhorst et al., 2004). Further investigation is currently under way to determine whether *S. pombe* Tsc2 targets SpRheb and serves as a GAP for SpRheb. The 156kDa *S. pombe* Tsc2 is smaller than the 190kDa HsTsc2 and 204kDa DmTsc2, but it has been shown to form a complex with *S. pombe* Tsc1(Matsumoto et al., 2002). *S. pombe* Tsc1/2 is not complemented by human Tsc1/2, and is believed to be involved in nutrient uptake and conjugation (Matsumoto et al., 2002). We are currently pursuing biochemical assays to provide direct evidence that Tsc2 in *S. pombe* is, in fact, a GAP for SpRheb.

3.2.3 Biosynthesis of SpRheb

SpRheb is farnesylated and the modification is critical for the function of SpRheb. A mutant form of SpRheb defective in farnesylation is incapable of complementing growth inhibition of the *rhb1*$^-$ mutant (Yang et al., 2000). In addition, non-farnesylated SpRheb was unable to rescue the arginine uptake defect of *rhb1*$^-$ mutants. Disruption of the *cpp1*$^-$ subunit in *S. pombe* results in a mutant farnesyltransferase. As a result, the mutant strain displays G0/G1 accumulation of the cell cycle as well as sensitivity to the toxic arginine analog, canavanine. These phenotypes can be rescued by expression of a mutant form of SpRheb that can be geranyl- geranylated (Yang et al., 2000).

4. CONCLUSIONS

Yeast serves as a powerful genetic system to study gene function. Much of what is known today about mammalian *ras* counterparts has been discovered by characterizing Ras1 and Ras2 from *S. cerevisiae* (Fraenkel et al., 1985). Data obtained from yeast studies have been extended to elucidate the function of ras in higher eukaryotic organisms. Continuing to study these Ras superfamily G-proteins in yeast will undoubtedly lead to additional findings as to what intracellular signaling pathways they are involved in, what proteins they interact with, and how theses molecules are in fact regulated in the cell. The ease in which genes can be disrupted in yeast also makes this system beneficial. The evolutionary closeness of yeast, especially the fission yeast, *Schizosaccharomyces pombe*, to higher eukaryotes allows for the possibility of functional conservation between yeast and human proteins. Thus, the future discoveries that are made with small GTPases in yeast have true potential in advancing what we know about these proteins in humans.

This chapter discussed the small GTPases Ras, Rsr1/Bud1, and Rheb in the yeasts *Saccharomyces cerevisiae* and *Schizosaccharomyces pombe*. We have provided a thorough discussion of the following areas: 1) general structure, 2) major functions and pathways in the cell, 3) how they are regulated in the cell, and 4) biosynthesis. As can be seen, these Ras superfamily G-proteins play a number of important functions in yeast from proliferation, mating, filamentous and invasive growth, budding, and nutrient uptake. Future work will bring us to a fuller understanding as to how these small G-proteins can affect so many diverse processes not only in yeast but in higher eukaryotic systems.

ACKNOWLEDGEMENTS

We would like to thank Dr. Douglas Johnson and Dr. Jun Urano for valuable comments about this manuscript. This work is supported by NIH grant CA41996.

REFERENCES

Arkinstall, S.J., Papasavvas, S.G. and Payton, M.A. (1991) Yeast alpha-mating factor receptor-linked G-protein signal transduction suppresses Ras-dependent activity. FEBS Lett. *284*, 123-128.

Arkowitz, R.A. (2001) Cell polarity: connecting to the cortex. Curr. Biol. *11*, R610-R612.

Bardin, A.J., Visintin, R. and Amon, A. (2000) A mechanism for coupling exit from mitosis to partitioning of the nucleus. Cell *102*, 21-31.

Bender, A. (1993) Genetic evidence for the roles of the bud-site-selection genes BUD5 and BUD2 in control of the Rsr1p (Bud1p) GTPase in yeast. Proc. Natl. Acad. Sci. U.S.A. *90*, 9926-9929.

Bender, A. and Pringle, J.R. (1989) Multicopy suppression of the cdc24 budding defect in yeast by CDC42 and three newly identified genes including the ras-related gene RSR1. Proc. Natl. Acad. Sci. U.S.A. *86*, 9976-9980.

Bourne, H.R., Sanders, D.A. and McCormick, F. (1990) The GTPase superfamily: a conserved switch for diverse cell functions. Nature *348*, 125-132.

Bourne, H.R., Sanders, D.A. and McCormick, F. (1991) The GTPase superfamily: conserved structure and molecular mechanism. Nature *349*, 117-127.

Boy-Marcotte, E., Buu, A., Soustelle, C., Poullet, P., Parmeggiani, A. and Jacquet, M. (1993) The C-terminal part of the CDC25 gene product has Ras-nucleotide exchange activity when present in a chimeric SDC25-CDC25 protein. Curr. Genet. *23*, 397-401.

Boy-Marcotte, E., Ikonomi, P. and Jacquet, M. (1996) SDC25, a dispensable Ras guanine nucleotide exchange factor of Saccharomyces cerevisiae differs from CDC25 by its regulation. Mol. Biol. Cell *7*, 529-539.

Boyartchuk, V.L., Ashby, M.N. and Rine, J. (1997) Modulation of Ras and a-factor function by carboxyl-terminal proteolysis. Science *275*, 1796-1800.

Breviario, D., Hinnebusch, A., Cannon, J., Tatchell, K. and Dhar, R. (1986) Carbon source regulation of RAS1 expression in Saccharomyces cerevisiae and the phenotypes of ras2- cells. Proc. Natl. Acad. Sci. U.S.A. *83*, 4152-4156.

Breviario, D., Hinnebusch, A.G. and Dhar, R. (1988) Multiple regulatory mechanisms control the expression of the RAS1 and RAS2 genes of Saccharomyces cerevisiae. EMBO J. *7*, 1805-1813.

Broek, D., Samiy, N., Fasano, O., Fujiyama, A., Tamanoi, F., Northup, J. and Wigler, M. (1985) Differential activation of yeast adenylate cyclase by wild-type and mutant RAS proteins. Cell *41*, 763-769.

Broek, D., Toda, T., Michaeli, T., Levin, L., Birchmeier, C., Zoller, M., Powers, S. and Wigler, M. (1987) The S. cerevisiae CDC25 gene product regulates the RAS/adenylate cyclase pathway. Cell *48*, 789-799.

Boy-Marcotte, D., Toda, T., Michaeli, T., Levin, L., Birchmeier, C., Zoller, M., Powers, S. and Wigler, M. (1987) The S. cerevisiae CDC25 gene product regulates the RAS/adenylate cyclase pathway. Cell *48*, 789-799.

Cabib, E., Drgonova, J. and Drgon, T. (1998) Role of small G proteins in yeast cell polarization and wall biosynthesis. Annu. Rev. Biochem. *67*, 307-333.

Camonis, J.H., Kalekine, M., Gondre, B., Garreau, H., Boy-Marcotte, E. and Jacquet, M. (1986) Characterization, cloning and sequence analysis of the CDC25 gene which controls the cyclic AMP level of Saccharomyces cerevisiae. EMBO J. *5*, 375-380.

Castro, A.F., Rebhun, J.F., Clark, G.J. and Quilliam, L.A. (2003) Rheb binds tuberous sclerosis complex 2 (TSC2) and promotes S6 kinase activation in a rapamycin- and farnesylation-dependent manner. J. Biol. Chem. *278*, 32493-32496.

Chang, E.C., Barr, M., Wang, Y., Jung, V., Xu, H.P. and Wigler, M.H. (1994) Cooperative interaction of S. pombe proteins required for mating and morphogenesis. Cell *79*, 131-141.

Chant, J. and Herskowitz, I. (1991) Genetic control of bud site selection in yeast by a set of gene products that constitute a morphogenetic pathway. Cell *65*, 1203-1212.

Chant, J., Corrado, K., Pringle, J.R. and Herskowitz, I. (1991) Yeast BUD5, encoding a putative GDP-GTP exchange factor, is necessary for bud site selection and interacts with bud formation gene BEM1. Cell *65*, 1213-1224.

Chant, J., Mischke, M., Mitchell, E., Herskowitz, I. and Pringle, J.R. (1995) Role of Bud3p in producing the axial budding pattern of yeast. J. Cell Biol. *129*, 767-778.

Chen, C.R., Li, Y.C., Chen, J., Hou, M.C., Papadaki, P. and Chang, E.C. (1999) Moe1, a conserved protein in Schizosaccharomyces pombe, interacts with a Ras effector, Scd1, to affect proper spindle formation. Proc. Natl. Acad. Sci. U.S.A. *96*, 517-522.

Chen, R.A., Michaeli, T., Van Aelst, L. and Ballester, R. (2000) A role for the noncatalytic N terminus in the function of Cdc25, a Saccharomyces cerevisiae Ras-guanine nucleotide exchange factor. Genetics *154*, 1473-1484.

Chen, T., Hiroko, T., Chaudhuri, A., Inose, F., Lord, M., Tanaka, S., Chant, J. and Fujita, A. (2000) Multigenerational cortical inheritance of the Rax2 protein in orienting polarity and division in yeast. Science *290*, 1975-1978.

Cherkasova, V.A., McCully, R., Wang, Y., Hinnebusch, A. and Elion, E.A. (2003) A novel functional link between MAP kinase cascades and the Ras/cAMP pathway that regulates survival. Curr. Biol. *13*, 1220-1226.

Clark, G.J., Kinch, M.S., Rogers-Graham, K., Sebti, S.M., Hamilton, A.D. and Der, C.J. (1997) The Ras-related protein Rheb is farnesylated and antagonizes Ras signaling and transformation. J. Biol. Chem. *272*, 10608-10615.

Crechet, J.B., Poullet, P., Bernardi, A., Fasano, O. and Parmeggiani, A. (1993) Properties of the SDC25 C-domain, a GDP to GTP exchange factor of RAS proteins and in vitro modulation of adenylyl cyclase. J. Biol. Chem. *268*, 14836-14841.

Damak, F., Boy-Marcotte, E., Le-Roscouet, D., Guilbaud, R. and Jacquet, M. (1991) SDC25, a CDC25-like gene which contains a RAS-activating domain and is a dispensable gene of Saccharomyces cerevisiae. Mol. Cell. Biol. *11*, 202-212.

Daniel, J., Becker, J.M., Enari, E. and Levitzki, A. (1987) The activation of adenylate cyclase by guanyl nucleotides in Saccharomyces cerevisiae is controlled by the CDC25 start gene product. Mol. Cell. Biol. *7*, 3857-3861.

Danjoh, I. and Fujiyama, A. (1996) Enzymic characterization of fission yeast farnesyl transferase: recognition of the -CAAL motif at the C-terminus. Eur. J. Biochem. *236*, 847-851.

DeFeo-Jones, D., Scolnick, E.M., Koller, R. and Dhar, R. (1983) ras-Related gene sequences identified and isolated from Saccharomyces cerevisiae. Nature *306*, 707-709.

DeFeo-Jones, D., Tatchell, K., Robinson, L.C., Sigal, I.S., Vass, W.C., Lowy, D.R. and Scolnick, E.M. (1985) Mammalian and yeast ras gene products: biological function in their heterologous systems. Science *228*, 179-184.

Del Villar, K., Mitsuzawa, H., Yang, W., Sattler, I. and Tamanoi, F. (1997) Amino acid substitutions that convert the protein substrate specificity of farnesyltransferase to that of geranylgeranyltransferase type I. J. Biol. Chem. *272*, 680-687.

Deschenes, R.J. and Broach, J.R. (1987) Fatty acylation is important but not essential for Saccharomyces cerevisiae RAS function. Mol. Cell. Biol. *7*, 2344-2351.

Downward, J. (1990) The ras superfamily of small GTP-binding proteins. Trends Biochem. Sci. *15*, 469-472.

Elion, E.A. (2000) Pheromone response, mating and cell biology. Curr. Opin. Microbiol. *3*, 573-581.

Elion, E.A., Brill, J.A. and Fink, G.R. (1991a) Functional redundancy in the yeast cell cycle: FUS3 and KSS1 have both overlapping and unique functions. Cold Spring Harb. Symp. Quant. Biol. *56*, 41-49.

Elion, E.A., Brill, J.A. and Fink, G.R. (1991b) FUS3 represses CLN1 and CLN2 and in concert with KSS1 promotes signal transduction. Proc. Natl. Acad. Sci. U.S.A. *88*, 9392-9396.

Fan, W.T., Koch, C.A., de Hoog, C.L., Fam, N.P. and Moran, M.F. (1998) The exchange factor Ras-GRF2 activates Ras-dependent and Rac-dependent mitogen-activated protein kinase pathways. Curr. Biol. *8*, 935-938.

Feig, L.A. (1999) Tools of the trade: use of dominant-inhibitory mutants of Ras-family GTPases. Nat. Cell Biol. *1*, E25-E27.

Finegold, A.A., Johnson, D.I., Farnsworth, C.C., Gelb, M.H., Judd, S.R., Glomset, J.A. and Tamanoi, F. (1991) Protein geranylgeranyltransferase of Saccharomyces cerevisiae is specific for Cys-Xaa-Xaa-Leu motif proteins and requires the CDC43 gene product but not the DPR1 gene product. Proc. Natl. Acad. Sci. U.S.A. *88*, 4448-4452.

Fraenkel, D.G. (1985) On ras gene function in yeast. Proc. Natl. Acad. Sci. U.S.A. *82*, 4740-4744.

Francois, J., Higgins, D.L., Chang, F. and Tatchell, K. (1991) Inhibition of glycogen synthesis in Saccharomyces cerevisiae by the mating pheromone alpha-factor. J. Biol. Chem. *266*, 6174-6180.

Freeman, N.L., Lila, T., Mintzer, K.A., Chen, Z., Pahk, A.J., Ren, R., Drubin, D.G. and Field, J. (1996) A conserved proline-rich region of the Saccharomyces cerevisiae cyclase-associated protein binds SH3 domains and modulates cytoskeletal localization. Mol. Cell. Biol. *16*, 548-556.

Freifelder, D. (1960) Bud position in Saccharomyces cerevisiae. J. Bacteriol. *80*, 567-568.

Fujita, A., Oka, C., Arikawa, Y., Katagai, T., Tonouchi, A., Kuhara, S. and Misumi, Y. (1994) A yeast gene necessary for bud-site selection encodes a protein similar to insulin-degrading enzymes. Nature *372*, 567-570.

Fujiyama, A. and Tamanoi, F. (1986) Processing and fatty acid acylation of RAS1 and RAS2 proteins in Saccharomyces cerevisiae. Proc. Natl. Acad. Sci. U.S.A. *83*, 1266-1270.

Fujiyama, A., Matsumoto, K. and Tamanoi, F. (1987) A novel yeast mutant defective in the processing of ras proteins: assessment of the effect of the mutation on processing steps. EMBO J. *6*, 223-228.

Fujiyama, A. and Tamanoi, F. (1990) RAS2 protein of Saccharomyces cerevisiae undergoes removal of methionine at N terminus and removal of three amino acids at C terminus. J. Biol. Chem. *265*, 3362-3368.

Fukui, Y., Kozasa, T., Kaziro, Y., Takeda, T. and Yamamoto, M. (1986) Role of a ras homolog in the life cycle of Schizosaccharomyces pombe. Cell *44*, 329-336.

Fukui, Y. and Yamamoto, M. (1988) Isolation and characterization of Schizosaccharomyces pombe mutants phenotypically similar to ras1. Mol. Gen. Genet. *215*, 26-31.

Garami, A., Zwartkruis, F.J., Nobukuni, T., Joaquin, M., Roccio, M., Stocker, H., Kozma, S.C., Hafen, E., Bos, J.L. and Thomas, G. (2003) Insulin activation of Rheb, a mediator of mTOR/S6K/4E-BP signaling, is inhibited by TSC1 and 2. Mol. Cell *11*, 1457-1466.

Garreau, H., Geymonat, M., Renault, G. and Jacquet, M. (1996) Membrane-anchoring domains of Cdc25p, a Saccharomyces cerevisiae ras exchange factor. Biol. Cell *86*, 93-102.

Garrett, S. and Broach, J. (1989) Loss of Ras activity in Saccharomyces cerevisiae is suppressed by disruptions of a new kinase gene, YAKI, whose product may act downstream of the cAMP-dependent protein kinase. Genes Dev. *3*, 1336-1348.

Gerst, J.E., Ferguson, K., Vojtek, A., Wigler, M. and Field, J. (1991) CAP is a bifunctional component of the Saccharomyces cerevisiae adenylyl cyclase complex. Mol. Cell. Biol. *11*, 1248-1257.

Goodman, L.E., Perou, C.M., Fujiyama, A. and Tamanoi, F. (1988) Structure and expression of yeast DPR1, a gene essential for the processing and intracellular localization of ras proteins. Yeast *4*, 271-281.

Goodman, L.E., Judd, S.R., Farnsworth, C.C., Powers, S., Gelb, M.H., Glomset, J.A. and Tamanoi, F. (1990) Mutants of Saccharomyces cerevisiae defective in the farnesylation of Ras proteins. Proc. Natl. Acad. Sci. U.S.A. *87*, 9665-9669.

Gromov, P.S., Madsen, P., Tomerup, N. and Celis, J.E. (1995) A novel approach for expression cloning of small GTPases: identification, tissue distribution and chromosome mapping of the human homolog of rheb. FEBS Lett. *377*, 221-226.

Gross, A., Winograd, S., Marbach, I. and Levitzki, A. (1999) The N-terminal half of Cdc25 is essential for processing glucose signaling in Saccharomyces cerevisiae. Biochemistry *38*, 13252-13262.

Hall, B.E., Yang, S.S., Boriack-Sjodin, P.A., Kuriyan, J. and Bar-Sagi, D. (2001) Structure-based mutagenesis reveals distinct functions for Ras switch 1 and switch 2 in Sos-catalyzed guanine nucleotide exchange. J. Biol. Chem. *276*, 27629-27637.

Halme, A., Michelitch, M., Mitchell, E.L. and Chant, J. (1996) Bud10p directs axial cell polarization in budding yeast and resembles a transmembrane receptor. Curr. Biol. *6*, 570-579.

Haney, S.A. and Broach, J.R. (1994) Cdc25p, the guanine nucleotide exchange factor for the Ras proteins of Saccharomyces cerevisiae, promotes exchange by stabilizing Ras in a nucleotide-free state. J. Biol. Chem. *269*, 16541-16548.

He, B., Chen, P., Chen, S.Y., Vancura, K.L., Michaelis, S. and Powers, S. (1991) RAM2, an essential gene of yeast, and RAM1 encode the two polypeptide components of the farnesyltransferase that prenylates a-factor and Ras proteins. Proc. Natl. Acad. Sci. U.S.A. *88*, 11373-11377.

Herskowitz, I., Park, H.O., Sanders, S., Valtz, N. and Peter, M. (1995) Programming of cell polarity in budding yeast by endogenous and exogenous signals. Cold Spring Harb. Symp. Quant. Biol. *60*, 717-727.

Hicks, J. and Fink, G.R. (1977) Identification of chromosomal location of yeast DNA from hybrid plasmid p Yeleu 10. Nature *269*, 265-267.

Hicks, J. and Strathern, J.N. (1977) Interconversion of mating type in S. cerevisiae and the Cassette model for gene transfer. Brookhaven Symp. Biol. 233-242.

Hrycyna, C.A. and Clarke, S. (1990) Farnesyl cysteine C-terminal methyltransferase activity is dependent upon the STE14 gene product in Saccharomyces cerevisiae. Mol. Cell. Biol. *10*, 5071-5076.

Hughes, D.A. (1995) Control of signal transduction and morphogenesis by Ras. Semin. Cell Biol. *6*, 89-94.

Hughes, D.A., Fukui, Y. and Yamamoto, M. (1990) Homologous activators of ras in fission and budding yeast. Nature *344*, 355-357.

Hughes, D.A., Ashworth, A. and Marshall, C.J. (1993) Complementation of byr1 in fission yeast by mammalian MAP kinase kinase requires coexpression of Raf kinase. Nature *364*, 349-352.

Hughes, D.A., Yabana, N. and Yamamoto, M. (1994) Transcriptional regulation of a Ras nucleotide-exchange factor gene by extracellular signals in fission yeast. J. Cell Sci. *107* (12), 3635-3642.

Im, E., von Lintig, F.C., Chen, J., Zhuang, S., Qui, W., Chowdhury, S., Worley, P.F., Boss, G.R. and Pilz, R.B. (2002) Rheb is in a high activation state and inhibits B-Raf kinase in mammalian cells. Oncogene *21*, 6356-6365.

Imai, Y., Miyake, S., Hughes, D.A. and Yamamoto, M. (1991) Identification of a GTPase-activating protein homolog in Schizosaccharomyces pombe. Mol. Cell. Biol. *11*, 3088-3094.

Inoki, K., Li, Y., Xu, T. and Guan, K.L. (2003) Rheb GTPase is a direct target of TSC2 GAP activity and regulates mTOR signaling. Genes Dev. *17*, 1829-1834.

Jiang, Y. and Ferro-Novick, S. (1994) Identification of yeast component A: reconstitution of the geranylgeranyltransferase that modifies Ypt1p and Sec4p. Proc. Natl. Acad. Sci. U.S.A. *91*, 4377-4381.

Jones, S., Vignais, M.L. and Broach, J.R. (1991) The CDC25 protein of Saccharomyces cerevisiae promotes exchange of guanine nucleotides bound to ras. Mol. Cell. Biol. *11*, 2641-2646.

Kang, P.J., Sanson, A., Lee, B. and Park, H.O. (2001) A GDP/GTP exchange factor involved in linking a spatial landmark to cell polarity. Science *292*, 1376-1378.

Kataoka, T., Powers, S., McGill, C., Fasano, O., Strathern, J., Broach, J. and Wigler, M. (1984) Genetic analysis of yeast RAS1 and RAS2 genes. Cell *37*, 437-445.

Kawai, M., Nakashima, A., Ueno, M., Ushimaru, T., Aiba, K., Doi, H. and Uritani, M. (2001) Fission yeast tor1 functions in response to various stresses including nitrogen starvation, high osmolarity, and high temperature. Curr. Genet. *39*, 166-174.

Kozminski, K.G., Beven, L., Angerman, E., Tong, A.H., Boone, C. and Park, H.O. (2003) Interaction between a Ras and a Rho GTPase couples selection of a growth site to the development of cell polarity in yeast. Mol. Biol. Cell *14*, 4958-4970.

Kuroda, Y., Suzuki, N. and Kataoka, T. (1993) The effect of posttranslational modifications on the interaction of Ras2 with adenylyl cyclase. Science *259*, 683-686.

Lai, C.C., Boguski, M., Broek, D. and Powers, S. (1993) Influence of guanine nucleotides on complex formation between Ras and CDC25 proteins. Mol. Cell. Biol. *13*, 1345-1352.

Lee, B.N. and Elion, E.A. (1999) The MAPKKK Ste11 regulates vegetative growth through a kinase cascade of shared signaling components. Proc. Natl. Acad. Sci. U.S.A. *96*, 12679-12684.

Leupold, U., Sipiczki, M. and Egel, R. (1991) Pheromone production and response in sterile mutants of fission yeast. Curr. Genet. *20*, 79-85.

Levitzki, A. (1996) Targeting signal transduction for disease therapy. Curr. Opin. Cell Biol. *8*, 239-244.

Li, Y.C., Chen, C.R. and Chang, E.C. (2000) Fission yeast Ras1 effector Scd1 interacts with the spindle and affects its proper formation. Genetics *156*, 995-1004.

Lobo, S., Greentree, W.K., Linder, M.E. and Deschenes, R.J. (2002) Identification of a Ras palmitoyltransferase in Saccharomyces cerevisiae. J. Biol. Chem. *277*, 41268-41273.

Mach, K.E., Furge, K.A. and Albright, C.F. (2000) Loss of Rhb1, a Rheb-related GTPase in fission yeast, causes growth arrest with a terminal phenotype similar to that caused by nitrogen starvation. Genetics *155*, 611-622.

Marshall, M.S., Gibbs, J.B., Scolnick, E.M. and Sigal, I.S. (1987) Regulatory function of the Saccharomyces cerevisiae RAS C-terminus. Mol. Cell. Biol. *7*, 2309-2315.

Marston, A.L., Chen, T., Yang, M.C., Belhumeur, P. and Chant, J. (2001) A localized GTPase exchange factor, Bud5, determines the orientation of division axes in yeast. Curr. Biol. *11*, 803-807.

Martegani, E., Baroni, M. and Wanoni, M. (1986) Interaction of cAMP with the CDC25-mediated step in the cell cycle of budding yeast. Exp. Cell Res. *162*, 544-548.

Masuda, T., Kariya, K., Shinkai, M., Okada, T. and Kataoka, T. (1995) Protein kinase Byr2 is a target of Ras1 in the fission yeast Schizosaccharomyces pombe. J. Biol. Chem. *270*, 1979-1982.

Matsumoto, K., Uno, I., Kato, K. and Ishikawa, T. (1985) Isolation and characterization of a phosphoprotein phosphatase-deficient mutant in yeast. Yeast *1*, 25-38.

Matsumoto, S., Bandyopadhyay, A., Kwiatkowski, D.J., Maitra, U. and Matsumoto, T. (2002) Role of the Tsc1-Tsc2 complex in signaling and transport across the cell membrane in the fission yeast Schizosaccharomyces pombe. Genetics *161*, 1053-1063.

Merla, A. and Johnson, D.I. (2001) The Schizosaccharomyces pombe Cdc42p GTPase signals through Pak2p and the Mkh1p-Pek1p-Spm1p MAP kinase pathway. Curr. Genet. *39*, 205-209.

Michelitch, M. and Chant, J. (1996) A mechanism of Bud1p GTPase action suggested by mutational analysis and immunolocalization. Curr. Biol. *6*, 446-454.

Mintzer, K.A. and Field, J. (1994) Interactions between adenylyl cyclase, CAP and RAS from Saccharomyces cerevisiae. Cell. Signal. *6*, 681-694.

Mintzer, K.A. and Field, J. (1999) The SH3 domain of the S. cerevisiae Cdc25p binds adenylyl cyclase and facilitates Ras regulation of cAMP signalling. Cell. Signal. *11*, 127-135.

Mitsuzawa, H., Esson, K. and Tamanoi, F. (1995) Mutant farnesyltransferase beta subunit of Saccharomyces cerevisiae that can substitute for geranylgeranyltransferase type I beta subunit. Proc. Natl. Acad. Sci. U.S.A. *92*, 1704-1708.

Mitts, M.R., Bradshaw-Rouse, J. and Heideman, W. (1991) Interactions between adenylate cyclase and the yeast GTPase-activating protein IRA1. Mol. Cell. Biol. *11*, 4591-4598.

Morillon, A., Springer, M. and Lesage, P. (2000) Activation of the Kss1 invasive-filamentous growth pathway induces Ty1 transcription and retrotransposition in Saccharomyces cerevisiae. Mol. Cell. Biol. *20*, 5766-5776.

Morishita, T., Mitsuzawa, H., Nakafuku, M., Nakamura, S., Hattori, S. and Anraku, Y. (1995) Requirement of Saccharomyces cerevisiae Ras for completion of mitosis. Science *270*, 1213-1215.

Mosch, H.U., Kubler, E., Krappmann, S., Fink, G.R. and Braus, G.H. (1999) Crosstalk between the Ras2p-controlled mitogen-activated protein kinase and cAMP pathways during invasive growth of Saccharomyces cerevisiae. Mol. Biol. Cell. *10*, 1325-1335.

Munder, T. and Kuntzel, H. (1989) Glucose-induced cAMP signaling in Saccharomyces cerevisiae is mediated by the CDC25 protein. FEBS Lett. *242*, 341-345.

Nadin-Davis, S.A., Yang, R.C., Narang, S.A. and Nasim, A. (1986) The cloning and characterization of a RAS gene from Schizosaccharomyces pombe. J. Mol. Evol. *23*, 41-51.

Nadin-Davis, S.A. and Nasim, A. (1990) Schizosaccharomyces pombe ras1 and byr1 are functionally related genes of the ste family that affect starvation-induced transcription of mating-type genes. Mol. Cell. Biol. *10*, 549-560.

Neiman, A.M., Stevenson, B.J., Xu, H.P., Sprague, G.F. Jr., Herskowitz, I., Wigler, M. and Marcus, S. (1993) Functional homology of protein kinases required for sexual differentiation in Schizosaccharomyces pombe and Saccharomyces cerevisiae suggests a conserved signal transduction module in eukaryotic organisms. Mol. Biol. Cell *4*, 107-120.

Ohya, Y., Goebl, M., Goodman, L.E., Petersen-Bjorn, S., Friesen, J.D., Tamanoi, F. and Anraku, Y. (1991) Yeast CAL1 is a structural and functional homologue to the DPR1 (RAM) gene involved in ras processing. J. Biol. Chem. *266*, 12356-12360.

Ottilie, S., Miller, P.J., Johnson, D.I., Creasy, C.L., Sells, M.A., Bagrodia, S., Forsburg, S.L. and Chernoff, J. (1995) Fission yeast pak1+ encodes a protein kinase that interacts with Cdc42p and is involved in the control of cell polarity and mating. EMBO J. *14*, 5908-5919.

Pan, X. and Heitman, J. (1999) Cyclic AMP-dependent protein kinase regulates pseudohyphal differentiation in Saccharomyces cerevisiae. Mol. Cell. Biol. *19*, 4874-4887.

Papadaki, P., Pizon, V., Onken, B. and Chang, E.C. (2002) Two ras pathways in fission yeast are differentially regulated by two ras guanine nucleotide exchange factors. Mol. Cell. Biol. *22*, 4598-4606.

Papasavvas, S., Arkinstall, S., Reid, J. and Payton, M. (1992) Yeast alpha-mating factor receptor and G-protein-linked adenylyl cyclase inhibition requires RAS2 and GPA2 activities. Biochem. Biophys. Res. Commun. *184*, 1378-1385.

Park, H.O., Chant, J. and Herskowitz, I. (1993) BUD2 encodes a GTPase-activating protein for Bud1/Rsr1 necessary for proper bud-site selection in yeast. Nature *365*, 269-274.

Park, H.O., Kang, P.J. and Rachfal, A.W. (2002) Localization of the Rsr1/Bud1 GTPase involved in selection of a proper growth site in yeast. J. Biol. Chem. *277*, 26721-26724.

Parrini, M.C., Jacquet, E., Bernardi, A., Jacquet, M. and Parmeggiani, A. (1995) Properties and regulation of the catalytic domain of Ira2p, a Saccharomyces cerevisiae GTPase-activating protein of Ras2p. Biochemistry *34*, 13776-13783.

Patel, P.H., Thapar, N., Guo, L., Martinez, M., Maris, J., Gau, C.L., Lengyel, J.A. and Tamanoi, F. (2003) Drosophila Rheb GTPase is required for cell cycle progression and cell growth. J. Cell Sci. *116*, 3601-3610.

Powers, S., Kataoka, T., Fasano, O., Goldfarb, M., Strathern, J., Broach, J. and Wigler, M. (1984) Genes in S. cerevisiae encoding proteins with domains homologous to the mammalian ras proteins. Cell *36*, 607-612.

Powers, S., Gonzales, E., Christensen, T., Cubert, J. and Broek, D. (1991) Functional cloning of BUD5, a CDC25-related gene from S. cerevisiae that can suppress a dominant-negative RAS2 mutant. Cell *65*, 1225-1231.

Pringle, J.R., Bi, E., Harkins, H.A., Zahner, J.E., De Virgilio, C., Chant, J., Corrado, K. and Fares, H. (1995) Establishment of cell polarity in yeast. Cold Spring Harb. Symp. Quant. Biol. *60*, 729-744.

Reuther, G.W. and. Der, C.J. (2000) The Ras branch of small GTPases: Ras family members don't fall far from the tree. Curr. Opin. Cell Biol. *12*, 157-165.

Robinson, L.C., Gibbs, J.B., Marshall, M.S., Sigal, I.S. and Tatchell, K. (1987) CDC25: a component of the RAS-adenylate cyclase pathway in Saccharomyces cerevisiae. Science *235*, 1218-1221.

Roemer, T., Madden, K., Chang, J. and Snyder, M. (1996) Selection of axial growth sites in yeast requires Axl2p, a novel plasma membrane glycoprotein. Genes Dev. *10*, 777-793.

Romano, J.D., Schmidt, W.K. and Michaelis, S. (1998) The Saccharomyces cerevisiae prenylcysteine carboxyl methyltransferase Ste14p is in the endoplasmic reticulum membrane. Mol. Biol. Cell *9*, 2231-2247.

Ruggieri, R., Bender, A., Matsui, Y., Powers, S., Takai, Y., Pringle, J.R. and Matsumoto, K. (1992) RSR1, a ras-like gene homologous to Krev-1 (smg21A/rap1A): role in the development of cell polarity and interactions with the Ras pathway in Saccharomyces cerevisiae. Mol. Cell. Biol. *12*, 758-766.

Sanders, S.L. and Herskowitz, I. (1996) The BUD4 protein of yeast, required for axial budding, is localized to the mother/BUD neck in a cell cycle-dependent manner. J. Cell Biol. *134*, 413-427.

Sattler, I. and Tamanoi, F. (1996) Prenylation of RAS and inhibitors of prenyltransferases. Regulation of the Ras Signaling Network. H. M. a. A. W. Burgess. Austin, R. G. Landes: 95-137.

Saucedo, L.J., Gao, X., Chiarelli, D.A., Li, L., Pan, D. and Edgar, B.A. (2003) Rheb promotes cell growth as a component of the insulin/TOR signalling network. Nat. Cell Biol. *5*, 566-571.

Schlessinger, J. and Bar-Sagi, D. (1994) Activation of Ras and other signaling pathways by receptor tyrosine kinases. Cold Spring Harb. Symp. Quant. Biol. *59*, 173-179.

Shama, S., Kirchman, P.A., Jiang, J.C. and Jazwinski, S.M. (1998) Role of RAS2 in recovery from chronic stress: effect on yeast life span. Exp. Cell Res. *245*, 368-378.

Shirayama, M., Matsui, Y., Tanaka, K. and Toh-e, A. (1994a) Isolation of a CDC25 family gene, MSI2/LTE1, as a multicopy suppressor of ira1. Yeast *10*, 451-461.

Shirayama, M., Matsui, Y. and Toh, E.A. (1994b) The yeast TEM1 gene, which encodes a GTP-binding protein, is involved in termination of M phase. Mol. Cell. Biol. *14*, 7476-7482.

Stocker, H., Radimerski, T., Schindelholz, B., Wittwer, F., Belawat, P., Daram, P., Breuer, S., Thomas, G. and Hafen, E. (2003) Rheb is an essential regulator of S6K in controlling cell growth in Drosophila. Nat. Cell Biol. *5*, 559-565.

Sugimoto, A., Iino, Y., Maeda, T., Watanabe, Y. and Yamamoto, M. (1991) Schizosaccharomyces pombe ste11+ encodes a transcription factor with an HMG motif that is a critical regulator of sexual development. Genes Dev. *5*, 1990-1999.

Sun, J., Kale, S.P., Childress, A.M., Pinswasdi, C. and Jazwinski, S.M. (1994) Divergent roles of RAS1 and RAS2 in yeast longevity. J. Biol. Chem. *269*, 18638-18645.

Tabancay, A.P. Jr., Gau, C.L., Machado, I.M., Uhlmann, E.J., Gutmann, D.H., Guo, L. and Tamanoi, F. (2003) Identification of dominant negative mutants of Rheb GTPase and their use to implicate the involvement of human Rheb in the activation of p70S6K. J. Biol. Chem. *278*, 39921-39930.

Tam, A., Schmidt, W.K. and Michaelis, S. (2001) The multispanning membrane protein Ste24p catalyzes CAAX proteolysis and NH2-terminal processing of the yeast a-factor precursor. J. Biol. Chem. *276*, 46798-46806.

Tanaka, K., Matsumoto, K. and Toh, E.A. (1989) IRA1, an inhibitory regulator of the RAS-cyclic AMP pathway in Saccharomyces cerevisiae. Mol. Cell. Biol. *9*, 757-768.

Tanaka, K., Nakafuku, M., Satoh, T., Marshall, M.S., Gibbs, J.B., Matsumoto, K., Kaziro, Y. and Toh-e, A. (1990a) S. cerevisiae genes IRA1 and IRA2 encode proteins that may be functionally equivalent to mammalian ras GTPase activating protein. Cell *60*, 803-807.

Tanaka, K., Nakafuku, M., Tamanoi, F., Kaziro, Y., Matsumoto, K. and Toh-e, A. (1990b) IRA2, a second gene of Saccharomyces cerevisiae that encodes a protein with a domain homologous to mammalian ras GTPase-activating protein. Mol. Cell. Biol. *10*, 4303-4313.

Tanaka, K., Lin, B.K., Wood, D.R. and Tamanoi, F. (1991) IRA2, an upstream negative regulator of RAS in yeast, is a RAS GTPase-activating protein. Proc. Natl. Acad. Sci. U.S.A. *88*, 468-472.

Tanaka, K., Wood, D.R., Lin, B.K., Khalil, M., Tamanoi, F. and Cannon, J.F. (1992) A dominant activating mutation in the effector region of RAS abolishes IRA2 sensitivity. Mol. Cell. Biol. *12*, 631-637.

Tatchell, K., Chaleff, D.T., DeFeo-Jones, D. and Scolnick, E.M. (1984) Requirement of either of a pair of ras-related genes of Saccharomyces cerevisiae for spore viability. Nature *309*, 523-527.

Tee, A.R., Manning, B.D., Roux, P.P., Cantley, L.C. and Blenis, J. (2003) Tuberous sclerosis complex gene products, Tuberin and Hamartin, control mTOR signaling by acting as a GTPase-activating protein complex toward Rheb. Curr. Biol. *13*, 1259-1268.

Thevelein, J.M. and de Winde, J.H. (1999) Novel sensing mechanisms and targets for the cAMP-protein kinase A pathway in the yeast Saccharomyces cerevisiae. Mol. Microbiol. *33*, 904-918.

Toda, T., Uno, I., Ishikawa, T., Powers, S., Kataoka, T., Broek, D., Cameron, S., Broach, J., Matsumoto, K. and Wigler, M. (1985) In yeast, RAS proteins are controlling elements of adenylate cyclase. Cell *40*, 27-36.

Toda, T., Cameron, S., Sass, P., Zoller, M. and Wigler, M. (1987) Three different genes in S. cerevisiae encode the catalytic subunits of the cAMP-dependent protein kinase. Cell *50*, 277-287.

Tratner, I., Fourticq-Esqueoute, A., Tillit, J. and Baldacci, G. (1997) Cloning and characterization of the S. pombe gene efc25+, a new putative guanine nucleotide exchange factor. Gene *193*, 203-210.

Trueblood, C.E., Ohya, Y. and Rine, J. (1993) Genetic evidence for in vivo cross-specificity of the CaaX-box protein prenyltransferases farnesyltransferase and geranylgeranyltransferase-I in Saccharomyces cerevisiae. Mol. Cell. Biol. *13*, 4260-4275.

Trueblood, C.E., Boyartchuk, V.L. and Rine, J. (1997) Substrate specificity determinants in the farnesyltransferase beta-subunit. Proc. Natl. Acad. Sci. U.S.A. *94*, 10774-10779.

Urano, J., Tabancay, A.P., Yang, W. and Tamanoi, F. (2000) The Saccharomyces cerevisiae Rheb G-protein is involved in regulating canavanine resistance and arginine uptake. J. Biol. Chem. *275*, 11198-11206.

Van Slegtenhorst, M., Carr, E., Stoyanova, R., Kruger, W. and Henske, E.P. (2004) Tsc1+ and tsc2+ regulate arginine uptake and metabolism in Schizosaccharomyces pombe. J. Biol. Chem.

Wang, Y., Boguski, M., Riggs, M., Rodgers, L. and Wigler, M. (1991a) sar1, a gene from Schizosaccharomyces pombe encoding a protein that regulates ras1. Cell Regul. *2*, 453-465.

Wang, Y., Xu, H.P., Riggs, M., Rodgers, L. and Wigler, M. (1991b) byr2, a Schizosaccharomyces pombe gene encoding a protein kinase capable of partial suppression of the ras1 mutant phenotype. Mol. Cell. Biol. *11*, 3554-3563.

Ward, M.P. and Garrett, S. (1994) Suppression of a yeast cyclic AMP-dependent protein kinase defect by overexpression of SOK1, a yeast gene exhibiting sequence similarity to a developmentally regulated mouse gene. Mol. Cell. Biol. *14*, 5619-5627.

Weisman, R. (2004) The fission yeast TOR proteins and the rapamycin response: an unexpected tale. Curr. Top. Microbiol. Immunol. *279*, 85-95.

Weisman, R. and Choder, M. (2001) The fission yeast TOR homolog, tor1+, is required for the response to starvation and other stresses via a conserved serine. J. Biol. Chem. *276*, 7027-7032.

Weissbach, L., Settleman, J., Kalady, M.F., Snijders, A.J., Murthy, A.E., Yan, Y.X. and Bernards, A. (1994) Identification of a human rasGAP-related protein containing calmodulin-binding motifs. J. Biol. Chem. *269*, 20517-20521.

Wilson, B.A., Khalil, M., Tamanoi, F. and Cannon, J.F. (1993) New activated RAS2 mutations identified in Saccharomyces cerevisiae. Oncogene *8*, 3441-3445.

Wood, D.R., Poullet, P., Wilson, B.A., Khalil, M., Tanaka, K., Cannon, J.F. and Tamanoi, F. (1994) Biochemical characterization of yeast RAS2 mutants reveals a new region of ras protein involved in the interaction with GTPase-activating proteins. J. Biol. Chem. *269*, 5322-5327.

Xu, H.P., White, M., Marcus, S. and Wigler, M. (1994) Concerted action of RAS and G proteins in the sexual response pathways of Schizosaccharomyces pombe. Mol. Cell. Biol. *14*, 50-58.

Yamagata, K., Sanders, L.K., Kaufmann, W.E., Yee, W., Barnes, C.A., Nathans, D. and Worley, P.F. (1994) rheb, a growth factor- and synaptic activity-regulated gene, encodes a novel Ras-related protein. J. Biol. Chem. *269*, 16333-16339.

Yang, W., Urano, J. and Tamanoi, F. (2000) Protein farnesylation is critical for maintaining normal cell morphology and canavanine resistance in Schizosaccharomyces pombe. J. Biol. Chem. *275*, 429-438.
Yang, W., Tabancay, A.P. Jr., Urano, J. and Tamanoi, F. (2001) Failure to farnesylate Rheb protein contributes to the enrichment of G0/G1 phase cells in the Schizosaccharomyces pombe farnesyltransferase mutant. Mol. Microbiol. *41*, 1339-1347.
Yee, W.M. and Worley, P.F. (1997) Rheb interacts with Raf-1 kinase and may function to integrate growth factor- and protein kinase A-dependent signals. Mol. Cell. Biol. *17*, 921-933.
Yoshida, S., Ichihashi, R. and Toh-e, A. (2003) Ras recruits mitotic exit regulator Lte1 to the bud cortex in budding yeast. J. Cell Biol. *161*, 889-897.
Zagulski, M., Becam, A.M., Grzybowska, E., Lacroute, F., Migdalski, A., Slonimski, P.P., Sokolowska, B. and Herbert, C.J. (1994) The sequence of 12.5 kb from the right arm of chromosome II predicts a new N-terminal sequence for the IRA1 protein and reveals two new genes, one of which is a DEAD-box helicase. Yeast *10*, 1227-1234.
Zahner, J.E., Harkins, H.A. and Pringle, J.R. (1996) Genetic analysis of the bipolar pattern of bud site selection in the yeast Saccharomyces cerevisiae. Mol. Cell. Biol. *16*, 1857-1870.
Zhang, Y., Gao, X., Saucedo, L.J., Ru, B., Edgar, B.A. and Pan, D. (2003) Rheb is a direct target of the tuberous sclerosis tumour suppressor proteins. Nat. Cell Biol. *5*, 578-581.
Zheng, Y., Bender, A. and Cerione, R.A. (1995) Interactions among proteins involved in bud-site selection and bud-site assembly in Saccharomyces cerevisiae. J. Biol. Chem. *270*, 626-630.

CHAPTER 11

COMPARISON OF THE EFFECTS OF RAS EFFECTOR MUTANTS AND RAS EFFECTORS ON TRANSFORMED AND TUMORIGENIC GROWTH OF HUMAN AND RODENT CELLS

KIAN-HUAT LIM AND CHRISTOPHER M. COUNTER
Duke University Medical Center, Durham NC

Abstract: How oncogenic Ras signaling leads to transformation of normal cells to malignant state has been under intense scrutiny over the last few decades. It is now well-appreciated that Ras induces an immensely complicated network of signaling cascades that can lead to very different outcomes depending on cell type, genetic background, phenotype assayed for, and so forth. On top of all these differences, mounting evidence suggests that there may even be differences in Ras-mediated oncogenesis between rodents, the primary model system used to study Ras oncogenesis, and humans. In this chapter, we will summarize what Ras effector pathways have been implicated in the most common and stringent phenotypes of Ras transformed cells, anchorage-independent growth using soft agar assay and in vivo xenograft tumorigenesis using immunocompromised mice, between mice and human cells

Keywords: Ras, RalGEF, PI3-kinase, Raf, human cells, transformation, tumorigenesis

1. INTRODUCTION

The Ras superfamily of monomeric GTP-binding proteins mediate a host of signals from activated growth factor receptors to the cell (Shields et al., 2000). Mutations that leave Ras in the constitutively active GTP-bound state (12V or 61L) have been found in one third of human tumors, arguing that inappropriate activation of this protein is a tumorigenic event (Bos, 1989). Much of what has been learned about the role of the oncogenic Ras in human cancer has been derived from manipulating the expression of this family of proteins, or their downstream targets, in cultured rodent cells. Proteins can be quickly and easily expressed in cultured cells, and a number of phenotypes characteristic of human cancer cells (transformed phenotypes) can be manifested and studied in culture. More recently, employing the same approaches it has been possible to extend these studies to human cells, with some interesting twists!

C. Der (ed.), RAS Family GTPases, 257–272.

2. SOFT AGAR AND TUMORIGENIC GROWTH

Oncogenic Ras expression can cause a host of alterations to cultured cells reminiscent of neoplastic phenotypes, although none of these phenotypes recapitulate the complex process of tumorigenic growth. Instead, most assays measure single transformed phenotypes, which can often be induced by many targets of Ras. The ability of cells to grow in an anchorage independent fashion in soft agar is, however, at least a consequence of multiple transformed phenotypes, and as such is considered one of the best in vitro correlates for tumorigenesis (Smets, 1980). While soft agar growth has the advantage of being a simple and fast assay that reflects tumor growth, cells can also be injected into immuno-compromised mice to assess actual tumor growth in vivo. These assays are commonly used to study cells from different tissues or species of animals, making it possible to compare transformation and tumorigenic potentials in different settings. Thus, for the sake of simplicity for this discussion, analysis of Ras oncogenesis will be limited to these two stringent assays: anchorage-independent growth and tumor formation in vivo.

3. RAS EFFECTOR MUTANTS AND CONSTITUTIVELY ACTIVE EFFECTORS

Ras is known to exert its oncogenic effects through the interaction of downstream proteins, termed effectors. The most studied of these effectors are Raf, PI3-kinase, and RalGEFs. The Raf family of proteins are serine/threonine kinases that are localized to the plasma membrane from the cytoplasm and activated upon binding to activated (GTP-bound) Ras, leading to a MAP kinase signal transduction cascade (Chong et al., 2003). Similarly, the p110 catalytic subunit of PI3-kinase can be activated via its interaction with GTP-Ras, leading to the phosphorylation of phosphoinositides, which in turn results in the activation of a host of proteins, the most celebrated being the serine/threonine kinase AKT/PKB (Cantley, 2002). Lastly, RalGEFs are a family of guanine nucleotide exchange factors (GEFs), four of which are known to be activated by recruitment to the plasma membrane by GTP-Ras, where they promote the Ral G-proteins to the active GTP-bound state (Feig, 2003; Wolthuis and Bos, 1999).

One tool that has proven extremely valuable in dissecting the contribution to the various effectors to Ras oncogenesis is a series of effector-binding mutants (White et al., 1995) that leave H-Ras capable of binding and activating the signaling cascade of primarily only one of the three major effector proteins (Fig. 1). Specifically, H-RasG12V with the mutation T_{35} to S (or 38E) binds to Raf1 and very weakly, if at all, to RalGDS (Khosravi-Far et al., 1996; Rodriguez-Viciana et al., 1997; Webb et al., 1998; White et al., 1995; White et al., 1996; Wolthuis et al., 1997). Mutant Y_{40} to C has the highest affinity for the p110 subunit of PI3-kinase (Rodriguez-Viciana et al., 1997) whereas mutant E_{37} to G binds specifically to RalGEF proteins (Rodriguez-Viciana et al., 1997; Webb et al., 1998; White et al., 1995; White et al., 1996; Wolthuis et al., 1997) in two-hybrid assays or co-immunoprecipitation assays using recombinant or over-expressed proteins. Correspondingly, RasG12V harboring the 35S mutation preferentially stimulates Raf1 kinase activity, leading to a measurable increase in the phosphorylation and activation of the downstream targets MEK and ERK in murine cells (Rodriguez-Viciana et al., 1997; Webb et al., 1998; White et al., 1995; Wolthuis et al., 1997). The 12V40C form of H-Ras is the only effector mutant capable of increasing PI3-kinase activity, as measured by elevated levels of PIP_3 (Rodriguez-Viciana et al., 1997). Lastly, the 12V37G mutant can increase the amount of GTP-Ral and leads to activation of the c-fos promoter, a target of the Ral pathway (Wolthuis and Bos, 1999; Wolthuis et al., 1997). The caveat of using these effector mutants is that they can still weakly activate other effectors, and bind to and possibly activate other known or yet to be discovered effectors of Ras aside from Raf, PI3-kinase and RalGEFs. Nevertheless, these mutants are useful in preferentially activating specific families of Ras effectors.

The second tool in dissecting Ras signaling has been expression of constitutively active versions of the three core effectors themselves. The most common means of recapitulating the activation of an effector by Ras is to fix the CAAX membrane targeting sequence of Ras to the C-terminus of the effector. Indeed, the p110α subunit of PI3-kinase or RalGEFs fused to the CAAX sequence (p110α-CAAX or RalGEF-CAAX) constitutively activate the downstream targets AKT or Ral, respectively (Matsubara et al., 1999; Rodriguez-Viciana et al., 1997; Urano et al., 1996; White et al., 1996; Wolthuis et al., 1997), although the p110α subunit can also be activated by fusion with portions of its regulatory subunit, p85 (Dhand et al., 1994; Klippel et al., 1994). Raf1 can be activated in this fashion (Raf1-CAAX) (Stokoe et al., 1994), or by deletions in the N terminus (ΔRaf1-22W, RafΔN, Raf-BXB) (Kerkhoff and Rapp, 1997; Stanton et al., 1989; Wasylyk et al., 1989), which negatively regulates the protein (Chong and Guan, 2003). While these types of mutants activate one arm of Ras signaling pathway (Fig. 1) with far more precision than Ras effector mutants, they do suffer from transmitting a signal using only one family member. Indeed, there are at least four RalGEF proteins (RalGDS, Rlf, Rgl1 and Rgl2) (Feig, 2003) and three Raf proteins (A-Raf,

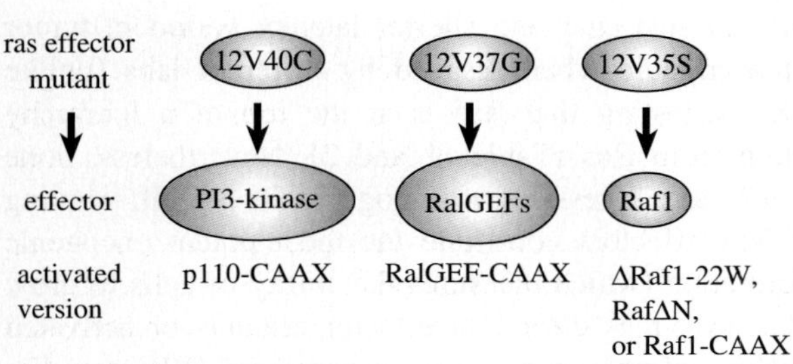

Figure 1. Ras effector mutants and effectors. Pictorial diagram to illustrate the three major effectors of Ras and the molecules commonly used to constitutively activate them

B-Raf and Raf1) (Chong et al., 2003) activated by Ras. Thus, a combined use of Ras effector mutants and active versions of the effectors have proven to be the most valuable approach at dissecting Ras function in many different settings (for example: Tables 1 and 2, references (Gille and Downward, 1999; Goi et al., 1999; Matsuguchi and Kraft, 1998; McFall et al., 2001; Peyssonnaux et al., 2000; Ramocki et al., 1998; Ward et al., 2001)). Moreover, because of the widespread use of these effector mutants and activated effectors, it is possible to directly compare the role of the Ras effectors between different phenotypes, cell types, or species.

4. RAS ONCOGENESIS IN NIH 3T3 CELLS

The foundation of Ras oncogenesis is based in large part on studies using NIH 3T3 cells, which are derived by a strict passaging protocol of primary murine fibroblasts to select for an immortal cell population. These cells teetered on becoming transformed, and hence have been extensively exploited to study oncogenesis. Ectopic expression of H-Ras12V causes NIH 3T3 cells to acquire a host of phenotypes characteristic of human tumor cells. Importantly for this review, upon the expression of Ras12V, NIH 3T3 cells will form colonies when plated in semi-solid medium and form tumor masses when injected into immuno-compromised mice.

Clues to how Ras exerts its oncogenic effects came from expressing the Ras effector mutants and activated effectors in NIH 3T3 cells. Specifically, expression of the Ras effector mutants 12V35S, 12V40C, or 12V37G, or the constitutively active versions the effectors themselves (Raf1 or its downstream target MEK1, various activated versions of PI3-kinase or AKT, or the oncogenic RalGEF, Rsc, or RalGDS fused to the CAAX sequence) all promoted anchorage-independent growth in vitro and tumor growth in vivo of NIH 3T3 cells (reviewed in Tables 1 and 2). Similar results were found when the spontaneous mouse tumor cell line EpH4 was transfected with Ras effector mutants and injected subcutaneously into immuno-compromised mice (Janda et al., 2002).

Collectively, these data argue that activation of any one of the MAP-kinase, PI3-kinase or RalGEF pathways is tumorigenic. However, activation of MAP-kinase generally yields more colonies in soft agar and shorter latency period in tumor assays, and consistently is tumorigenic when assayed by different labs (unlike the other effector pathways), suggesting that Raf is at the top of a hierarchy in oncogenic signals emanating from Ras (Tables 1 and 2). Nevertheless, none of these pathways are as potent as expression of oncogenic Ras itself, arguing that the united collective of Ras effectors constitute the most potent oncogenic signal. Indeed, using focus formation (which measures the ability of cells to grow at low density) or soft agar assays, it is clear that effector mutants or activated effectors promote a more transformed phenotype when co-expressed (Khosravi-Far et al., 1996; Rodriguez-Viciana et al., 1997; White et al., 1996; Yang et al., 1998).

Table 1. Soft agar growth of cells expressing Ras effectors of effector mutants

	% soft agar growth							Reference
		RalGEF pathway		PI3-kinase pathway		MAP-kinase pathway		
cell line	12V	12V37G	RalGEF activation[1]	12V40C	PI3-K/AKT activation [2]	12V35S	MAP-kinase activation [3]	
NIH 3T3	0.8-14.7%[4,5]	0.1-2.0%[4,5]	0.2%	0.02-2.3%[4,5]	similar to Ras12V[4]	0.2-5.7%[4,5]	0.8-7.9%[4,5]	Cheng et al., 1997; Cowley et al., 1994; Hamad et al., 2002; Khosravi-Far et al., 1996; Klippel et al., 1998; Mansour et al., 1994; Samuels et al., 1993; Sun et al., 2001; Ward et al., 2001; Yang et al., 1998
Rat1a	10%[6]	0%[6]		0.4%[6]		1.0%[6]	Up to 0.4%	Pritchard et al., 1995; Samuels et al., 1993; Tang et al., 1999
RIE	2-20%						0%	McFall et al., 2001; Oldham et al., 1996; Sheng et al., 2001
Rat-6	1.9%	1.5%[4]		1.5%[4]		1.5%[4]		Yang et al., 1998
Rat thyroid cells[2]		0[5]				Similar to Ras12V[5]		Miller et al., 1998

Activation by [1](Rlf-CAAX), or [2](p110α-CAAX, DBp110, or myr-AKT) or [3](ΔRaf1-22W, RafΔN or derivatives, Raf1-CAAX, MEKΔN3+S222D or MEK 218D+S222D). [4]Yang et al. (Yang et al., 1998) detected only microscopic colonies using Ras effector mutants. [5]No quantitation was provided by Samuels et al. (Samuels et al., 1993) or Miller et al. (Miller et al., 1998), [6]As estimated from graphs.

Table 2. Tumor growth of cells expressing Ras effectors or effector mutants

	% animals with subcutaneous tumors (latency in days)							Reference
		RalGEF pathway		PI3-kinase pathway		MAP-kinase pathway		
cell line	12V	12V37G	RalGEF activation[1]	12V40C	PI3-K/AKT activation[2]	12V35S	MAP-kinase activation[3]	
NIH 3T3	100% (10-15) 80-100%[4]	90-100% (20-28) 0-100%[4]	100% 100%[4]	100%(21) 0%[4]	100%(7-28)	100%[4] 100% (12-21)	100%(15-25) 70-100%[4]	Cheng et al., 1997; Cowley et al., 1994; D'Adamo et al., 1997; Khosravi-Far et al., 1996; Mansour et al., 1998; Sun et al., 2001; Ward et al., 2001; Webb et al., 1998
Murine tumor line EpH4	100% (7)			100%(7)		100%(7)		Janda et al., 2002
Rat mammary tissue in situ		100% (105)		100% (112)		100%(35)	90%(49)	McFarlin and Gould, 2003; McFarlin et al., 2003

Activation by expression of [1](Rlf-CAAX), [2](p110α-CAAX, DBp110, myr-AKT or *v-P3k*) or [3](ΔRaf1-22W, RafΔN, Raf1-CAAX, MEKΔN3+S222D or MEK 218*D*+*S*222D). [4]tail vein injections.

5. RAS ONCOGENESIS IN RATS

The next most studied animal of Ras oncogenesis is the rat. Like murine NIH 3T3 fibroblasts, Rat1a or Rat-6 fibroblasts are easily induced to grow in soft agar upon the expression of most or all of the Ras effector mutants or, when tested, activated Ras effectors, suggesting that Ras oncogenesis may be conserved in rodents (Table 1) (Pritchard et al., 1995; Samuels et al., 1993; Tang et al., 1999; Yang et al., 1998). Rat epithelial cells, on the other hand, are more resistant to transformation, and can not be induced to grow in semi-solid media simply by expression of activated Raf alone, indicating that cell type can affect what effector pathways are required for oncogenic transformation (McFall et al., 2001; Oldham et al., 1996; Sheng et al., 2001). In vivo, the MAP-kinase pathway also appears to be the dominant oncogenic signal of Ras as infection of rat mammary glands in situ with retroviruses encoding different Ras effector mutants leads to tumor growth with different latency periods. Glands infected with a Ras12V38E or RafΔN-expressing retrovirus formed tumors with nearly identical kinetics as glands infected with Ras12V-expressing retrovirus. However, glands expressing the 12V40C or 12V37G mutants took an additional 10 weeks to form tumors (McFarlin and Gould, 2003; McFarlin et al., 2003). It can therefore be concluded that Ras oncogenesis is quite similar between mice and rats, relying primarily on the MAP-kinase pathway.

6. RAS ONCOGENESIS IN HUMAN CELLS

Far less is known about Ras oncogenesis in human cells. Part of the problem stems from a lack of a cell system to study Ras. For example, expression of oncogenic Ras induces growth arrest in normal human fibroblasts (Serrano et al., 1997), unless very early passage cells are used (Benanti and Galloway, 2004). Even when Ras can be stably introduced into normal human cells without causing an immediate growth arrest, as in the case of human thyroid epithelial cells, the cells retain a differentiated phenotype and eventually growth arrest, suggesting that they are not highly transformed (Lemoine et al., 1990). Instead, most studies of Ras oncogenesis have relied on loss-of-function experiments using human tumor cell lines. For example, the small molecular weight MEK inhibitor PD184352 and the c-Raf1 anti-sense phosphorothioate oligodeoxynucleotide ISIS 5132 have both been shown to reduce the growth of the human colon carcinoma cell line HT-26 or the human lung carcinoma cell line A549, respectively, in xenograft mouse models (Monia et al., 1996; Sebolt-Leopold et al., 1999). Moreover, pulmonary metastatic tumors derived from the human melanoma cell line A375M actually regressed in immuno-compromised mice upon a single dose of PD184352 (Collisson et al., 2003). Treatment of primary AML blast samples with the small molecular weight MEK inhibitors PD184352 and PD98059 was also found to specifically reduce the proliferation of the AML cells in vitro, while having no measurable effect on normal hematopoietic cells (Milella et al., 2001). Thus, pharmacological inhibition of the MAP kinase pathway can curb human tumor cell growth in vivo.

The PI3-kinase pathway is also important for human tumor growth. Deregulation of this pathway is very common in human malignancies and can occur by several mechanisms such as mutation of EGFR (Moscatello et al., 1998), amplification of the p110α catalytic subunit of the PI3-kinase (Samuels and Velculescu, 2004; Shayesteh et al., 1999), mutation of the p85-regulatory subunit leading to constitutive activation of the enzyme, amplification or overexpression of the major downstream target AKT, or most commonly, functional loss of the tumor suppressor PTEN (Whang et al., 1998; Wu et al., 1998). In cultured cells, the oncogenic function of v-src (Fukui and Hanafusa, 1989), abl (Varticovski et al., 1991) and polyoma middle T-antigens (Courtneidge and Heber, 1987; Whitman et al., 1985) all depend upon associating with PI3-kinase. Similarly, the transforming ability of TC21 (or R-Ras2), a Ras-related GTPase, and EGFRvIII, a naturally occurring mutant form of epidermal growth factor receptor, both appear to be mediated through activation of the PI3-kinase pathway (Adnane et al., 2002; Moscatello et al., 1998; Murphy et al., 2002; Rosario et al., 2001). Many studies have also demonstrated the requirement of the PI3-kinase pathway in potentiating the transforming ability of other oncogenes, such as c-Myc and Ras (Kauffmann-Zeh et al., 1997). Activation of this pathway has been frequently associated with aggressive tumor phenotypes such as increased survival, cell growth and metabolism, cell cycle progression, angiogenesis, and resistance to apoptosis induced by radiation or chemotherapeutic agents (Vivanco and Sawyers, 2002). Correspondingly, small molecule PI3-kinase inhibitors such as wortmannin and LY294002 have been shown to induce apoptosis of many human cancer cell lines such as ovarian cancers (Shayesteh et al., 1999), Ewing's sarcomas (Toretsky et al., 1999), cervical cancers (Ma et al., 2000), and sensitize glioblastomas (Kubota et al., 2000), lung cancers (Brognard et al., 2001) and pancreatic cancers (Ng et al., 2000) to radiation and chemotherapeutic agents in vitro. Like wise, inhibition of a major downstream target of the PI3-kinase pathway, mTOR (mammalian target of Rapamycin), using rapamycin or its derivative CCI-779 has shown promising anti-tumor activity against cancers harboring deregulation of this pathway (Bjornsti and Houghton, 2004). Thus, PI3-kinase is not only a key Ras effector, but potentiates oncogenesis independent of Ras.

7. DIRECT COMPARISON OF RAS ONCOGENESIS BETWEEN HUMAN AND MOUSE CELLS

Comparisons of Ras oncogenesis between mice and humans are difficult to interpret as the method to induce tumors, cell type, genetic background, and means to dissect the targets of Ras all vary. To address this comparison, the most ideal situation would be to compare Ras oncogenesis in human and murine cells using rigorous assays, such as growth in soft agar and tumor growth, in a system whereby the only difference is that the cells come from either humans or mice.

In an effort to directly compare Ras oncogenesis in human and murine cells, we developed a cell-based approach whereby normal human or murine cells could be converted to a tumorigenic state by the enforced expression of four genes, one of them being Ras12V. Specifically, normal human somatic cells can be converted to a tumorigenic state (Hahn et al., 1999; Hamad et al., 2002; Rich et al., 2001), as assessed in immunocompromised mice, by ectopic expression of two SV40 viral proteins: *T-Ag*, which disrupt the p53 and Rb pathways that are altered in virtually in all cancers (Levine et al., 1991; Livingston, 1992), and t-Ag, which increases c-Myc stability (Yeh et al., 2004), a change often detected in human cancers (Henriksson and Luscher, 1996), and expression of two mammalian genes: $H\text{-}Ras^{G12V}$ and the *hTERT* catalytic telomerase subunit, which is activated in 90% of human cancers to promote unlimited proliferation (Shay and Bacchetti, 1997). Using this system, normal murine and human fibroblasts expressing these four proteins were created, with the exception being that Ras was replaced with one of the three Ras effector mutants (Figure 1). The murine cells were, as expected, transformed by all the Ras effector mutants, but most readily by 12V35S, which activates the MAP-kinase pathway. Surprisingly, it was found that human fibroblasts could not be induced to grow in soft agar via the activation of Raf, even in the presence of activated PI3-kinase, and instead grew only if the RalGEF pathway was activated. This effect was reproduced in human epithelial cells and astrocytes, and when the effector mutants were replaced with the activated versions of the effectors- ΔRaf1-22W, p110α-CAAX or Rlf-CAAX (Figure 2). Thus, unlike murine cells, human cells rely heavily on the RalGEF pathway for anchorage-independent growth. But what about tumorigenesis? Again, it was shown that unless human cells expressed the Ras effector mutant 12V37G, which activates RalGEFs, the cells failed to grow as xenografts in immuno-compromised mice (Hamad et al., 2002).

The prediction from these experiments is that activation of the MAP-kinase pathway should play a smaller role in human cancer than it does in rodents.

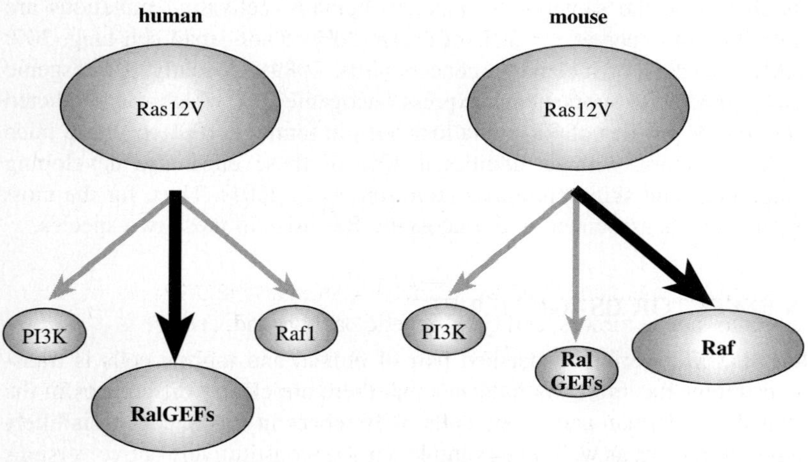

Figure 2. Ras effector usage model. Pictorial diagram to illustrate the ability of the three major effectors of Ras to promote anchorage-independent growth of human or murine fibroblasts expressing telomerase, T and T-Ag

However, it is clear that pharmacological inhibitors of the MAP kinase pathway can curb human tumor cell growth in vivo. However, MEK inhibitors have had no objective response in four independent phase II clinical trials on prostate, ovarian, colorectal and non-small cell lung cancers (Coudert et al., 2001; Cripps et al., 2002; Oza et al., 2003; Tolcher et al., 2002).

Many cancers are also now known to harbor activating mutations in the *BRAF* gene, which would argue against the notion that Ras-induced tumorigenesis in humans relies less on the MAP-kinase pathway. Specifically, *BRAF* activating mutations, primarily E599V, occur in 50-70% of human malignant melanomas (Brose et al., 2002; Davies et al., 2002; Dong et al., 2003), 25-45% thyroid cancers (Cohen et al., 2003; Kimura et al., 2003; Namba et al., 2003), being particularly high in the papillary subtype (Cohen et al., 2003; Kimura et al., 2003), 2-50% (~10% on average) of colorectal cancers (all stages) (Chan et al., 2003; Davies et al., 2002; Oliveira et al., 2003; Rajagopalan et al., 2002; Wang et al., 2003; Yuen et al., 2002), 14% of ovarian cancers (Davies et al., 2002), and less than 5% in lung, head and neck, pancreatic, and (rarely) gastric cancers (Brose et al., 2002; Calhoun et al., 2003; Cohen et al., 2003; Davies et al., 2002; Ishimura et al., 2003; Kim et al., 2003; Lee et al., 2003; Naoki et al., 2002; Zhao et al., 2004). However, it is becoming clear that an activated BRAF protein does not functionally replace Ras in vivo. Activating mutations in both these genes have been found in the same tumor samples from a variety of cancers (Davies et al., 2002; Ishimura et al., 2003; Lee et al., 2003; Yuen et al., 2002) and knock-down or knock-out of BRAF expression has no effect on the proliferation or tumorigenicity of Ras-transformed human or mouse cells (Kim et al., 2004; Wellbrock et al., 2004).

If Ras oncogenesis was indeed identical between humans and mice, it is reasonable to assume that the types of tumors associated with activating Ras mutations in humans should be the same as in mice. In humans, activating mutations are detected in ~90% of pancreatic, ~50% of colon, 30% of non small cell lung, 30% of ADS/AML and 20% of melanoma cancers (Bos, 1989). Recently, a transgenic mouse was engineered to spontaneous express oncogenic Ras, which was predicted to model the tumor profiles of Ras mutations seen in humans. 100% of these mice developed lung tumors, with an additional 30% of these mice also developing thymic lymphomas and skin papillomas (Johnson et al., 2001). Thus, for the most part, different cancers associated with oncogenic Ras arise in these two species.

8. AN EFFECTOR USAGE MODEL

The finding that in a perfectly matched pair of human and murine cells is transformed by different Ras effectors indicates that there are clearly differences in the way Ras transforms human and rodent cells. Differences in Ras oncogenesis likely occur in other organisms as well. For example, viral or constitutively active versions of PI3-kinase and AKT readily transform chicken embryo fibroblasts (CEF) and promote tumor growth in chickens (Akagi et al., 2000; Aoki et al., 2001; Aoki et al., 2000; Chang et al., 1997; Nguyen et al., 2000) whereas expression of v-mil, a viral version of Raf in CEF, was reported to be inefficient at these processes (Bechade et al., 1988; Palmieri and Vogel, 1987). It is nevertheless clear that effectors like Raf, which are less important in tissue culture models of Ras oncogenesis, are still important for neoplastic growth in vivo. To reconcile these differences, we propose that while the various Ras effectors each play important roles in Ras oncogenesis, the relative contribution of each effector to this process may vary from species to species (Figure 2). It should be stressed that this is a simplified view of a highly complex process, and that there are likely to be other Ras effectors that may affect oncogenesis, and that the oncogenic signal will be interpreted differently depending on cell types and genetic background. Nevertheless, this represents a starting point to explore differences in transformation processes between different models of cancer.

ACKNOWLEDGEMENTS

This work was supported by NCI grant CA94184 (CMC). K.-H.L. is a Department of Defense Breast Cancer Research Pre-doctoral Scholar; C.M.C. is a Leukemia and Lymphoma Scholar.

REFERENCES

Adnane, J., Seijo, E., Chen, Z., Bizouarn, F., Leal, M., Sebti, S.M. and Munoz-Antonia, T. (2002) RhoB, not RhoA, represses the transcription of the transforming growth factor beta type II receptor by a mechanism involving activator protein 1. J. Biol. Chem. *277*, 8500-8507.

Akagi, T., Shishido, T., Murata, K. and Hanafusa, H. (2000) v-Crk activates the phosphoinositide 3-kinase/AKT pathway in transformation. Proc. Natl. Acad. Sci. U.S.A. *97*, 7290-7295.

Aoki, M., Schetter, C., Himly, M., Batista, O., Chang, H.W. and Vogt, P.K. (2000) The catalytic subunit of phosphoinositide 3-kinase: requirements for oncogenicity. J. Biol. Chem. *275*, 6267-6275.

Aoki, M., Blazek, E. and Vogt, P.K. (2001) A role of the kinase mTOR in cellular transformation induced by the oncoproteins P3k and Akt. Proc. Natl. Acad. Sci. U.S.A. *98*, 136-141.

Bechade, C., Dambrine, G., David-Pfeuty, T., Esnault, E. and Calothy, G. (1988) Transformed and tumorigenic phenotypes induced by avian retroviruses containing the v-mil oncogene. J. Virol. *62*, 1211-1218.

Benanti, J.A. and Galloway, D.A. (2004) Normal human fibroblasts are resistant to RAS-induced senescence. Mol. Cell. Biol. *24*, 2842-2852.

Bjornsti, M.A. and Houghton, P.J. (2004) The TOR pathway: a target for cancer therapy. Nat. Rev. Cancer *4*, 335-348.

Bos, J.L. (1989) ras oncogenes in human cancer: a review. Cancer Res. *49*, 4682-4689.

Brognard, J., Clark, A.S., Ni, Y. and Dennis, P.A. (2001) Akt/protein kinase B is constitutively active in non-small cell lung cancer cells and promotes cellular survival and resistance to chemotherapy and radiation. Cancer Res. *61*, 3986-3997.

Brose, M.S., Volpe, P., Feldman, M., Kumar, M., Rishi, I., Gerrero, R., Einhorn, E., Herlyn, M., Minna, J., Nicholson, A. et al. (2002) BRAF and RAS mutations in human lung cancer and melanoma. Cancer Res. *62*, 6997-7000.

Calhoun, E.S., Jones, J.B., Ashfaq, R., Adsay, V., Baker, S.J., Valentine, V., Hempen, P.M., Hilgers, W., Yeo, C.J., Hruban, R.H. and Kern, S.E. (2003) BRAF and FBXW7 (CDC4, FBW7, AGO, SEL10) mutations in distinct subsets of pancreatic cancer: potential therapeutic targets. Am. J. Pathol. *163*, 1255-1260.

Cantley, L.C. (2002) The phosphoinositide 3-kinase pathway. Science *296*, 1655-1657.

Chan, T.L., Zhao, W., Leung, S.Y., Yuen, S.T. and Cancer Genome, P. (2003) BRAF and KRAS mutations in colorectal hyperplastic polyps and serrated adenomas. Cancer Res. *63*, 4878-4881.

Chang, H.W., Aoki, M., Fruman, D., Auger, K.R., Bellacosa, A., Tsichlis, P.N., Cantley, L.C., Roberts, T.M. and Vogt, P.K. (1997) Transformation of chicken cells by the gene encoding the catalytic subunit of PI 3-kinase. Science *276*, 1848-1850.

Cheng, J.Q., Altomare, D.A., Klein, M.A., Lee, W.C., Kruh, G.D., Lissy, N.A. and Testa, J.R. (1997) Transforming activity and mitosis-related expression of the AKT2 oncogene: evidence suggesting a link between cell cycle regulation and oncogenesis. Oncogene *14*, 2793-2801.

Chong, H. and Guan, K.L. (2003) Regulation of Raf through phosphorylation and N terminus–C terminus interaction. J. Biol. Chem. *278*, 36269-36276.

Chong, H., Vikis, H.G. and Guan, K.L. (2003) Mechanisms of regulating the Raf kinase family. Cell. Signal. *15*, 463-469.

Cohen, Y., Xing, M., Mambo, E., Guo, Z., Wu, G., Trink, B., Beller, U., Westra, W.H., Ladenson, P.W. and Sidransky, D. (2003) BRAF mutation in papillary thyroid carcinoma. J. Natl. Cancer Inst. *95*, 625-627.

Collisson, E.A., De, A., Suzuki, H., Gambhir, S.S. and Kolodney, M.S. (2003) Treatment of metastatic melanoma with an orally available inhibitor of the Ras-Raf-MAPK cascade. Cancer Res. *63*, 5669-5673.

Coudert, B., Anthoney, A., Fiedler, W., Droz, J.P., Dieras, V., Borner, M., Smyth, J.F., Morant, R., de Vries, M.J., Roelvink, M. et al. (2001) Phase II trial with ISIS 5132 in patients with small-cell (SCLC) and non-small cell (NSCLC) lung cancer. A European Organization for Research and Treatment of Cancer (EORTC) Early Clinical Studies Group report. Europ. J. Cancer *37*, 2194-2198.

Courtneidge, S.A. and Heber, A. (1987) An 81 kd protein complexed with middle T antigen and pp60c-src: a possible phosphatidylinositol kinase. Cell *50*, 1031-1037.

Cowley, S., Paterson, H., Kemp, P. and Marshall, C.J. (1994) Activation of MAP kinase kinase is necessary and sufficient for PC12 differentiation and for transformation of NIH 3T3 cells. Cell *77*, 841-852.

Cripps, M.C., Figueredo, A.T., Oza, A.M., Taylor, M.J., Fields, A.L., Holmlund, J.T., McIntosh, L.W., Geary, R.S. and Eisenhauer, E.A. (2002) Phase II randomized study of ISIS 3521 and ISIS 5132 in patients with locally advanced or metastatic colorectal cancer: a National Cancer Institute of Canada clinical trials group study. Clin. Cancer Res. *8*, 2188-2192.

D'Adamo, D.R., Novick, S., Kahn, J.M., Leonardi, P. and Pellicer, A. (1997) rsc: a novel oncogene with structural and functional homology with the gene family of exchange factors for Ral. Oncogene *14*, 1295-1305.

Davies, H., Bignell, G.R., Cox, C., Stephens, P., Edkins, S., Clegg, S., Teague, J., Woffendin, H., Garnett, M.J., Bottomley, W. et al. (2002) Mutations of the BRAF gene in human cancer [see comment]. Nature *417*, 949-954.

Dhand, R., Hara, K., Hiles, I., Bax, B., Gout, I., Panayotou, G., Fry, M.J., Yonezawa, K., Kasuga, M. and Waterfield, M.D. (1994) PI 3-kinase: structural and functional analysis of intersubunit interactions. EMBO J. *13*, 511-521.

Dong, J., Phelps, R.G., Qiao, R., Yao, S., Benard, O., Ronai, Z. and Aaronson, S.A. (2003) BRAF oncogenic mutations correlate with progression rather than initiation of human melanoma. Cancer Res. *63*, 3883-3885.

Feig, L.A. (2003) Ral-GTPases: approaching their 15 minutes of fame. Trends Cell Biol. *13*, 419-425.

Fukui, Y. and Hanafusa, H. (1989) Phosphatidylinositol kinase activity associates with viral p60src protein. Mol. Cell. Biol. *9*, 1651-1658.

Gille, H. and Downward, J. (1999) Multiple ras effector pathways contribute to G(1) cell cycle progression. J. Biol. Chem. *274*, 22033-22040.

Goi, T., Rusanescu, G., Urano, T. and Feig, L.A. (1999) Ral-specific guanine nucleotide exchange factor activity opposes other Ras effectors in PC12 cells by inhibiting neurite outgrowth. Mol. Cell. Biol. *19*, 1731-1741.

Hahn, W.C., Counter, C.M., Lundberg, A.S., Beijersbergen, R.L., Brooks, M.W. and Weinberg, R.A. (1999) Creation of human tumour cells with defined genetic elements. Nature *400*, 464-468.

Hamad, N.M., Elconin, J.H., Karnoub, A.E., Bai, W., Rich, J.N., Abraham, R.T., Der, C.J. and Counter, C.M. (2002) Distinct requirements for Ras oncogenesis in human versus mouse cells. Genes Dev. *16*, 2045-2057.

Henriksson, M. and Luscher, B. (1996) Proteins of the Myc network: essential regulators of cell growth and differentiation. Adv. Cancer Res. *68*, 109-182.

Ishimura, N., Yamasawa, K., Karim Rumi, M.A., Kadowaki, Y., Ishihara, S., Amano, Y., Nio, Y., Higami, T. and Kinoshita, Y. (2003) BRAF and K-ras gene mutations in human pancreatic cancers. Cancer Lett. *199*, 169-173.

Janda, E., Lehmann, K., Killisch, I., Jechlinger, M., Herzig, M., Downward, J., Beug, H. and Grunert, S. (2002) Ras and TGF[beta] cooperatively regulate epithelial cell plasticity and metastasis: dissection of Ras signaling pathways. J. Cell Biol. *156*, 299-313.

Johnson, L., Mercer, K., Greenbaum, D., Bronson, R.T., Crowley, D., Tuveson, D.A. and Jacks, T. (2001) Somatic activation of the K-ras oncogene causes early onset lung cancer in mice. Nature *410*, 1111-1116.

Kauffmann-Zeh, A., Rodriguez-Viciana, P., Ulrich, E., Gilbert, C., Coffer, P., Downward, J. and Evan, G. (1997) Suppression of c-Myc-induced apoptosis by Ras signalling through PI(3)K and PKB. Nature *385*, 544-548.

Kerkhoff, E. and Rapp, U.R. (1997) Induction of cell proliferation in quiescent NIH 3T3 cells by oncogenic c-Raf-1. Mol. Cell. Biol. *17*, 2576-2586.

Khosravi-Far, R., White, M.A., Westwick, J.K., Solski, P.A., Chrzanowska-Wodnicka, M., Van Aelst, L., Wigler, M.H. and. Der, C.J. (1996) Oncogenic Ras activation of Raf/mitogen-activated protein kinase-independent pathways is sufficient to cause tumorigenic transformation. Mol. Cell. Biol. *16*, 3923-3933.

Kim, I.J., Park, J.H., Kang, H.C., Shin, Y., Park, H.W., Park, H.R., Ku, J.L., Lim, S.B. and Park, J.G. (2003) Mutational analysis of BRAF and K-ras in gastric cancers: absence of BRAF mutations in gastric cancers. Hum. Genet. *114*, 118-120.

Kim, J.S., Lee, C., Foxworth, A. and Waldman, T. (2004) B-Raf is dispensable for K-Ras-mediated oncogenesis in human cancer cells. Cancer Res. *64*, 1932-1937.

Kimura, E.T., Nikiforova, M.N., Zhu, Z., Knauf, J.A., Nikiforov, Y.E. and Fagin, J.A. (2003) High prevalence of BRAF mutations in thyroid cancer: genetic evidence for constitutive activation of the RET/PTC-RAS-BRAF signaling pathway in papillary thyroid carcinoma. Cancer Res. *63*, 1454-1457.

Klippel, A., Escobedo, J.A., Hirano, M. and Williams, L.T. (1994) The interaction of small domains between the subunits of phosphatidylinositol 3-kinase determines enzyme activity. Mol. Cell. Biol. *14*, 2675-2685.

Klippel, A., Escobedo, M.A., Wachowicz, M.S., Apell, G., Brown, T.W., Giedlin, M.A., Kavanaugh, W.M. and Williams, L.T. (1998) Activation of phosphatidylinositol 3-kinase is sufficient for cell cycle entry and promotes cellular changes characteristic of oncogenic transformation. Mol. Cell. Biol. *18*, 5699-5711.

Kubota, N., Okada, S., Inada, T., Ohnishi, K. and Ohnishi, T. (2000) Wortmannin sensitizes human glioblastoma cell lines carrying mutant and wild type TP53 gene to radiation. Cancer Lett. *161*, 141-147.

Lee, S.H., Lee, J.W., Soung, Y.H., Kim, H.S., Park, W.S., Kim, S.Y., Lee, J.H., Park, J.Y., Cho, Y.G., Kim, C.J. et al. (2003) BRAF and KRAS mutations in stomach cancer. Oncogene *22*, 6942-6945.

Lemoine, N.R., Staddon, S., Bond, J., Wyllie, F.S., Shaw, J.J. and Wynford-Thomas, D. (1990) Partial transformation of human thyroid epithelial cells by mutant Ha-ras oncogene. Oncogene *5*, 1833-1837.

Levine, A.J., Momand, J. and Finlay, C.A. (1991) The p53 tumour suppressor gene. Nature *351*, 453-456.

Livingston, D.M. (1992) Functional analysis of the retinoblastoma gene product and of RB-SV40 T antigen complexes. Cancer Surv. *12*, 153-160.

Ma, Y.Y., Wei, S.J., Lin, Y.C., Lung, J.C., Chang, T.C., Whang-Peng, J., Liu, J.M., Yang, D.M., Yang, W.K. and Shen, C.Y. (2000) PIK3CA as an oncogene in cervical cancer. Oncogene *19*, 2739-2744.

Mansour, S.J., Matten, W.T., Hermann, A.S., Candia, J.M., Rong, S., Fukasawa, K., Vande Woude, G.F. and Ahn, N.G. (1994) Transformation of mammalian cells by constitutively active MAP kinase kinase. Science *265*, 966-970.

Mansour, S.L., Thomas, K.R. and Capecchi, M.R. (1988) Disruption of the proto-oncogene int-2 in mouse embryo-derived stem cells: a general strategy for targeting mutations to non-selectable genes. Nature *336*, 348-352.

Matsubara, K., Kishida, S., Matsuura, Y., Kitayama, H., Noda, M. and Kikuchi, A. (1999) Plasma membrane recruitment of RalGDS is critical for Ras-dependent Ral activation. Oncogene *18*, 1303-1312.

Matsuguchi, T. and Kraft, A.S. (1998) Regulation of myeloid cell growth by distinct effectors of Ras. Oncogene *17*, 2701-2709.

McFall, A., Ulku, A., Lambert, Q.T., Kusa, A., Rogers-Graham, K. and Der, C.J. (2001) Oncogenic Ras blocks anoikis by activation of a novel effector pathway independent of phosphatidylinositol 3-kinase. Mol. Cell. Biol. *21*, 5488-5499.

McFarlin, D.R. and Gould, M.N. (2003) Rat mammary carcinogenesis induced by in situ expression of constitutive Raf kinase activity is prevented by tethering Raf to the plasma membrane. Carcinogenesis *24*, 1149-1153.

McFarlin, D.R., Lindstrom, M.J. and Gould, M.N. (2003) Affinity with Raf is sufficient for Ras to efficiently induce rat mammary carcinomas. Carcinogenesis *24*, 99-105.

Milella, M., Kornblau, S.M., Estrov, Z., Carter, B.Z., Lapillonne, H., Harris, D., Konopleva, M., Zhao, S., Estey, E. and Andreeff, M. (2001) Therapeutic targeting of the MEK/MAPK signal transduction module in acute myeloid leukemia. J. Clin. Invest. *108*, 851-859.

Miller, M.J., Rioux, L., Prendergast, G.V., Cannon, S., White, M.A. and Meinkoth, J.L. (1998) Differential effects of protein kinase A on Ras effector pathways. Mol. Cell. Biol. *18*, 3718-3726.

Monia, B.P., Johnston, J.F., Geiger, T., Muller, M. and Fabbro, D. (1996) Antitumor activity of a phosphorothioate antisense oligodeoxynucleotide targeted against C-raf kinase. Nat. Med. *2*, 668-675.

Moscatello, D.K., Holgado-Madruga, M., Emlet, D.R., Montgomery, R.B. and Wong, A.J. (1998) Constitutive activation of phosphatidylinositol 3-kinase by a naturally occurring mutant epidermal growth factor receptor. J. Biol. Chem. *273*, 200-206.

Murphy, G.A., Graham, S.M., Morita, S., Reks, S.E., Rogers-Graham, K., Vojtek, A., Kelley, G.G. and Der, C.J. (2002) Involvement of phosphatidylinositol 3-kinase, but not RalGDS, in TC21/R-Ras2-mediated transformation. J. Biol. Chem. *277*, 9966-9975.

Namba, H., Nakashima, M., Hayashi, T., Hayashida, N., Maeda, S., Rogounovitch, T.I., Ohtsuru, A., Saenko, V.A., Kanematsu, T. and Yamashita, S. (2003) Clinical implication of hot spot BRAF mutation, V599E, in papillary thyroid cancers. J. Clin. Endocrinol. Metab. *88*, 4393-4397.

Naoki, K., Chen, T.H., Richards, W.G., Sugarbaker, D.J. and Meyerson, M. (2002) Missense mutations of the BRAF gene in human lung adenocarcinoma. Cancer Res. *62*, 7001-7003.

Ng, S.S.W., Tsao, M.S., Chow, S. and Hedley, D.W. (2000) Inhibition of phosphatidylinositide 3-kinase enhances gemcitabine-induced apoptosis in human pancreatic cancer cells. Cancer Res. *60*, 5451-5455.

Nguyen, K.T., Wang, W.J., Chan, J.L. and Wang, L.H. (2000) Differential requirements of the MAP kinase and PI3 kinase signaling pathways in Src- versus insulin and IGF-1 receptors-induced growth and transformation of rat intestinal epithelial cells. Oncogene *19*, 5385-5397.

Oldham, S.M., Clark, G.J., Gangarosa, L.M., Coffey, R.J., Jr. and. Der, C.J. (1996) Activation of the Raf-1/MAP kinase cascade is not sufficient for Ras transformation of RIE-1 epithelial cells. Proc. Natl. Acad. Sci. U.S.A. *93*, 6924-6928.

Oliveira, C., Pinto, M., Duval, A., Brennetot, C., Domingo, E., Espin, E., Armengol, M., Yamamoto, H., Hamelin, R., Seruca, R. and Schwartz, S., Jr. (2003) BRAF mutations characterize colon but not gastric cancer with mismatch repair deficiency. Oncogene *22*, 9192-9196.

Oza, A.M., Elit, L., Swenerton, K., Faught, W., Ghatage, P., Carey, M., McIntosh, L., Dorr, A., Holmlund, J.T., Eisenhauer, E. and Study, N.C.T.G. (2003) Phase II study of CGP 69846A (ISIS 5132) in recurrent epithelial ovarian cancer: an NCIC clinical trials group study (NCIC IND.116). Gynecol. Oncol. *89*, 129-133.

Palmieri, S. and Vogel, M.L. (1987) Fibroblast transformation parameters induced by the avian v-mil oncogene. J. Virol. *61*, 1717-1721.

Peyssonnaux, C., Provot, S., Felder-Schmittbuhl, M.P., Calothy, G. and Eychene, A. (2000) Induction of postmitotic neuroretina cell proliferation by distinct Ras downstream signaling pathways. Mol. Cell. Biol. *20*, 7068-7079.

Pritchard, C.A., Samuels, M.L., Bosch, E. and McMahon, M. (1995) Conditionally oncogenic forms of the A-Raf and B-Raf protein kinases display different biological and biochemical properties in NIH 3T3 cells. Mol. Cell. Biol. *15*, 6430-6442.

Rajagopalan, H., Bardelli, A., Lengauer, C., Kinzler, K.W., Vogelstein, B. and Velculescu, V.E. (2002) Tumorigenesis: RAF/RAS oncogenes and mismatch-repair status. Nature *418*, 934.

Ramocki, M.B., White, M.A., Konieczny, S.F. and Taparowsky, E.J. (1998) A role for RalGDS and a novel Ras effector in the Ras-mediated inhibition of skeletal myogenesis. J. Biol. Chem. *273*, 17696-17701.

Rich, J.N., Guo, C., McLendon, R.E., Bigner, D.D., Wang, X.F. and Counter, C.M. (2001) A genetically tractable model of human glioma formation. Cancer Res. *61*, 3556-3560.

Rodriguez-Viciana, P., Warne, P.H., Khwaja, A., Marte, B.M., Pappin, D., Das, P., Waterfield, M.D., Ridley, A. and Downward, J. (1997) Role of phosphoinositide 3-OH kinase in cell transformation and control of the actin cytoskeleton by Ras. Cell *89*, 457-467.

Rosario, M., Paterson, H.F. and Marshall, C.J. (2001) Activation of the Ral and phosphatidylinositol 3′ kinase signaling pathways by the ras-related protein TC21. Mol. Cell. Biol. *21*, 3750-3762.

Samuels, M.L., Weber, M.J., Bishop, J.M. and McMahon, M. (1993) Conditional transformation of cells and rapid activation of the mitogen-activated protein kinase cascade by an estradiol-dependent human raf-1 protein kinase. Mol. Cell. Biol. *13*, 6241-6252.

Samuels, Y. and Velculescu, V.E. (2004) Oncogenic mutations of PIK3CA in human cancers. Cell Cycle *3*, 1221-1224.

Sebolt-Leopold, J.S., Dudley, D.T., Herrera, R., Van Becelaere, K., Wiland, A., Gowan, R.C., Tecle, H., Barrett, S.D., Bridges, A., Przybranowski, S. et al. (1999) Blockade of the MAP kinase pathway suppresses growth of colon tumors in vivo. Nat. Med. *5*, 810-816.

Serrano, M., Lin, A.W., McCurrach, M.E., Beach, D. and Lowe, S.W. (1997) Oncogenic ras provokes premature cell senescence associated with accumulation of p53 and p16INK4a. Cell *88*, 593-602.

Shay, J.W. and Bacchetti, S. (1997) A survey of telomerase activity in human cancer. Eur. J. Cancer *33*, 787-791.

Shayesteh, L., Lu, Y., Kuo, W.L., Baldocchi, R., Godfrey, T., Collins, C., Pinkel, D., Powell, B., Mills, G.B. and Gray, J.W. (1999) PIK3CA is implicated as an oncogene in ovarian cancer. Nat. Genet. *21*, 99-102.

Sheng, H., Shao, J. and DuBois, R.N. (2001) Akt/PKB activity is required for Ha-Ras-mediated transformation of intestinal epithelial cells. J. Biol. Chem. *276*, 14498-14504.

Shields, J.M., Pruitt, K., McFall, A., Shaub, A. and Der, C.J. (2000) Understanding Ras: 'it ain't over' til it's over'. Trends Cell Biol. *10*, 147-154.

Smets, L.A. (1980) Cell transformation as a model for tumor induction and neoplastic growth. Biochim. Biophys. Acta *605*, 93-111.

Stanton, V.P., Nichols, D.W., Laudano, A.P. and Cooper, G.M. (1989) Definition of the human raf amino-terminal regulatory region by deletion mutagenesis. Mol. Cell. Biol. *9*, 639-647.

Stokoe, D., Macdonald, S.G., Cadwallader, K., Symons, M. and Hancock, J.F. (1994) Activation of Raf as a result of recruitment to the plasma membrane. Science *264*, 1463-1467.

Sun, M., Wang, G., Paciga, J.E., Feldman, R.I., Yuan, Z.Q., Ma, X.L., Shelley, S.A., Jove, R., Tsichlis, P.N., Nicosia, S.V. and Cheng, J.Q. (2001) AKT1/PKBalpha kinase is frequently elevated in human cancers and its constitutive activation is required for oncogenic transformation in NIH3T3 cells. Am. J. Pathol. *159*, 431-437.

Tang, Y., Yu, J. and Field, J. (1999) Signals from the Ras, Rac, and Rho GTPases converge on the Pak protein kinase in Rat-1 fibroblasts. Mol. Cell. Biol. *19*, 1881-1891.

Tolcher, A.W., Reyno, L., Venner, P.M., Ernst, S.D., Moore, M., Geary, R.S., Chi, K., Hall, S., Walsh, W., Dorr, A. and Eisenhauer, E. (2002) A randomized phase II and pharmacokinetic study of the antisense oligonucleotides ISIS 3521 and ISIS 5132 in patients with hormone-refractory prostate cancer. Clin. Cancer Res. *8*, 2530-2535.

Toretsky, J.A., Thakar, M., Eskenazi, A.E. and Frantz, C.N. (1999) Phosphoinositide 3-hydroxide kinase blockade enhances apoptosis in the Ewing's sarcoma family of tumors. Cancer Res. *59*, 5745-5750.

Urano, T., Emkey, R. and Feig, L.A. (1996) Ral-GTPases mediate a distinct downstream signaling pathway from Ras that facilitates cellular transformation. EMBO J. *15*, 810-816.

Varticovski, L., Daley, G.Q., Jackson, P., Baltimore, D. and Cantley, L.C. (1991) Activation of phosphatidylinositol 3-kinase in cells expressing abl oncogene variants. Mol. Cell. Biol. *11*, 1107-1113.

Vivanco, I. and Sawyers, C.L. (2002) The phosphatidylinositol 3-Kinase AKT pathway in human cancer. Nat. Rev. Cancer *2*, 489-501.

Wang, L., Cunningham, J.M., Winters, J.L., Guenther, J.C., French, A.J., Boardman, L.A., Burgart, L.J., McDonnell, S.K., Schaid, D.J. and Thibodeau, S.N. (2003) BRAF mutations in colon cancer are not likely attributable to defective DNA mismatch repair. Cancer Res. *63*, 5209-5212.

Ward, Y., Wang, W., Woodhouse, E., Linnoila, I., Liotta, L. and Kelly, K. (2001) Signal pathways which promote invasion and metastasis: critical and distinct contributions of extracellular signal-regulated kinase and Ral-specific guanine exchange factor pathways. Mol. Cell. Biol. *21*, 5958-5969.

Wasylyk, C., Wasylyk, B., Heidecker, G., Huleihel, M. and Rapp, U.R. (1989) Expression of raf oncogenes activates the PEA1 transcription factor motif. Mol. Cell. Biol. *9*, 2247-2250.

Webb, C.P., Van Aelst, L., Wigler, M.H. and Woude, G.F. (1998) Signaling pathways in Ras-mediated tumorigenicity and metastasis. Proc. Natl. Acad. Sci. U.S.A. *95*, 8773-8778.

Wellbrock, C., Ogilvie, L., Hedley, D., Karasarides, M., Martin, J., Niculescu-Duvaz, D., Springer, C.J. and Marais, R. (2004) V599EB-RAF is an oncogene in melanocytes. Cancer Res. *64*, 2338-2342.

Whang, Y.E., Wu, X. and Sawyers, C.L. (1998) Identification of a pseudogene that can masquerade as a mutant allele of the PTEN/MMAC1 tumor suppressor gene. J. Natl. Cancer Inst. *90*, 859-861.

White, M.A., Nicolette, C., Minden, A., Polverino, A., Van Aelst, L., Karin, M. and Wigler, M.H. (1995) Multiple Ras functions can contribute to mammalian cell transformation. Cell *80*, 533-541.

White, M.A., Vale, T., Camonis, J.H., Schaefer, E. and Wigler, M.H. (1996) A role for the Ral guanine nucleotide dissociation stimulator in mediating Ras-induced transformation. J. Biol. Chem. *271*, 16439-16442.

Whitman, M., Kaplan, D.R., Schaffhausen, B., Cantley, L. and Roberts, T.M. (1985) Association of phosphatidylinositol kinase activity with polyoma middle-T competent for transformation. Nature *315*, 239-242.

Wolthuis, R.M., de Ruiter, N.D., Cool, R.H. and Bos, J.L. (1997) Stimulation of gene induction and cell growth by the Ras effector Rlf. EMBO J. *16*, 6748-6761.

Wolthuis, R.M. and Bos, J.L. (1999) Ras caught in another affair: the exchange factors for Ral. Curr. Opin. Genet. Dev. *9*, 112-117.

Wu, X., Senechal, K., Neshat, M.S., Whang, Y.E. and Sawyers, C.L. (1998) The PTEN/MMAC1 tumor suppressor phosphatase functions as a negative regulator of the phosphoinositide 3-kinase/Akt pathway. Proc. Natl. Acad. Sci. U.S.A. *95*, 15587-15591.

Yang, J.J., Kang, J.S. and Krauss, R.S. (1998) Ras signals to the cell cycle machinery via multiple pathways to induce anchorage-independent growth. Mol. Cell. Biol. *18*, 2586-2595.

Yeh, E., Cunningham, M., Arnold, H., Chasse, D., Monteith, T., Ivaldi, G., Hahn, W.C., Stukenberg, P.T., Shenolikar, S., Uchida, T. et al. (2004) A signalling pathway controlling c-Myc degredation that impacts oncogenic transformation of human cells. Nat. Cell Biol. *6*, 308-318.

Yuen, S.T., Davies, H., Chan, T.L., Ho, J.W., Bignell, G.R., Cox, C., Stephens, P., Edkins, S., Tsui, W.W., Chan, A.S. et al. (2002) Similarity of the phenotypic patterns associated with BRAF and KRAS mutations in colorectal neoplasia. Cancer Res. *62*, 6451-6455.

Zhao, W., Chan, T.L., Chu, K.M., Chan, A.S., Stratton, M.R., Yuen, S.T. and Leung, S.Y. (2004) Mutations of BRAF and KRAS in gastric cancer and their association with microsatellite instability. Int. J. Cancer *108*, 167-169.

CHAPTER 12

GENETICALLY ENGINEERED MICE HARBORING RAS MUTATIONS AS MODELS OF HUMAN CANCER: IN MEDIAS RAS[1]

TAL Z. ZAKS, MICHAEL A. JACOBETZ, JR., AND DAVID A. TUVESON

Abramson Cancer Research Institute and Avramson Cancer Center, Department of Medicine and Cancer Biology, University of Pennsylvania, Philadelphia, PA 19104

Abstract: The molecular and histopathological consequences of expressing a mutated *ras* gene have been the focus of intense studies for the past 30 years. Initially, introducing *H-ras* mutations as transgenes under the control of a heterologous promoter was shown to cause neoplastic changes where expressed, proving *in-vivo* the relevance of *ras* mutations to cancer. Lesions usually developed after a long latent period, and the additional mutation of tumor suppressor genes (such as *p*53) resulted in a more aggressive phenotype. In an attempt to better mimic genetic lesions found in human cancer, more recent studies have used the *Cre-Lox* recombination system to tissue-specific directed mutations in the endogenous *K-ras* allele. When expressed in the lungs, mice develop lesions that include hyperplasias, adenomas, invasive carcinomas, and occasionally metastatic lesions. When expressed in the pancreas, early pancreatic intraepithelial neoplastic lesions (PanIN) are induced, with progression of histological atypia that appears indistinguishable from human disease. Moreover, when crossed to a conditional Ink4a/ARF background (as is frequently found in human pancreatic adenocarcinoma), animals succumb to both local and distant disease within 12 weeks. Similarly, when the endogenous *K-ras* is mutated in hematopoietic cells, all mice die from a myeloproliferative disorder. By faithfully recapitulating both the genetic and pathophysiology of the cognate human malignancy, these models provide a means to dissect the relevant molecular pathways leading to cancer susceptibility, formation and progression; and to design and test prevention, early detection, and treatment strategies for human cancers that are driven by *ras* mutations

Keywords: ras; Ink4a; p53; endogenous mutation; transgenic; embryonic stem cells

[1]From Latin "in media res" – in the midst of things

C. Der (ed.), RAS Family GTPases, 273–293.

1. INTRODUCTION

Cancer is the second leading cause of death in the United States, despite the tremendous investment of resources and considerable scientific investigations over the past three decades. Substantial progress has been made towards identifying the numerous genetic alterations present in malignancy. However, the precise role of these genes in regulating the biochemical pathways and cellular behavior of tumor cells is largely unknown. Animal models of malignancy that closely resemble the cognate human disease offer the opportunity to dissect the molecular requirements of neoplasia and evaluate potential therapeutic and detection strategies. Neoplasia has been studied most prominently in mice, due to the existence of genetic inbred mouse strains, the relative ease of animal husbandry, and the well-established genetic methods to produce mutant mice. The importance of the *ras* genes in human disease, as expounded upon elsewhere in this volume, has spurred numerous attempts to accurately model human cancer by genetically manipulating *ras* expression in the mouse. Genetically engineered *ras* mutant mice have provided key insights into the role of *ras* in tumor formation and growth, and have provided model systems to evaluate both cancer-promoting (carcinogens) and cancer-regressing (i.e., therapeutic) agents.

This review will briefly discuss the methods used to generate *ras* mutant mice, followed by a description of the various models that have been created and their use in the study of Ras-specific therapy.

2. GENETICALLY ENGINEERED MUTANT MOUSE METHODS

2.1 Conventional Transgenics

The initial intent of developing genetically engineered mice was to discern the involvement of oncogenes in tumorigenesis. Transgenic mice are produced by pro-nuclear injection of cDNA constructs that contain ectopic and/or endogenous regulatory sequences (i.e., promoters and enhancers). The injected cDNA integrates randomly by non-homologous recombination generally in a tandem fashion in multiple sites (Hanahan 1988) (Figure 1). Several of these one-cell embryos are implanted into a pseudo-pregnant female and allowed to develop. The random recombination events result in mice that are genetically unique with regard to the precise integration site of the foreign DNA and the copy number of the transgene. Viable offspring are genotyped, by PCR or Southern blot analysis, to confirm the introduction of the transgene. Those harboring the transgene are then classified as founders and are further propagated to create a lineage of transgenics that are genetically similar.

2.2 Embryonic Stem Cell Targeting

In the mid 1980s, gene targeting approaches in murine embryonic stem (ES) cells were successfully reported by Cappechi and colleagues (Thomas and Capecchi 1987). Targeting vectors consisted of complementary genomic DNA fragments that

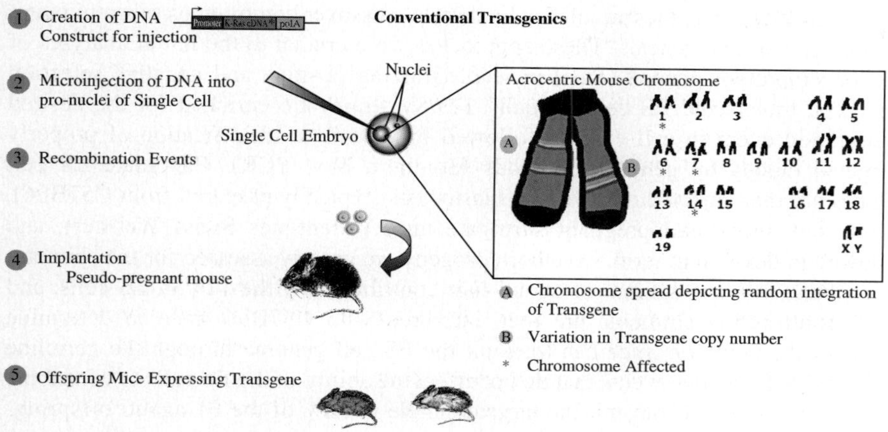

Figure 1. Conventional Transgenic

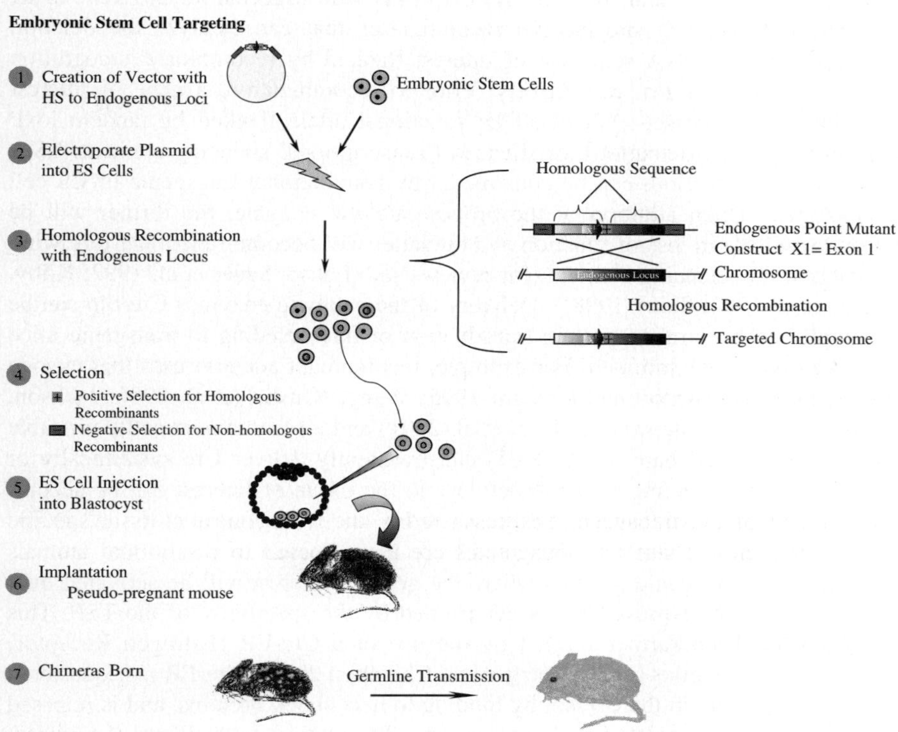

Figure 2. Embryonic Stem Cell Targeting

mutagenized the endogenous alleles by double crossover homologous recombination or insertional mutagenesis. These approaches were crucial to the initial analyses of tumor suppressor genes (TSGs) *in vivo*. Utilizing positive and negative selection markers, transfected ES cells (usually 129/Sv strain) are enriched for the desired integration event in cell culture, followed by definitive identification of properly targeted clones by genomic methods (Southern blot, PCR). Candidate ES cell clones are then injected into day 2.5 blastocysts (typically prepared from C57Bl/6), implanted into pseudopregnant surrogate mice (oftentimes Swiss Webster), and allowed to develop to term. Live born progeny are grossly assessed for the chimeric contribution by the amount of agouti hair contributed by the 129/Sv ES cells, and high contribution chimeras are then interbred with C57Bl/6 mice to determine whether the chimeric mice can transmit the ES cell genome through the germline (Figure 2). Targeting events that do not affect the ability of the ES cells to contribute to the germline will transmit the targeted allele in 50% of the F1 agouti offspring.

2.3 Conditional Control of Genetic Expression

Conditional gene targeting affords the spatial and temporal control of gene inactivation and activation. This is most often accomplished by the use of the viral (Phage P1, Cre) (Le and Sauer 2001) or prokaryotic (Saccharomyces cerevisiae, Flp) (Sadowski 1995) site-specific recombinases that can catalyze the deletion or inversion of a DNA sequence of interest flanked by recombinase recognition sites called loxP or Frt, respectively. Prior to recombination, a gene of interest can either be expressed ("flox" allele: functional allele flanked by tandem loxP sites in the same orientation), or silenced (transcriptional silencing element; LSL: loxP STOP loxP). Both can be constructed by conventional transgenic or ES cell technologies. Upon addition of the proper catalytic enzyme, the former will be deleted resulting in loss of function and the latter will become functional following removal of the silencing element (for reviews see (Lakso, Sauer et al. 1992; Kilby, Snaith et al. 1993; Sauer 1998)). Delivery of the catalytic enzymes Cre/Flp can be accomplished by viral or protein transduction or interbreeding to transgenic mice that express the recombinase. For example, recombinant adenoviruses that encode cre recombinase (Anton and Graham 1995; Wang, Krushel et al. 1996; Jackson, Willis et al. 2001; Meuwissen, Linn et al. 2001) and a recombinant cell permeable Cre protein (Jo, Nishabi et al., 2001) can transiently deliver Cre systemically or directly to the organ site. Genetic delivery to the tissue of interest can be accomplished with transgenic mice that express Cre/Flp under the control of tissue specific promoters (TSP). These Cre transgenics are then crossed to conditional animals to produce bi-transgenic animals where the gene of interest will be activated only in the tissues that express Cre, as determined by the specificity of the TSP. This approach has been further refined by the use of a Cre-ER (Estrogen Receptor) fusion protein (Babinet 2000; Metzger and Chambon 2001). Cre-ER is sequestered in an inactive state in the cytosol by binding to heat shock proteins, and is released upon binding of agonists (estrogen analogs, Tamoxifen) to facilitate the migration of the active Cre-ER protein to the nucleus. This enables the gene of interest to be activated only in the tissue that expresses Cre (providing spatial control) and only when Tamoxifen is administered systemically (allowing for temporal control).

2.4 Inducible and Reversible Transgene Expression

An additional approach to generate genetically engineered mice is the tetracycline-operon/repressor system (Gossen and Bujard 1992; Furth, St Onge et al. 1994) with a gene of interest placed downstream of tet. Bi-transgenic tet-op; tet repressor mice are produced by either injecting both constructs into one pro-nucleus (Kucera, Bortner et al. 1996) or by crossing transgenics expressing the tet-gene of interest to mice expressing tet transcriptional regulatory elements (inducible (rtTA) and repressible (tTA) factors) that are controlled by the exogenous administration of small molecules such as tetracycline or its analog doxycycline (Gossen and Bujard 1993; Chrast-Balz and Hooft van Huijsduijnen 1996; Kistner, Gossen et al. 1996). The bio-availability of these compounds enables them to be orally administered in water or food. This approach enables both a gene-dose effect, by varying the amount of doxycycline, and the cessation of transgene expression by withholding tetracycline.

3. EARLY TRANSGENIC MODELS

In the mid-80s it was shown that skin tumors induced by the chemical carcinogen dimethlybenzathracene (DMBA) contained high protein or RNA levels of oncogenic v-*H-ras* that were capable of transforming NIH3T3 fibroblasts in culture. This was shown to be a frequent event in both early benign papillomas as well as in more invasive carcinomas (Balmain, Ramsden et al. 1984). When a mutated human Ha-*ras* was placed under the suprabasal keratin 10 promoter (expressed in differentiating epidermal cells of the skin), multiple benign papillomas and hyperkeratosis were observed, with papilloma formation being restricted to sites of injury and wound healing. Significantly, human Ha-*ras* was detected in these lesions, but not in adjacent normal skin or in other organs (Bailleul, Surani et al. 1990). When oncogenic human *H-ras* was placed under the rat elastase I promoter in transgenic mice (targeting expression to the exocrine pancreas), massive pancreatic acinar tumors formed soon after pancreatic development, but without metastatic spread (Quaife, Pinkert et al. 1987) (see below). An additional model involved creating transgenic mice with *H-ras*^{G12V} under the promoter of the murine whey acidic protein (wap) gene, which is expressed in mammary epithelial cells in response to lactogenic hormones. Tumors developed after a long latency only in tissues expressing *H-ras*^{G12V} — which were mammary glands in females and salivary glands in males. It appeared that oncogenic *ras* was sufficient for an early initiation of neoplasia, but not for tumor progression; and that tissue specificity, at least in experimental systems, was a function of the promoter.

The issue of tissue specificity of *ras* effects was addressed initially by creating mice with transgenic human *H-ras* under their own promoter region; specifically, the transgene was not mutated and was incapable of transforming NIH3T3 cells. Two transgenic strains, containing 5-6 tandem repeats of the transgene, were studied. Approximately 50% of mice, in both lines, had cancers when examined at 18 months. The major histologies were angiosarcomas followed by lung adenocarcinomas (Saitoh, Kimura et al. 1990). While the transgene was expressed in all tissues, only the cancers contained (de-novo) point mutations. Most of the mutations were a glutamine to leucine change at position 61 (G61L), with the exception of glycine to valine changes at position 12 (G12V) that occurred in skin papillomas. Neoplasia of blood vessels and lungs were an unexpected finding in this study. In a separate study, transgenic mice carrying the entire coding regions of mutated human *H-ras* (with an Ig enhancer and the SV40 early gene promoter) developed predominantly lung tumors that were multicentric, well differentiated, and with limited, if any, capacity to form metastasis (Suda, Aizawa et al. 1987). In yet another example, transgenic mice harboring a mutant human Ha-*ras* linked to the murine albumin enhancer and promoter region, in an attempt to target expression to the liver, developed very similar lung tumors (Maronpot, Palmiter et al. 1991). In all these cases transgene expression was not limited to the tumors, suggesting additional mutational events had occurred, and mice died from pulmonary insufficiency with very little evidence of metastatic spread. A number of scientific concerns regarding the validity of these mice as models of human disease remain, including the artificially high levels of mutant *ras* transgene expression conferred by heterologous promoters, the integration of multiple tandem transgene copies, and the non-physiologic temporal and spatial expression. Despite these concerns, it appeared that mutated *ras*, when broadly over-expressed in transgenic mice, has a propensity for forming early lung tumors.

3.1 Cooperation between *ras*, Oncogenes, and Tumor Suppressor Genes

Most *in vitro* studies have shown that oncogenic *ras* alone was incapable of transforming primary fibroblasts cells, but could cooperate with a number of other proto-oncogenes, such as c-*myc*, to transform cells *in vitro* (Land, Parada et al. 1983). These observations were validated when separate mutant mouse strains harboring v-Ha-*ras*$^{(G12R,\ A59T)}$ and c-*myc*, both under the control of the murine mammary tumor virus (MMTV) promoter, were interbred. While it took almost a year for MMTV/c-*myc* females to develop mammary tumors, and about 6 months for MMTV/v-Ha-*ras*$^{(G12R,\ A59T)}$ females to develop tumors, 50% of the hybrid mice carrying both transgenes had tumors by day 100, and none were tumor free at day 200 (Sinn, Muller et al. 1987). Histologically, most of the tumors were mammary adenocarcinomas, with a minority developing harderian and salivary gland adenocarcinomas and lymphomas (the latter expected as transgenic mice expressing c-*myc* under the immunoglobulin gene enhancer had been shown to develop lymphomas).

In this system, v-Has-*ras*$^{(G12R,\ A59T)}$ alone was not sufficient for a fully malignant phenotype, but was synergistic with c-*myc*, recapitulating oncogenic cooperation *in vivo*.

The ability to interbreed mice harboring different transgenes has been employed often to show the *in vivo* relevance of cooperation both between different oncogenes and between oncogenes and tumor suppressor genes. Thus, MMTV/v-Ha-*ras*$^{(G12R,\ A59T)}$ mice develop tumors much more rapidly on a *p53*$^{-/-}$ background (Hundley, Koester et al. 1997).

Mice deficient in the tumor suppressor gene *Ink4a/ARF* do not develop melanomas, but when crossed to founders that express oncogenic *H-ras* (G12V) under the control of the tyrosinase promoter (Tyr-*ras*), more than 50% of Tyr-*ras; Ink4a/ARF*$^{-/-}$ mice develop melanoma within 6 months (versus few in the Tyr-*ras*; *Ink4a/ARF*$^{+/-}$ animals) (Chin, Pomerantz et al. 1997), characterized by multiple lesions that do not metastasize. As both INK4a/ARF and the RAS protein pathways are implicated in human melanoma, this provides both a confirmation and a relevant model of the genetic interaction between the two. Similarly, while *Ink4a/ARF*$^{-/-}$ mice do not develop spontaneous glioblastomas (despite a high frequency of deletion of *Ink4a* in human glioblastomas), infecting them with a retroviral construct targeting oncogenic *K-ras* to neural progenitors or astrocytes *in vivo* was sufficient to induce glioblastomas in the majority of the animals studied (Uhrbom, Dai et al. 2002). More recently, a mutation of the endogenous *K-ras* allele in the pancreas has been shown to cooperate with loss of Ink4a/ARF in the formation of pancreatic ductal cancer (see below).

3.2 Limitations of Early Models

These early studies recapitulated the role of an activated *ras* pathway in oncogenesis, as well as the cooperation between genetic alterations in oncogenes and tumor suppressor genes. However, several characteristics limit their ability to mimic human disease. First, expression of an oncogenic mutation throughout ontogeny is not characteristic of human cancer, where sporadic *de novo* mutations are thought to occur as stochastic events in single cells. It has been suggested that the timing of a mutational event during development can influence the type of tumor that ultimately develops (Knudson 1993). For example, in the study of melanoma cited above, most of the Tyr-*H-ras*$^{(G12V)}$ founders were non-viable (Chin, Pomerantz et al. 1997), raising the question of what was different about transgene expression in the viable animals. Related to this issue, transgenic expression simultaneously by many or all cells leads to a “field cancerization effect” (Garcia, Park et al., 1999).

An additional concern relates to the identity of the *ras* paralogue being studied. Not only are there significant post-translational modification differences between the different *ras* isoforms (Hancock, Magee et al. 1989; Magee and Marshall 1999), but both *H-ras* and *N-ras* are dispensable for normal murine development (Esteban, Vicario-Abejon et al. 2001). While the vast majority of these earlier studies were conducted with *H-ras*, it became apparent that the *ras* gene that is most commonly mutated in human malignancy is *K-ras* (Bos 1989; Khosravi-Far and Der 1994). Moreover, in an inducible system where *myc* was transiently overexpressed from the MMTV promoter (with a conditional tetracycline regulatory system), half the tumors that failed to regress upon withdrawal of *myc* harbored spontaneous *ras* mutations: the majority of these (75%) were *K-ras*, with the remainder being *N-ras* and no evidence of *H-ras* mutations (D'Cruz, Gunther et al., 2001). To better mimic human sporadic cancers, *K-ras* had to be targeted in a controllable temporal, spatial, and quantitive fashion. In the following sections, modern murine models of *ras*-induced neoplasia will be discussed according to tissue of origin.

4. CURRENT TRANSGENIC MODELS

4.1 Lung Cancer

Lung cancer is the leading cause of cancer death in the United States, with most cases presenting late in the course and the vast majority of patients dying of their illness. Most disease occurs in current or former users of tobacco, and multiple chromosomal and genetic alterations are found throughout the airways. Despite a seemingly unifying cause, tobacco smoke contains many different carcinogenic elements, and the pattern of DNA changes in lung cancer is not unique to this histology. Of the more common alterations found in human lung cancer, *K-ras* is mutated in 20-50%; *p53* is deleted or mutated in 50%; 60% have deletions or reduced expression of $p16^{Ink4a}$, and up to 30% show deletion or reduced expression of *Rb* (see (Tuveson and Jacks 1999) and references therein). Mice develop spontaneous lung cancers at a varying rate depending on their background, ranging from 3% in wild-type to 100% by 18-24 months in the sensitive A/J strain. Of note, while the majority of murine lung tumors are histologically early hyperplasias and adenomas, they share some of the molecular alterations of their human counterparts, including a high prevalence of mutated *K-ras* (80-90%) as well as deletion or reduced levels of expression of $p16^{Ink4a}$ in late lesions. While many of the genetically engineered models mentioned above manifested early lung tumors in response to constitutive over-expression of oncogenic *H-ras*, the involvement of *K-ras* was recently studied by a number of groups using an inducible targeted transgene. In an effort to mimic somatic mutations that would be subject to physiological controls, a latent *K-ras* allele carrying a mutant exon 1 upstream of a normal exon 1 was introduced into ES cells, with the expectation that an infrequent second recombination event would occur *in vivo*, yielding an oncogenic form half the time (Johnson, Mercer et al. 2001). A second strain, harboring two mutant exon 1 copies in tandem, was also constructed, with the expectation that every subsequent recombination would result in an oncogenic *K-ras* (see (Johnson, Mercer et al. 2001). By day 300 animals developed multifocal tumors at various stages of progression. Serial analysis revealed a temporal pattern of histological progression from small alveolar adenomatous hyperplasias to solid papillary adenomas with glandular formation; a fraction of the older mice demonstrated invasive adenocarcinomas with metastasis to thoracic lymph nodes, kidney, and other visceral organs occasionally noted. Additional tumors, including skin papillomas (in areas of abrasion) and thymic lymphomas, neither of which progressed, were seen in 30% of the animals. PCR amplification following Laser Capture Microdissection was used to show that recombination had occurred in tumors. It is unclear why tumors developed preferentially in the lungs (and thymus) in this model. Either these tissues are more sensitive to oncogenic *K-ras*^{G12D} or the frequency of recombination in them was higher. The latent *K-ras* alleles cooperated with *p53* loss, with mice succumbing earlier to tumors with more aggressive histological features. In addition to lung tumors, 30% of double mutant animals developed sarcomas; all mice developed non-progressing aberrant crypt foci (ACF); however, no pancreatic tumors were observed. As these mice succumbed relatively early to multiple tumors of varying histologies, the model was refined by introducing an oncogenic *K-ras*^{G12D} preceded by a cre/LoxP conditioned "STOP" element, so that exogenous administration of Cre (via an adenovirus encoding Cre) would result in removal of the STOP and transcription (under normal regulatory elements) of the oncogene (Jackson, Willis et al. 2001). By intranasal instillation of recombinant adenovirus encoding cre recombinase (Adeno-Cre), both the timing and location (e.g., lungs) could be controlled; by varying the dose of AdenoCre, mice with discrete synchronous tumors could be studied for histological progression. At 6 weeks, atypical alveolar adenomatous hyperplasia (AAH) and epithelial hyperplasia of the bronchioles and adenomas were seen. Of note, AAH are dysplastic lesions believed to be a precursor of human pulmonary adenocarcinoma (Kerr 2001). By 12 weeks post infection, large adenomas outnumbered AAH lesions, and by 16 weeks overt adenocarcinomas were observed with near complete disappearance of AAH lesions. Similar results were obtained by a different group, who generated a single copy transgenic mouse that conditionally expressed the human *K-ras*^{G12V} allele downstream of the β-actin promoter (Meuwissen, Linn et al. 2001). In this mutant mouse GFP is expressed in all transgenic cells, while following Cre recombination GFP is excised and *K-ras*^{G12V} is expressed. Here again, Adeno-Cre was administered to the lungs (at levels shown to infect only sporadic cells both in bronchial epithelium and distal alveoli). At 5-6 weeks post-infection all mice developed AAH and by 12-13 weeks all had papillary tumors that caused cachexia and respiratory difficulty due to bronchial obstruction, with sporadic occurrence of macroscopic metastasis to lymph nodes and kidneys. Only neoplastic cells, but not surrounding normal appearing cells (as determined by Laser Capture Microdissection followed by PCR) had undergone Cre-mediated recombination, suggesting that neoplasia was not a product of a "field effect" of *K-ras*^{G12V} mutant cells but in each case a direct consequence of *K-ras*^{G12V} expression in that cell. These models demonstrated that mutant *K-ras*^{G12V} is sufficient to induce hyperplasia with a short latency, and that histologic progression from AAH to frank adenocarcinoma can be shown. More recently, mice harboring a *K-ras*^{G12V} preceded by a "floxed" stop cassette and followed by an internal ribosomal entry site (IRES) and a by a β-Geo cassette (to enable monitoring of the mutant allele expression) have been described (Guerra, Mijimolle et al. 2003). When these

K-ras$^{G12V\text{-}IRES\text{-}BGeo}$ mice were crossed with mice expressing Cre systemically under the CMV promoter, multifocal adenomas and adenocarcinomas developed in the lungs. Additionally, when Cre was activated for a short period post-natally in mice harboring Cre-ERT2 (see methods), temporally controlling the activation of the mutant allele, only lung lesions were observed. It is unclear whether these results indicate that bronchoalveolar cells are uniquely susceptible to mutant *K-ras*, or whether other features of this model account for the lack of penetrance in other organs (for example, a bicistronic allele may alter the levels of either transcript, and protein levels of ras were not directly compared). However, an additional finding merits mention: murine embryonal fibroblasts exposed to Cre recombinase *in-vitro* were immortalized, and no senescence was observed (as is observed when mutant *ras* driven by an exogenous promoter is introduced into fibroblasts (Serrano, Lin et al. 1997)). Thus, both *in-vivo* and *in-vitro*, the single genetic event of a *ras* mutation is sufficient to induce unchecked proliferation.

It is expected that the ability to induce *K-ras* mutations synchronously in the lungs will make these valuable tools for studying the interactions of *K-ras* with other genetic and epigenetic lesions, as well as for prevention and treatment studies. It is unclear to what extent additional mutations are required for a more metastatic phenotype (e.g., a significant proportion of human lung cancer patients die of distant metastasis rather than a local tumor burden), and what these mutations might be. Finally, these models did not enable "withdrawal" of the oncogenic *K-ras*, to assess for ongoing dependence of the tumor on this pathway, a critical question that relates to the therapeutic implication of interfering with oncogenic *ras*. This last question was answered by placing *K-ras*4b^{G12D} in an inducible manner under the control of the reverse tetracycline transactivator in alveolar type II pneumocytes (Fisher, Wellen et al. 2001). When provided doxycycline in drinking water, these mice developed tumors with similar kinetics to the previously described models, with hyperplastic lesions throughout the lungs by day 14 and solid adenomas and adenocarcinomas by 2-3 months, with mutant *K-ras*^{G12D} readily detectable after (but not prior to) doxycycline administration. These lesions were totally dependent on continuous oncogene expression; three days after withdrawal of doxycycline from their diet, the lung surface appeared pitted with reduced cellular density and widespread apoptosis; by one month post-withdrawal no tumors were evident and lung weight was nearly back to baseline. These studies showed that *ras* is required not only for the initiation but also for the continued maintenance of the transformed phenotype (in contrast to the effect of ectopic expression of *myc* cited earlier (D'Cruz, Gunther et al. 2001)). Furthermore, as was expected, *ras* cooperated with *p53* and *Ink4a/ARF* to decrease tumor latency and increase histological and nuclear atypia. However, even in these cases withdrawal of doxycycline (and thus, of activated *K-ras*) still led to complete tumor regression.

The above studies all suggest that expression of oncogenic *K-ras*, alone, is sufficient to drive tumor cell formation. This contradicts the prevailing hypothesis that human tumors derive from a combination of activated oncogenes and deactivated tumor suppressor genes (Kinzler and Vogelstein 1996; Hanahan and Weinberg

2000). The relative role that the normal *K-ras* allele plays in these transgenic systems is unclear, and it is worth mentioning that wild-type *K-ras* can inhibit lung cancer in mice treated with agents that cause mutations in *K-ras* (Zhang, Wang et al. 2001). Furthermore, *K-ras* and *p53* do cooperate *in vivo*, in that tumors form more rapidly and are associated with increased disorganization both at the cellular (nuclear) and histological levels.

4.2 Skin Cancer

The relevance of oncogenic *ras* to skin tumors was initially based on studies of chemical carcinogenesis, and subsequently confirmed by early transgenic models (see above). Directing *ras* expression to superficial skin with either murine keratin 10 or human keratin K1 promoters yielded only limited benign growth (Bailleul, Surani et al. 1990; Greenhalgh, Rothnagel et al. 1993). Similar results were described in an inducible model of *ras* activation, where a conditionally activated *H-ras*^{G12V} was targeted to the skin under the human keratin14 promoter. Embryonic expression was avoided by inducing *H-ras*^{G12V} expression with a cre-Estrogen receptor (cre-ER) fusion protein activated in response to the topical application of tamoxifen (Tarutani, Cai et al. 2003). As noted above, *H-ras*^{G12V} cooperates with *Ink4a/ARF* deletion to form non-metastasizing melanoma (Chin, Pomerantz et al. 1997). Similar results were obtained when *H-ras*^{G12V} was targeted to melanocytes in an inducible tetracycline-responsive model; and similar to *K-ras*^{G12D} lung tumors, complete tumor regression was found upon tetracycline withdrawal, leading to the conclusion that continued tumor survival or "maintenance" remained dependent on the expression of mutated *H-ras*^{G12V} (Chin, Tam et al. 1999).

4.3 Colorectal Cancer

While mutated *ras* is found in up to 50% of colorectal cancer (Fearon and Vogelstein 1990), the majority of transgenic models of systemic activation of *ras* fail to demonstrate colon neoplasia. This has been true whether *ras* was expressed from its own promoter (Saitoh, Kimura et al. 1990), the albumin promoter (Maronpot, Palmiter et al. 1991), or in the spontaneous recombination model described earlier, although potentially pre-malignant aberrant crypt foci (ACF) were occasionally seen (Johnson, Mercer et al. 2001). Furthermore, directing *K-ras*^{G12V} to intestinal villous enterocytes did not produce a neoplastic phenotype (Kim, Roth et al. 1993). In contrast to these results, transgenic expression of human *K-ras4b*^{G12V} directed by a villin regulatory element that is expressed in both mature and immature crypt and villi cells resulted in tumor formation in 80% of animals by the age of 9 months. There were an average of 2.5 tumors per mouse, with a histology ranging from ACF to tubular adenoma to malignant adenocarcinomas; however, no distant metastases were seen. Four separate lines were studied, and they varied in the frequency of tumor formation, which correlated with transgene copy number.

The tumors that formed were found to be diploid, though 3 of 7 tested had alterations in *p53*. The discrepancy between these results may be due to multiple factors. Beyond possible strain, line, and transgenic integration site effects, different promoters will vary in expression levels not only by cell type but also temporally in ontogeny and development as well as in response to the cellular and extra-cellular milieu.

Ras mutations are thought to collaborate with *APC* mutations in human colorectal cancer (Fearon and Vogelstein 1990), but the same may not appear to be true of these transgenic models of *ras* activation. The murine model of *APC* loss, APC^{min}, shares the polyposis phenotype with the cognate human condition, however the distribution of polyps is enriched in the small intestine in mice. However, when either the enterocyte-*ras* (Kim, Roth et al. 1993) or the latent-recombination *K-ras* (Johnson, Mercer et al. 2001) were crossed with APC^{min} mice, no effect on tumor formation or progression was seen, and no spontaneous mutations in *APC* were found in tumors in the villin-*ras* mice (Janssen, el-Marjou et al. 2002). While it always difficult to explain a lack of an observed effect, the bulk of these studies suggest that in murine colorectal epithelia, the ability of *ras* mutations to cause or promote neoplastic changes may be relatively limited to expression in a specific developmental niche (i.e., stem cells) or in cells in which other genetic lesions had occurred (such as loss of *p53*).

4.4 Pancreatic Cancer

In one of the first transgenic mice to be created, normal or activated human *H-ras*$^{\mathrm{G12V}}$ was placed under the rat elastase promoter and enhancer, which is expressed in pancreatic acinar cells beginning day 14 of development. The majority of mice died either in-utero or as newborns from pancreatic tumors; several were mosaics and survived to adulthood before succumbing to pancreatic cancer. Histologically, acinar cells were formed but failed to complete their differentiation in late gestation, leading to dysplasia, large cysts, and an overall increase in size, but without invasion of either adjacent or distant sites. In animals that were transgenic for the proto-oncogenic *H-ras*, anaplasia and histological anomalies were frequently noted when the mice were sacrificed at 12 months, but no tumors developed. These results established the ability of oncogenic *ras*, as a single lesion, to induce early tumors in the pancreas. However, the ras isoform most frequently mutated in human pancreatic cancer is *K-ras*, and carcinomas are proposed to arise in the ductal epithelium rather the acini. Two recent models have addressed *K-ras* mutations in the exocrine pancreas. In one, *K-ras*^{G12D} is targeted to pancreatic acini by the elastase promoter (Grippo, Nowlin et al. 2003), and in the second *K-ras*^{G12V} is targeted specifically to pancreatic ductal cells by the cytokeratin 19 promoter (Brembeck, Schreiber et al. 2003). Both of these mice display non-progressing phenotypes (i.e., "early" ductal hyperplasia) despite clear expression of both the *ras* transgene and elevated RAS-GTP protein levels. Possible explanations include non-physiological levels of *ras* expression, a need for additional cooperating mutations (e.g., *p53, Ink4a/ARF*, etc), or targeting of a different cell type (e.g., expression in a cell with a differentiation state nonpermissive to the initiation of neoplasia). The findings of infiltrating CD4+ T cells in the cytokeratin 19 model are intriguing, and suggest that host immune mechanisms may limit neoplastic growth (Brembeck, Schreiber et al. 2003). However, similar findings have not been described in other models.

Two important caveats hamper the interpretation of these results as accurate models of *KRAS* mutations in human disease. First, active cellular proliferation, as occurs in the pancreas during ontogeny, may be a prerequisite for the oncogenic effect of *ras*; as such, these transgenic models may not reflect the effect of a sporadic mutation occurring later in life, as is assumed to occur in human disease. Second, expression of mutant *ras* driven by an "exogenous" promoter (rather than its own) may not mimic the physiological levels or response to various stimuli that may affect the endogenous gene. For example, as alluded to earlier, murine embryonal fibroblasts that express mutant *ras* from the endogenous promoter do not undergo senescence, a well-studied effect of *ras* when driven by heterologous promoters (such as when introduced into MEFs via retroviral vectors) (Guerra, Mijimolle et al. 2003). We have recently described mice that express an endogenous *K-ras*^{G12D} only in the pancreas (as a result of crossing a mutant mouse strain with a "Floxed" stop cassette in front of *K-ras*^{G12D} with founders that express Cre from the pancreatic-specific promoters *pdx* or *p48*) (Hingorani, Petricoin et al. 2004). These mice develop early pre-invasive neoplasia with complete penetrance, which are histologically identical to the PanIN lesions (Pancreatic Intraepithelial Neoplasia) that are seen in human pancreata. Human PanIN are characterized by different degrees of dysplasia leading to carcinoma *in-situ*, with an increasing number of lesions harboring mutant *KRAS* as the atypia progresses. In these *K-ras*^{G12D} mice, these lesions histologically progress, with spontaneous invasive and metastatic adenocarcinoma arising at a low frequency. This model demonstrated that *K-ras*^{G12D}, as a single genetic event, can induce preinvasive cancer and predisposes to pancreatic adenocarcinoma. An additional important finding in this study has been the identification of a serum proteomic profile in mice that harbor PanIN. While the frequency of PanIN in humans is unknown due to the inability to image or sample pancreatic tissue from asymptomatic patients, once pancreatic cancer occurs, it is nearly uniformly fatal even if detected at the earliest stages. Thus, the potential ability to identify PanIN by screening sera of asymptomatic individuals may allow detection and prevention, as is routinely accomplished with colorectal cancer, for example.

Given that *K-ras*^{G12D} expression alone was sufficient to induce PanIN that progressed spontaneously to pancreatic ductal adenocarcinoma (PDA), it was predictable that the concomitant loss of tumor suppressor genes known to be important in human PDA would accelerate tumor progression in this model. Indeed, the concomitant loss of the Ink4a/ARF locus cooperated with *K-ras*^{G12D} to produce a model of advanced PDA, with all mice succumbing by 12 weeks of age of locally advanced invasive PDA (Aguierre, Bardeesy et al. 2003). Interestingly, only micrometastatic disease was noted, whereas when PanIN spontaneously progresses to PDA in the model of Hingorani et al, gross metastatic disease is evident and the clear cause of death. It is likely that mutations in additional tumor suppressor genes will also cooperate with *K-ras*^{G12D} to produce invasive and metastatic disease, and this remains the current topic of intensive research.

4.5 Astrocytoma

Astrocytomas are the most common primary brain tumor in adults, with the majority of them being histologically high grade and rapidly lethal. The most common genetic abnormality detected in astrocytomas involves the EGFR, with amplification, overexpression, and mutations occurring in up to 50% of tumors. Ras is one of the downstream targets of activated epidermal growth factor receptor (EGFR), and while astrocytomas lack primary *ras* mutations, elevated levels of RAS-GTP have been observed, suggesting that RAS-GTP may be important to their formation or continued growth ((Ding, Roncari et al. 2001) and references therein). Moreover, this elevation is seen without a decrease in negative regulatory molecules such as *ras*-GAP, suggesting that they result from increased activation via RTKs such as EGFR. In mice, astrocytomas are induced when both oncogenic *K-ras*^{G12D} and *akt* are transferred into neural progenitor cells, with a high-grade histology similar to that seen in human disease (Holland, Celestino et al. 2000). Expressing activated *K-ras* or *akt* on their own was not sufficient, but *K-ras* could co-operate *in vivo* with loss of *Ink4a/Arf* (Uhrbom, Dai et al. 2002). However, the tumorigenic effect in astrocytoma formation may be dose-dependent. In a separate investigation, mutated *H-ras* ^{G12V} was placed under the glial fibrillary acidic protein promoter in chimeric transgenic mice. Mice expressing high levels of the transgene all died of astrocytomas within 2 weeks. Those expressing lower doses of the oncogenic *ras* could produce offspring, but ultimately succumbed to similar intermediate to high-grade astrocytomas (which were multifocal in 20% of the animals) between 3 and 4 months of age. Of note, homozygous progeny died between 2 and 7 weeks, recapitulating the gene-dose effect seen in the chimeras (Ding, Roncari et al. 2001). Finally, the tumor cells exhibited a wide range of additional molecular alterations, with aneuploidy and decreased expression of *p16*, *p19*, and *PTEN* (as well as overexpression of *MDM2, CDK4*, and *EGFR*), further mimicking human disease. Interestingly, the transgenic expression of an activated *EGFR* allele alone (using the exon 2-7 deletion mutant commonly found in human astrocytomas) is insufficient to form astrocytomas; however, this mutation did cooperate with the GFAP-*H-ras*^{G12V} transgenic mice described above, to significantly shorten their survival (Ding, Shannon et al. 2003). These studies suggested that activated *ras* is capable of initiating astrocytoma formation; that the effect may be dependent on either levels of expression or alteration in additional pathways; and that this activation may model human disease both in the observed histology and subsequent genetic alterations.

4.6 Ovarian Cancer

The progression of ovarian cancer from early lesions to disseminated metastatic disease has been recently modeled in transgenic mice by utilizing the avian leucosis virus (ALV) retroviral vector/receptor (RCAS/TVA) system to introduce oncogenes into ovarian cells *ex vivo*, followed by intraperitoneal or subcutaneous transplantation (Orsulic, Li et al. 2002). Ovaries from mice that expressed TVA under either the β-actin or keratin 5 promoter (either wild-type or on a *p53*$^{-/-}$ background) were harvested and subjected to *in vitro* infection with ALV packaged with RCAS vectors containing human c-*myc*, activated *K-ras*^{G12D}, myristoylated murine *Akt1*, or various combinations of these genes (the promoter for keratin 5 was chosen as it is only active in ovarian surface epithelium). When transferred back into nude mice, ovarian cells transduced with only one oncogene failed to form tumors, regardless of *p53* status or promoter origin (although *p53*$^{-/-}$ cells infected with any one oncogene led to latent tumor formation when mice were followed for up to 6 months). However, *p53*$^{-/-}$ cells infected with any combination of two or all three of these oncogenes were able to form subcutaneous tumors. These results demonstrate that for ovarian cancer, multiple mutations both in oncogenes and a tumor suppressor gene are required for *in vivo* tumor formation. In addition, the cell of origin of these tumors could be identified. While it could be anticipated that in the keratin-5 TVA derived tumors only the epithelial tumor cells expressed TVA, this was also the case for the β-actin derived tumors, indicating that the histology of origin of ovarian cancer, at least in this model, is the ovarian epithelium. To better mimic human disease, infected cells were also implanted under the ovarian bursa; this led to a palpable solid tumor in the ovary within two weeks and culminated in intraperitoneal spread to the contralateral ovary, intestines and omentum, liver, pancreas, kidney, and diaphragm, with a histology ranging from a poorly differentiated carcinoma to a more organized papillary serous carcinoma.

Human studies have indicated that ovarian cancer cells harbor multiple genetic mutations, including c-*myc*, *K-ras*, *Akt*, *Her-2/neu*, *p53*, and *BRCA1/2*, among others. This model recapitulates the importance of multiple oncogenic lesions. However, it should be noted that tumors were formed by implanting cells that expressed supraphysiological levels of oncogenes and that had undergone short-term *in vitro* propagation and multiple proviral insertions, potentially altering their properties (as well as that of the microenvironment in which they were ultimately able to grow *in vivo*).

4.7 Myeloid Malignancies

N-ras and *K-ras* mutations occur in ~30% of acute myeloid leukemias, myeloproliferative disorders, and myelodysplastic syndromes (Bos 1989). While many genetic lesions associated with leukemia are specific for certain disease phenotypes, oncogenic *N-ras* and *K-ras* are found across the spectrum of myeloid malignancies, suggesting that *ras* may not initiate disease but can cooperate with additional genetic events to promote proliferation; this is consistent with the lack of myeloid disease in

CMV-Cre;K-ras$^{V12IRES\text{-}BGeo}$ mice. However, this notion has been recently challenged by the generation of mice that express an endogenous *K-ras*^{G12D} in hematopoietic cells that respond to interferon (by driving the expression of Cre recombinase from the Interferon-inducible Mx1 promoter). Two groups have simultaneously reported nearly identical results, the occurrence of a myeloid hyperproliferative syndrome that is uniformly fatal by ~4 months of age (Braun, Tuveson et al. 2004; Chan, Kutok et al. 2004). These myeloid cells bear phenotypic markers of mature cells and thus correspond to the adult spectrum of myeloproliferative disorders (as opposed to acute myeloid leukemia).

4.8 Disregulated Signal Transduction Pathways

As has been discussed, the phenotype of mice harboring an endogenous mutation in *K-ras* more closely resembles the cognate human cancer than do the earlier *ras* transgenic models, and MEFs from these mice proliferate rather than senesce. Consequently, the downstream biochemical signaling of an endogenous mutation can be expected to differ from that of transgenic *ras* mutations; moreover, the targets of an endogenous *K-ras* mutation are likely to be more relevant to therapy of human cancer. What these key differences are is the subject of current investigations in our laboratory and others. Of the known ras effector pathways, it is difficult to see activation of either the PI3K/Akt or the Raf/MEK/ERK MAP kinase cascades in *K-ras*^{G12D} MEFs by quantifying the phosphorylated intermediates, although increased transcriptional activity of Elk-1 was evident (Tuveson et al., in press). In MEFs harboring *K-ras*$^{G12V\text{-}IRES\text{-}BGeo}$ paradoxically lower steady-state levels of both phosphorylated Akt and Erk have been seen (Guerra, Mijimolle et al. 2003). Finally, the *K-ras*^{G12D} bone marrow cells described above were not different than *K-ras*^{wt} controls in levels of phosphorylated Akt and MEK. In the latter system, the markedly increased proliferative response to extracellular growth factor stimulation suggests that *K-ras*^{G12D} may function to increase the biochemical sensitivity rather than lock a constant "on" signal as is commonly believed. Much more work will be needed to clarify the biochemical consequences of having an endogenous *K-ras* mutation.

5. RAS TRANSGENIC MODELS IN EVALUATING CARCINOGENICITY AND THERAPY

Tumor prone mice can be instrumental in assessing the impact of environmental factors such as chemicals or diet on tumor formation, progression, or repression. For example, transgenic mice carrying the human *c-Ha-ras* develop pulmonary adenomas when injected intraperitoneally with urethane (Umemura, Kodama et al. 1999), and rats harboring *c-Ha-ras* develop bladder cancer when exposed to nitrosamine compounds in drinking water (Ota, Asamoto et al. 2000). These studies have particular relevance as *ras* mutations are commonly found in lung and bladder cancers. In contrast, MMTV-*v-Ha-ras* have often been employed as a model because of the predictable kinetics of tumor formation. Thus, to cite one example, dietary restriction of fats has been shown to decrease tumor incidence (DeWille, Waddell et al. 1993; Fernandes, Chandrasekar et al. 1995). However, the relevance of these findings to sporadic human breast cancer, which rarely harbor *ras* mutations, is unclear.

Genetically defined mice harboring mutations in oncogenes or tumor suppressor genes can serve as an *in vivo* system to evaluate potential cancer therapies, and several groups have utilized *ras* transgenic mice to test novel therapies that may be relevant to human disease (Nielsen, Gurnani et al. 1992). These studies can be particularly informative when the tumors mimic both the histological site and the molecular pathogenesis of human malignancy. Mice that express oncogenic *ras* have been used most extensively to study the pharmacological inhibition of RAS by inhibitors of farnesyl protein transferase (FTPase). RAS proteins require farnesylation of the C-terminal CAAX motif to localize to the plasma membrane. A number of compounds have been developed that are able to block this action, and can cause regression of established tumors in MMTV-*v-Ha-ras* mice (Kohl, Omer et al. 1995; Liu, Bryant et al. 1998; Mangues, Corral et al. 1998); these results had paved the way for clinical trials of FTPase inhibitors. Their ultimate failure to affect a response in human disease may be due to differences between *K-ras* and *Ha-ras* in requirements for farnesylation, and the presences of other key targets of FTPs, such as Rho B. (reviewed in (Gibbs and Oliff 1997)).

6. SUMMARY AND FUTURE DIRECTIONS

The striking technical advances in manipulating mouse genetics over the past two decades has enabled targeting of specific genetic lesions to chosen tissues in an attempt to recapitulate human cancer; the study of the *ras* oncogene has and continues to be at the forefront of these efforts. Taken together, the results described above suggest that *ras* is critical to early tumor formation, be it adenomas in the lung, papillomas in the skin, aberrant crypt foci in the colon, or PanIN in the pancreas. Furthermore, they also suggest that *ras* is insufficient on its own to produce fully invasive and metastatic disease, and point the way to elucidating the other lesions (i.e., loss of tumor suppressor genes) that are relevant for *ras*-induced cancer *in vivo* in a given histological location. In addition, some studies suggest that the continued presence of an activated *ras* allele is crucial to the viability of the cancer cells, strengthening the role of mutant *ras* as a therapeutic target for human disease. We are now at the point where these murine genetic models faithfully reproduce the histological and pathophysiological features of the cognate human diseases. The modern genetically engineered mice hold the promise of enabling us to study prevention, early detection, and treatment of the molecular pathology caused by a mutation in *ras* in ways that have not been previously possible. Having this tool is especially critical today, when the rapid development of multiple therapeutic approaches requires preclinical models that will rapidly and faithfully predict for responses in patients. In addition, the ability to manipulate mice both genetically and pharmacologically can enable a better understanding of the basic biological questions of where and how *ras* mutations transform cells. Is there a cancer stem cell that is the target? Why are *ras* mutations found in some histologies but not in others? Are mutations in *ras* restricted to a certain differentiation state of the tissue, and if so, why? What is the role of secondary genetic lesions? Of the extracellular milieu? These questions and many others will be the focus of future work. However, several important limitations of these murine models should be acknowledged, both theoretical and practical. Spontaneously arising lesions in mice may take many months to manifest, and the "turn-around" time for any given set of experiments may take several years. While progress is being made in non-invasive imaging of mice to assess disease burden, current methods of assessing histology still rely on sacrificing mice, enabling only a single "end-point" of efficacy. These factors, coupled with the cost of caring for the animals, make "high-throughput" screening of candidate drugs difficult with current technology. Finally, although it appears that orthologic genetic lesions can lead to similar disease in mice and humans, we must acknowledge that important genetic differences (such as the role of telomerase and p16, for example) remain between rodent cancer and human disease; these may inherently limit the ability of genetically manipulated mice to predict response to therapies in humans. Despite these caveats, the striking histological similarity to human pre-invasive and invasive cancer caused by introducing the exact *ras* genetic lesions found in human disease leaves us optimistic that the study of genetically engineered mice will, in the foreseeable future, translate into better prevention, detection, and treatment of our patients.

REFERENCES

Aguirre, A.J., Bardeesy, N. et al. (2003) Activated Kras and Ink4a/Arf deficiency cooperate to produce metastatic pancreatic ductal adenocarcinoma. Genes Dev. *17* (24), 3112-3126.

Anton, M. and Graham, F.L. (1995) Site-specific recombination mediated by an adenovirus vector expressing the Cre recombinase protein: a molecular switch for control of gene expression. J. Virol. *69* (8), 4600-4606.

Babinet, C. (2000) Transgenic mice: an irreplaceable tool for the study of mammalian development and biology. J. Am. Soc. Nephrol. *11* (Suppl. 16), S88-S94.

Bailleul, B., Surani, M.A. et al. (1990) Skin hyperkeratosis and papilloma formation in transgenic mice expressing a ras oncogene from a suprabasal keratin promoter. Cell *62* (4), 697-708.

Balmain, A., Ramsden, M. et al. (1984) Activation of the mouse cellular Harvey-ras gene in chemically induced benign skin papillomas. Nature *307* (5952), 658-660.

Bos, J.L. (1989) ras oncogenes in human cancer: a review. Cancer Res. *49* (17), 4682-4689.

Braun, B.S., Tuveson, D.A. et al. (2004) Somatic activation of oncogenic Kras in hematopoietic cells initiates a rapidly fatal myeloproliferative disorder. Proc. Natl. Acad. Sci. U.S.A. *101* (2), 597-602.

Brembeck, F.H., Schreiber, F.S. et al. (2003) The mutant K-ras oncogene causes pancreatic periductal lymphocytic infiltration and gastric mucous neck cell hyperplasia in transgenic mice. Cancer Res. *63* (9), 2005-2009.

Chan, I.T., Kutok, J.L. et al. (2004) Conditional expression of oncogenic K-ras from its endogenous promoter induces a myeloproliferative disease. J. Clin. Invest. *113* (4), 528-538.

Chin, L., Pomerantz, J. et al. (1997) Cooperative effects of INK4a and ras in melanoma susceptibility in vivo. Genes Dev. *11* (21), 2822-2834.

Chin, L., Tam, A. et al. (1999) Essential role for oncogenic Ras in tumour maintenance. Nature *400* (6743), 468-472.

Chrast-Balz, J. and Hooft van Huijsduijnen, R. (1996) Bi-directional gene switching with the tetracycline repressor and a novel tetracycline antagonist. Nucleic Acids Res. *24* (15), 2900-2904.

D'Cruz, C.M., Gunther, E.J. et al. (2001) c-MYC induces mammary tumorigenesis by means of a preferred pathway involving spontaneous Kras2 mutations. Nat. Med. *7* (2), 235-239.

DeWille, J.W., Waddell, K. et al. (1993) Dietary fat promotes mammary tumorigenesis in MMTV/v-Ha-ras transgenic mice. Cancer Lett. *69* (1), 59-66.

Ding, H., Roncari, L. et al. (2001) Astrocyte-specific expression of activated p21-ras results in malignant astrocytoma formation in a transgenic mouse model of human gliomas. Cancer Res. *61* (9), 3826-3836.

Ding, H., Shannon, P. et al. (2003) Oligodendrogliomas result from the expression of an activated mutant epidermal growth factor receptor in a RAS transgenic mouse astrocytoma model. Cancer Res. *63* (5), 1106-1113.

Esteban, L.M., Vicario-Abejon, C. et al. (2001) Targeted genomic disruption of H-ras and N-ras, individually or in combination, reveals the dispensability of both loci for mouse growth and development. Mol. Cell. Biol. *21* (5), 1444-1452.

Fearon, E.R. and Vogelstein, B. (1990) A genetic model for colorectal tumorigenesis. Cell *61* (5), 759-767.

Fernandes, G., Chandrasekar, B. et al. (1995) Dietary lipids and calorie restriction affect mammary tumor incidence and gene expression in mouse mammary tumor virus/v-Ha-ras transgenic mice. Proc. Natl. Acad. Sci. U.S.A. *92* (14), 6494-6498.

Fisher, G.H., Wellen, S.L. et al. (2001) Induction and apoptotic regression of lung adenocarcinomas by regulation of a K-Ras transgene in the presence and absence of tumor suppressor genes. Genes Dev. *15* (24), 3249-3262.

Furth, P.A., St Onge, L. et al. (1994) Temporal control of gene expression in transgenic mice by a tetracycline-responsive promoter. Proc. Natl. Acad. Sci. U.S.A. *91* (20), 9302-9306.

Garcia, S.B., Park, H.S. et al. (1999) Field cancerization, clonality, and epithelial stem cells: the spread of mutated clones in epithelial sheets. J. Pathol. *187* (1), 61-81.

Gibbs, J.B. and Oliff, A. (1997) The potential of farnesyltransferase inhibitors as cancer chemotherapeutics. Annu. Rev. Pharmacol. Toxicol. *37*, 143-166.

Gossen, M. and Bujard, H. (1992) Tight control of gene expression in mammalian cells by tetracycline-responsive promoters. Proc. Natl. Acad. Sci. U.S.A. *89* (12), 5547-5551.

Gossen, M. and Bujard, H. (1993) Anhydrotetracycline, a novel effector for tetracycline controlled gene expression systems in eukaryotic cells. Nucleic Acids Res. *21* (18), 4411-4412.

Greenhalgh, D.A., Rothnagel, J.A. et al. (1993) Induction of epidermal hyperplasia, hyperkeratosis, and papillomas in transgenic mice by a targeted v-Ha-ras oncogene. Mol. Carcinog. *7* (2), 99-110.

Grippo, P.J., Nowlin, P.S. et al. (2003) Preinvasive pancreatic neoplasia of ductal phenotype induced by acinar cell targeting of mutant Kras in transgenic mice. Cancer Res. *63* (9), 2016-2019.

Guerra, C., Mijimolle, N. et al. (2003) Tumor induction by an endogenous K-ras oncogene is highly dependent on cellular context. Cancer Cell *4* (2), 111-120.

Hanahan, D. (1988) Dissecting multistep tumorigenesis in transgenic mice. Annu. Rev. Genet. *22*, 479-519.

Hanahan, D. and Weinberg, R.A. (2000) The hallmarks of cancer. Cell *100* (1), 57-70.

Hancock, J.F., Magee, A.I. et al. (1989) All ras proteins are polyisoprenylated but only some are palmitoylated. Cell *57* (7), 1167-1177.

Hingorani, S.R., Petricoin, E.F. et al. (2004) Preinvasive and invasive ductal pancreatic cancer and its early detection in the mouse. Cancer Cell *5* (1), 103.

Holland, E.C., Celestino, J. et al. (2000) Combined activation of Ras and Akt in neural progenitors induces glioblastoma formation in mice. Nat. Genet. *25* (1), 55-57.

Hundley, J.E., Koester, S.K. et al. (1997) Increased tumor proliferation and genomic instability without decreased apoptosis in MMTV-ras mice deficient in p53. Mol. Cell. Biol. *17* (2), 723-731.

Jackson, E.L., Willis, N. et al. (2001) Analysis of lung tumor initiation and progression using conditional expression of oncogenic K-ras. Genes Dev. *15* (24), 3243-3248.

Janssen, K.P., el-Marjou, F. et al. (2002) Targeted expression of oncogenic K-ras in intestinal epithelium causes spontaneous tumorigenesis in mice. Gastroenterology *123* (2), 492-504.

Jo, D., Nashabi, A. et al. (2001) Epigenetic regulation of gene structure and function with a cell-permeable Cre recombinase [Comment]. Nat. Biotechnol. *19* (10), 929-933.

Johnson, L., Mercer, K. et al. (2001) Somatic activation of the K-ras oncogene causes early onset lung cancer in mice. Nature *410* (6832), 1111-1116.

Kerr, K.M. (2001) Pulmonary preinvasive neoplasia. J. Clin. Pathol. *54* (4), 257-271.

Khosravi-Far, R. and Der, C.J. (1994) The Ras signal transduction pathway. Cancer Metastasis Rev. *13* (1), 67-89.

Kilby, N.J., Snaith, M.R. et al. (1993) Site-specific recombinases: tools for genome engineering. Trends Genet. *9* (12), 413-421.

Kim, S.H., Roth, K.A. et al. (1993) Transgenic mouse models that explore the multistep hypothesis of intestinal neoplasia. J. Cell Biol. *123* (4), 877-893.

Kinzler, K.W. and Vogelstein, B. (1996) Lessons from hereditary colorectal cancer. Cell *87* (2), 159-170.

Kistner, A., Gossen, M. et al. (1996) Doxycycline-mediated quantitative and tissue-specific control of gene expression in transgenic mice. Proc. Natl. Acad. Sci. U.S.A. *93* (20), 10933-10938.

Knudson, A.G. (1993) Antioncogenes and human cancer. Proc. Natl. Acad. Sci. U.S.A. *90* (23), 10914-10921.

Kohl, N.E., Omer, C.A. et al. (1995) Inhibition of farnesyltransferase induces regression of mammary and salivary carcinomas in ras transgenic mice. Nat. Med. *1* (8), 792-797.

Kucera, G.T., Bortner, D.M. et al. (1996) Overexpression of an Agouti cDNA in the skin of transgenic mice recapitulates dominant coat color phenotypes of spontaneous mutants. Dev. Biol. *173* (1), 162-173.

Lakso, M., Sauer, B. et al. (1992) Targeted oncogene activation by site-specific recombination in transgenic mice. Proc. Natl. Acad. Sci. U.S.A. *89* (14), 6232-6236.

Land, H., Parada, L.F. et al. (1983) Tumorigenic conversion of primary embryo fibroblasts requires at least two cooperating oncogenes. Nature *304* (5927), 596-602.

Le, Y. and Sauer, B. (2001) Conditional gene knockout using Cre recombinase. Mol. Biotechnol. *17* (3), 269-275.

Liu, M., Bryant, M.S. et al. (1998) Antitumor activity of SCH 66336, an orally bioavailable tricyclic inhibitor of farnesyl protein transferase, in human tumor xenograft models and wap-ras transgenic mice. Cancer Res. *58* (21), 4947-4956.

Magee, T. and Marshall, C. (1999) New insights into the interaction of Ras with the plasma membrane. Cell *98* (1), 9-12.

Mangues, R., Corral, T. et al. (1998) Antitumor effect of a farnesyl protein transferase inhibitor in mammary and lymphoid tumors overexpressing N-ras in transgenic mice. Cancer Res. *58* (6), 1253-1259.

Maronpot, R.R., Palmiter, R.D. et al. (1991) Pulmonary carcinogenesis in transgenic mice. Exp. Lung Res. *17* (2), 305-320.

Metzger, D. and Chambon, P. (2001) Site- and time-specific gene targeting in the mouse. Methods *24* (1), 71-80.

Meuwissen, R., Linn, S.C. et al. (2001) Mouse model for lung tumorigenesis through Cre/lox controlled sporadic activation of the K-Ras oncogene. Oncogene *20* (45), 6551-6558.

Nielsen, L.L., Gurnani, M. et al. (1992) Evaluation of the wap-ras transgenic mouse as a model system for testing anticancer drugs. Cancer Res. *52* (13), 3733-3738.

Orsulic, S., Li, Y. et al. (2002) Induction of ovarian cancer by defined multiple genetic changes in a mouse model system. Cancer Cell *1* (1), 53-62.

Ota, T., Asamoto, M. et al. (2000) Transgenic rats carrying copies of the human c-Ha-ras proto-oncogene exhibit enhanced susceptibility to N-butyl-N-(4-hydroxybutyl) nitrosamine bladder carcinogenesis. Carcinogenesis *21* (7), 1391-1396.

Quaife, C.J., Pinkert, C.A. et al. (1987) Pancreatic neoplasia induced by ras expression in acinar cells of transgenic mice. Cell *48* (6), 1023-1034.

Sadowski, P.D. (1995) The Flp recombinase of the 2-microns plasmid of Saccharomyces cerevisiae. Prog. Nucleic Acid Res. Mol. Biol. *51*, 53-91.

Saitoh, A., Kimura, M. et al. (1990) Most tumors in transgenic mice with human c-Ha-ras gene contained somatically activated transgenes. Oncogene *5* (8), 1195-1200.

Sauer, B. (1998) Inducible gene targeting in mice using the Cre/lox system. Methods *14* (4), 381-392.

Serrano, M., Lin, A.W. et al. (1997) Oncogenic ras provokes premature cell senescence associated with accumulation of p53 and p16INK4a. Cell *88* (5), 593-602.

Sinn, E., Muller, W. et al. (1987) Coexpression of MMTV/v-Ha-ras and MMTV/c-myc genes in transgenic mice: synergistic action of oncogenes in vivo. Cell *49* (4), 465-475.

Suda, Y., Aizawa, S. et al. (1987) Driven by the same Ig enhancer and SV40 T promoter ras induced lung adenomatous tumors, myc induced pre-B cell lymphomas and SV40 large T gene a variety of tumors in transgenic mice. EMBO J. *6* (13), 4055-4065.

Tarutani, M., Cai, T. et al. (2003) Inducible activation of Ras and Raf in adult epidermis. Cancer Res. *63* (2), 319-323.

Thomas, K.R. and Capecchi, M.R. (1987) Site-directed mutagenesis by gene targeting in mouse embryo-derived stem cells. Cell *51* (3), 503-512.

Tuveson, D.A. and Jacks, T. (1999) Modeling human lung cancer in mice: similarities and shortcomings. Oncogene *18* (38), 5318-5324.

Uhrbom, L., Dai, C. et al. (2002) Ink4a-Arf loss cooperates with KRas activation in astrocytes and neural progenitors to generate glioblastomas of various morphologies depending on activated Akt. Cancer Res. *62* (19), 5551-5558.

Umemura, T., Kodama, Y. et al. (1999) Susceptibility to urethane carcinogenesis of transgenic mice carrying a human prototype c-Ha-ras gene (rasH2 mice) and its modification by butylhydroxytoluene. Cancer Lett. *145* (1–2), 101-106.

Wang, Y., Krushel, L.A. et al. (1996) Targeted DNA recombination in vivo using an adenovirus carrying the cre recombinase gene. Proc. Natl. Acad. Sci. U.S.A. *93* (9), 3932-3936.

Zhang, Z., Wang, Y. et al. (2001) Wildtype Kras2 can inhibit lung carcinogenesis in mice. Nat. Genet. *29* (1), 25-33.

CHAPTER 13

RAS FAMILY PROTEINS

Rap, Ral, R-Ras, Rheb, Rit/Rin, and others

JEAN DE GUNZBURG

Inserm U528, Institut Curie-Section de Recherche, 26 rue d'Ulm, 75005, Paris, France

Abstract: The Ras branch of the Ras superfamily of GTPases comprises 20 proteins that can be classified in 7 subgroups (Ras, Rap, Ral, R-Ras, Rit/Rin, Rheb, ARRHI/Di-Ras) according to sequence homology. Most of them act as molecular switches that alternate between an inactive GDP-bound and an active GTP-bound conformation, except for ARHI/Di-Ras that remain complexed to GTP. Each of these proteins may be activated by several GEFs (guanine nucleotide exchange factors) in response to various extracellular stimuli, and interacts with several downstream effectors, many of which have been characterized in great detail. Proteins of the Ras family are involved in a great array of biological functions such as the control of cellular proliferation, differentiation, integrin-dependent adhesion, cell-cell junctions, motility and intracellular trafficking. This chapter attempts to review our current understanding of the biology of the proteins, other than H-, K- and N-Ras, of the Ras family

Keywords: Rap, Ral, R-Ras, Rit, Rin, Rheb, ARHI, Di-Ras

1. INTRODUCTION

The Ras branch of the Ras superfamily presently comprises 20 proteins, that belong to various subgroups: Ras (H-Ras, K-Ras with two alternatively spliced variants expressing the A or B fourth exon, and N-Ras), Rap (with the Rap1 A and B proteins, and Rap2 A, B and C proteins), Ral (A and B), R-Ras (comprising the R-Ras, R-Ras2/TC-21 and R-Ras3/M-Ras proteins), Rit/Rin, Rheb, Di-Ras (1 and 2) and ARHI proteins. Except for the K-Ras 4A and 4B variants that are generated by alternative splicing, all of the other proteins are the products of distinct genes.

An alignment of the sequences of these proteins (Fig. 1), as well as a phylogenetic tree (Fig. 2) presented below show that they all share the common hallmarks of GDP/GTP binding proteins with closely related regions involved in interactions with the phosphate moieties (PM1-PM3) and guanine base (G1-G3). With the exception

C. Der (ed.), RAS Family GTPases, 295–339.

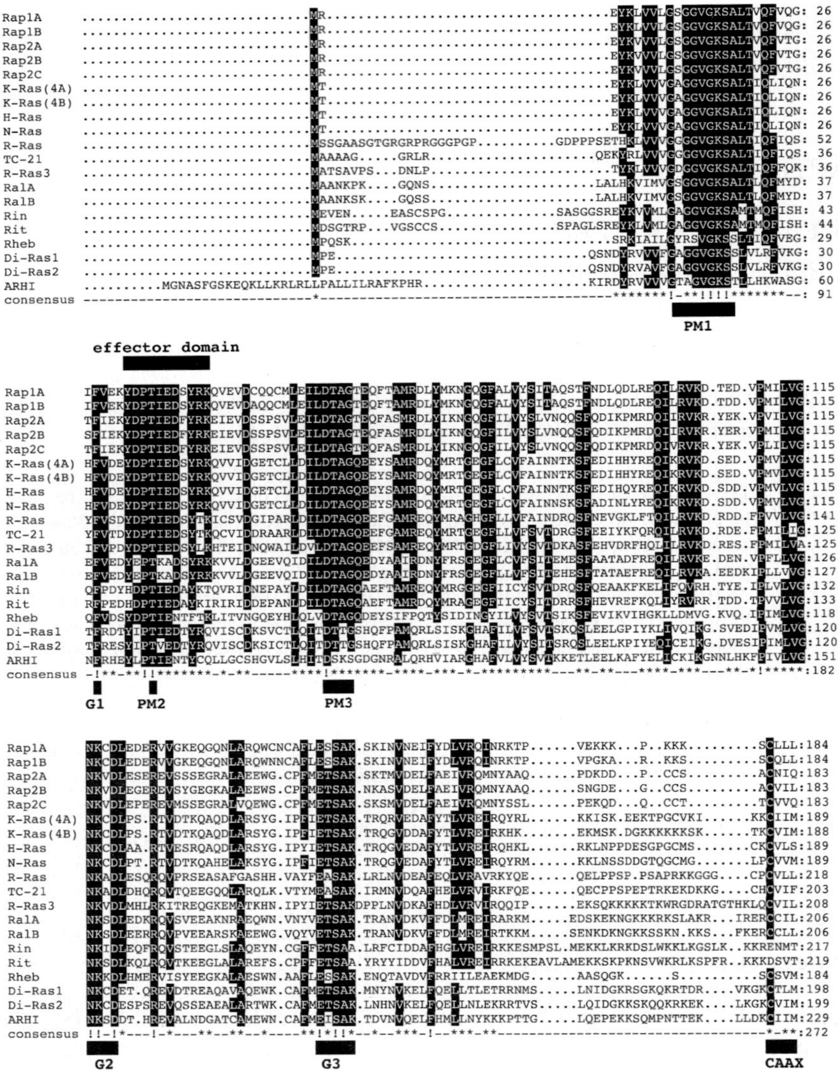

Figure 1. Sequence alignment of Ras family proteins

of Rit and Rin, they all present a C-terminal CAAX box, and are predicted to be farnesylated (the four Ras proteins, Rap2 A and C, R-Ras2/TC-21, Rheb, Di-Ras 1 and 2, ARHI) or geranylgeranylated (Rap1 A and B, Rap2B, Ral A and B, R-Ras and R-Ras3/M-Ras). They form a branch that is easily distinguishable, on the basis of sequence identity (in particular in the switch 1 region also termed effector domain, as well as specific features), from the Rho branch (exhibiting a T instead

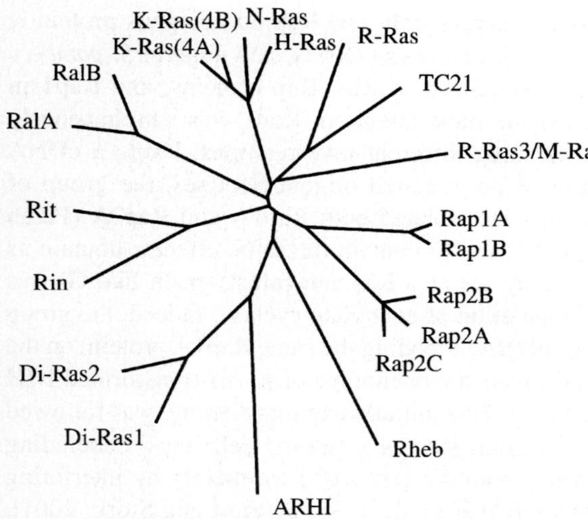

Figure 2. Phylogenetic tree of Ras family proteins

of S at position 17, a W preceding the conserved switch 2 DTAGQE sequence, and a 13 residue insert starting at residue 122 that is involved in binding Rho-GDIs), the Rab branch (these proteins also contain a W preceding the conserved switch 2 DTAGQE sequence, and most of them exhibit a C-terminal CC or CxC sequence instead of the CAAX box characteristic of proteins from the Ras and Rho branches), Ran, the Rad/Gem/Rem branch (that contains long N- and C-terminal extensions, is devoid of the residue equivalent to T35 in Ras, a modified switch 2 region, and no C-terminal prenylation sequences), and the Arf branch (that presents many distinctive features).

As is described lower in the text, these proteins exert a large variety of physiological functions, only some of which have been to date clearly demonstrated. This chapter aims to sum up a vast body of information, mainly gathered in the last ten years and sometimes controversial, concerning the biology of these proteins with the exception of H-, K- and N-Ras to which several chapters of this book are devoted, ranging from biochemistry and structure, to interacting partners and cellular functions.

2. RAP PROTEINS

The Rap group of GTPases comprises two subgroups of proteins, Rap1 and Rap2, that altogether share close to 60% sequence identity. The Rap1 group contains the 95% identical Rap1A and Rap1B proteins. The Rap2 group contains the Rap2A, Rap2B and Rap2B proteins that share 85-90% sequence identity. These two groups are found in all multicellular eukaryotes, with a single member of each represented in *C. elegans* and *D. melanogaster*, whereas only one Rap homologous protein is found in *Dictyostelium* (Dd Rap) and *S. cerevisiae* (Rsr1), and none in *S. pombe*.

Amongst the proteins most closely related to Ras, Rap proteins, and Rap1 in particular, have undoubtedly raised the most attention. Rap1 was simultaneously discovered by two laboratories, pursuing different investigations. Using a cDNA cloning strategy based on the use of degenerated oligonucleotides, the group of Pierre Chardin in Armand Tavitian's lab isolated both Rap1A and Rap2A (Pizon et al., 1988); the fact that Rap1 was found to contain the same effector domain as Ras prompted speculations that it may act as a Ras antagonist, much like Gi was thought to antagonize Gs for the regulation of adenylate cyclase. Indeed, the group of Makoto Noda cloned the K-*rev*1 cDNA, encoding the same Rap1A protein, on the basis that its overexpression could revert the phenotype of K-*ras*-transformed NIH 3T3 fibroblasts (Kitayama et al., 1989). This initially exciting finding, was followed by reports, mainly based on transfection studies in model cell lines, concluding that Rap1 counteracted the growth-promoting effect of Ras, mainly by interfering with activation of the ERK pathway (Cook et al., 1993; Schmitt and Stork, 2001). This idea has now been seriously revisited, and though it may still be valid in certain physiological situations, other functions, such as the control of inside-out integrin signaling and establishment/maintenance of adherens junctions, have been attributed to Rap1. No clear role has to date been attributed to Rap2.

Since several recent reviews have addressed various aspects of Rap signaling in great detail (Bos, 1998; Bos, 2003; Bos et al., 2003; Bos et al., 2001; Caron, 2003; Stork, 2003; Stork and Schmitt, 2002), only the most salient features of Rap's biology will be addressed below.

2.1 Biochemistry and GAPs

One of the hallmarks of Rap proteins is that Glutamine 61 of the DTAGQE sequence in the switch 2 region, which is conserved in proteins of the Ras, Rho and Rab branches of the Ras superfamily, as well as in Ran, is replaced by a Threonine (Pizon et al., 1988). This residue plays an important role in the catalytic mechanism of the GTPase reaction (Scheffzek et al., 1997); despite this substitution, Rap proteins retain an intrinsic GTPase activity, as well as other GDP and GTP binding characteristics, comparable to those of Ras (Lerosey et al., 1991a; Quilliam et al., 1990).

A protein, with GAP activity specific to Rap was purified as a cytosolic protein, with a higher molecular weight membrane-associated isoform (Nice et al., 1992; Polakis et al., 1991), and the cDNA corresponding to the soluble form was cloned (Rubinfeld et al., 1991). RapGAP is unable to stimulate the GTPase activity of Ras, as p120RasGAP is unable to stimulate the GTPase activity of Rap, although the two latter proteins were shown to interact; several studies showed that this specificity of GAPs for their cognate GTPase was attributable to residues in the switch 2 region and its C-terminal vicinity extending until position 65 (Frech et al., 1990; Hart and Marshall, 1990; Holden et al., 1991; Maruta et al., 1991; Zhang et al., 1991). The catalytic domain of RapGAP, that shares no homology with that of RasGAPs such as p120 RasGAP or Neurofibromin, is also found in other proteins with GAP activity towards Rap: SPA-1, whose expression is restricted to lymphoid tissues (Kurachi et al., 1997) and the related protein SPAL/SPAR/E6TP1 found in neurons (Gao et al., 1999; Pak et al., 2001; Roy et al., 2002). This latter protein is a large scaffolding molecule consisting in two actin regulatory domains termed Act 1 and 2, a GAP domain specific for Rap, a PDZ domain, and a C-terminal region that binds the guanylate kinase domain of the post-synaptic density protein PSD-95 (Pak et al., 2001). Through its Act2 domain, SPAR binds a Polo-related kinase, SNK, that phosphorylates SPAR and targets it to degradation via the ubiquitin-dependent proteasome pathway; the induction of SNK induced in brain via electric stimulation or drug-induced seizures results via this SPAR-dependent pathway in the decrease of post-synaptic structures, and hence is proposed to regulate the morphology and activity of synapses (Pak and Sheng, 2003). Whether the degradation of SPAR results in a locally enhanced Rap activity that could downmodulate receptor trafficking in those postsynaptic structures (Zhu et al., 2002), or on the contrary Rap could affect the stability of SPAR and/or its interaction with SNK has not yet been assessed.

Homology searches in sequence databases led to the identification of several other proteins containing a RapGAP domain, however their functional relevance has not been investigated (Bernards, 2003).

An in depth biochemical study recently showed that RapGAP activates the GTPase activity of Rap by a mechanism distinct from that of GAPs for Ras (Ahmadian et al., 1997), Rho (Rittinger et al., 1997) and Rab (Albert et al., 1999) that stabilize Glutamine 61 and contribute a critical Arginine (in a structure referred to as the Arginine finger) to the active site in order to complement the catalytic machinery (Vetter and Wittinghofer, 2001). Hence Rap proteins, despite their high degree of similarity with Ras, exhibit a unique catalytic mechanism for their GAP-stimulated GTPase activity which may ensure that the negative regulation of their signaling pathways is controlled with great specificity, and in particular no interference with the negative regulation of Ras.

2.2 Activators and Upstream Pathways

The development of an assay to measure the activity level of Rap in cell lysates (i.e. the relative amount of Rap-GTP complexes) constituted a critical step towards the understanding of Rap signaling pathways. The laboratory of J. L. Bos used the property of the Rap binding domain of RalGDS to bind Rap1-GTP with nanomolar affinity, and high specificity as compared with Rap-GDP, to trap active Rap-GTP complexes, via a GST-fusion protein, on Glutathione-covered beads; the detection of Rap is then ensured by Western blotting with antibodies specific of Rap1 or Rap2 (Franke et al., 1997). Assays of this type were subsequently developed for many other GTPases such as Ras (de Rooij and Bos, 1997), Ral (Bauer et al., 1999b), Rho/Rac/Cdc42 (Ren and Schwartz, 2000; Sander et al., 1999), and Arf6

(Niedergang et al., 2003); they have advantageously replaced the cumbersome traditional assay that required metabolic labeling of cellular nucleotide pools with ^{32}P-orthophosphate, followed by immunoprecipitation of GTPases and analysis of associated nucleotides by thin layer chromatography.

Only a short description of Rap activators will be given here since it constitutes the subject of another chapter in this book.

2.2.1 Smg-GDS

SmgGDS was the first protein to be biochemically identified which promoted nucleotide exchange on Rap1 (called Smg p21 by the laboratory of Yoshimi Takai, hence SmgGDS for the activity that stimulates the dissociation of GDP) (Yamamoto et al., 1990). SmgGDS was found to act on several GTPases with polybasic stretches at their C-termini, i.e. K-Ras, Rap1, RhoA and Rac1, and to require their post-translational processing (Hiroyoshi et al., 1991; Mizuno et al., 1991). Cloning of its cDNA revealed that this protein contains two Armadillo repeats similar to those found in β-catenin, and presented no sequence homology with known GEFs for GTPases or the Ras superfamily (Kaibuchi et al., 1991). In addition to its GEF activity, SmgGDS is able to extract Rap1 from membranes (similarly to the action of Rho-GDIs on Rho family proteins), and the activities of SmgGDS on Rap1 (both membrane extraction and nucleotide exchange) are enhanced by the prior phosphorylation of Rap1 by cAMP-dependent protein kinase (PKA) on a serine residue of its C-terminal domain (Ser179 of Rap1B) (Hata et al., 1991; Itoh et al., 1991). Despite more recent studies, the physiological role of SmgGDS remains elusive (Shimizu et al., 1996; Strassheim et al., 2000; Vikis et al., 2002).

2.2.2 C3G

C3G was originally identified as a protein associated with the SH3 domain of the Crk adaptor (Tanaka et al., 1994). It contains a region homologous to the catalytic domain of the RasGEF Cdc25 (Cdc25 homology region), and was shown to act *in vitro* as well as in cells specifically on Rap1 (Gotoh et al., 1995; van den Berghe et al., 1997). Expression of C3G is essential for mouse early development since homozygous C3G-deficient mice die before E 7.5; derived MEFs display defects in adhesion and spreading that are ablated by the expression of active Rap or other RapGEFs (Ohba et al., 2001).

C3G couples Rap1 to many receptors via Crk family adaptors, and its activity requires its phosphorylation on Tyrosine 504, induced by Crk, via a yet unidentified kinase (Ichiba et al., 1999a). Several different molecular cascades have been described to link growth factor and hormone receptors to C3G. For instance, Insulin and EGF activate Rap1 in CHO cells expressing the corresponding receptors via the regulation of a CrkII-C3G complex (Okada and Pessin, 1997). Similarly, bombesin activates Rap1 via Crk/CrkL-C3G complexes in Swiss 3T3 and primary mouse embryo fibroblasts (Posern et al., 2000), and HGF activates Rap1 in HEK 293 cells via the association of Gab1 with CrkL that recruits C3G (Sakkab et al., 2000). The treatment of human NB-4 acute promyelocytic leukemia cells with IFN-γ leads to

Rap1 activation via the phosphorylation of c-Cbl that recruits a CrkL-C3G complex (Alsayed et al., 2000). Growth hormone also activates Rap in NIH 3T3 cells via the phosphorylation of C3G complexed to CrkII, which requires the combined activity of JAK2 and c-Src kinases (Ling et al., 2003). Conversely, the E3 ubiquitin ligases Cbl and Cbl-b negatively regulate Rap1 activation by promoting the ubiquitination and subsequent degradation of CrkL (Shao et al., 2003; Zhang et al., 2003a).

In PC12 cells, EGF transiently activates Rap1 by assembling a short-lived Crk-C3G complex on the receptor itself, whereas NGF leads to a prolonged Rap1 activation via the phosphorylation of FRS2 that scaffolds the assembly of a stable Crk-C3G-Rap1 complex (Kao et al., 2001). Such a persistent Rap1 activation following the stimulation of PC12 cells with NGF has also been described in endosomal compartments via the induction of a long lived TrkA/Gab2/Shp2/CrkL/C3G complex (Wu et al., 2001). Several lines of evidence suggest that the activation of Rap1 via pathways involving C3G indeed requires endocytosis of the activated receptor, such as TrkA in PC12 cells (York et al., 2000), or the EFG receptor in COS cells (Mochizuki et al., 2001; Ohba et al., 2003).

2.2.3 cAMP-activated GEFs (Epac/cAMP-GEF family)

Early work had shown that Rap1 (both A and B isoforms) could be phosphorylated by PKA, and that it resulted in translocation of the protein from a membrane-bound compartment to the cytosol (Kawata et al., 1989; Lapetina et al., 1989; Lerosey et al., 1991b). It was thereafter shown that the treatment of cells with agonists elevating intracellular cAMP levels resulted in the activation of Rap1B (Altschuler et al., 1995).

Two groups simultaneously identified a family of RapGEFs, called Epacs or cAMP-GEFs, that are directly activated by cAMP (de Rooij et al., 1998; Kawasaki et al., 1998a). These proteins contain one (Epac/cAMP-GEF I) or two (Epac2/cAMP-GEF II) cAMP binding domains that present a high degree of similarity with those of the regulatory subunits of PKA. In the absence of cAMP, the cAMP-binding domains bind with high affinity to the GEF moiety and repress its catalytic activity; the binding of cAMP releases this inhibition, hence leading to the activation of the GEF (de Rooij et al., 2000). A third member of this family, called Repac or GFR, is devoid of cAMP binding sequences and thus thought to be constitutively active (de Rooij et al., 2000; Ichiba et al., 1999b). These factors are specific for Rap (they activate both Rap1 and Rap2), and have no effect on Rap-related GTPases such as Ras or R-Ras. The structure of their cAMP-binding domains was established (Rehmann et al., 2003a). Its comparison with those of the regulatory subunits of PKA enabled to propose a mechanism for the cAMP-dependent activation of Epac (Rehmann et al., 2003a; Rehmann et al., 2003b), and to rationally design analogs of cAMP that selectively activate Epac, but not PKA (Christensen et al., 2003; Enserink et al., 2002) thus opening the way for a pharmacological investigation of the cAMP-Rap1 pathway (Maillet et al., 2003). In addition to cAMP-binding and GEF catalytic domains, Epacs contain a DEP domain involved in membrane localization of the protein (de Rooij et al., 2000), and an RA domain (for Ras Association, see below) within the C-terminal catalytic domain that should be able to bind activated proteins of the Ras, R-Ras or Rap groups. In fact, Repac, also called MR-GEF, specifically binds activated M-Ras/R-Ras3 through its RA, and this interaction has been shown to inhibit MR-GEF's ability to activate Rap1 (Rebhun et al., 2000a).

The discovery of Epacs/cAMP-GEFs constituted a major breakthrough in showing that, contrarily to what had been thought for many years, cAMP does not only act in cells via the control of PKA or ion channels, but also via the Ras-related Rap GTPases (Bos, 2003). It should be noted that Rap1 may also be indirectly activated in cells by cAMP via the action of PKA: such a cascade has recently been described in NIH 3T3 fibroblasts where cAMP activates PKA to phosphorylate c-Src on Serine 17, leading to its ability to phosphorylate c-Cbl on tyrosine thus enabling the recruitment of the Crk-C3G complex and subsequent activation of Rap1 (Schmitt and Stork, 2002b).

2.2.4 CalDAG-GEFs

Another family of GEFs, carrying binding domains for Ca^{2+} ions (EF Hands) and diacylglycerol, was discovered simultaneously with Epacs. It is composed of 4 proteins called GRP/CalDAG-GEF that exhibit distinct specificities vis à vis Ras family GTPases: whereas Ras-GRP1/CalDAG-GEF II acts on Ras (Ebinu et al., 1998; Kawasaki et al., 1998b), CalDAG-GEF I is specific of Rap (Rap1 and Rap2) (Kawasaki et al., 1998b), GRP2 acts on both Ras and Rap (Clyde-Smith et al., 2000), and CalDAG-GEF III activates Ras, Rap and R-Ras (Yamashita et al., 2000). So far, their contribution to the activation of Rap1 in response to elevations of intracellular Ca^{2+} and/or diacylglycerol has not yet been demonstrated; it is however likely, especially in hematopoietic cells such as platelets (Franke et al., 1997; Franke et al., 2000), neutrophils (M'Rabet et al., 1998) or B lymphocytes (McLeod and Gold, 2001), in which the elevation of intracellular Ca^{2+} and/or diacylglycerol elicits the activation of Rap1.

2.2.5 PDZ-GEFs/RA-GEFs

Two closely related GEFs contain both a PDZ domain and an RA domain, and were named PDZ-GEF I and II (de Rooij et al., 1999) or RA-GEF-1 and 2 (Gao et al., 2001; Liao et al., 1999). They are also closely related to Epacs, in that they contain at their N-terminus a domain homologous to cAMP-binding domains; similarly to Epacs, this domain acts as a negative regulator of the GEF activity, however it does not bind cyclic nucleotides with a physiologically relevant affinity and is therefore probably regulated by other signals (de Rooij et al., 1999; Kuiperij et al., 2003). Both RA-GEFs/PDZ-GEFs activate Rap1 and Rap2, but they exhibit distinct tissue distributions. A major functional difference between the two factors is that the RA domain of PDZ-GEF I/RA-GEF-1 binds active Rap1 which induces the translocation of the GEF to the perinuclear region and the local activation of Raps; in contrast, the RA domain of PDZ-GEF II/RA-GEF-2 binds active M-Ras, thereby recruiting the GEF to the plasma membrane and activating Raps at that location (Gao et al., 2001; Liao et al., 2001). This suggests a mechanism by which M-Ras, acting at the plasma membrane, and Rap1 in the perinuclear region, respectively control the activation of discrete, and possibly functionally distinct, pools of Rap GTPases resident in these different subcellular compartments.

2.2.6 *Shep/AND-34/Chat*

Ephrins, signaling through Eph receptor tyrosine kinases, lead to the activation of Rap1 and R-Ras GTPases, but neither H-Ras nor Ral. This is mediated by SHEP1, a protein that binds activated EphB2 receptors via its N-terminal SH2 domain, and contains a C-terminal GEF domain (Dodelet et al., 1999). The C-terminal GEF domain of the closely related protein AND-34 (the murine homologue of BCAR3) is active on Rap, R-Ras and Ral, but not on H-Ras; it also binds the focal adhesion docking protein p130Cas in the region containing the GEF domain, and overexpression studies suggest that p130Cas binding inhibits the GEF activity of AND-34, hence linking Src and adhesion signaling with the activation of Rap, R-Ras and Ral (Gotoh et al., 2000). A third closely related protein termed Chat (Cas/HEF1-associated signal transducer) has been shown to lead to Rap1 activation, probably by upregulating the Cas-Crk-C3G pathway rather than by exerting a GEF activity (Sakakibara et al., 2002).

2.2.7 *DOCK4*

DOCK4 is a recently identified member of the CDM family (for Ced-5, DOCK180 and Myoblast city) of proteins that act as activators of the Rac GTPase (Yajnik et al., 2003). These proteins, that contain close to 2000 amino acids, are devoid of known GEF domains (such as DH or Cdc25-homology); they contain an N-terminal SH3 domain, a C-terminal proline-rich region, and are highly related to each other as well as to the unconventional Cdc42 activator Zizimin (Meller et al., 2002) in the region encompassing residues 100-1700. In contrast with other CDM proteins, DOCK4 acts as an activator of Rap1, associates with the N-terminal SH3 domain of the adaptor CrkII, and is involved in the formation of adherens junctions. Mutations in the DOCK4 gene are found in various tumors, and a recurrent missense mutation identified in prostate and ovarian cancers (Pro1718Leu) leads to a protein that is defective in Rap1 activation, and is instead capable of activating Rac and Cdc42 (Yajnik et al., 2003).

2.3 Effectors and Downstream Pathways

2.3.1 *Rap effectors*

Similarly to other GTPases of the Ras superfamily, Rap proteins are able to interact with several distinct effector proteins that each mediate some of their biological activities. Due to the similarity between their effector domains, Ras and Rap1 share a subset of common potential effector proteins.

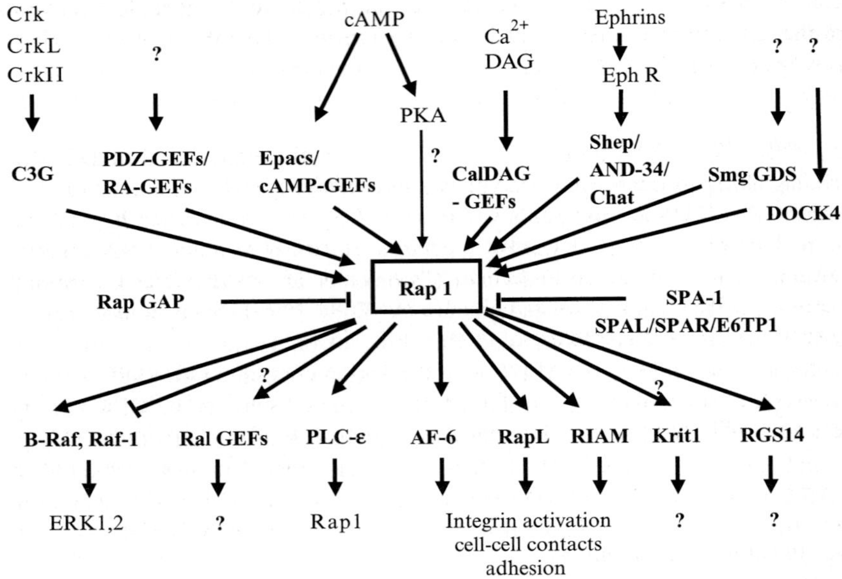

Figure 3. The Rap pathway

2.3.1.1 Raf kinases Similarly to Ras, Rap1 interacts with Raf kinases through their N-terminal Ras Binding Domain (RBD). The interaction of Rap1A with the RBD of Raf-1 has been characterized in detail, and the structure of the complex has been solved by X-ray crystallography (Nassar et al., 1995). Though both Ras and Rap1 bind to the Raf-1 RBD, H-Ras does so with a much higher affinity than Rap1 (20 nM and 1 μM respectively) (Herrmann et al., 1996), a difference that rests principally on the ability of Glu 31 in Ras (changed to Lys in Rap) to interact with Lys 84 of Raf-1 (Nassar et al., 1996). However, Ras and Rap1 also interact with the cysteine-rich domain (CRD) of Raf kinases, and the strength of this interaction has been suggested to determine whether the kinase activity will be activated or not. In essence, while Rap1 interacts with both Raf-1 and B-Raf, it is only able to activate B-Raf *in vitro* as well as in cells. The weaker respective interactions of Ras and Rap1 with the CRDs of Raf-1 and B-Raf is thought to enable the activation of the kinases (Ohtsuka et al., 1996; Okada et al., 1999). In contrast, the stronger interaction of Rap1 with the CRD of Raf-1 would prevent activation of the kinase by Rap1, and would also sequester it away from its activator, Ras; this Rap1-Raf interaction was reported to be inhibited by the phosphorylation of Rap1A on S180 by PKA (Hu et al., 1999). Raf-1 itself is subject to multiple regulatory phosphorylation events, and its phosphorylation at Y341 and S338 have been reported to be essential for Ras-induced Raf-1 activation. A recent report suggests that Ras, but not Rap1, is competent to recruit Raf-1 to cholesterol-rich rafts where it is phosphorylated on Y341, which enables its subsequent phosphorylation at S338 and activation (Carey et al., 2003); the presence of such rafts has been shown to favor mitogenic signaling downstream of Ras to the ERK MAP kinase cascade (Kranenburg et al., 2001). This could provide a biochemical mechanism, based on their differential subcellular localizations, to segregate the different downstream effects of Ras and Rap on Raf.

2.3.1.2 RalGEFs Rap proteins also interact with the RalGEFs RalGDS, Rlf and RGL via their C-terminal Ras interacting region (Nancy et al., 1999b; Wolthuis et al., 1996). Though this domain presents little sequence conservation with the RBD of Raf kinases, it presents a similar three dimensional structure, consisting in a ubiquitin fold, and interaction surface with Ras/Rap GTPases (Esser et al., 1998b; Huang et al., 1998; Huang et al., 1997b; Vetter et al., 1999). Such domains have been termed RA for Ras Association domains.

Especially remarkable is the fact that Rap1 binds RalGDS with high affinity, comparable with that of the Ras-Raf interaction, whereas it binds Rlf and RGL with a lower affinity similarly to their interaction with Ras (Herrmann et al., 1996). However, the interaction of Rap proteins with RalGEFs does not lead to the activation of Ral in mammalian cells (Urano et al., 1996; Zwartkruis et al., 1998). This may be due to the distinct subcellular localizations of Rap and Ral GTPases: whereas Rap proteins recruit RalGEFs to their endomembrane resident compartments, Ral is mainly found at or in the vicinity of the plasma membrane where it is activated downstream from Ras (Nancy et al., 1999b).

2.3.1.3 AF-6 AF-6, the mammalian homologue of Canoe in Drosophila, is a multidomain protein that contains two potential Ras binding domains at its N-terminus, as well as a PDZ domain towards its C-terminus, and plays a role in the establishment or maintenance of cell-cell junctions (Boettner et al., 2000). AF-6 binds Rap1 with higher affinity than Ras, and experiments in mammalian cells as well as in flies suggest that Rap1 regulates the positioning of adherens junctions and adhesion between epithelial cells via its interaction with AF-6/Canoe (Boettner et al., 2000; Knox and Brown 2002). In addition to its role as an effector, AF-6 may also recruit the RapGAP SPA-1 via its PDZ domain, hence also contributing to locally regulate the level of active Rap1 at cell attachment sites (Su et al., 2003).

2.3.1.4 PLC-ε PLC-ε is a recently uncovered form of phosphoinositide-specific phospholipase C that contains a CDC25-like GEF domain N-terminal to the PLC catalytic region, and a C-terminal RA domain. On one hand, PLC-ε acts as an effector through its interaction with the activated form of Ras, Rap1 and Rap2 (Schmidt et al., 2001; Song et al., 2001). It also acts as a GEF for Rap1, exclusive of any other Ras family GTPase including Ras and Rap2, and acts to amplify or prolong Rap1-dependent signaling (Jin et al., 2001; Song et al., 2002).

2.3.1.5 Other potential Rap effectors Several other proteins have been shown to interact with GTP-bound Rap proteins, and may constitute potential Rap effectors. Rap1 was reported to bind the p110 catalytic subunit of PI3 kinase, without leading to its activation, a case which is reminiscent of the interaction of Rap1 with

Raf-1 (Bos, 1998). Rap1 has also been reported to bind Krit1, an Ankyrin-repeat containing protein (Serebriiskii et al., 1997) whose loss of expression is responsible for hereditary cavernous angiomas (Laberge-le Couteulx et al., 1999; Sahoo et al., 1999); however, the involvement of the Rap1-Krit1 interaction in the generation of the pathology, or in the normal process of capillary angiogenesis, has not yet been substantiated. Rap proteins also interact with a protein carrying an RGS domain, RGS14, that may act to coordinate pathways involving Rap and heterotrimeric Gi/Go proteins (Traver et al., 2000). Finally, a database survey of the human genome reveals that may proteins are predicted to carry RA domains, and hence may bind active Rap proteins. One such recently uncovered proteins, that is essentially composed of a single RA domain, was termed RapL (Katagiri et al., 2003), an alternatively spliced form of the RASSF5/Nore 1 protein that is closely related to the RASSF1 tumor suppressor protein (Shivakumar et al., 2002). Whether other RA domain containing proteins, including members of the Nore/RASSF family actually constitute Rap effectors is under investigation in several laboratories.

2.3.2 Rap1, Raf kinases and the ERK MAP kinase pathway

Since it was known that Rap bound Raf-1 without causing its activation (see higher), and that the overexpression of Rap1 lead to the phenotypic reversion of Ras-transformed cells (Kitayama et al., 1989), experiments showing that the ectopic expression of constitutively activated Rap1 antagonized the Ras-dependent activation of ERK1/2 by LPA and EGF in Rat fibroblasts came as no surprise (Cook et al., 1993). However, there is still considerable debate whether the inhibition of Ras-dependent Raf-1 activation is an artefact of Rap1 overexpression, or whether it actually occurs in vivo under physiological conditions (Okada et al., 1998; Schmitt and Stork, 2001; Zwartkruis et al., 1998) (see lower).

2.3.2.1 The cAMP – Rap1 – B-Raf module An early report had shown that cAMP triggered the activation of the ERK 1/2 cascade in PC12 cells (Frodin et al., 1994). A further study by the group of Stork suggested that cAMP, acting through PKA, led to the phosphorylation and activation of Rap1B, which in turn associated with and activated B-Raf which was responsible for switching on the ERK 1/2 cascade (Vossler et al., 1997). This work, along with another study (Dugan et al., 1999), suggested that the ability of Rap1 to activate ERK was dependent on the presence of high molecular weight (95 kD) isoforms of B-Raf, which are mainly expressed in cells of neuronal or neuroendocrine origin (Barnier et al., 1995). Such a Rap1 – B-Raf cascade has since been found in PC12 cells stimulated by cAMP-elevating agents, NGF, and agonists of the M1 muscarinic acetylcholine receptor acting through CalDAG-GEF1 (Grewal et al., 2000; Guo et al., 2001; Vossler et al., 1997; York et al., 1998), HEK293 cells stimulated by cAMP-elevating β2-adrenergic agonists (Schmitt and Stork, 2000; Schmitt and Stork, 2002a), neuronal cells stimulated by the cAMP-elevating peptide PACAP (Bouschet et al., 2003), differentiating male germ cells (Berruti, 2000) and stimulation by thrombopoietin of a megakaryoblastic cell line expressing the thrombopoietin receptor (Garcia et al., 2001). Besides the presence of B-Raf isoforms, existence of this pathway may also depend on the expression of 14-3-3 molecules whose association with B-Raf have been suggested to protect it from inhibition by PKA (Qiu et al., 2000).

2.3.2.2 Differential phosphorylation of Rap1A and Rap1B by PKA Phosphorylation of Rap1 by PKA may influence its ability to activate cell signaling pathways. Rap1A carries a single site for phosphorylation by PKA at S180, which is also conserved in Rap1B (Hu et al., 1999; Lerosey et al., 1991b); however, Rap1B is phosphorylated *in vivo* at S179, a site which is not conserved in Rap1A that carries instead a positively charged Lys residue at the corresponding position (Hata et al., 1991). Phosphorylation of Rap1B on S179 was reported to be required for activation by cAMP of the B-Raf - ERK cascade in PC12 cells (Vossler et al., 1997), as well as for the mitogenic effects and inhibition of Akt/PKB by cAMP via Rap1B in thyroid follicular cells (Lou et al., 2002; Ribeiro-Neto et al., 2002). Hence it is possible that PKA, acting through the phosphorylation of Rap1B at S179, specifically enables Rap1B, but not the 95% identical isoform Rap1A, to activate a subset of downstream pathways such as the B-Raf – ERK cascade. This would also explain why analogues of cAMP that specifically activate Epacs/cAMP-GEFs, but not PKA, fail to enable the Rap1 pathway to activate the B-Raf – ERK cascade (Enserink et al., 2002), and require the concomitant activation of PKA to maximally promote the extension of neurites by PC12 cells (Christensen et al., 2003). In contrast, the phosphorylation of Rap1A at S180 by PKA was reported to weaken its interaction with Raf-1 (Hu et al., 1999). The hypothesis that the selective phosphorylation of Rap1A or Rap1B isoforms by PKA could distinguish their physiological effects on the ERK MAP kinase cascade remains to be experimentally challenged.

2.3.2.3 Inhibition of the Raf-1 – ERK cascade by cAMP and Rap1 cAMP is known to inhibit the proliferation of many cell types, and has been shown to prevent the Ras-dependent activation of the Raf-1 – ERK pathway (Burgering et al., 1993; Cook and McCormick, 1993; Wu et al., 1993). This latter effect may be in part consequential to the phosphorylation of Raf-1 by PKA (Dumaz and Marais, 2003). Activation and phosphorylation of Rap1B by PKA was also suggested to be responsible for this effect, in cells devoid of B-Raf, by enabling Rap1B to sequester Raf-1 and prevent its activation by Ras (Schmitt and Stork, 2001; Schmitt and Stork, 2002b; Vossler et al., 1997). This is in apparent contradiction with experiments showing that the activation of endogenous Rap1 fails to interfere with activation of the Ras-Raf-ERK cascade (Enserink et al., 2002; Zwartkruis et al., 1998). One possibility to reconcile these results is to postulate, as developed higher, that phosphorylation by PKA might confer to Rap1A and Rap1B different signaling capacities. Inhibition of the Ras-dependent Raf-1 activation by Rap1 would only occur in certain cell lines, or specific experimental conditions, where Rap1B is expressed and phosphorylated on S179 by PKA. On the contrary, under circumstances where Rap1A is predominantly expressed, or preferentially targeted by PKA over Rap1B, it would not be able to prevent the activation of Raf by Ras since its phosphorylation at S180 by PKA weakens the Rap1 – Raf-1 interaction (Hu et al., 1999).

2.3.3 *Spatio-temporal aspects of Rap signaling*

In order to fully comprehend signaling, it is necessary to take into account its spatio-temporal aspects. For instance, many cells respond to the stimulation of tyrosine kinase receptors by a rapid and transient activation of ERKs, followed by a second more sustained ERK response; those two waves of MAP kinase activation are thought to be responsible for different biological effects (Marshall, 1995). In view of this, it should be noted that in several biological models, the Rap1 – B-Raf – ERK cascade was shown to exhibit a sustained response, contrarily to the rapid and transient one elicited by the Ras-dependent pathway (Bouschet et al., 2003; Garcia et al., 2001; Kao et al., 2001; York et al., 1998).

Among spatial aspects that should be taken into consideration is the fact that Rap1 has been localized to the perinuclear region ascribed to the Golgi or late endocytotic compartment depending on studies and cell types (Beranger et al., 1991; Pizon et al., 1994). Indeed, FRET-based experiments have shown that Rap1 is activated, in response to EGF, in a perinuclear compartment (Mochizuki et al., 2001; Ohba et al., 2003). Such a spatial pattern of the response may be related to the subcellular localization of Rap1, as well as to evidence showing that its activation in response to tyrosine kinase receptors triggered by NGF in PC12 cells (York et al., 2000) or EGF in COS-1 cells (Mochizuki et al., 2001; Ohba et al., 2003) requires endocytosis of the receptors, in contrast with the activation of Ras by EGF that is endocytosis-independent. This pattern may also be determined by the localization, or local activation of GAPs, as suggested by a recent study (Ohba et al., 2003).

Such a perinuclear activation of the Rap pathway may however not be universal. For instance in mouse forebrain neurons, Rap1 has been shown to be involved, in response to cAMP and Ca^{2+}, in the activation of a membrane-associated pool of ERK1/2 that plays a role in learning and memory (Morozov et al., 2003).

These few examples illustrate that the kinetics and subcellular location of Rap activation may heavily influence biological outcomes, and that these aspects will require detailed attention in future signaling studies.

2.3.4 *Rap1 and adhesion*

The first unrecognized (at the time) evidence suggesting the involvement of Rap1 signaling in adhesion was the selection of Krev-1 (encoding Rap1A) as a cDNA whose overexpression could reverse the phenotype of *ras*-transformed NIH 3T3 fibroblasts to their original adherent and fully spread morphology (Kitayama et al., 1989). A further hint was published ten years later in a report showing that the expression of the Rap-GAP SPA-1 inhibited, whereas C3G enhanced, cell spreading (Tsukamoto et al., 1999). The idea that Rap1 positively regulates integrin-mediated cell adhesion was finally acknowledged through a series of reports showing that Rap1 activates several integrins such as β1 (VLA-4) and β2 (LFA-1) in B or T lymphocytes (Arai et al., 2001; Katagiri et al., 2000; Reedquist et al., 2000), αMβ2 in macrophages (Caron et al., 2000), or α2β3 in platelets and megakaryocytes (Bertoni et al., 2002; de Bruyn et al., 2003). Increasing integrin-mediated adhesion through inside-out signaling may actually constitute one of the major biological effects of the physiological activation of Rap1 through stimuli such as β-adrenergic agonists (Rangarajan et al., 2003) or chemokines (Shimonaka et al., 2003). Stimulation of the T cell receptor also activates Rap1 which in turn regulates the interaction of T cells with antigen presenting cells and enhances the responses of T lymphocytes to antigen via integrin-dependent pathways (Katagiri et al., 2002; Sebzda et al., 2002).

Genetic evidence also supports the notion that the Rap pathway positively regulates cell adhesion, and in particular during development: C3G deficient mice die *in utero*, and derived MEFs display impaired adhesion and spreading (Ohba et al., 2001). Similarly, Rap1 plays a role in morphogenesis and cell migration processes in drosophila (Asha et al., 1999). At the cellular level, Rap1 is necessary for the formation and/or maintenance of adherens junctions in drosophila and mice epithelial cells (Knox and Brown, 2002; Yajnik et al., 2003). The pathways downstream of Rap1 leading to these effects are however not yet understood. Recent evidence suggests that some of them are mediated via its effectors AF-6, and RapL and RIAM (Boettner et al., 2000; Katagiri et al., 2003; Su et al., 2003, Lafuente et al., 2004); another possible pathway proceeds via the potential Rap effector Krit-1 that interacts with ICAP-1, a protein that binds the cytoplasmic domain of β1 integrins (Zawistowski et al., 2002; Zhang et al., 2001).

2.3.5 *Rap1 and trafficking*

Rap1 may also play a role in intracellular trafficking. It was found to play a role in the removal of AMPA receptors from the membrane during synaptic plasticity in brain (Zhu et al., 2002). Rap1 also regulates the secretion of the non-amyloidogenic soluble form of amyloid precursor protein via a pathway involving the Rac GTPase (Maillet et al., 2003); however the means by which Rap activates Rac, and whether this is a general feature of Rap signaling, or restricted to certain cell types, remains to be established.

2.3.6 *Rap2-dependent pathways*

Rap2 presents many functional similarities with Rap1, including the ability to be regulated by the same GAPs and GEFs (Janoueix-Lerosey et al., 1992; Ohba et al., 2000), as well as to interact with common putative effectors such as RalGEFs and RGS14 (Nancy et al., 1999b; Traver et al., 2000). However, Rap2 does not appear to mediate the same cellular functions as Rap1, and may therefore act through distinct pathways. Indeed, Rap2 binds to a specific effector, RPIP8, that is selectively expressed in cells of neuronal and neuro-endocrine origin (Janoueix-Lerosey et al., 1998); its cellular function has not yet been elucidated. In addition, Rap2B has been suggested to mediate the activation of PLC-ε mediated by pathways elevating the intracellular levels of cAMP (Evellin et al., 2002; Schmidt et al., 2001). Much work is still required to decipher the respective functions of the A, B and C Rap2 isoforms.

2.4 Rap and Cancer

There is now converging evidence suggesting that deregulated Rap1 signaling participates in oncogenesis. Early reports had shown that the overexpression of Rap1 promoted growth and transformation of Swiss 3T3 cells (Altschuler and Ribeiro-Neto, 1998; Yoshida et al., 1992). Two recent studies show that interfering with RapGAP function, hence increasing the cellular levels of active Rap, may lead to cancer. The E6 protein from the high risk human papilloma virus type 16, that plays a role in tumor formation, promotes the degradation of the E6TP1/SPAL/SPAR protein carrying a RapGAP function, resulting in higher Rap-GTP levels (Singh et al., 2003). Similarly, the targeted disruption of the RapGAP protein SPA-1 in mice leads to myeloproliferative stem cell disorders associated with the elevation of Rap-GTP levels (Ishida et al., 2003). Conversely, the Rap1 activator DOCK4 acts as a tumor suppressor in mouse osteosarcoma cells, and carries a point mutation in ovary and prostate tumor cell lines that switches its biochemical properties to a form that no longer activates Rap1, but instead Rac and Cdc42 (Yajnik et al., 2003). Hence, depending on the tissue, the Rap1 pathway may display oncogenic or tumor suppressor properties; however, no clear pattern emerges enabling to develop a rationale for these apparently contradictory effects.

3. RAL PROTEINS

Ral was the first protein of the Ras superfamily to be isolated by a strategy purposefully designed to search for genes encoding proteins related to Ras (Chardin and Tavitian, 1986), a strategy that subsequently led to the discovery of Rab and Rap proteins (Pizon et al., 1988; Touchot et al., 1987). The Ral group consists of two closely related proteins, RalA and RalB (Chardin and Tavitian, 1989), that are 82% identical exhibiting only 6 different residues amongst the 110 N-terminal amino acids. Compared with Ras, both Ral proteins contain a 11 residue N-terminal extension, whose integrity is required for their interaction with one of their downstream effectors, phospholipase D (PLD) (Jiang et al., 1995). At their C-termini, Ral proteins contain a CAAL box that ensures that they are geranylgeranylated. A notable feature is that these boxes contain two cysteines (CCIL and CCLL for RalA and RalB respectively) and it is therefore possible that both residues could be geranylgeranylated, similarly to Rab proteins that contain C-terminal CC or CxC sequences (Nancy et al., 2002). There is still some debate concerning the precise subcellular localization of Ral proteins, at the plasma membrane or on vesicular structures close to the membrane. Several factors could contribute to these localizations such as the presence upstream from the CAAX box of a stretch enriched in basic residues (11 and 13 out of 27 in RalA and RalB respectively), and the interaction of Ral with components of the exocyst complex (Brymora et al., 2001; Moskalenko et al., 2002; Sugihara et al., 2001).

So far, a Ral-GAP activity of Mr greater than 10^6 has been partially purified from cytosolic fractions, yet not characterized at the molecular level (Emkey et al., 1991).

3.1 Pathways Leading to Ral Activation

Several pathways leading to the physiological activation of Ral have been identified. First and foremost, Ral is activated downstream from Ras, and the activity of Ral participates in the transformation and growth-promoting activity of Ras (Gille and Downward, 1999; Urano et al., 1996; White et al., 1996). This cascade is based on the action of a family of related RalGEFs, RalGDS (Albright et al., 1993; Hofer et al., 1994), RGL (RalGDS like) (Kikuchi et al., 1994), Rlf (RalGDS like factor, also termed RGL2) (Peterson et al., 1996; Wolthuis et al., 1996), Rgr (D'Adamo et al., 1997) and RGL3 (Shao and Andres, 2000). These proteins contain in their central region a canonical catalytic Cdc25 homology domain, also found in RasGEFs such as GRF and Sos, consisting of tandem N-GEF (also called LTE1-homology region or REM for Ras Exchange Motif) – GEF (the *stricto sensu* catalytic domain) domains, and in their C-terminal region a regulatory RA (for Ras Association) domain that binds to Ras and Rap. This latter region shares only limited sequence homology with the Ras binding domain of Raf kinases, yet both exhibit very similar three-dimensional structures (see lower) (Esser et al., 1998a; Geyer et al., 1997; Huang et al., 1997a; Kigawa et al., 1998). RalGEFs of this family are only catalytically active on Ral, and interaction with the GTP-bound form of Ras is necessary for their catalytic activity in cells. In an *in vitro* system where prenylated GTPases are reconstituted on artificial liposomes, both Rap and Ras can stimulate the activation of Ral by RalGDS (Hinoi et al., 1996; Kishida et al., 1997). However, the colocalisation of Ras and Ral, as well as recruitment of RalGDS by Ras to the plasma membrane are necessary for the activation of Ral downstream of Ras (Matsubara et al., 1999). Active Rap1 and Rap2 can also bind RalGDS and related factors, and recruit them to their respective resident compartments (perinuclear Golgi or endosomal for Rap1, and endoplasmic reticulum for Rap2); however, these compartments do not contain Ral, which might explain why the Ras->Ral cascade, but not the Rap->Ral cascade, actually occurs in the mammalian cell lines studied to date (Nancy et al., 1999a; Zwartkruis et al., 1998). Hence, Ras-GTP displays a dual action on RalGEFs of the RalGDS family by enhancing their catalytic activity and recruiting them in the vicinity of their substrate, Ral. An additional regulatory step involving PDK1, downstream of PI3 kinase, may ensure a coordination of the Ras->Ral and Ras->PI3K pathways downstream from Ras (Rosario et al., 2001b).

More recently, another family of RalGEFs was described, called RalGPS or RalGEF2 (de Bruyn et al., 2000; Rebhun et al., 2000b), consisting of RalGPS 1 (with two forms resulting from alternative splicing named 1A and 1B) and RalGPS 2. They contain the catalytic Cdc25 domain at their N-terminus, followed by proline-rich sequences that bind the SH3 domains of adaptor proteins Grb2 and Nck, and a C-terminal PH domain. These factors, that lack the RA domain characteristic of the RalGDS family, are thought to activate Ral downstream from growth factor receptors, independently of the activation of Ras.

Since the development of a pull-down assay that measures the activation state of Ral in cells, several physiological stimuli and pathways leading to the *in vivo* activation of Ral have been described (Wolthuis et al., 1998a; Wolthuis et al., 1998b).

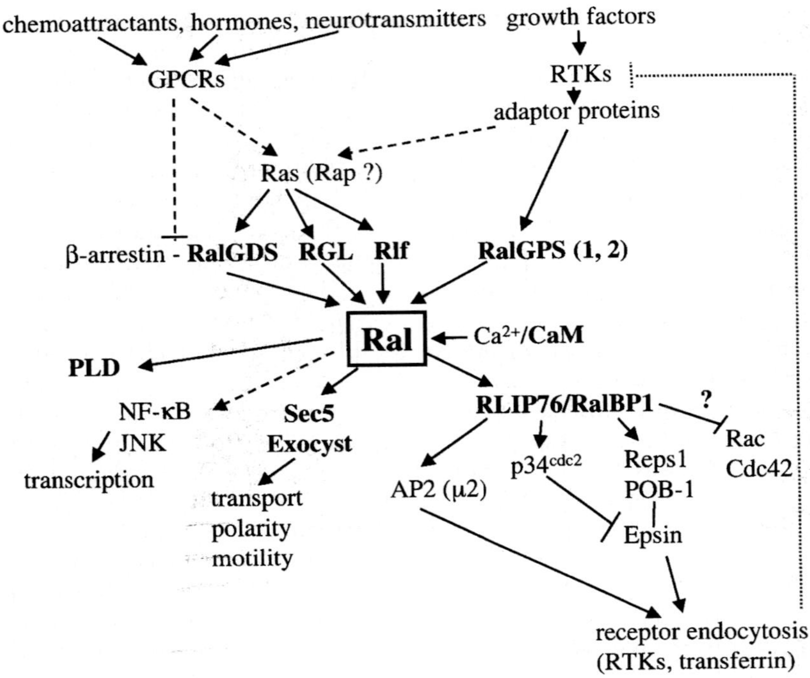

Figure 4. The Ral pathway

Ral is rapidly activated in platelets by alpha-thrombin as well as in neutrophils in response to fMet-Leu-Phe, PAF or GM-CSF (M'Rabet et al., 1998; Wolthuis et al., 1998a). It has been suggested that fMet-Leu-Phe, acting through a serpentine receptor, would induce the dissociation of a cytoplasmic RalGDS-β-arrestin complex and translocation of RalGDS to the membrane, where it could be free to activate Ral (Bhattacharya et al., 2002). In platelets and neutrophils, the elevation of intracellular Ca^{2+} levels is sufficient to induce activation of Ral (M'Rabet et al., 1998; Wolthuis et al., 1998a). Ca^{2+} participates in the activation of Ral via Calmodulin (CaM) independently from Ras (Hofer et al., 1998); indeed, both RalA and RalB contain a Ca^{2+}-dependent Calmodulin binding site in their basic-rich C-terminal region (Wang et al., 1997; Wang and Roufogalis, 1999), and the interaction of Ca^{2+}-Calmodulin with Ral is necessary for its activation by thrombin (Clough et al., 2002).

3.2 Pathways Downstream from Ral

3.2.1 2.1 Phospholipase D

One of the responses of cells to transformation by v-Src is an increase in phospholipase D (PLD) activity, leading to the generation of biologically active second messengers such as phosphatidic acid, lysophosphatidic acid and diacylglycerol.

This effect is mediated by the Ras-RalGDS-Ral pathway and results from a direct interaction of Ral with PLD (Jiang et al., 1995). This interaction involves the 11 residue N-terminal extension of Ral, and not the classical effector domain; nevertheless, Ral-mediated activation of PLD requires the active form of Ral since it is inhibited by expression of dominant-negative RalA. RalA associates with the PLD1 enzyme, which is dependent on PIP2 for its activity and is activated by ARF1 (Luo et al., 1997). It appears that RalA and ARF1 both associate with PLD, via different sites, and that they act synergistically to activate PLD (Kim et al., 1998; Luo et al., 1998). A recent report shows that ARF6 is also able to cooperate with RalA to activate PLD activity; interestingly, this pathway is active in H- but not in K-Ras transformed cells and seems to result from an H-Ras specific ability to activate RalA and to elevate ARF6 levels in NIH 3T3 cells (Xu et al., 2003).

3.2.2 RLIP76/RalBP1

Searches for Ral binding proteins led to the identification of RLIP76/RalBP1 (Cantor et al., 1995; Jullien-Flores et al., 1995; Park and Weinberg, 1995), a protein that specifically binds the active form of Ral (complexed with GTP) via its effector domain (Bauer et al., 1999a), and hence has the characteristics of a Ral effector. RLIP76/RalBP1 carries a RhoGAP domain that is able to stimulate the GTPase activity of Cdc42, and to a lesser extent Rac, but not Rho; whether Ral regulates the activity of Cdc42 and Rac via RLIP76/RalBP1, and the physiological role of this putative pathway, remain to be established.

RLIP76/RalBP1 is a multidomain protein that interacts with several cellular partners involved in endocytosis. Its C-terminal region binds two proteins containing an Eps15 homology region (EH), Reps1 (Yamaguchi et al., 1997) and POB1 (Ikeda et al., 1998), both of which interact with the SH3 domain of the growth factor receptor binding/adaptor protein Grb2. POB1 itself interacts with Epsin and Eps15, two proteins involved in the endocytotic machinery, and is necessary for the ligand-dependent internalization of insulin and EGF receptors (Nakashima et al., 1999). The N-terminal region of RLIP76/RalBP1 interacts with $\mu 2$, the medium chain of the AP2 complex involved in clathrin-dependent endocytosis; there is evidence that Ral acts through this pathway to regulate endocytosis of transferrin and EGF receptors (Jullien-Flores et al., 2000). RLIP76/RalBP1 also binds a novel PDZ domain-containing protein, ARIP2, that acts in the endocytotic pathway and links the endocytosis of activin type II receptors to the activation level of Ral (Matsuzaki et al., 2002).

Several reports suggest that the Ral-RLIP76/RalBP1-POB1-Epsin-Eps15-AP2 endocytotic pathway is inhibited by phosphorylation during mitosis. Indeed, RLIP76/RalBP1, also called cytocentrin, transiently associates with the mitotic spindle poles in early prophase and dissociates from them after completion of mitosis (Quaroni and Paul, 1999). RLIP76/RalBP1, POB1, Epsin and Eps15 are phosphorylated during mitosis, and the phosphorylation of Epsin by $p34^{cdc2}$ prevents endocytosis by disrupting its association with POB1 (Kariya et al., 2000). The proposed role for RLIP76/RalBP1 in mitotic cells is to interact with the $p34^{cdc2}$/cyclinB1

(Cdk1) complex and act as a platform to facilitate the phosphorylation of Epsin hence switching off endocytosis during this phase of the cell cycle (Rossé et al., 2003).

3.2.3 The exocyst

The active form of Ral also interacts with Sec5, a member of the multiprotein exocyst complex (Brymora et al., 2001; Moskalenko et al., 2002; Sugihara et al., 2001) through which Ral could exert a regulatory control on events requiring delivery of membrane material or cargo to the plasma membrane. This pathway is involved in cytoskeletal events such as filopod formation in response to inflammatory cytokines in Swiss 3T3 fibroblasts (Sugihara et al., 2001). It also regulates the targeting of basolateral proteins in epithelial cells, secretagogue-dependent exocytosis in neuroendocrine (Moskalenko et al., 2002) and endothelial cells (de Leeuw et al., 2001), and the re-filling of the readily releasable pool of synaptic vesicles in neurons (Polzin et al., 2002).

3.3 Physiological Effects of the RalGEF-Ral Pathway

Ras has several potential downstream effector pathways, amongst which three main ones (Raf-MAP kinase, PI3 kinase, and RalGEF-Ral) are thought to mediate most of Ras' physiological effects and have been investigated in great detail (see other chapters in this book). Partial loss of function effector mutants of Ras were developed that enable the selective activation of each of these three pathways: Ras (G12V, T35S) activates the Raf-MAPK pathway, Ras (V12G, E37G) targets RalGEFs, and Ras (G12V, Y40C) activates the PI3 kinase pathway (Joneson et al., 1996; Khosravi-Far et al., 1996; White et al., 1995). They have been instrumental to determine the respective roles of these pathways in Ras signaling, however, one must remain cautious when interpreting the results from experiments using such mutants. In particular, while it is well documented that Ras (V12G, E37G) activates RalGEFs, it is possible that these proteins, in addition to activating Ral, could engage other downstream pathways. Another frequently used tool to document the involvement of the RalGEF-Ral pathway is the ectopic expression of the dominant-negative RalS28N mutant. However, one should be careful not to over interpret these results as demonstrating that the activity of Ral is required in a given signaling pathway: indeed, RalS28N acts by titering endogenous RalGEFs, but in doing so, it may also inactivate them or target them for degradation, hence curtailing all of their effects, including those that are independent of Ral.

There is in fact ample experimental evidence to sustain the hypothesis that RalGEFs do not solely act through the activation of Ral, and may do so via their interaction with other proteins. For instantance, active Raps bind RalGEFs with high affinity (Herrmann et al., 1996) and recruit them to their resident compartments (Nancy et al., 1999a); despite that, Raps do not activate Ral in most mammalian cell lines (Nancy et al., 1999a; Urano et al., 1996; Zwartkruis et al., 1998) hence raising the possibility that RalGEFs might activate alternative pathways downstream from

Rap. Another example, recently documented, is the direct interaction of RalGDS with β-arrestin which could mediate a cross-talk between Ras and GPCR-controlled pathways (Bhattacharya et al., 2002).

Additional evidence comes from the fact that many of the effects observed upon the ectopic expression of RalGEFs or Ras (G12V, E37G) cannot be reproduced by a constitutively active Ral mutant (Hamad et al., 2002; Ward et al., 2001; Wolthuis et al., 1997). One explanation for such results is that Ral could need to cycle between GDP and GTP-bound forms to exert some of its effects; alternatively, RalGEFs could also activate pathways that do not involve Ral. For instance, investigators have used a constitutively active mutant of Rlf, made by targeting the protein to membranes by the addition to its C-terminus of the C-terminal sequence of K-Ras (Rlf-CAAX) (Wolthuis et al., 1997), to decipher its cellular roles. However, Rlf contains a consensus SH3-binding domain in its N-terminal region; a plausible hypothesis (that has not yet been experimentally addressed) is that the recruitment of SH3 domain-containing proteins (adaptors for instance) to membranes via Rlf-CAAX could activate several signaling pathways downstream from these proteins, and hence recapitulate some of their physiological effects. This could explain why results obtained with Rlf-CAAX on the induction of transcription and transformation of human cells were not reproduced using constitutively active Ral (Hamad et al., 2002; Wolthuis et al., 1997), and only poorly using RalGDS-CAAX (Channing Der, personal communication). Since the activation of Ras recruits RalGEFs to the plasma membrane, it is also possible that these Ral-independent pathways are involved in the physiological activation of Rlf downstream from Ras.

3.3.1 Development

Developmental systems have often been used as a means of investigating signal transduction pathways. For instance, RalB has been shown to be involved in the early development of xenopus (Iouzalen et al., 1998). The Ras-RalGEF-Ral pathway is conserved in drosophila, and is therefore amenable to genetic analysis; indeed, transgenic flies overexpressing wild type or activated Ral exhibit developmental abnormalities (Mirey et al., 2003; Sawamoto et al., 1999b). One of these investigations has surprisingly shown that contrarily to the case of mammalian cell lines where Ral acts downstream from Ras, it acts downstream from Rap during development of the drosophila wing and notum (Mirey et al., 2003). This observation urges investigators to be cautious when transposing results between biological systems, as it suggests that the conservation of proteins and their interactions through evolution does not necessarily mean that biological functions remain identical.

3.3.2 Motility

Motility is a complex multistep process requiring cytoskeletal as well as membrane remodeling events. During oogenesis in drosophila, the Ras-Ral pathway is necessary for the migration of border cells of the follicle (Lee et al., 1996b). In mammalian cells, Ral participates via its interaction with filamin in the formation of filopodia, a frequent early step of chemotactic responses (Ohta et al., 1999). It is also involved in cell migration in various biological models such as skeletal myoblasts (Suzuki et al., 2000b), bladder carcinoma cell lines (Gildea et al., 2002) and NRK cells where it acts via the exocyst complex (Rossé et al., 2006).

3.3.3 Transcription

All three Ras-dependent RalGEFs, RalGDS, RGL and Rlf, have some ability to activate the c-*fos* promoter, and this effect is greatly enhanced in the presence of active Ras (Murai et al., 1997; Okazaki et al., 1997; Wolthuis et al., 1997). In the case of RalGDS, its transcription-promoting activity strongly synergizes with the concomitant activation of the Raf pathway (Okazaki et al., 1997). Targeting Rlf to membranes by the addition of the C-terminal sequence of K-Ras (Rlf-CAAX) enables it to act independently of Ras to activate transcription from the c-*fos* promoter via the SRE, and to induce the proliferation of NIH 3T3 cells under low serum conditions (Wolthuis et al., 1997). However, neither the synergistic effects of RalGDS with Raf, nor those of Rlf-CAAX can be replicated by replacing RalGEFs with a constitutively active mutant of Ral, suggesting the involvement of additional downstream pathways (see higher and (Okazaki et al., 1997; Wolthuis et al., 1997). Nevertheless, the concomitant activation of Ral may also be necessary since deletion of one of the catalytic motifs in Rlf-CAAX abolishes its effects (Wolthuis et al., 1997).

Some transcriptional effects of the RalGEF-Ral pathway clearly involve Ral. Downstream from Ras, the RalGEF-Ral pathway regulates the phosphorylation of c-Jun via JNK in human as well as in Drosophila, and this effect requires the activity of Ral (De Ruiter et al., 2001; Sawamoto et al., 1999a). Moreover, activated Ral induces the transcription of cyclin D1 via the activation of NF-κB (Henry et al., 2000). Finally, both Ral and PKB/Akt independently contribute to regulate the activity of the Forkhead transcription factor AFX via phosphorylation (Kops et al., 1999), and the activation of Ral has been shown to enhance the activity of AFX (De Ruiter et al., 2001).

3.3.4 Growth, transformation and invasion

Several reports have suggested a role for the Ral pathway in cancer. Ral participates in the Src transduction cascade, and is necessary for transformation by v-Src (Aguirre-Ghiso et al., 1999; Goi et al., 2000; Jiang et al., 1995). Furthermore, the activity of Ral is required for the conversion of breast cancer cells overexpressing EGF receptor family members to an estrogen-independent state (Yu and Feig, 2002).

Oncogenic Ras can cooperate with SV40 large T and telomerase to direct the tumorigenic conversion of normal human epithelial and fibroblast cells (Hahn et al., 1999); in depth analysis of this biological model led to the surprising result that in human, but not in mouse cells, the oncogenic effect of Ras could be recapitulated by the ectopic expression of Ras (G12V, E37G) or Rlf-CAAX (Hamad et al., 2002), suggesting an oncogenic role in human cells for signaling through the RalGEF-Ral pathway. The RalGEF pathway may also play a role in invasion: a recent investigation shows that the RalGEF pathway synergizes with ERK to experimentally promote an invasive phenotype in NIH 3T3 and breast cancer cells (Ward et al., 2001). Here again, the invasive phenotype was a consequence of the expression of Ras (G12V, E37G) or RalGDS-CAAX, but could not be reproduced by the constitutively active Ral (G23V) mutant.

However suggestive they may be, these results do not actually demonstrate the involvement of the RalGEF-Ral pathway in human oncogenesis, or the development of metastasis. The experiments reported above use ectopically expressed proteins that do not mimic the physiological situation because they are usually overexpressed, frequently leading to improper subcellular localization, and in some cases constitutively activated, as opposed to the transient activation of their endogenous counterparts. One would have to use conditionally expressed knock-in mutant animals to demonstrate the role of these proteins in the development of cancer, or conditional knock-out mutants to show that the abrogation of their expression reverts the transformed phenotype of tumor cells.

Challenges for the future years will be to actually demonstrate, in the appropriate animal models, the role of the RalGEF-Ral pathway in cancer, as well as to decipher the Ral-independent pathways triggered downstream of Ras by activated RalGEFs.

4. THE R-RAS GROUP

This group contains 3 closely related, yet functionally different, proteins: R-Ras (Lowe et al., 1987; Lowe and Goeddel, 1987), R-Ras2/TC-21 (Chan et al., 1994; Drivas et al., 1990) and R-Ras3/M-Ras (Kimmelman et al., 1997). Their core effector domain is conserved relative to Ras, and is flanked at the N-terminal side by an acidic residue, which may play a role in the interaction with certain effectors. All three proteins contain an N-terminal extension relative to Ras (26 residues for R-Ras, and 10 for TC-21 and M-Ras) to which no function has yet been ascribed. TC-21 is farnesylated at its C-terminal extremity and contains a consensus palmitoylation site, similarly to N-Ras. In contrast, both R-Ras and M-Ras are geranylgeranylated; TC-21 and R-Ras contain multiple basic residues in their C-terminal region.

4.1 R-Ras

R-Ras was originally discovered by low stringency hybridization with a v-H-*ras* probe, and considerable efforts were deployed to investigate whether, similarly to H-, K- and N-Ras, R-Ras carried transformation activity (Lowe et al., 1987; Lowe and Goeddel, 1987). Overexpression of the constitutively activated protein in NIH-3T3 cells induces malignant transformation, characterized by anchorage-independent growth and tumorigenicity in nude mice, but not the morphological changes characteristic of *ras*-transformed cells (Cox et al., 1994). The interaction of R-Ras with the Ras binding domains of Ras targets Raf and RalGDS was shown to be weak and non-specific (Herrmann et al., 1996), and the expression of activated R-Ras does not lead to the activation of Raf or ERK kinases. In contrast, R-Ras stimulates the activity of the PI3 kinase pathway, and the downstream kinase PKB/Akt (Marte et al., 1997; Osada, 1999, #13). This latter activity may be responsible for the positive effect of R-Ras on the differentiation of C2C12 skeletal myoblasts into myotubes (Suzuki et al., 2000a), the induction of estrogen-independent proliferation of beast cancer MCF-7 cells (Yu and Feig, 2002), as well as the integrin-independent formation of focal adhesions via FAK (focal adhesion kinase) and p130Cas (Kwong et al., 2003).

An early report showing that R-Ras, through its C-terminal region, interacts with Bcl-2, suggested that R-Ras could play a role in apoptosis (Fernandez-Sarabia and Bischoff, 1993). It was proposed that active R-Ras acted to promote cell death in growth factor deprived cells (Wang et al., 1995). However, this report is in contradiction with the antiapoptotic effects played by R-Ras via activation of PI3 kinase and induction of passive leakage of Ca^{2+} from the endoplasmic reticulum (Koopman et al., 2003).

The principal biological role identified for R-Ras so far consists in the induction of integrin activation (Zhang et al., 1996). Several pathways are reported to be involved in this activity, including antagonizing the Ras/Raf /ERK pathway whose activation suppresses integrin activation (Sethi et al., 1999), PI3 kinase-dependent (Keely et al., 1999; Osada et al., 1999) or independent pathways (Kinashi et al., 2000; Oertli et al., 2000), and a yet uncharacterized mechanism that involves the further downstream activation of Rap1 (Caron et al., 2000; Self et al., 2001). Another recently described pathway involves a proline-rich SH3-binding domain in the C-terminal region of R-Ras, that may act through the adaptor protein Nck, and that is both necessary and sufficient to induce integrin activation (Hansen et al., 2002; Wang et al., 2000). R-Ras was also found to localize to focal adhesion sites and to promote their assembly, a process that is not dependent on its interaction with Nck (Furuhjelm and Peranen, 2003). Conversely, it was recently described that the EphB2 tyrosine kinase receptor, as well as Src, reduce integrin-dependent cell adhesion via phosphorylation of R-Ras in the effector domain at tyrosine 66 (Zou et al., 2002; Zou et al., 1999), suggesting that the phosphorylation and subsequent inactivation of R-Ras may constitute a mechanism through which the activation of tyrosine kinases enhances the invasive phenotype of tumor cells.

4.2 R-Ras2/TC21

TC21, which is closely related to R-Ras, was found to be constitutively activated by a single point mutation and to carry under those conditions a high transforming activity, similar to that of *ras* oncogenes (Chan et al., 1994; Graham et al., 1994).

However, unlike Ras, such mutations have only been occasionally found in cell lines derived from human tumors and therefore probably constitute rare oncogenic events (Barker and Crompton, 1998; Chan et al., 1994). In contrast, the increased expression of normal TC21 may contribute more frequently to oncogenesis since an elevated level of TC21 protein has been found in several breast tumor lines (Clark et al., 1996), and *TC-21* was recently shown to be the most frequently deregulated gene in mouse T-cell lymphomas induced by the SL3-3 retrovirus (Kim et al., 2003).

However, the physiological function of TC21 has not yet been established, and its downstream pathways are still a matter of debate. It was initially suggested that TC21 could activate the Ral pathway via RalGDS (Lopez-Barahona et al., 1996), and that, similarly to Ras, its transforming ability was largely attributable to its capacity to activate the Raf-MAP kinase pathway (Movilla et al., 1999; Rosario et al., 1999). However, more recent reports disputed this finding, by showing that TC21 transforms cells, or affects their differentiation, via Raf-independent pathways (Graham et al., 1999; Graham et al., 1996). TC21 has also been reported to activate PLCε, and in some cases Ral; however these latter studies were performed by overexpressing constitutively activated forms of TC21, and may therefore have limited physiological significance (Murphy et al., 2002; Rosario et al., 2001a). It now emerges from several converging studies that TC21 activates the PI3 kinase pathway, and that this activity is required for its transforming effects (Murphy et al., 2002; Rong et al., 2002; Rosario et al., 2001a).

4.3 R-Ras3/M-Ras

Little is known on the function of the most recently described member of this group of GTPases, R-Ras3/M-Ras, that exhibits an expression pattern restricted to brain and heart (Kimmelman et al., 1997). In cells of neuronal origin such as PC12, R-Ras3/M-Ras is a potent activator of the PI3 kinase/Akt pathway through which it may play a role in the survival of neural-derived cells (Kimmelman et al., 2000; Kimmelman et al., 2002).

R-Ras3/M-Ras is able to bind to AF-6 (Quilliam et al., 1999), a putative effector of Rap, as well as a Ral activator related to Rlf termed RGL3 (Ehrhardt et al., 2001). Furthermore, the GTP-bound form of R-Ras3/M-Ras interacts with the Rap-specific GEF MR-GEF (Rebhun et al., 2000a), also termed RA-GEF2 (Gao et al., 2001); however, these two studies report contradictory results as to whether this interaction leads to an activation or an inhibition of Rap downstream from M-Ras.

5. RHEB

Rheb was originally discovered as the product of a gene whose expression is rapidly and transiently induced in brain by receptor-dependent synaptic activity; this gene is quite highly expressed in developing brain, adult cerebral cortex as well as several peripheral tissues such as lung and intestine (Yamagata et al., 1994). Rheb is a farnesylated protein whose overexpression antagonizes Ras transformation (Clark et al., 1997) through its interaction with Raf kinases. On one hand, the interaction of Rheb with Raf-1 is potentiated by its phosphorylation on Ser 43 by cAMP-dependent protein kinase; this reaction reciprocally decreases its interaction with H-Ras, thereby downmodulating the activation of Raf-1 downstream from Ras (Yee and Worley, 1997). Similarly, Rheb has also been reported to bind B-Raf and to inhibit its kinase activity (Im et al., 2002).

A highly conserved ortholog of Rheb is found in unicellular organisms such as yeast and fungi. In these organisms, Rheb is involved in the response to nitrogen-containing nutrients; Rheb is induced by nitrogen starvation, and is involved in regulating the uptake of lysine and arginine (Panepinto et al., 2002; Urano et al., 2000). In fission yeast, the absence of Rheb arrests cell growth and division with a phenotype analogous to nitrogen starvation (Mach et al., 2000), and the presence of farnesylated Rheb is critical to the exit of cells from the G0/G1 phase of the cell cycle (Yang et al., 2001). These data suggest that in yeast, Rheb links cell cycle entry to extracellular nitrogen levels.

A series of recent reports have shed new light on the biological activity of Rheb and its signaling pathway. Tuberous sclerosis (TSC) syndrome is an autosomal-dominant genetic disorder that can cause severe pathological consequences (including mental retardation, epilepsy, autism, cardiac pulmonary and renal failure) arising from the development of hamartomas, largely manifested as benign tumors, but that in rare case progress to renal cell carcinoma; this syndrome is characterized by mutations in either the *TSC1* or *TSC2* gene, whose respective protein products, Hamartin (*TSC1*) and Tuberin (*TSC2*) form together a putative tumor suppressor complex (Sparagana and Roach, 2000). It had also been known from studies in Drosophila and in mammals that TSC1/2 acts in the pathway through which Insulin and the deficiency in nutrients (amino acids and energy) inhibit cell growth via mTOR (Target of Rapamycin), S6K (ribosomal protein S6 kinase) and the translation regulator 4E-PB (that regulates the activity of the translation initiation factor eIF4E). Several studies in Drosophila and in mammalian cells now show that Rheb promotes cell growth by enhancing the activity of mTOR, and that Tuberin (in the Tuberin/Hamartin complex) inhibits the pathway by acting as a GAP for Rheb (Castro et al., 2003; Garami et al., 2003; Patel et al., 2003; Saucedo et al., 2003; Stocker et al., 2003; Tabancay et al., 2003; Tee et al., 2003; Zhang et al., 2003b). Tuberin proteins that carry point mutations from TSC patients have lost their ability to downregulate the level of Rheb-GTP in cells. Hence, Tuberin acts as a GAP for Rheb, and not for Rap1 as originally thought (Wienecke et al., 1995). Insulin activates the mTOR pathway through PI3 kinase, which phosphorylates TSC2 and inhibits its activity, thereby resulting in an elevated level of Rheb-GTP; this pathway requires the proper post-translational modification (farnesylation) and membrane localization of Rheb. The way in which nutrients regulate the cascade if however not yet known. It is presently clear that Rheb signals via mTOR and not, as earlier suggested via the ERK cascade, but how Rheb activates mTOR remains to be established.

6. THE RIT/RIN GROUP

This group contains the human proteins Rit and Rin (Lee et al., 1996a; Wes et al., 1996). Only an orthologue for Rin, called Ric (Wes et al., 1996), is found in Drosophila; members of this branch are not found in unicellular organisms such as yeasts. The most striking feature of these proteins is that they are devoid of C-terminal prenylation motifs, but contain in that region a cluster of basic residues through which Rin binds calmodulin (Lee et al., 1999). The central part of their effector domain is conserved relative to Ras, which is consistent with their reported interaction with a subset of Ras effectors such as AF-6 as well as the Ral-GEFs Ral-GDS, but neither Raf kinases nor PI3 kinase (Shao et al., 1999).

Rin is only expressed in neural tissues (Lee et al., 1996a; Wes et al., 1996), and more specifically in adult neurons (Spencer et al., 2002b). It is activated (exchange of GDP for GTP) downstream of Ras in neuronal-like PC6 pheochromocytoma cells by factors such as EGF and NGF, and this activation is thought to be necessary for the Ras-dependent extension of neurites in response to NGF since it is inhibited by the expression of dominant-negative Rin (Spencer et al., 2002b). Furthermore, the ectopic expression of Rin (wild type as well as constitutively activated) induces neurite outgrowth in PC12 cells, acting through a Ras-independent pathway that involves the activation of Rac and Cdc42 and requires the interaction of Rin with Calmodulin; siRNA depletion experiments suggest that endogenous Rin function is necessary for the Calcium/Calmodulin-mediated extension of neurites in PC12 cells (Hoshino and Nakamura, 2003).

Rit, in contrast, is ubiquitously expressed. It is associated with membranes, which requires the presence of the C-terminal basic stretch (Lee et al., 1996a). Little is know concerning its function. When ectopically expressed, Rit is able to confer anchorage-independence and tumorigenicity (but not focus formation) to NIH 3T3 cells (Rusyn et al., 2000) as well as to induce neurite extension and survival in response to low serum conditions in pheochromocytoma cells (Spencer et al., 2002a). Activated Rit signals to Ras-responsive elements, but fails to activate ERK, JNK, p38 and Akt/PKB pathways. It interacts with and activates a novel Ral-GEF termed RGL3, that can also be activated by Ras and Rap (Shao and Andres, 2000). This suggests that Rit may act, at least in part, by signaling through the Ral-GEF/Ral pathway.

7. RECENTLY IDENTIFIED MEMBERS: ARHI AND DI-RAS

ARHI (originally called *NOEY2*) was identified as a gene normally expressed in breast and ovarian epithelial cells, but not in ovarian and breast cancers (Yu et al., 1999). It behaves as a tumor suppressor gene, and re-expression of the ARHI protein in tumor cells prevents their growth and promotes their apoptosis via a caspase-independent and calpain-dependent pathway (Bao et al., 2002). Most remarquable is the fact that this gene is maternally imprinted, and therefore monoallelically expressed; 41% of breast and ovarian cancers were reported to exhibit a loss of heterozygosity at this locus, which was associated in most cases with a deletion of the non-imprinted functional allele (Yu et al., 1999). Moreover, the *ARHI* gene is within the region most commonly affected in deletions of the 1p31 region, a frequent chromosomal alteration in breast and ovarian cancers (Peng et al., 2000). The ARHI protein exhibits 60% identity to Ras and Rap proteins. It carries a low intrinsic GTPase activity and is maintained in a constitutively activated GTP-bound form in cells (Luo et al., 2003). This may be attributable to changes, as compared with Ras, in the GTP-binding regions of the protein. In the P-loop, the residue equivalent to Glycine 12 of Ras is changed to Alanine; more importantly, the conserved DTAGQE sequence of the switch 2 region is changed to DSKSGD, in which the Glutamine 61 residue important for the GTPase activity is changed to a Glycine, a substitution which abolishes the intrinsic GTPase activity of H-Ras (Der et al., 1986). In addition, ARHI contains a 34 residue N-terminal extension as compared with Ras; this region was shown to be essential for the growth inhibitory effect of ARHI (Luo et al., 2003). No evidence as to its possible physiological role has yet been reported.

A recent report describes the occurrence of two closely related proteins, Di-Ras 1 and 2, that are specifically expressed in brain (D-Ras1) or heart and brain (Di-Ras2). Similarly to ARHI, they exhibit a reduced intrinsic GTPase activity and are predominantly GTP-bound in cells, which may be ascribed to the change in the switch 2 region of Glutamine 61 to Serine. The only functional indication is that its overexpression in HEK 293 cells leads to the formation of large vacuoles (Kontani et al., 2002).

8. CONCLUSION

Since the discovery of Ras family GTPases, considerable efforts have been devoted to identify their physiological function as well as to characterize their regulators (GEFs and GAPs) and effectors. The molecular mechanisms by which GEFs and GAPs control the activation and GTPase activities of Ras family proteins have been described, often to the submolecular level, thanks to the fascinating progress of structural studies. The extracellular cues to which many of these GTPases respond, and pathways leading to their activation have been identified thanks to the development of convenient assays to selectively detect the GTP-bound form of Ras family proteins. Finally, the development of powerful methods to investigate protein-protein interactions has allowed the identification of effector proteins, and many of their downstream pathways.

Several major challenges now lie ahead of us, such as to decipher the spatio-temporal aspects of signaling controlled by these GTPases, which begin to be accessible to analysis using emerging imaging technologies such as FRET-based time-lapse microscopy on living cells. We must also understand, beyond the present descriptive level, the cross-talks and cross-regulations between signaling pathways that appear to be linked in a giant cell-wide network. Finally, we will need to integrate all of this information to conceptualize the mechanisms by which a cell, or a group of cells, is able to react with exquisite specificity to extracellular cues.

ACKNOWLEDGEMENTS

I wish to thank my colleagues from the Inserm U528 laboratory, and in particular Jacques Camonis, for many fruitful discussions. Portions of this work were supported by grants from the Association pour la Recherche contre le Cancer and Inserm.

REFERENCES

Aguirre-Ghiso, J.A., Frankel, P., Farias, E.F., Lu, Z.M., Jiang, H., Olsen, A., Feig, L.A., Joffe, E.B. and Foster, D.A. (1999) RalA requirement for v-Src- and v-Ras-induced tumorigenicity and overproduction of urokinase-type plasminogen activator: involvement of metalloproteases. Oncogene *18*, 4718-4725.

Ahmadian, M.R., Stege, P., Scheffzek, K. and Wittinghofer, A. (1997) Confirmation of the arginine-finger hypothesis for the GAP-stimulated GTP-hydrolysis reaction of Ras. Nat. Struct. Biol. *4*, 686-689.

Albert, S., Will, E. and Gallwitz, D. (1999) Identification of the catalytic domains and their functionally critical arginine residues of two yeast GTPase-activating proteins specific for Ypt/Rab transport GTPases. EMBO J. *18*, 5216-5225.

Albright, C.F., Giddings, B.W., Liu, J., Vito, M. and Weinberg, R.A. (1993) Characterization of a guanine nucleotide dissociation stimulator for a ras-related GTPase. EMBO J. *12*, 339-347.

Alsayed, Y., Uddin, S., Ahmad, S., Majchrzak, B., Druker, B.J., Fish, E.N. and Platanias, L.C. (2000) IFN-gamma activates the C3G/Rap1 signaling pathway. J. Immunol. *164*, 1800-1806.

Altschuler, D.L., Peterson, S.N., Ostrowski, M.C. and Lapetina, E.G. (1995) Cyclic AMP-dependent activation of Rap1b. J. Biol. Chem. *270*, 10373-10376.

Altschuler, D.L. and Ribeiro-Neto, F. (1998) Mitogenic and oncogenic properties of the small G protein Rap1b. Proc. Natl. Acad. Sci. U.S.A. *95*, 7475-7479.

Arai, A., Nosaka, Y., Kanda, E., Yamamoto, K., Miyasaka, N. and Miura, O. (2001) Rap1 is activated by erythropoietin or interleukin-3 and is involved in regulation of beta1 integrin-mediated hematopoietic cell adhesion. J. Biol. Chem. *276*, 10453-10462.

Asha, H., de Ruiter, N.D., Wang, M.G. and Hariharan, I.K. (1999) The Rap1 GTPase functions as a regulator of morphogenesis in vivo. EMBO J. *18*, 605-615.

Bao, J.J., Le, X.F., Wang, R.Y., Yuan, J., Wang, L., Atkinson, E.N., LaPushin, R., Andreeff, M., Fang, B., Yu, Y. and Bast, R.C., Jr. (2002) Reexpression of the tumor suppressor gene ARHI induces apoptosis in ovarian and breast cancer cells through a caspase-independent calpain-dependent pathway. Cancer Res. *62*, 7264-7272.

Barker, K.T. and Crompton, M.R. (1998) Ras-related TC21 is activated by mutation in a breast cancer cell line, but infrequently in breast carcinomas in vivo. Br. J. Cancer. *78*, 296-300.

Barnier, J.V., Papin, C., Eychene, A., Lecoq, O. and Calothy, G. (1995) The mouse B-raf gene encodes multiple protein isoforms with tissue-specific expression. J. Biol. Chem. *270*, 23381-23389.

Bauer, B., Mirey, G., Vetter, I.R., Garcia-Ranea, J.A., Valencia, A., Wittinghofer, A., Camonis, J.H. and Cool, R.H. (1999a) Effector recognition by the small GTP-binding proteins Ras and Ral. J. Biol. Chem. *274*, 17763-17770.

Bauer, B., Mirey, G., Vetter, I.R., Garcia-Ranea, J.A., Valencia, A., Wittinghofer, A., Camonis, J.H. and Cool, R.H. (1999b) Effector recognition by the small GTP-binding proteins Ras and Ral. J. Biol. Chem. *274*, 17763-17770.

Beranger, F., Goud, B., Tavitian, A. and de Gunzburg, J. (1991) Association of the Ras-antagonistic Rap1/Krev-1 proteins with the Golgi complex. Proc. Natl. Acad. Sci. U.S.A. *88*, 1606-1610.

Bernards, A. (2003) GAPs galore! A survey of putative Ras superfamily GTPase activating proteins in man and Drosophila. Biochim. Biophys. Acta *1603*, 47-82.

Berruti, G. (2000) A novel rap1/B-Raf/14-3-3 theta protein complex is formed in vivo during the morphogenetic differentiation of postmeiotic male germ cells. Exp. Cell Res. *257*, 172-179.

Bertoni, A., Tadokoro, S., Eto, K., Pampori, N., Parise, L.V., White, G.C. and Shattil, S.J. (2002) Relationships between Rap1b, affinity modulation of integrin alpha IIbbeta 3, and the actin cytoskeleton. J. Biol. Chem. *277*, 25715-25721.

Bhattacharya, M., Anborgh, P.H., Babwah, A.V., Dale, L.B., Dobransky, T., Benovic, J.L., Feldman, R.D., Verdi, J.M., Rylett, R.J. and Ferguson, S.S. (2002) Beta-arrestins regulate a Ral-GDS Ral effector pathway that mediates cytoskeletal reorganization. Nat. Cell Biol. *4*, 547-555.

Boettner, B., Govek, E., Cross, J. and Van Aelst, L. (2000) The junctional multidomain protein AF-6 is a binding partner of the Rap1A GTPase and associates with the actin cytoskeletal regulator profilin. Proc. Natl. Acad. Sci. U.S.A. *97*, 9064-9069.

Bos, J.L. (1998) All in the family? New insights and questions regarding interconnectivity of Ras, Rap1 and Ral. EMBO J. *17*, 6776-6782.

Bos, J.L. (2003) Epac: a new cAMP target and new avenues in cAMP research. Nat. Rev. Mol. Cell. Biol. *4*, 733-738.

Bos, J.L., de Rooij, J. and Reedquist, K.A. (2001) Rap1 signalling: adhering to new models. Nat. Rev. Mol. Cell. Biol. *2*, 369-377.

Bos, J.L., De Bruyn, K., Enserink, J., Kuiperij, B., Rangarajan, S., Rehmann, H., Riedl, J., De Rooij, J., Van Mansfeld, F. and Zwartkruis, F. (2003) The role of Rap1 in integrin-mediated cell adhesion. Biochem. Soc. Trans. *31*, 83-86.

Bouschet, T., Perez, V., Fernandez, C., Bockaert, J., Eychene, A. and Journot, L. (2003) Stimulation of the ERK pathway by GTP-loaded Rap1 requires the concomitant activation of Ras, protein kinase C, and protein kinase A in neuronal cells. J. Biol. Chem. *278*, 4778-4785.

Brymora, A., Valova, V.A., Larsen, M.R., Roufogalis, B.D. and Robinson, P.J. (2001) The brain exocyst complex interacts with RalA in a GTP-dependent manner: identification of a novel mammalian Sec3 gene and a second Sec15 gene. J. Biol. Chem. *276*, 29792-29797.

Burgering, B.M.T., Pronk, G.J., van Weeren, P.C., Chardin, P. and Bos, J. (1993) cAMP antagonizes $p21^{ras}$-directed activation of extracellular signal-regulated kinase 2 and phosphorylation of mSos nucleotide exchange factor. EMBO J. *12*, 4211-4220.

Cantor, S.B., Urano, T. and Feig, L.A. (1995) Identification and characterization of Ral-binding protein 1, a potential downstream target of Ral GTPases. Mol. Cell. Biol. *15*, 4578-4584.

Carey, K.D., Watson, R.T., Pessin, J.E. and Stork, P.J.S. (2003) The requirement of specific membrane domains for Raf-1 phosphorylation and activation. J. Biol. Chem. *278*, 3185-3196.

Caron, E. (2003) Cellular functions of the Rap1 GTP-binding protein: a pattern emerges. J. Cell Sci. *116*, 435-440.

Caron, E., Self, A.J. and Hall, A. (2000) The GTPase Rap1 controls functional activation of macrophage integrin alphaMbeta2 by LPS and other inflammatory mediators. Curr. Biol. *10*, 974-978.

Castro, A.F., Rebhun, J.F., Clark, G.J. and Quilliam, L.A. (2003) Rheb binds tuberous sclerosis complex 2 (TSC2) and promotes S6 kinase activation in a rapamycin- and farnesylation-dependent manner. J. Biol. Chem. *278*, 32493-32496.

Chan, A.M., Miki, T., Meyers, K.A. and Aaronson, S.A. (1994) A human oncogene of the RAS superfamily unmasked by expression cDNA cloning. Proc. Natl. Acad. Sci. U.S.A. *91*, 7558-7562.

Chardin, P. and Tavitian, A. (1986) The ral gene: a new ras related gene isolated by the use of a synthetic probe. EMBO J. *5*, 2203-2208.

Chardin, P. and Tavitian, A. (1989) Coding sequences of human ralA and ralB cDNAs. Nucl. Acids Res. *17*, 4380.

Christensen, A.E., Selheim, F., de Rooij, J., Dremier, S., Schwede, F., Dao, K.K., Martinez, A., Maenhaut, C., Bos, J.L., Genieser, H.G. and Doskeland, S.O. (2003) cAMP analog mapping of Epac1 and cAMP kinase. Discriminating analogs demonstrate that Epac and cAMP kinase act synergistically to promote PC-12 cell neurite extension. J. Biol. Chem. *278*, 35394-35402.

Clark, G.J., Kinch, M.S., Gilmer, T.M., Burridge, K. and Der, C.J. (1996) Overexpression of the Ras-related TC21/R-Ras2 protein may contribute to the development of human breast cancers. Oncogene *12*, 169-176.

Clark, G.J., Kinch, M.S., Rogers-Graham, K., Sebti, S.M., Hamilton, A.D. and Der, C.J. (1997) The Ras-related protein Rheb is farnesylated and antagonizes Ras signaling and transformation. J. Biol. Chem. *272*, 10608-10615.

Clough, R.R., Sidhu, R.S. and Bhullar, R.P. (2002) Calmodulin binds RalA and RalB and is required for the thrombin-induced activation of Ral in human platelets. J. Biol. Chem. *277*, 28972-28980.

Clyde-Smith, J., Silins, G., Gartside, M., Grimmond, S., Etheridge, M., Apolloni, A., Hayward, N. and Hancock, J.F. (2000) Characterization of RasGRP2, a plasma membrane-targeted, dual specificity Ras/Rap exchange factor. J. Biol. Chem. *275*, 32260-32267.

Cook, S.J. and McCormick, F. (1993) Inhibition by cAMP of Ras-dependent activation of Raf. Science *262*, 1069-1072.

Cook, S.J., Rubinfeld, B., Albert, I. and McCormick, F. (1993) RapV12 antagonizes Ras-dependent activation of ERK1 and ERK2 by LPA and EGF in Rat-1 fibroblasts. EMBO J. *12*, 3475-3485.

Cox, A.D., Brtva, T.R., Lowe, D.G. and Der, C.J. (1994) R-Ras induces malignant, but not morphologic, transformation of NIH3T3 cells. Oncogene *9*, 3281-3288.

D'Adamo, D.R., Novick, S., Kahn, J.M., Leonardi, P. and Pellicer, A. (1997) rsc: a novel oncogene with structural and functional homology with the gene family of exchange factors for Ral. Oncogene *14*, 1295-1305.

de Bruyn, K.M., de Rooij, J., Wolthuis, R.M., Rehmann, H., Wesenbeek, J., Cool, R.H., Wittinghofer, A.H. and Bos, J.L. (2000) RalGEF2, a pleckstrin homology domain containing guanine nucleotide exchange factor for Ral. J. Biol. Chem. *275*, 29761-29766.

de Bruyn, K.M.T., Zwartkruis, F.T.J., de Rooij, J., Akkerman, J.-W.N. and Bos, J.L. (2003) The small GTPase Rap1 is activated by turbulence and involved in integrin alpha IIbbeta 3-mediated cell adhesion in human megakaryocytes. J. Biol. Chem. M212036200.

de Leeuw, H.P., Fernandez-Borja, M., Reits, E.A., Romani de Wit, T., Wijers-Koster, P.M., Hordijk, P.L., Neefjes, J., van Mourik, J.A. and Voorberg, J. (2001) Small GTP-binding protein Ral modulates regulated exocytosis of von Willebrand factor by endothelial cells. Arterioscler. Thromb. Vasc. Biol. *21*, 899-904.

de Rooij, J. and Bos, J.L. (1997) Minimal Ras-binding domain of Raf1 can be used as an activation-specific probe for Ras. Oncogene *14*, 623-625.

de Rooij, J., Zwartkruis, F.J., Verheijen, M.H., Cool, R.H., Nijman, S.M., Wittinghofer, A. and Bos, J.L. (1998) Epac is a Rap1 guanine-nucleotide-exchange factor directly activated by cyclic AMP. Nature *396*, 474-477.

de Rooij, J., Boenink, N.M., van Triest, M., Cool, R.H., Wittinghofer, A. and Bos, J.L. (1999) PDZ-GEF1, a guanine nucleotide exchange factor specific for Rap1 and Rap2. J. Biol. Chem. *274*, 38125-38130.

de Rooij, J., Rehmann, H., van Triest, M., Cool, R.H., Wittinghofer, A. and Bos, J.L. (2000) Mechanism of regulation of the Epac family of cAMP-dependent RapGEFs. J. Biol. Chem. *275*, 20829-20836.

De Ruiter, N.D., Burgering, B.M. and Bos, J.L. (2001) Regulation of the Forkhead transcription factor AFX by Ral-dependent phosphorylation of threonines 447 and 451. Mol. Cell. Biol. *21*, 8225-8235.

Der, C.J., Finkel, T. and Cooper, G.M. (1986) Biological and biochemical properties of human rasH genes mutated at codon 61. Cell *44*, 167-176.

Dodelet, V.C., Pazzagli, C., Zisch, A.H., Hauser, C.A. and Pasquale, E.B. (1999) A novel signaling intermediate, SHEP1, directly couples Eph receptors to R-Ras and Rap1A. J. Biol. Chem. *274*, 31941-31946.

Drivas, G.T., Shih, A., Coutavas, E., Rush, M.G. and D'Eustachio, P. (1990) Characterization of four novel ras-like genes expressed in a human teratocarcinoma cell line. Mol. Cell. Biol. *10*, 1793-1798.

Dugan, L.L., Kim, J.S., Zhang, Y., Bart, R.D., Sun, Y., Holtzman, D.M. and Gutmann, D.H. (1999) Differential effects of cAMP in neurons and astrocytes. Role of B-raf. J. Biol. Chem. *274*, 25842-25848.

Dumaz, N. and Marais, R. (2003) Protein Kinase A Blocks Raf-1 Activity by Stimulating 14-3-3 Binding and Blocking Raf-1 Interaction with Ras. J. Biol. Chem. *278*, 29819-29823.

Ebinu, J.O., Bottorff, D.A., Chan, E.Y., Stang, S.L., Dunn, R.J. and Stone, J.C. (1998) RasGRP, a Ras guanyl nucleotide-releasing protein with calcium- and diacylglycerol-binding motifs. Science *280*, 1082-1086.

Ehrhardt, G.R., Korherr, C., Wieler, J.S., Knaus, M. and Schrader, J.W. (2001) A novel potential effector of M-Ras and p21 Ras negatively regulates p21 Ras-mediated gene induction and cell growth. Oncogene *20*, 188-197.

Emkey, R., Freedman, S. and Feig, L.A. (1991) Characterization of a GTPase-activating protein for the Ras-related Ral protein. J. Biol. Chem. *266*, 9703-9706.

Enserink, J.M., Christensen, A.E., de Rooij, J., van Triest, M., Schwede, F., Genieser, H.G., Doskeland, S.O., Blank, J.L. and Bos, J.L. (2002) A novel Epac-specific cAMP analogue demonstrates independent regulation of Rap1 and ERK. Nat. Cell Biol. *4*, 901-906.

Esser, D., Bauer, B., Wolthuis, R.M., Wittinghofer, A., Cool, R.H. and Bayer, P. (1998a) Structure determination of the Ras-binding domain of the Ral-specific guanine nucleotide exchange factor Rlf. Biochemistry *37*, 13453-13462.

Esser, D., Bauer, B., Wolthuis, R.M.F., Wittinghofer, A., Cool, R.H. and Bayer, P. (1998b) Structure determination of the Ras-binding domain of the Ral-specific guanine nucleotide exchange factor Rlf. Biochemistry *37*, 13453-13462.

Evellin, S., Nolte, J., Tysack, K., vom Dorp, F., Thiel, M., Weernink, P.A.O., Jakobs, K.H., Webb, E.J., Lomasney, J.W. and Schmidt, M. (2002) Stimulation of phospholipase C-epsilon by the M3 muscarinic acetylcholine receptor mediated by cyclic AMP and the GTPase Rap2B. J. Biol. Chem. *277*, 16805-16813.

Fernandez-Sarabia, M.J. and Bischoff, J.R. (1993) Bcl-2 associates with the ras-related protein R-ras p23. Nature *366*, 274-275.

Franke, B., Akkerman, J.W. and Bos, J.L. (1997) Rapid Ca2+-mediated activation of Rap1 in human platelets. EMBO J. *16*, 252-259.

Franke, B., van Triest, M., de Bruijn, K.M., van Willigen, G., Nieuwenhuis, H.K., Negrier, C., Akkerman, J.W. and Bos, J.L. (2000) Sequential regulation of the small GTPase Rap1 in human platelets. Mol. Cell. Biol. *20*, 779-785.

Frech, M., John, J., Pizon, V., Chardin, P., Tavitian, A., Clark, R., McCormick, F. and Wittinghofer, A. (1990) Inhibition of GTPase activating protein stimulation of Ras-p21 GTPase by the Krev-1 gene product. Science *249*, 169-171.

Frodin, M., Peraldi, P. and Van Obberghen, E. (1994) Cyclic AMP activates the mitogen-activated protein kinase cascade in PC12 cells. J. Biol. Chem. *269*, 6207-6214.

Furuhjelm, J. and Peranen, J. (2003) The C-terminal end of R-Ras contains a focal adhesion targeting signal. J. Cell Sci. *116*, 3729-3738.

Gao, Q., Srinivasan, S., Boyer, S.N., Wazer, D.E. and Band, V. (1999) The E6 oncoproteins of high-risk papillomaviruses bind to a novel putative GAP protein, E6TP1, and target It for Degradation. Mol. Cell. Biol. *19*, 733-744.

Gao, X., Satoh, T., Liao, Y., Song, C., Hu, C.D., Kariya Ki, K. and Kataoka, T. (2001) Identification and characterization of RA-GEF-2, a Rap guanine nucleotide exchange factor that serves as a downstream target of M-Ras. J. Biol. Chem. *276*, 42219-42225.

Garami, A., Zwartkruis, F.J., Nobukuni, T., Joaquin, M., Roccio, M., Stocker, H., Kozma, S.C., Hafen, E., Bos, J.L. and Thomas, G. (2003) Insulin activation of Rheb, a mediator of mTOR/S6K/4E-BP signaling, is inhibited by TSC1 and 2. Mol. Cell *11*, 1457-1466.

Garcia, J., de Gunzburg, J., Eychene, A., Gisselbrecht, S. and Porteu, F. (2001) Thrombopoietin-mediated sustained activation of extracellular signal-regulated kinase in UT7-Mpl cells requires both Ras-Raf-1- and Rap1-B-Raf-dependent pathways. Mol. Cell. Biol. *21*, 2659-2670.

Geyer, M., Herrmann, C., Wohlgemuth, S., Wittinghofer, A. and Kalbitzer, H.R. (1997) Structure of the Ras-binding domain of RalGEF and implications for Ras binding and signalling. Nat. Struct. Biol. *4*, 694-699.

Gildea, J.J., Harding, M.A., Seraj, M.J., Gulding, K.M. and Theodorescu, D. (2002) The role of Ral A in epidermal growth factor receptor-regulated cell motility. Cancer Res. *62*, 982-985.

Gille, H. and Downward, J. (1999) Multiple Ras effector pathways contribute to G1 cell cycle progression. J. Biol. Chem. *274*, 22033-22040.

Goi, T., Shipitsin, M., Lu, Z., Foster, D.A., Klinz, S.G. and Feig, L.A. (2000) An EGF receptor/Ral-GTPase signaling cascade regulates c-Src activity and substrate specificity. EMBO J. *19*, 623-630.

Gotoh, T., Hattori, S., Nakamura, S., Kitayama, H., Noda, M., Takai, Y., Kaibuchi, K., Matsui, H., Hatase, O., Takahashi, H. et al. (1995) Identification of Rap1 as a target for the Crk SH3 domain-binding guanine nucleotide-releasing factor C3G. Mol. Cell. Biol. *15*, 6746-6753.

Gotoh, T., Cai, D., Tian, X., Feig, L.A. and Lerner, A. (2000) p130Cas regulates the activity of AND-34, a novel Ral, Rap1, and R-Ras guanine nucleotide exchange factor. J. Biol. Chem. *275*, 30118-30123.

Graham, S.M., Cox, A.D., Drivas, G., Rush, M.G., D'Eustachio, P. and Der, C.J. (1994) Aberrant function of the Ras-related protein TC21/R-Ras2 triggers malignant transformation. Mol. Cell. Biol. *14*, 4108-4115.

Graham, S.M., Vojtek, A.B., Huff, S.Y., Cox, A.D., Clark, G.J., Copper, J.A. and Der, C.J. (1996) TC21 causes transformation by Raf-independent signaling pathways. Mol. Cell. Biol. *16*, 6132-6140.

Graham, S.M., Oldham, S.M., Martin, C.B., Drugan, J.K., Zohn, I.E., Campbell, S. and Der, C.J. (1999) TC21 and Ras share indistinguishable transforming and differentiating activities. Oncogene *18*, 2107-2116.

Grewal, S.S., Horgan, A.M., York, R.D., Withers, G.S., Banker, G.A. and Stork, P.J. (2000) Neuronal calcium activates a Rap1 and B-Raf signaling pathway via the cyclic adenosine monophosphate-dependent protein kinase. J. Biol. Chem. *275*, 3722-3728.

Guo, F.-F., Kumahara, E. and Saffen, D. (2001) A CalDAG-GEFI/Rap1/B-Raf cassette couples M1 muscarinic acetylcholine receptors to the activation of ERK1/2. J. Biol. Chem. *276*, 25568-25581.

Hahn, W.C., Counter, C.M., Lundberg, A.S., Beijersbergen, R.L., Brooks, M.W. and Weinberg, R.A. (1999) Creation of human tumour cells with defined genetic elements. Nature *400*, 464-468.

Hamad, N.M., Elconin, J.H., Karnoub, A.E., Bai, W., Rich, J.N., Abraham, R.T., Der, C.J. and Counter, C.M. (2002) Distinct requirements for Ras oncogenesis in human versus mouse cells. Genes Dev. *16*, 2045-2057.

Hansen, M., Rusyn, E.V., Hughes, P.E., Ginsberg, M.H., Cox, A.D. and Willumsen, B.M. (2002) R-Ras C-terminal sequences are sufficient to confer R-Ras specificity to H-Ras. Oncogene *21*, 4448-4461.

Hart, P.A. and Marshall, C.J. (1990) Amino acid 61 is a determinant of sensitivity of rap proteins to the ras GTPase activating protein. Oncogene *5*, 1099-1101.

Hata, Y., Kaibuchi, K., Kawamura, S., Hiroyoshi, M., Shirataki, H. and Takai, Y. (1991) Enhancement of the actions of smg p21 GDP/GTP exchange protein by the protein kinase A-catalyzed phosphorylation of smg p21. J. Biol. Chem. *266*, 6571-6577.

Henry, D.O., Moskalenko, S.A., Kaur, K.J., Fu, M., Pestell, R.G., Camonis, J.H. and White, M.A. (2000) Ral GTPases contribute to regulation of cyclin D1 through activation of NF-kappaB. Mol. Cell. Biol. *20*, 8084-8092.

Herrmann, C., Horn, G., Spaargaren, M. and Wittinghofer, A. (1996) Differential interaction of the ras family GTP-binding proteins H-Ras, Rap1A, and R-Ras with the putative effector molecules Raf kinase and Ral-guanine nucleotide exchange factor. J. Biol. Chem. *271*, 6794-6800.

Hinoi, T., Kishida, S., Koyama, S., Ikeda, M., Matsuura, Y. and Kikuchi, A. (1996) Post-translational modifications of Ras and Ral are important for the action of Ral GDP dissociation stimulator. J. Biol. Chem. *271*, 19710-19716.

Hiroyoshi, M., Kaibuchi, K., Kawamura, S., Hata, Y. and Takai, Y. (1991) Role of the C-terminal region of smg p21, a ras-like small GTP-binding protein, in membrane and smg p21 GDP/GTP exchange protein interactions. J. Biol. Chem. *266*, 2962-2969.

Hofer, F., Fields, S., Schneider, C. and Martin, G.S. (1994) Activated Ras interacts with the Ral guanine nucleotide dissociation stimulator. Proc. Natl. Acad. Sci. U.S.A. *91*, 11089-11093.

Hofer, F., Berdeaux, R. and Martin, G.S. (1998) Ras-independent activation of Ral by a Ca(2+)-dependent pathway. Curr. Biol. *8*, 839-842.

Holden, J.L., Nur, E.K.M.S., Fabri, L., Nice, E., Hammacher, A. and Maruta, H. (1991) Rsr1 and Rap1 GTPases are activated by the same GTPase-activating protein and require threonine 65 for their activation. J. Biol. Chem. *266*, 16992-16995.

Hoshino, M. and Nakamura, S. (2003) Small GTPase Rin induces neurite outgrowth through Rac/Cdc42 and calmodulin in PC12 cells. J. Cell Biol. *163*, 1067-1076.

Hu, C.D., Kariya, K., Okada, T., Qi, X., Song, C. and Kataoka, T. (1999) Effect of phosphorylation on activities of Rap1A to interact with Raf-1 and to suppress Ras-dependent Raf-1 activation. J. Biol. Chem. *274*, 48-51.

Huang, L., Weng, X., Hofer, F., Martin, G.S. and Kim, S.H. (1997a) Three-dimensional structure of the Ras-interacting domain of RalGDS. Nat. Struct. Biol. *4*, 609-615.

Huang, L., Weng, X.W., Hofer, F., Martin, G.S. and Kim, S.H. (1997b) Three-dimensional structure of the Ras-interacting domain of RalGDS. Nat. Struct. Biol. *4*, 609-615.

Huang, L., Hofer, F., Martin, G.S. and Kim, S.H. (1998) Structural basis for the interaction of Ras with RalGDS. Nat. Struct. Biol. *5*, 422-426.

Ichiba, T., Hashimoto, Y., Nakaya, M., Kuraishi, Y., Tanaka, S., Kurata, T., Mochizuki, N. and Matsuda, M. (1999a) Activation of C3G guanine nucleotide exchange factor for Rap1 by phosphorylation of tyrosine 504. J. Biol. Chem. *274*, 14376-14381.

Ichiba, T., Hoshi, Y., Eto, Y., Tajima, N. and Kuraishi, Y. (1999b) Characterization of GFR, a novel guanine nucleotide exchange factor for Rap1. FEBS Lett. *457*, 85-89.

Ikeda, M., Ishida, O., Hinoi, T., Kishida, S. and Kikuchi, A. (1998) Identification and characterization of a novel protein interacting with Ral-binding protein 1, a putative effector protein of Ral. J. Biol. Chem. *273*, 814-821.

Im, E., von Lintig, F.C., Chen, J., Zhuang, S., Qui, W., Chowdhury, S., Worley, P.F., Boss, G.R. and Pilz, R.B. (2002) Rheb is in a high activation state and inhibits B-Raf kinase in mammalian cells. Oncogene *21*, 6356-6365.

Iouzalen, N., Camonis, J. and Moreau, J. (1998) Identification and characterization in Xenopus of XsmgGDS, a RalB binding protein. Biochem. Biophys. Res. Commun. *250*, 359-363.

Ishida, D., Kometani, K., Yang, H., Kakugawa, K., Masuda, K., Iwai, K., Suzuki, M., Itohara, S., Nakahata, T., Hiai, H., Kawamoto, H., Hattori, M. and Minato, N. (2003) Myeloproliferative stem cell disorders by deregulated Rap1 activation in SPA-1-deficient mice. Cancer Cell *4*, 55-65.

Itoh, T., Kaibuchi, K., Sasaki, T. and Takai, Y. (1991) The smg GDS-induced activation of smg p21 is initiated by cyclic AMP-dependent protein kinase-catalyzed phosphorylation of smg p21. Biochem. Biophys. Res. Commun. *177*, 1319-1324.

Janoueix-Lerosey, I., Polakis, P., Tavitian, A. and de Gunzburg, J. (1992) Regulation of the GTPase activity of the ras-related rap2 protein. Biochem. Biophys. Res. Comm. *189*, 455-464.

Janoueix-Lerosey, I., Pasheva, E., de Tand, M.F., Tavitian, A. and de Gunzburg, J. (1998) Identification of a specific effector of the small GTP-binding protein Rap2. Eur. J. Biochem. *252*, 290-298.

Jiang, H., Luo, J.Q., Urano, T., Frankel, P., Lu, Z., Foster, D.A. and Feig, L.A. (1995) Involvement of Ral GTPase in v-Src-induced phospholipase D activation. Nature *378*, 409-412.

Jin, T.G., Satoh, T., Liao, Y., Song, C., Gao, X., Kariya, K., Hu, C.D. and Kataoka, T. (2001) Role of the CDC25 homology domain of phospholipase Cepsilon in amplification of Rap1-dependent signaling. J. Biol. Chem. *276*, 30301-30307.

Joneson, T., White, M.A., Wigler, M.H. and Bar-Sagi, D. (1996) Stimulation of membrane ruffling and MAP kinase activation by distinct effectors of RAS. Science *271*, 810-812.

Jullien-Flores, V., Dorseuil, O., Romero, F., Letourneur, F., Saragosti, S., Berger, R., Tavitian, A., Gacon, G. and Camonis, J.H. (1995) Bridging Ral GTPase to Rho pathways. RLIP76, a Ral effector with CDC42/Rac GTPase-activating protein activity. J. Biol. Chem. *270*, 22473-22477.

Jullien-Flores, V., Mahe, Y., Mirey, G., Leprince, C., Meunier-Bisceuil, B., Sorkin, A. and Camonis, J.H. (2000) RLIP76, an effector of the GTPase Ral, interacts with the AP2 complex: involvement of the Ral pathway in receptor endocytosis. J. Cell Sci. *113*, 2837-2844.

Kaibuchi, K., Mizuno, T., Fujioka, H., Yamamoto, T., Kishi, K., Fukumoto, Y., Hori, Y. and Takai, Y. (1991) Molecular cloning of the cDNA for stimulatory GDP/GTP exchange protein for smg p21s (ras p21-like small GTP-binding proteins) and characterization of stimulatory GDP/GTP exchange protein. Mol. Cell. Biol. *11*, 2873-2880.

Kao, S., Jaiswal, R.K., Kolch, W. and Landreth, G.E. (2001) Identification of the mechanisms regulating the differential activation of the mapk cascade by epidermal growth factor and nerve growth factor in PC12 cells. J. Biol. Chem. *276*, 18169-18177.

Kariya, K., Koyama, S., Nakashima, S., Oshiro, T., Morinaka, K. and Kikuchi, A. (2000) Regulation of complex formation of POB1/epsin/adaptor protein complex 2 by mitotic phosphorylation. J. Biol. Chem. *275*, 18399-18406.

Katagiri, K., Hattori, M., Minato, N., Irie, S., Takatsu, K. and Kinashi, T. (2000) Rap1 is a potent activation signal for leukocyte function-associated antigen 1 distinct from protein kinase C and phosphatidylinositol-3-OH kinase. Mol. Cell. Biol. *20*, 1956-1969.

Katagiri, K., Hattori, M., Minato, N. and Kinashi, T. (2002) Rap1 Functions as a Key Regulator of T-Cell and Antigen–Presenting Cell Interactions and Modulates T-Cell Responses. Mol. Cell. Biol. *22*, 1001-1015.

Katagiri, K., Maeda, A., Shimonaka, M. and Kinashi, T. (2003) RAPL, a Rap1-binding molecule that mediates Rap1-induced adhesion through spatial regulation of LFA-1. Nat. Immunol. *4*, 741-748.

Kawasaki, H., Springett, G.M., Mochizuki, N., Toki, S., Nakaya, M., Matsuda, M., Housman, D.E. and Graybiel, A.M. (1998a) A family of cAMP-binding proteins that directly activate Rap1. Science *282*, 2275-2279.

Kawasaki, H., Springett, G.M., Toki, S., Canales, J.J., Harlan, P., Blumenstiel, J.P., Chen, E.J., Bany, I.A., Mochizuki, N., Ashbacher, A., Matsuda, M., Housman, D.E. and Graybiel, A.M. (1998b) A Rap guanine nucleotide exchange factor enriched highly in the basal ganglia. Proc. Natl. Acad. Sci. U.S.A. *95*, 13278-13283.

Kawata, M., Kikuchi, A., Hoshijima, M., Yamamoto, K., Hashimoto, E., Yamamura, H. and Takai, Y. (1989) Phosphorylation of smg p21, a ras p21-like GTP-binding protein, by cyclic AMP-dependent protein kinase in a cell-free system and in response to prostaglandin E1 in intact human platelets. J. Biol. Chem. *264*, 15688-15695.

Keely, P.J., Rusyn, E.V., Cox, A.D. and Parise, L.V. (1999) R-Ras signals through specific integrin alpha cytoplasmic domains to promote migration and invasion of breast epithelial cells. J. Cell Biol. *145*, 1077-1088.

Khosravi-Far, R., White, M.A., Westwick, J.K., Solski, P.A., Chrzanowska-Wodnicka, M., Van Aelst, L., Wigler, M.H. and Der, C.J. (1996) Oncogenic Ras activation of Raf/mitogen-activated protein kinase-independent pathways is sufficient to cause tumorigenic transformation. Mol. Cell. Biol. *16*, 3923-3933.

Kigawa, T., Endo, M., Ito, Y., Shirouzu, M., Kikuchi, A. and Yokoyama, S. (1998) Solution structure of the Ras-binding domain of RGL. FEBS Lett. *441*, 413-418.

Kikuchi, A., Demo, S.D., Ye, Z.H., Chen, Y.W. and Williams, L.T. (1994) ralGDS family members interact with the effector loop of ras p21. Mol. Cell. Biol. *14*, 7483-7491.

Kim, J.H., Lee, S.D., Han, J.M., Lee, T.G., Kim, Y., Park, J.B., Lambeth, J.D., Suh, P.G. and Ryu, S.H. (1998) Activation of phospholipase D1 by direct interaction with ADP-ribosylation factor 1 and RalA. FEBS Lett. *430*, 231-235.

Kim, R., Trubetskoy, A., Suzuki, T., Jenkins, N.A., Copeland, N.G. and Lenz, J. (2003) Genome-based identification of cancer genes by proviral tagging in mouse retrovirus-induced T-cell lymphomas. J. Virol. *77*, 2056-2062.

Kimmelman, A., Tolkacheva, T., Lorenzi, M.V., Osada, M. and Chan, A.M.L. (1997) Identification and characterization of R-ras3: a novel member of the RAS gene family with a non-ubiquitous pattern of tissue distribution. Oncogene *15*, 2675-2685.

Kimmelman, A.C., Osada, M. and Chan, A.M. (2000) R-Ras3, a brain-specific Ras-related protein, activates Akt and promotes cell survival in PC12 cells. Oncogene *19*, 2014-2022.

Kimmelman, A.C., Rodriguez, N.N. and Chan, A.M.-L. (2002) R-Ras3/M-Ras Induces Neuronal Differentiation of PC12 Cells through Cell-Type-Specific Activation of the Mitogen-Activated Protein Kinase Cascade. Mol. Cell. Biol. *22*, 5946-5961.

Kinashi, T., Katagiri, K., Watanabe, S., Vanhaesebroeck, B., Downward, J. and Takatsu, K. (2000) Distinct mechanisms of alpha 5beta 1 integrin activation by Ha-Ras and R-Ras. J. Biol. Chem. *275*, 22590-22596.

Kishida, S., Koyama, S., Matsubara, K., Kishida, M., Matsuura, Y. and Kikuchi, A. (1997) Colocalization of Ras and Ral on the membrane is required for Ras-dependent Ral activation through Ral GDP dissociation stimulator. Oncogene *15*, 2899-2907.

Kitayama, H., Sugimoto, Y., Matsuzaki, T., Ikawa, Y. and Noda, M. (1989) A ras-related gene with transformation suppressor activity. Cell *56*, 77-84.

Knox, A.L. and Brown, N.H. (2002) Rap1 GTPase regulation of adherens junction positioning and cell adhesion. Science *295*, 1285-1288.

Kontani, K., Tada, M., Ogawa, T., Okai, T., Saito, K., Araki, Y. and Katada, T. (2002) Di-Ras: A distinct subgroup of Ras-family GTPases with unique biochemical properties. J. Biol. Chem. M202150200.

Koopman, W.J.H., Bosch, R.R., van Emst-de Vries, S.E., Spaargaren, M., De Pont, J.J.H.H.M. and Willems, P.H.G.M. (2003) R-RAS alters CA2+ homeostasis by increasing the CA2+ leak across the endoplasmic reticular membrane. J. Biol. Chem. M211256200.

Kops, G.J., de Ruiter, N.D., De Vries-Smits, A.M., Powell, D.R., Bos, J.L. and Burgering, B.M. (1999) Direct control of the Forkhead transcription factor AFX by protein kinase B. Nature *398*, 630-634.

Kranenburg, O., Verlaan, I. and Moolenaar, W.H. (2001) Regulating c-Ras function; cholesterol depletion affects caveolin association, GTP loading, and signaling. Curr. Biol. *11*, 1880-1884.

Kuiperij, H.B., de Rooij, J., Rehmann, H., van Triest, M., Wittinghofer, A., Bos, J.L. and Zwartkruis, F.J. (2003) Characterisation of PDZ-GEFs, a family of guanine nucleotide exchange factors specific for Rap1 and Rap2. Biochim. Biophys. Acta *1593*, 141-149.

Kurachi, H., Wada, Y., Tsukamoto, N., Maeda, M., Kubota, H., Hattori, M., Iwai, K. and Minato, N. (1997) Human SPA-1 gene product selectively expressed in lymphoid tissues is a specific GTPase-activating protein for Rap1 and Rap2 – segregate expression profiles from a rap1GAP gene product. J. Biol. Chem. *272*, 28081-28088.

Kwong, L., Wozniak, M.A., Collins, A.S., Wilson, S.D. and Keely, P.J. (2003) R-Ras promotes focal adhesion formation through focal adhesion kinase and p130Cas by a novel mechanism that differs from integrins. Mol. Cell. Biol. *23*, 933-949.

Laberge-ie Couteulx, S., Jung, H.H., Labauge, P., Houtteville, J.P., Lescoat, C., Cecillon, M., Marechal, E., Joutel, A., Bach, J.F. and Tournier-Lasserve, E. (1999) Truncating mutations in CCM1, encoding KRIT1, cause hereditary cavernous angiomas. Nat. Genet. *23*, 189-193.

Lafuente, E.M., van Puijebroek, A.A., Krause, M., Carman, C.V., Freedman, G.J., Berezovskaya, A., Constantine, E., Springer, T.A., Gertler, F.B. and Boussiotis, V.A. (2004) RIAM, an Ena/VASP and Profilin Ligand, interacts with Rap1-GTP and mediates Rap1-induced adhesion. Dev. Cell *7*, 585-595.

Lapetina, E.G., Lacal, J.C., Reep, B.R. and Molina y Vedia, L. (1989) A ras-related protein is phosphorylated and translocated by agonists that increase cAMP levels in human platelets. Proc. Natl. Acad. Sci. U.S.A. *86*, 3131-3134.

Lee, A.C., Fenster, B.E., Ito, H., Takeda, K., Bae, N.S., Hirai, T., Yu, Z.X., Ferrans, V.J., Howard, B.H. and Finkel, T. (1999) Ras proteins induce senescence by altering the intracellular levels of reactive oxygen species. J. Biol. Chem. *274*, 7936-7940.

Lee, C.-H.J., Della, N.G., Chew, C.E. and Zack, D.J. (1996a) Rin, a Neuron-specific and calmodulin-binding small G-protein, and Rit define a novel subfamily of Ras proteins. J. Neurosci. *16*, 6784-6794.

Lee, T., Feig, L. and Montell, D.J. (1996b) Two distinct roles for Ras in a developmentally regulated cell migration. Development *122*, 409-418.

Lerosey, I., Chardin, P., de Gunzburg, J. and Tavitian, A. (1991a) The product of the rap2 gene, member of the ras superfamily. Biochemical characterization and site-directed mutagenesis. J. Biol. Chem. *266*, 4315-4321.

Lerosey, I., Pizon, V., Tavitian, A. and de Gunzburg, J. (1991b) The cAMP-dependent protein kinase phosphorylates the rap1 protein in vitro as well as in intact fibroblasts, but not the closely related rap2 protein. Biochem. Biophys. Res. Commun. *175*, 430-436.

Liao, Y., Kariya, K., Hu, C.D., Shibatohge, M., Goshima, M., Okada, T., Watari, Y., Gao, X., Jin, T.G., Yamawaki-Kataoka, Y. and Kataoka, T. (1999) RA-GEF, a novel Rap1A guanine nucleotide exchange factor containing a Ras/Rap1A-associating domain, is conserved between nematode and humans. J. Biol. Chem. *274*, 37815-37820.

Liao, Y., Satoh, T., Gao, X., Jin, T.-G., Hu, C.-D. and Kataoka, T. (2001) RA-GEF-1, a guanine nucleotide exchange factor for Rap1, is activated by translocation induced by association with Rap1·GTP and enhances Rap1-dependent B-Raf activation. J. Biol. Chem. *276*, 28478-28483.

Ling, L., Zhu, T. and Lobie, P.E. (2003) Src-CrkII-C3G dependent activation of Rap1 switches growth hormone stimulated p44/42 MAP kinase and JNK/SAPK activities. J. Biol. Chem. *278*, 27301-27311.

Lopez-Barahona, M., Bustelo, X.R. and Barbacid, M. (1996) The TC21 oncoprotein interacts with the Ral guanosine nucleotide dissociation factor. Oncogene *12*, 463-470.

Lou, L., Urbani, J., Ribeiro-Neto, F. and Altschuler, D.L. (2002) cAMP inhibition of Akt is mediated by activated and phosphorylated Rap1b. J. Biol. Chem. *277*, 32799-32806.

Lowe, D.G. and Goeddel, D.V. (1987) Heterologous expression and characterization of the human R-ras gene product. Mol. Cell. Biol. *7*, 2845-2856.

Lowe, D.G., Capon, D.J., Delwart, E., Sakaguchi, A.Y., Naylor, S.L. and Goeddel, D.V. (1987) Structure of the human and murine R-ras genes, novel genes closely related to ras proto-oncogenes. Cell *48*, 137-146.

Luo, J.Q., Liu, X., Hammond, S.M., Colley, W.C., Feig, L.A., Frohman, M.A., Morris, A.J. and Foster, D.A. (1997) RalA interacts directly with the Arf-responsive, PIP2-dependent phospholipase D1. Biochem. Biophys. Res. Commun. *235*, 854-859.

Luo, J.Q., Liu, X., Frankel, P., Rotunda, T., Ramos, M., Flom, J., Jiang, H., Feig, L.A., Morris, A.J., Kahn, R.A. and Foster, D.A. (1998) Functional association between Arf and RalA in active phospholipase D complex. Proc. Natl. Acad. Sci. U.S.A. *95*, 3632-3637.

Luo, R.Z., Fang, X., Marquez, R., Liu, S.Y., Mills, G.B. and Liao, W.S., Yu, Y. and Bast, R.C. (2003) ARHI is a Ras-related small G-protein with a novel N-terminal extension that inhibits growth of ovarian and breast cancers. Oncogene *22*, 2897-2909.

M'Rabet, L., Coffer, P., Zwartkruis, F., Franke, B., Segal, A.W., Koenderman, L. and Bos, J.L. (1998) Activation of the small GTPase rap1 in human neutrophils. Blood *92*, 2133-2140.

Mach, K.E., Furge, K.A. and Albright, C.F. (2000) Loss of Rhb1, a Rheb-related GTPase in fission yeast, causes growth arrest with a terminal phenotype similar to that caused by nitrogen starvation. Genetics *155*, 611-622.

Maillet, M., Robert, S.J., Cacquevel, M., Gastineau, M., Vivien, D., Bertoglio, J., Zugaza, J.L., Fischmeister, R. and Lezoualc'h, F. (2003) Crosstalk between Rap1 and Rac regulates secretion of sAPPalpha. Nat. Cell Biol. *5*, 633-639.

Marshall, C.J. (1995) Specificity of receptor tyrosine kinase signaling: transient versus sustained extracellular signal-regulated kinase activation. Cell *80*, 179-185.

Marte, B.M., RodriguezViciana, P., Wennstrom, S., Warne, P.H. and Downward, J. (1997) R-Ras can activate the phosphoinositide 3-kinase but not the MAP kinase arm of the Ras effector pathways. Curr. Biol. *7*, 63-70.

Maruta, H., Holden, J., Sizeland, A. and D'Abaco, G. (1991) The residues of ras and rap that determine their GAP specificities. J. Biol. Chem. *266*, 11661-11668.

Matsubara, K., Kishida, S., Matsuura, Y., Kitayama, H., Noda, M. and Kikuchi, A. (1999) Plasma membrane recruitment of RalGDS is critical for Ras-dependent Ral activation. Oncogene *18*, 1303-1312.

Matsuzaki, T., Hanai, S., Kishi, H., Liu, Z., Bao, Y., Kikuchi, A., Tsuchida, K. and Sugino, H. (2002) Regulation of endocytosis of activin type II receptors by a novel PDZ protein through Ral/Ral-binding protein 1-dependent pathway. J. Biol. Chem. *277*, 19008-19018.

McLeod, S.J. and Gold, M.R. (2001) Activation and function of the Rap1 GTPase in B lymphocytes. Int. Rev. Immunol. *20*, 763-789.

Meller, N., Irani-Tehrani, M., Kiosses, W.B., Del Pozo, M.A. and Schwartz, M.A. (2002) Zizimin1, a novel Cdc42 activator, reveals a new GEF domain for Rho proteins. Nat. Cell Biol. *4*, 639-647.

Mirey, G., Balakireva, M., L'Hoste, S., Rosse, C., Voegeling, S. and Camonis, J. (2003) A Ral guanine exchange factor-Ral pathway is conserved in Drosophila melanogaster and sheds new light on the connectivity of the Ral, Ras, and Rap pathways. Mol. Cell. Biol. *23*, 1112-1124.

Mizuno, T., Kaibuchi, K., Yamamoto, T., Kawamura, M., Sakoda, T., Fujioka, H., Matsuura, Y. and Takai, Y. (1991) A stimulatory GDP/GTP exchange protein for smg p21 is active on the post-translationally processed form of c-Ki-ras p21 and rhoA p21. Proc. Natl. Acad. Sci. U.S.A. *88*, 6442-6446.

Mochizuki, N., Yamashita, S., Kurokawa, K., Ohba, Y., Nagai, T., Miyawaki, A. and Matsuda, M. (2001) Spatio-temporal images of growth-factor-induced activation of Ras and Rap1. Nature *411*, 1065-1068.

Morozov, A., Muzzio, I.A., Bourtchouladze, R., Van-Strien, N., Lapidus, K., Yin, D., Winder, D.G., Adams, J.P., Sweatt, J.D. and Kandel, E.R. (2003) Rap1 couples cAMP signaling to a distinct pool of p42/44MAPK regulating excitability, synaptic plasticity, learning, and memory. Neuron *39*, 309-325.

Moskalenko, S., Henry, D.O., Rosse, C., Mirey, G., Camonis, J.H. and White, M.A. (2002) The exocyst is a Ral effector complex. Nat. Cell Biol. *4*, 66-72.

Movilla, N., Crespo, P. and Bustelo, X.R. (1999) Signal transduction elements of TC21, an oncogenic member of the R-Ras subfamily of GTP-binding proteins. Oncogene *18*, 5860-5869.

Murai, H., Ikeda, M., Kishida, S., Ishida, O., Okazaki-Kishida, M., Matsuura, Y. and Kikuchi, A. (1997) Characterization of Ral GDP dissociation stimulator-like (RGL) activities to regulate c-fos promoter and the GDP/GTP exchange of Ral. J. Biol. Chem. *272*, 10483-10490.

Murphy, G.A., Graham, S.M., Morita, S., Reks, S.E., Rogers-Graham, K., Vojtek, A., Kelley, G.G. and Der, C.J. (2002) Involvement of phosphatidylinositol 3-kinase, but not RalGDS, in TC21/R-Ras2-mediated transformation. J. Biol. Chem. *277*, 9966-9975.

Nakashima, S., Morinaka, K., Koyama, S., Ikeda, M., Kishida, M., Okawa, K., Iwamatsu, A., Kishida, S. and Kikuchi, A. (1999) Small G protein Ral and its downstream molecules regulate endocytosis of EGF and insulin receptors. EMBO J. *18*, 3629-3642.

Nancy, V., Wolthuis, R.M., de Tand, M.F., Janoueix-Lerosey, I., Bos, J.L. and de Gunzburg, J. (1999a) Identification and characterization of potential effector molecules of the Ras-related GTPase Rap2. J. Biol. Chem. *274*, 8737-8745.

Nancy, V., Wolthuis, R.M.F., de Tand, M.-F., Janoueix-Lerosey, I., Bos, J.L. and de Gunzburg, J. (1999b) Identification and characterization of potential effector molecules of the Ras-related GTPase Rap2. J. Biol. Chem. *274*, 8737-8745.

Nancy, V., Callebaut, I., El Marjou, A. and de Gunzburg, J. (2002) The delta subunit of retinal rod cGMP phosphodiesterase regulates the membrane association of Ras and Rap GTPases. J. Biol. Chem. *277*, 15076-15084.

Nassar, N., Horn, G., Herrmann, C., Scherer, A., McCormick, F. and Wittinghofer, A. (1995) The 2.2 A crystal structure of the Ras-binding domain of the serine/threonine kinase c-Raf1 in complex with Rap1A and a GTP analogue. Nature *375*, 554-560.

Nassar, N., Horn, G., Herrmann, C., Block, C., Janknecht, R., and Wittinghofer, A. (1996) Ras/Rap effector specificity determined by charge reversal. Nat. Struct. Biol. *3*, 723-729.

Nice, E.C., Fabri, L., Hammacher, A., Holden, J., Simpson, R.J. and Burgess, A.W. (1992) The purification of a Rap1 GTPase-activating protein from bovine brain cytosol. J. Biol. Chem. *267*, 1546-1553.

Niedergang, F., Colucci-Guyon, E., Dubois, T., Raposo, G. and Chavrier, P. (2003) ADP ribosylation factor 6 is activated and controls membrane delivery during phagocytosis in macrophages. J. Cell Biol. *161*, 1143-1150.

Oertli, B., Han, J., Marte, B.M., Sethi, T., Downward, J., Ginsberg, M. and Hughes, P.E. (2000) The effector loop and prenylation site of R-Ras are involved in the regulation of integrin function. Oncogene *19*, 4961-4969.

Ohba, Y., Mochizuki, N., Matsuo, K., Yamashita, S., Nakaya, M., Hashimoto, Y., Hamaguchi, M., Kurata, T., Nagashima, K. and Matsuda, M. (2000) Rap2 as a slowly responding molecular switch in the Rap1 signaling cascade. Mol. Cell. Biol. *20*, 6074-6083.

Ohba, Y., Ikuta, K., Ogura, A., Matsuda, J., Mochizuki, N., Nagashima, K., Kurokawa, K., Mayer, B.J., Maki, K., Miyazaki, J. and Matsuda, M. (2001) Requirement for C3G-dependent Rap1 activation for cell adhesion and embryogenesis. EMBO J. *20*, 3333-3341.

Ohba, Y., Kurokawa, K. and Matsuda, M. (2003) Mechanism of the spatio-temporal regulation of Ras and Rap1. EMBO J. *22*, 859-869.

Ohta, Y., Suzuki, N., Nakamura, S., Hartwig, J.H. and Stossel, T.P. (1999) The small GTPase RalA targets filamin to induce filopodia. Proc. Natl. Acad. Sci. U.S.A. *96*, 2122-2128.

Ohtsuka, T., Shimizu, K., Yamamori, B., Kuroda, S. and Takai, Y. (1996) Activation of brain B-Raf protein kinase by Rap1B small GTP-binding protein. J. Biol. Chem. *271*, 1258-1261.

Okada, S. and Pessin, J.E. (1997) Insulin and epidermal growth factor stimulate a conformational change in Rap1 and dissociation of the CrkII-C3G complex. J. Biol. Chem. *272*, 28179-28182.

Okada, S., Matsuda, M., Anafi, M., Pawson, T. and Pessin, J.E. (1998) Insulin regulates the dynamic balance between Ras and Rap1 signaling by coordinating the assembly states of the Grb2-SOS and CrkII-C3G complexes. EMBO J. *17*, 2554-2565.

Okada, T., Hu, C.D., Jin, T.G., Kariya, K., Yamawaki-Kataoka, Y. and Kataoka, T. (1999) The strength of interaction at the Raf cysteine-rich domain is a critical determinant of response of Raf to Ras family small GTPases. Mol. Cell. Biol. *19*, 6057-6064.

Okazaki, M., Kishida, S., Hinoi, T., Hasegawa, T., Tamada, M., Kataoka, T. and Kikuchi, A. (1997) Synergistic activation of c-fos promoter activity by Raf and Ral GDP dissociation stimulator. Oncogene *14*, 515-521.

Osada, M., Tolkacheva, T., Li, W.Q., Chan, T.O., Tsichlis, P.N., Saez, R., Kimmelman, A.C. and Chan, A.M.L. (1999) Differential roles of Akt, Rac, and Ral in R-Ras-mediated cellular transformation, adhesion, and survival. Mol. Cell. Biol. *19*, 6333-6344.

Pak, D.T., Yang, S., Rudolph-Correia, S., Kim, E. and Sheng, M. (2001) Regulation of dendritic spine morphology by SPAR, a PSD-95-associated RapGAP. Neuron *31*, 289-303.

Pak, D.T.S. and Sheng, M. (2003) Targeted protein degradation and synapse remodeling by an inducible protein kinase. Science *302*, 1368-1373.

Panepinto, J.C., Oliver, B.G., Amlung, T.W., Askew, D.S. and Rhodes, J.C. (2002) Expression of the Aspergillus fumigatus rheb homologue, rhbA, is induced by nitrogen starvation. Fungal Genet Biol. *36*, 207-214.

Park, S.H. and Weinberg, R.A. (1995) A putative effector of Ral has homology to Rho/Rac GTPase activating proteins. Oncogene *11*, 2349-2355.

Patel, P.H., Thapar, N., Guo, L., Martinez, M., Maris, J., Gau, C.-L., Lengyel, J.A. and Tamanoi, F. (2003) Drosophila Rheb GTPase is required for cell cycle progression and cell growth. J. Cell Sci. *116*, 3601-3610.

Peng, H., Xu, F., Pershad, R., Hunt, K.K., Frazier, M.L., Berchuck, A., Gray, J.W., Hogg, D., Bast, R.C., Jr. and Yu, Y. (2000) ARHI is the center of allelic deletion on chromosome 1p31 in ovarian and breast cancers. Int. J. Cancer *86*, 690-694.

Peterson, S.N., Trabalzini, L., Brtva, T.R., Fischer, T., Altschuler, D.L., Martelli, P., Lapetina, E.G., Der, C.J. and White, G.C.N. (1996) Identification of a novel RalGDS-related protein as a candidate effector for Ras and Rap1. J. Biol. Chem. *271*, 29903-29908.

Pizon, V., Chardin, P., Lerosey, I., Olofsson, B. and Tavitian, A. (1988) Human cDNAs rap1 and rap2 homologous to the Drosophila gene Dras3 encode proteins closely related to ras in the 'effector' region. Oncogene *3*, 201-204.

Pizon, V., Desjardins, M., Bucci, C., Parton, R.G. and Zerial, M. (1994) Association of Rap1a and Rap1b proteins with late endocytic/phagocytic compartments and Rap2a with the Golgi complex. J. Cell Sci. *107*, 1661-1670.

Polakis, P.G., Rubinfeld, B., Evans, T. and McCormick, F. (1991) Purification of a plasma membrane-associated GTPase-activating protein specific for rap1/Krev-1 from HL60 cells. Proc. Natl. Acad. Sci. U.S.A. *88*, 239-243.

Polzin, A., Shipitsin, M., Goi, T., Feig, L.A. and Turner, T.J. (2002) Ral-GTPase influences the regulation of the readily releasable pool of synaptic vesicles. Mol. Cell. Biol. *22*, 1714-1722.

Posern, G., Rapp, U.R. and Feller, S.M. (2000) The Crk signaling pathway contributes to the bombesin-induced activation of the small GTPase Rap1 in Swiss 3T3 cells. Oncogene *19*, 6361-6368.

Qiu, W., Zhuang, S., von Lintig, F.C., Boss, G.R. and Pilz, R.B. (2000) Cell type-specific regulation of B-Raf kinase by cAMP and 14-3-3 proteins. J. Biol. Chem. *275*, 31921-31929.

Quaroni, A. and Paul, E.C. (1999) Cytocentrin is a Ral-binding protein involved in the assembly and function of the mitotic apparatus. J. Cell Sci. *112*, 707-718.

Quilliam, L.A., Der, C.J., Clark, R., O'Rourke, E.C., Zhang, K., McCormick, F. and Bokoch, G.M. (1990) Biochemical characterization of baculovirus-expressed rap1A/Krev-1 and its regulation by GTPase-activating proteins. Mol. Cell. Biol. *10*, 2901-2908.

Quilliam, L.A., Castro, A.F., RogersGraham, K.S., Martin, C.B., Der, C.J. and Bi, C. (1999) M-Ras/R-Ras3, a transforming Ras protein regulated by Sos1, GRF1, and p120 Ras GTPase-activating protein, interacts with the putative Ras effector AF6. J. Biol. Chem. *274*, 23850-23857.

Rangarajan, S., Enserink, J.M., Kuiperij, H.B., de Rooij, J., Price, L.S., Schwede, F. and Bos, J.L. (2003) Cyclic AMP induces integrin-mediated cell adhesion through Epac and Rap1 upon stimulation of the β2-adrenergic receptor. J. Cell Biol. *160*, 487-493.

Rebhun, J.F., Castro, A.F. and Quilliam, L.A. (2000a) Identification of guanine nucleotide exchange factors (GEFs) for the Rap1 GTPase. Regulation of MR-GEF by M-Ras–GTP interaction. J. Biol. Chem. *275*, 34901-34908.

Rebhun, J.F., Chen, H. and Quilliam, L.A. (2000b) Identification and characterization of a new family of guanine nucleotide exchange factors for the Ras-related GTPase Ral. J. Biol. Chem. *275*, 13406-13410.

Reedquist, K.A., Ross, E., Koop, E.A., Wolthuis, R.M., Zwartkruis, F.J., van Kooyk, Y., Salmon, M., Buckley, C.D. and Bos, J.L. (2000) The small GTPase, Rap1, mediates CD31-induced integrin adhesion. J. Cell Biol. *148*, 1151-1158.

Rehmann, H., Prakash, B., Wolf, E., Rueppel, A., De Rooij, J., Bos, J.L. and Wittinghofer, A. (2003a) Structure and regulation of the cAMP-binding domains of Epac2. Nat. Struct. Biol. *10*, 26-32.

Rehmann, H., Rueppel, A., Bos, J.L. and Wittinghofer, A. (2003b) Communication between the regulatory and the catalytic domain of a cAMP responsive guanine nucleotide exchange factor Epac. J. Biol. Chem. M301680200.

Ren, X.D. and Schwartz, M.A. (2000) Determination of GTP loading on Rho. Meth. Enzymol. *325*, 264-272.

Ribeiro-Neto, F., Urbani, J., Lemee, N., Lou, L. and Altschuler, D.L. (2002) On the mitogenic properties of Rap1b: cAMP-induced G1/S entry requires activated and phosphorylated Rap1b. PNAS *99*, 5418-5423.

Rittinger, K., Walker, P.A., Eccleston, J.F., Smerdon, S.J. and Gamblin, S.J. (1997) Structure at 1.65 A of RhoA and its GTPase-activating protein in complex with a transition-state analogue. Nature *389*, 758-762.

Rong, R., He, Q., Liu, Y., Sheikh, M.S. and Huang, Y. (2002) TC21 mediates transformation and cell survival via activation of phosphatidylinositol 3-kinase/Akt and NF-kappaB signaling pathway. Oncogene *21*, 1062-1070.

Rosario, M., Paterson, H.F. and Marshall, C.J. (1999) Activation of the Raf/MAP kinase cascade by the Ras-related protein TC21 is required for the TC21-mediated transformation of NIH 3T3 cells. EMBO J. *18*, 1270-1279.

Rosario, M., Paterson, H.F. and Marshall, C.J. (2001a) Activation of the Ral and phosphatidylinositol 3′ kinase signaling pathways by the Ras-related protein TC21. Mol. Cell. Biol. *21*, 3750-3762.

Rosario, M., Paterson, H.F. and Marshall, C.J. (2001b) Activation of the Ral and phosphatidylinositol 3′ kinase signaling pathways by the ras-related protein TC21. Mol. Cell. Biol. *21*, 3750-3762.

Rossé, C., Hatzoglou, A., Parrini, M.C., White, M.A., Chavrier, P. and Camonis, J. (2006) RalB mobilizes the exocyst to drive cell migration. Mol. Cell. Biol. *26*, 727-734.

Rossé, C., L'Hoste, S., Offner, N., Picard, A. and Camonis, J.H. (2003) RLIP, an effector of the Ral GTPases, is a platform for Cdk1 to phosphorylate Epsin during the switch off of endocytosis in mitosis. J. Biol. Chem. M302191200.

Roy, B.C., Kohu, K., Matsuura, K., Yanai, H. and Akiyama, T. (2002) SPAL, a Rap-specific GTPase activating protein, is present in the NMDA receptor-PSD-95 complex in the hippocampus. Genes Cells. *7*, 607-617.

Rubinfeld, B., Munemitsu, S., Clark, R., Conroy, L., Watt, K., Crosier, W.J., McCormick, F. and Polakis, P. (1991) Molecular cloning of a GTPase activating protein specific for the Krev-1 protein p21rap1. Cell *65*, 1033-1042.

Rusyn, E.V., Reynolds, E.R., Shao, H., Grana, T.M., Chan, T.O., Andres, D.A. and Cox, A.D. (2000) Rit, a non-lipid-modified Ras-related protein, transforms NIH3T3 cells without activating the ERK, JNK, p38 MAPK or PI3K/Akt pathways. Oncogene *19*, 4685-4694.

Sahoo, T., Johnson, E.W., Thomas, J.W., Kuehl, P.M., Jones, T.L., Dokken, C.G., Touchman, J.W., Gallione, C.J., Lee-Lin, S.Q., Kosofsky, B., Kurth, J.H., Louis, D.N., Mettler, G., Morrison, L., Gil-Nagel, A., Rich, S.S., Zabramski, J.M., Boguski, M.S., Green, E.D. and Marchuk, D.A. (1999) Mutations in the gene encoding KRIT1, a Krev-1/rap1a binding protein, cause cerebral cavernous malformations (CCM1). Hum. Mol. Genet. *8*, 2325-2333.

Sakakibara, A., Ohba, Y., Kurokawa, K., Matsuda, M. and Hattori, S. (2002) Novel function of Chat in controlling cell adhesion via Cas-Crk-C3G-pathway-mediated Rap1 activation. J. Cell Sci. *115*, 4915-4924.

Sakkab, D., Lewitzky, M., Posern, G., Schaeper, U., Sachs, M., Birchmeier, W. and Feller, S.M. (2000) Signaling of hepatocyte growth factor/scatter factor (HGF) to the small GTPase Rap1 via the large docking protein Gab1 and the adapter protein CRKL. J. Biol. Chem. *275*, 10772-10778.

Sander, E.E., tenKlooster, J.P., vanDelft, S., vanderKammen, R.A. and Collard, J.G. (1999) Rac down-regulates Rho activity: Reciprocal balance between both GTPases determines cellular morphology and migratory behavior. J. Cell Biol. *147*, 1009-1021.

Saucedo, L.J., Gao, X., Chiarelli, D.A., Li, L., Pan, D. and Edgar, B.A. (2003) Rheb promotes cell growth as a component of the insulin/TOR signalling network. Nat. Cell Biol. *5*, 566-571.

Sawamoto, K., Winge, P., Koyama, S., Hirota, Y., Yamada, C., Miyao, S., Yoshikawa, S., Jin, M.H., Kikuchi, A. and Okano, H. (1999a) The Drosophila Ral GTPase regulates developmental cell shape changes through the Jun NH(2)-terminal kinase pathway. J. Cell Biol. *146*, 361-372.

Sawamoto, K., Yamada, C., Kishida, S., Hirota, Y., Taguchi, A., Kikuchi, A. and Okano, H. (1999b) Ectopic expression of constitutively activated Ral GTPase inhibits cell shape changes during Drosophila eye development. Oncogene *18*, 1967-1974.

Scheffzek, K., Ahmadian, M.R., Kabsch, W., Wiesmuller, L., Lautwein, A., Schmitz, F. and Wittinghofer, A. (1997) The Ras-RasGAP complex: Structural basis for GTPase activation and its loss in oncogenic Ras mutants. Science *277*, 333-338.

Schmitt, J.M. and Stork, P.J. (2000) beta 2-adrenergic receptor activates extracellular signal-regulated kinases (ERKs) via the small G protein rap1 and the serine/threonine kinase B-Raf. J. Biol. Chem. *275*, 25342-25350.

Schmitt, J.M. and Stork, P.J. (2001) Cyclic AMP-mediated inhibition of cell growth requires the small G protein Rap1. Mol. Cell. Biol. *21*, 3671-3683.

Schmitt, J.M. and Stork, P.J. (2002a) Galpha and Gbeta gamma require distinct Src-dependent pathways to activate Rap1 and Ras. J. Biol. Chem. *277*, 43024-43032.

Schmitt, J.M. and Stork, P.J. (2002b) PKA phosphorylation of Src mediates cAMP's inhibition of cell growth via Rap1. Mol. Cell *9*, 85-94.

Schmidt, M., Evellin, S., Weernink, P.A., von Dorp, F., Rehmann, H., Lomasney, J.W. and Jakobs, K.H. (2001) A new phospholipase-C-calcium signalling pathway mediated by cyclic AMP and a Rap GTPase. Nat. Cell Biol. *3*, 1020-1024.

Sebzda, E., Bracke, M., Tugal, T., Hogg, N. and Cantrell, D.A. (2002) Rap1A positively regulates T cells via integrin activation rather than inhibiting lymphocyte signaling. Nat. Immunol. *3*, 251-258.

Self, A.J., Caron, E., Paterson, H.F. and Hall, A. (2001) Analysis of R-Ras signalling pathways. J. Cell Sci. *114*, 1357-1366.

Serebriiskii, I., Estojak, J., Sonoda, G., Testa, J.R. and Golemis, E.A. (1997) Association of Krev-1/rap1a with Krit1, a novel ankyrin repeat-containing protein encoded by a gene mapping to 7q21–22. Oncogene *15*, 1043-1049.

Sethi, T., Ginsberg, M.H., Downward, J. and Hughes, P.E. (1999) The small GTP-binding protein R-Ras can influence integrin activation by antagonizing a Ras/Raf-initiated integrin suppression pathway. Mol. Biol. Cell. *10*, 1799-1809.

Shao, H., Kadono-Okuda, K., Finlin, B.S. and Andres, D.A. (1999) Biochemical characterization of the Ras-related GTPases Rit and Rin. Arch. Biochem. Biophys. *371*, 207-219.

Shao, H. and Andres, D.A. (2000) A novel RalGEF-like protein, RGL3, as a candidate effector for Rit and Ras. J. Biol. Chem. *275*, 26914-26924.

Shao, Y., Elly, C. and Liu, Y.C. (2003) Negative regulation of Rap1 activation by the Cbl E3 ubiquitin ligase. EMBO Rep. *4*, 425-431.

Shimizu, K., Kawabe, H., Minami, S., Honda, T., Takaishi, K., Shirataki, H. and Takai, Y. (1996) SMAP, an Smg GDS-associating protein having arm repeats and phosphorylated by Src tyrosine kinase. J. Biol. Chem. *271*, 27013-27017.

Shimonaka, M., Katagiri, K., Nakayama, T., Fujita, N., Tsuruo, T., Yoshie, O. and Kinashi, T. (2003) Rap1 translates chemokine signals to integrin activation, cell polarization, and motility across vascular endothelium under flow. J. Cell Biol. *161*, 417-427.

Shivakumar, L., Minna, J., Sakamaki, T., Pestell, R. and White, M.A. (2002) The RASSF1A tumor suppressor blocks cell cycle progression and inhibits cyclin D1 accumulation. Mol. Cell. Biol. *22*, 4309-4318.

Singh, L., Gao, Q., Kumar, A., Gotoh, T., Wazer, D.E., Band, H., Feig, L.A. and Band, V. (2003) The high-risk human papillomavirus type 16, E6 counters the GAP function of E6TP1 toward small Rap G proteins. J. Virol. *77*, 1614-1620.

Song, C., Hu, C.D., Masago, M., Kariyai, K., Yamawaki-Kataoka, Y., Shibatohge, M., Wu, D., Satoh, T. and Kataoka, T. (2001) Regulation of a novel human phospholipase C, PLCepsilon, through membrane targeting by Ras. J. Biol. Chem. *276*, 2752-2757.

Song, C., Satoh, T., Edamatsu, H., Wu, D., Tadano, M., Gao, X. and Kataoka, T. (2002) Differential roles of Ras and Rap1 in growth factor-dependent activation of phospholipase C epsilon. Oncogene *21*, 8105-8113.

Sparagana, S.P. and Roach, E.S. (2000) Tuberous sclerosis complex. Curr. Opin. Neurol. *13*, 115-119.

Spencer, M.L., Shao, H. and Andres, D.A. (2002a) Induction of neurite extension and survival in pheochromocytoma cells by the Rit GTPase. J. Biol. Chem. *277*, 20160-20168.

Spencer, M.L., Shao, H., Tucker, H.M. and Andres, D.A. (2002b) Nerve growth factor-dependent activation of the small GTPase Rin. J. Biol. Chem. *277*, 17605-17615.

Stocker, H., Radimerski, T., Schindelholz, B., Wittwer, F., Belawat, P., Daram, P., Breuer, S., Thomas, G. and Hafen, E. (2003) Rheb is an essential regulator of S6K in controlling cell growth in Drosophila. Nat. Cell Biol. *5*, 559-565.

Stork, P.J. (2003) Does Rap1 deserve a bad Rap? Trends Biochem. Sci. *28*, 267-275.

Stork, P.J. and Schmitt, J.M. (2002) Crosstalk between cAMP and MAP kinase signaling in the regulation of cell proliferation. Trends Cell Biol. *12*, 258-266.

Strassheim, D., Porter, R.A., Phelps, S.H. and Williams, C.L. (2000) Unique in vivo associations with SmgGDS and RhoGDI and different guanine nucleotide exchange activities exhibited by RhoA, dominant negative RhoAAsn-19, and activated RhoAVal-14. J. Biol. Chem. *275*, 6699-6702.

Su, L., Hattori, M., Moriyama, M., Murata, N., Harazaki, M., Kaibuchi, K. and Minato, N. (2003) AF-6 controls integrin-mediated cell adhesion by regulating Rap1 activation through the specific recruitment of Rap1GTP and SPA-1. J. Biol. Chem. *278*, 15232-15238.

Sugihara, K., Asano, S., Tanaka, K., Iwamatsu, A., Okawa, K. and Ohta, Y. (2001) The exocyst complex binds the small GTPase RalA to mediate filopodia formation. Nat. Cell Biol. *4*, 73-78.

Suzuki, J., Kaziro, Y. and Koide, H. (2000a) Positive regulation of skeletal myogenesis by R-Ras. Oncogene *19*, 1138-1146.

Suzuki, J., Yamazaki, Y., Li, G., Kaziro, Y. and Koide, H. (2000b) Involvement of Ras and Ral in chemotactic migration of skeletal myoblasts. Mol. Cell. Biol. *20*, 4658-4665.

Tabancay, A.P., Jr., Gau, C.-L., Machado, I.M.P., Uhlmann, E.J., Gutmann, D.H., Guo, L. and Tamanoi, F. (2003) Identification of dominant negative mutants of Rheb GTPase and their use to implicate the involvement of human Rheb in the activation of p70S6K. J. Biol. Chem. *278*, 39921-39930.

Tanaka, S., Morishita, T., Hashimoto, Y., Hattori, S., Nakamura, S., Shibuya, M., Matuoka, K., Takenawa, T., Kurata, T., Nagashima, K. and Matsuda, M. (1994) C3G, a guanine nucleotide-releasing protein expressed ubiquitously, binds to the Src homology 3 domains of CRK and GRB2/ASH proteins. Proc. Natl. Acad. Sci. U.S.A. *91*, 3443-3447.

Tee, A.R., Manning, B.D., Roux, P.P., Cantley, L.C. and Blenis, J. (2003) Tuberous sclerosis complex gene products, tuberin and hamartin, control mTOR signaling by acting as a GTPase-activating protein complex toward Rheb. Curr. Biol. *13*, 1259-1268.

Touchot, N., Chardin, P. and Tavitian, A. (1987) Four additional members of the ras gene superfamily isolated by an oligonucleotide strategy: molecular cloning of YPT-related cDNAs from a rat brain library. Proc. Natl. Acad. Sci. U.S.A. *84*, 8210-8214.

Traver, S., Bidot, C., Spassky, N., Baltauss, T., De Tand, M.F., Thomas, J.L., Zalc, B., Janoueix-Lerosey, I. and Gunzburg, J.D. (2000) RGS14 is a novel Rap effector that preferentially regulates the GTPase activity of Galphao. Biochem. J. *350*, 19-29.

Tsukamoto, N., Hattori, M., Yang, H., Bos, J.L. and Minato, N. (1999) Rap1 GTPase-activating protein SPA-1 negatively regulates cell adhesion. J. Biol. Chem. *274*, 18463-18469.

Urano, J., Tabancay, A.P., Yang, W. and Tamanoi, F. (2000) The Saccharomyces cerevisiae Rheb G-protein is involved in regulating canavanine resistance and arginine uptake. J. Biol. Chem. *275*, 11198-11206.

Urano, T., Emkey, R. and Feig, L.A. (1996) Ral-GTPases mediate a distinct downstream signaling pathway from Ras that facilitates cellular transformation. EMBO J. *15*, 810-816.

van den Berghe, N., Cool, R.H., Horn, G. and Wittinghofer, A. (1997) Biochemical characterization of C3G: an exchange factor that discriminates between Rap1 and Rap2 and is not inhibited by Rap1A (S17N). Oncogene *15*, 845-850.

Vetter, I.R., Linnemann, T., Wohlgemuth, S., Geyer, M., Kalbitzer, H.R., Herrmann, C. and Wittinghofer, A. (1999) Structural and biochemical analysis of Ras-effector signaling via RalGDS. FEBS Lett. *451*, 175-180.

Vetter, I.R. and Wittinghofer, A. (2001) The guanine nucleotide-binding switch in three dimensions. Science *294*, 1299-1304.

Vikis, H.G., Stewart, S. and Guan, K.L. (2002) SmgGDS displays differential binding and exchange activity towards different Ras isoforms. Oncogene *21*, 2425-2432.

Vossler, M.R., Yao, H., York, R.D., Pan, M.G., Rim, C.S. and Stork, P.J. (1997) cAMP activates MAP kinase and Elk-1 through a B-Raf- and Rap1-dependent pathway. Cell *89*, 73-82.

Wang, B.C., Zou, J.X., EkRylander, B. and Ruoslahti, E. (2000) R-Ras contains a proline-rich site that binds to SH3 domains and is required for integrin activation by R-Ras. J. Biol. Chem. *275*, 5222-5227.

Wang, H.G., Millan, J.A., Cox, A.D., Der, C.J., Rapp, U.R., Beck, T., Zha, H. and Reed, J.C. (1995) R-Ras promotes apoptosis caused by growth factor deprivation via a Bcl-2 suppressible mechanism. J. Cell Biol. *129*, 1103-1114.

Wang, K.L., Khan, M.T. and Roufogalis, B.D. (1997) Identification and characterization of a calmodulin-binding domain in Ral-A, a Ras-related GTP-binding protein purified from human erythrocyte membrane. J. Biol. Chem. *272*, 16002-16009.

Wang, K.L. and Roufogalis, B.D. (1999) Ca2+/calmodulin stimulates GTP binding to the ras-related protein ral-A. J. Biol. Chem. *274*, 14525-14528.

Ward, Y., Wang, W., Woodhouse, E., Linnoila, I., Liotta, L. and Kelly, K. (2001) Signal pathways which promote invasion and metastasis: critical and distinct contributions of extracellular signal-regulated kinase and ral-specific guanine exchange factor pathways. Mol. Cell. Biol. *21*, 5958-5969.

Wes, P., Yu, M. and Montell, C. (1996) RIC, a calmodulin-binding Ras-like GTPase. EMBO J. *15*, 5839-5848.

White, M.A., Nicolette, C., Minden, A., Polverino, A., Van Aelst, L., Karin, M. and Wigler, M.H. (1995) Multiple Ras functions can contribute to mammalian cell transformation. Cell *80*, 533-541.

White, M.A., Vale, T., Camonis, J.H., Schaefer, E. and Wigler, M.H. (1996) A role for the Ral guanine nucleotide dissociation stimulator in mediating Ras-induced transformation. J. Biol. Chem. *271*, 16439-16442.

Wienecke, R., Konig, A. and DeClue, J.E. (1995) Identification of tuberin, the tuberous sclerosis-2 product. Tuberin possesses specific Rap1GAP activity. J. Biol. Chem. *270*, 16409-16414.

Wolthuis, R.M., Bauer, B., van't Veer, L.J., de Vries-Smits, A.M., Cool, R.H., Spaargaren, M., Wittinghofer, A., Burgering, B.M. and Bos, J.L. (1996) RalGDS-like factor (Rlf) is a novel Ras and Rap 1A-associating protein. Oncogene *13*, 353-362.

Wolthuis, R.M., de Ruiter, N.D., Cool, R.H. and Bos, J.L. (1997) Stimulation of gene induction and cell growth by the Ras effector Rlf. EMBO J. *16*, 6748-6761.

Wolthuis, R.M., Franke, B., van Triest, M., Bauer, B., Cool, R.H., Camonis, J.H., Akkerman, J.W. and Bos, J.L. (1998a) Activation of the small GTPase Ral in platelets. Mol. Cell. Biol. *18*, 2486-2491.

Wolthuis, R.M., Zwartkruis, F., Moen, T.C. and Bos, J.L. (1998b) Ras-dependent activation of the small GTPase Ral. Curr. Biol. *8*, 471-474.

Wu, C., Lai, C.F. and Mobley, W.C. (2001) Nerve growth factor activates persistent Rap1 signaling in endosomes. J. Neurosci. *21*, 5406-5416.

Wu, J., Dent, P., Jelinek, T., Wolfman, A., Weber, M.J. and Sturgill, T.W. (1993) Inhibition of the EGF-activated MAP kinase signaling pathway by adenosine 3′, 5′-monophosphate. Science *262*, 1065-1069.

Xu, L., Frankel, P., Jackson, D., Rotunda, T., Boshans, R.L., D'Souza-Schorey, C. and Foster, D.A. (2003) Elevated phospholipase D activity in H-Ras- but not K-Ras-transformed cells by the synergistic action of RalA and ARF6. Mol. Cell. Biol. *23*, 645-654.

Yajnik, V., Paulding, C., Sordella, R., McClatchey, A.I., Saito, M., Wahrer, D.C., Reynolds, P., Bell, D.W., Lake, R., van den Heuvel, S., Settleman, J. and Haber, D.A. (2003) DOCK4, a GTPase activator, is disrupted during tumorigenesis. Cell *112*, 673-684.

Yamagata, K., Sanders, L.K., Kaufmann, W.E., Yee, W., Barnes, C.A., Nathans, D. and Worley, P.F. (1994) rheb, a growth factor- and synaptic activity-regulated gene, encodes a novel Ras-related protein. J. Biol. Chem. *269*, 16333-16339.

Yamaguchi, A., Urano, T., Goi, T. and Feig, L.A. (1997) An Eps homology (EH) domain protein that binds to the Ral-GTPase target, RalBP1. J. Biol. Chem. *272*, 31230-31234.

Yamamoto, T., Kaibuchi, K., Mizuno, T., Hiroyoshi, M., Shirataki, H. and Takai, Y. (1990) Purification and characterization from bovine brain cytosol of proteins that regulate the GDP/GTP exchange reaction of smg p21s, ras-like GTP-binding proteins. J. Biol. Chem. *265*, 16626-16634.

Yamashita, S., Mochizuki, N., Ohba, Y., Tobiume, M., Okada, Y., Sawa, H., Nagashima, K. and Matsuda, M. (2000) CalDAG-GEFIII activation of Ras, R-Ras, and Rap1. J. Biol. Chem. *275*, 25488-25493.

Yang, W., Tabancay, A.P. Jr., Urano, J. and Tamanoi, F. (2001) Failure to farnesylate Rheb protein contributes to the enrichment of G0/G1 phase cells in the Schizosaccharomyces pombe farnesyltransferase mutant. Mol. Microbiol. *41*, 1339-1347.

Yee, W.M. and Worley, P.F. (1997) Rheb interacts with Raf-1 kinase and may function to integrate growth factor- and protein kinase A-dependent signals. Mol. Cell. Biol. *17*, 921-933.

York, R.D., Yao, H., Dillon, T., Ellig, C.L., Eckert, S.P., McCleskey, E.W. and Stork, P.J. (1998) Rap1 mediates sustained MAP kinase activation induced by nerve growth factor. Nature *392*, 622-626.

York, R.D., Molliver, D.C., Grewal, S.S., Stenberg, P.E., McCleskey, E.W. and Stork, P.J. (2000) Role of phosphoinositide 3-kinase and endocytosis in nerve growth factor-induced extracellular signal-regulated kinase activation via Ras and Rap1. Mol. Cell. Biol. *20*, 8069-8083.

Yoshida, Y., Kawata, M., Miura, Y., Musha, T., Sasaki, T., Kikuchi, A. and Takai, Y. (1992) Microinjection of smg/rap1/Krev-1 p21 into Swiss 3T3 cells induces DNA synthesis and morphological changes. Mol. Cell. Biol. *12*, 3407-3414.

Yu, Y., Xu, F., Peng, H., Fang, X., Zhao, S., Li, Y., Cuevas, B., Kuo, W.L., Gray, J.W., Siciliano, M., Mills, G.B. and Bast, R.C., Jr. (1999) NOEY2 (ARHI), an imprinted putative tumor suppressor gene in ovarian and breast carcinomas. Proc. Natl. Acad. Sci. U.S.A. *96*, 214-219.

Yu, Y. and Feig, L.A. (2002) Involvement of R-Ras and Ral GTPases in estrogen-independent proliferation of breast cancer cells. Oncogene *21*, 7557-7568.

Zawistowski, J.S., Serebriiskii, I.G., Lee, M.F., Golemis, E.A. and Marchuk, D.A. (2002) KRIT1 association with the integrin-binding protein ICAP-1: a new direction in the elucidation of cerebral cavernous malformations (CCM1) pathogenesis. Hum. Mol. Genet. *11*, 389-396.

Zhang, J., Clatterbuck, R.E., Rigamonti, D., Chang, D.D. and Dietz, H.C. (2001) Interaction between krit1 and icap1alpha infers perturbation of integrin beta1-mediated angiogenesis in the pathogenesis of cerebral cavernous malformation. Hum. Mol. Genet. *10*, 2953-2960.

Zhang, K., Papageorge, A.G., Martin, P., Vass, W.C., Olah, Z., Polakis, P.G., McCormick, F. and Lowy, D.R. (1991) Heterogeneous amino acids in Ras and Rap1A specifying sensitivity to GAP proteins. Science *254*, 1630-1634.

Zhang, W., Shao, Y., Fang, D., Huang, J., Jeon, M.S. and Liu, Y.C. (2003a) Negative regulation of TCR-mediated Crk-L-C3G signaling and cell adhesion by Cbl-b. J. Biol. Chem. *278*, 23978-23983.

Zhang, Y., Gao, X., Saucedo, L.J., Ru, B., Edgar, B.A. and Pan, D. (2003b) Rheb is a direct target of the tuberous sclerosis tumour suppressor proteins. Nat. Cell Biol. *5*, 578-581.

Zhang, Z., Vuori, K., Wang, H., Reed, J.C. and Ruoslahti, E. (1996) Integrin activation by R-ras. Cell *85*, 61-69.

Zhu, J.J., Qin, Y., Zhao, M., Van Aelst, L. and Malinow, R. (2002) Ras and Rap control AMPA receptor trafficking during synaptic plasticity. Cell *110*, 443-455.

Zou, J.X., Wang, B., Kalo, M.S., Zisch, A.H., Pasquale, E.B. and Ruoslahti, E. (1999) An Eph receptor regulates integrin activity through R-Ras. Proc. Natl. Acad. Sci. U.S.A. *96*, 13813-13818.

Zou, J.X., Liu, Y., Pasquale, E.B. and Ruoslahti, E. (2002) Activated Src oncogene phosphorylates R-Ras and suppresses integrin activity. J. Biol. Chem. *277*, 1824-1827.

Zwartkruis, F.J., Wolthuis, R.M., Nabben, N.M., Franke, B. and Bos, J.L. (1998) Extracellular signal-regulated activation of Rap1 fails to interfere in Ras effector signalling. EMBO J. *17*, 5905-5912.

CHAPTER 14

κB-RAS: A SMALL GTPASE THAT INFLUENCES NF-κB SIGNALING

TOM HUXFORD, ANU K. MOORTHY, AND GOURISANKAR GHOSH
Department of Chemistry and Biochemistry, University of California, San Diego

Abstract: κB-Ras1 and κB-Ras2 are two small proteins that display similarity at the amino acid level to Ras-like small GTPases. Although little is known about the function of the κB-Ras proteins, they have been shown to interfere with activation of transcription factor NF-κB. They accomplish this by binding to IκB proteins, natural inhibitors of NF-κB, and delaying their stimulus-dependent degradation. In this chapter, we consider the κB-Ras proteins in light of their NF-κB regulatory properties. Three fundamental questions about κB-Ras function are addressed: 1) Does κB-Ras regulate NF-κB *in vivo*? 2) Does κB-Ras selectively regulate specific NF-κB/IκB complexes? 3) Does κB-Ras function as a true GTPase, that is, with molecular switching properties that correlate with the phosphorylation state of bound guanine nucleotide? Finally, we detail the methods currently used to study the κB-Ras proteins as regulators of NF-κB activation

Keywords: κB-Ras, NF-κB, IκB, GTPase, 26S proteasome, protein phosphorylation, signal transduction, transcription factor, ubiquitin

1. INTRODUCTION

κB-Ras1 and κB-Ras2 are relatively new members of the small GTPase superfamily. κB-Ras1 was originally identified by yeast two-hybrid screening with IκBβ, a known inhibitor of transcription factor NF-κB, as bait (Fenwick et al., 2000). κB-Ras2 was discovered in an EST database by virtue of its homology to κB-Ras1 (71% identity and 85% similarity). A sequence encoding an ortholog in *Drosophila melanogaster* was also discovered by database searching and its cDNA was isolated (dmκB-Ras). The dmκB-Ras sequence bears 40% identity and 68% similarity to the human κB-Ras1 sequence.

Both κB-Ras1 and -2 inhibit the activity of transcription factor NF-κB. However, the mechanism of this inhibition process has yet to be fully explored. To date, only three articles on κB-Ras proteins have been published, which highlight only a few aspects of their biochemical and biological properties (Chen et al., 2004; Chen

C. Der (ed.), RAS Family GTPases, 341–352.

et al., 2003; Fenwick et al., 2000). Here, we summarize these findings, explore the phylogeny of κB-Ras proteins, and propose possible regulatory mechanisms employed by κB-Ras in the NF-κB signaling pathway.

2. SEQUENCE AND STRUCTURAL FEATURES OF κB-RAS

Proteins of the κB-Ras family bear signature features of the small GTPases superfamily (Bourne et al., 1990; Bourne et al., 1991). κB-Ras1 and κB-Ras2 are most homologous to the Ras GTPase sub-family displaying nearly 45% sequence similarity. Three of the four conserved structural features observed in small GTPases

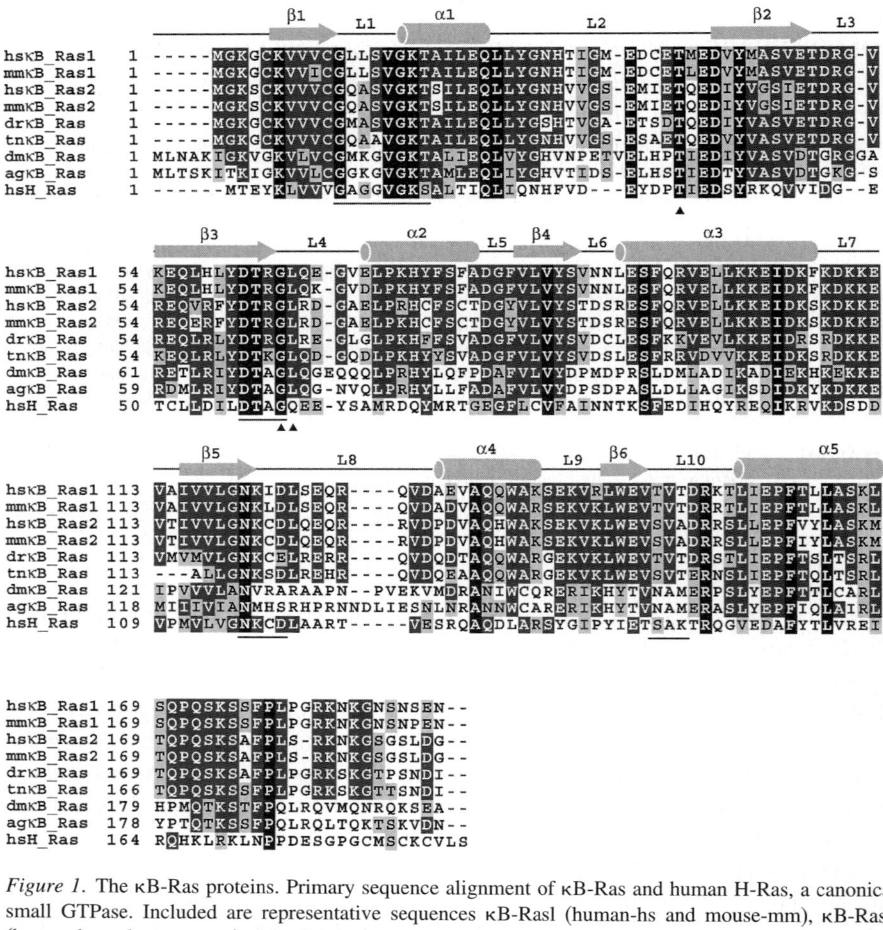

Figure 1. The κB-Ras proteins. Primary sequence alignment of κB-Ras and human H-Ras, a canonical small GTPase. Included are representative sequences κB-Rasl (human-hs and mouse-mm), κB-Ras2 (human-hs and mouse-mm), fish κB-Ras (zebrafish-dr and pufferfish-tn), and insect κB-Ras (fly-dm and mosquito-ag). Secondary structure elements are indicated schematically and labeled above the sequences. Conserved elements of primary structure including the GXXXXGKS/T, DXXG, NKXD and SAK motifs are underlined. Key residues mentioned in the text (Gly12, Gly13, Thr35, Gly60 and Gln 61) are labeled with small triangles below their position in the human H-Ras sequence

are present in κB-Ras proteins (Figure 1). These include the N/TKXD motif that recognizes the guanine nucleotide base, the GXXXXGKS/T motif of the P-loop that contacts the β- and γ-phosphates of GTP, and the DXXG motif that contributes to guanine nucleotide binding specificity (Saraste et al., 1990; Vetter and Wittinghofer, 2001). The SAK motif, present between amino acids 145 and 147 of human H-Ras and also implicated in guanine nucleotide specificity, is less well conserved. Presence of most of the defining G nucleotide binding features in κB-Ras1 and κB-Ras2 clearly suggests that, like all other GTPases, κB-Ras proteins are capable of binding GTP and GDP. Biochemical experiments have confirmed this to be the case (Chen et al., 2004).

The conversion of small GTPases from their GTP- to GDP-bound states requires structural changes within two regions, known as switch I and switch II (Milburn et al., 1990; Vetter and Wittinghofer, 2001). This conformational change has a profound effect on the ability of the GTPase proteins to influence signal transduction (Bourne et al., 1990; Bourne et al., 1991; Milburn et al., 1990). In their triphosphate bound states main chain NH groups of two highly conserved amino acids, Thr35 from switch I and Gly60 from switch II (human H-Ras numbering), donate two hydrogen bonds to the γ-phosphate oxygens of the bound GTP (Pai et al., 1990). These two residues remain conserved in κB-Ras and it is anticipated that they make similar contacts and induce analogous conformational changes (Figure 1).

However, several other features of prototypical small GTPases are markedly absent in κB-Ras. Three residues that are involved in GTP hydrolysis, a defining character of all GTPases, are altered in κB-Ras (Frech et al., 1994; Li and Zhang, 2004). Glycine at positions 12 and 13 in the P loop and glutamine at position 61 in H-Ras are replaced by a leucine/alanine, serine and leucine, respectively (Figure 1). Although these residues are not involved in nucleotide binding, their critical role in GTP hydrolysis has been demonstrated in several cases. Due to these alterations, it is possible that unlike most other members of the sub-family, κB-Ras may not function as a true GTPase. Also, κB-Ras proteins are generally more basic than the other Ras proteins and do not contain a recognizable signals for lipid modification or proteolysis at their C-termini (Casey et al., 1989; Fujiyama and Tamanoi, 1990; Gutierrez et al., 1989; Hancock et al., 1989). This suggests that their cellular localization and biochemical properties also differ from those exhibited by classical Ras proteins.

3. REGULATION OF NF-κB BY IκB

NF-κB represents a family of transcription factors that activate the expression of genes involved in innate and adaptive immunity, inflammation, cell growth, adhesion, development, and death (Baldwin et al., 1996; Ghosh et al., 1998). Functional dimeric NF-κB proteins assemble combinatorially from five subunits: p50, p52, RelA (p65), c-Rel, and RelB. All these proteins contain the Rel homology region (RHR), a highly homologous segment of approximately 300 amino acids in length, which is responsible for DNA binding, subunit dimerization and nuclear localization. Whereas, NF-κB dimers that contain RelB can function either to activate or repress target gene transcription, RelA and c-Rel homo- and heterodimers function exclusively as transcriptional activators. The p50 and p52 subunits do not contain activation domains. As a result, homodimers of p50 and p52 function as repressors of NF-κB target gene transcription.

Due to their inherent potential to activate gene expression, NF-κB dimers that contain RelA and c-Rel subunits are kept under tight regulation by a class of inhibitor proteins known as IκB. Two of these, IκBα and IκBβ, are known to function as inhibitors of NF-κB dimers that contain at least one RelA or c-Rel subunit by forming stable complexes with them. IκB binding masks the nuclear localization signals present in all NF-κB subunits, thereby sequestering the dimer to the cytoplasm.

Conversion of inactive NF-κB to its active state is carefully regulated by the NF-κB signaling pathway (Figure 2). A diverse array of inducing signals has been shown to lead to activation of a kinase known as IκB kinase (IKK). Active IKK phosphorylates the NF-κB-bound IκB leading to the latter's degradation by the ubiquitin

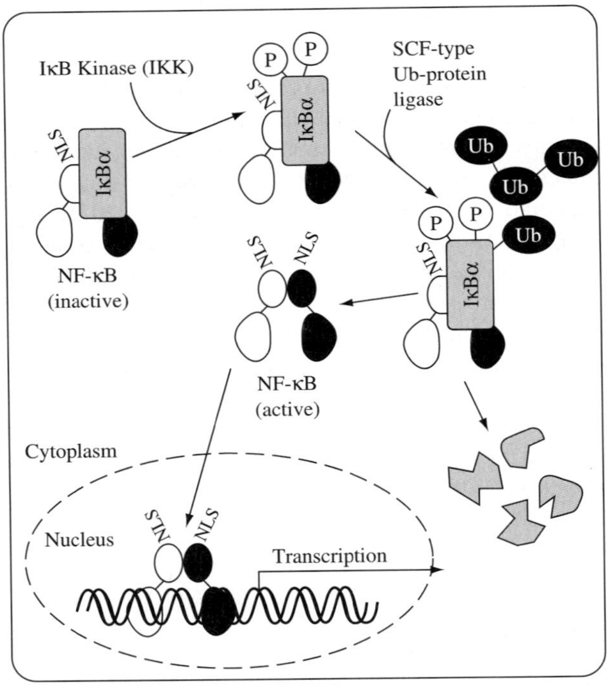

Figure 2. The NF-κB signaling pathway. NF-κB exists in resting cells as an inactive cytoplasmic complex with a member of the IκB inhibitor protein family. Inducing signals activate the IκB kinase (IKK) leading to the dual phosphorylation of the NF-κB associated IκB and its subsequent polyubiquitinylation, which is catalyzed by a specific SCF-type ubiquitin-protein ligase. Phosphorylated and ubiquitinylated IκB is rapidly degraded by the 26 S proteasome. Removal of IκB unmasks nuclear localization signals (NLS) present on NF-κB subunits. The freed NF-κB migrates to the nucleus where it binds to specific target gene promoters and activates transcription

and 26S proteasome-dependent pathway (Karin and Ben-Neriah, 2000). Free RelA and c-Rel dimers can then enter the nucleus, bind to specific DNA sequences within the promoter elements of target genes, and activate gene transcription.

4. NF-κB/IκBα AND NF-κB/IκBβ COMPLEXES ARE DIFFERENTIALLY REGULATED

In vitro experiments have shown that RelA and c-Rel dimers fail to bind DNA when bound to IκB (Malek et al., 1998; Phelps et al., 2000). This suggests that NF-κB dimers form mutually exclusive complexes with IκB and DNA. However, by virtue of the higher affinity, NF-κB/IκB complex formation is favored. This ability of IκB proteins to inhibit NF-κB DNA binding is particularly important in biological regulation. It has recently been shown that inactive NF-κB/IκB complexes show differing localization profiles in resting cells. Inactive NF-κB/IκBα complexes undergo steady state shuttling between the nucleus and cytoplasm in resting cells while inactive NF-κB/IκBβ complexes are exclusively cytoplasmic (Huang et al., 2000; Johnson et al., 1999; Malek et al., 2001; Tam and Sen, 2001). The gene encoding IκBα, but not IκBβ, lies under NF-κB transcriptional control and newly synthesized IκBα is capable of removing active NF-κB from DNA. Finally, it has been known since its discovery that IκBβ mediates a prolonged NF-κB response while IκBα is responsible for rapid transient activation of the transcription factor. These observations provide the first line of evidence that NF-κB dimers with transcriptional activation potential are regulated differently by their IκBα and IκBβ binding partners.

Further evidence in support of the differential inhibition strategies employed by the two inhibitors comes from structural studies of the NF-κB/IκB complexes. X-ray crystal structures of the NF-κB p50/RelA heterodimer bound to IκBα and the NF-κB RelA/RelA homodimer bound to IκBβ have been elucidated (Huxford et al., 1989; Jacobs and Harrison, 1998; Malek et al., 2003). These structures and their accompanying *in vitro* biochemical experiments revealed that both IκB proteins rely primarily upon their ankyrin repeat-containing domains (ARD) to interact with the RHR of NF-κB dimers. The ARDs of both IκBα and IκBβ are flanked by a flexible N-terminal signal response region and a flexible C-terminal tail rich in the amino acids proline, glutamic acid, serine, and threonine (PEST). Several serine and threonine residues within the C-terminal PEST region are phosphorylated *in vivo* by the protein kinase CK2 (casein kinase II), which renders this region highly acidic (Barroga et al., 1995; McElhinny et al., 1996). This PEST region from IκBβ was used in the original discovery of κB-Ras1 by yeast two-hybrid screening (Fenwick et al., 2000).

There are two significant differences in the domain organization of IκBα and IκBβ proteins. IκBβ contains an insertion of 40 amino acid residues between its third and fourth ankyrin repeats and its N-terminal signal response region is shorter by 15 residues. The insert of IκBβ is disordered and does not contact NF-κB in the binary complex. However, its location and unique primary sequence make it tempting to speculate that additional regulatory proteins might use the IκBβ insert to discriminate between NF-κB/IκBα and NF-κB/IκBβ complexes. κB-Ras might be one such protein (Chen et al., 2003).

5. REGULATION OF NF-κB/IκB COMPLEXES BY κB-RAS

Cell-based experiments such as co-immunoprecipitation from endogenous or transiently expressed sources have shown that κB-Ras proteins associate predominantly with IκBβ and to a lesser extent with IκBα (Fenwick et al., 2000). Pull-down experiments from endogenous sources also revealed that κB-Ras proteins associate with the RelA/IκBβ complex. These results suggest that at least a pool of NF-κB/IκB exists in a ternary complex with κB-Ras. The observation that IκB and κB-Ras can form binary complexes independent of NF-κB suggests that IκB proteins interact with both NF-κB and κB-Ras through two independent surfaces.

Transfection experiments have shown that κB-Ras inhibits stimulus-dependent activation of NF-κB by blocking the degradation of IκB proteins (Chen et al., 2004). As κB-Ras does not directly bind to or inhibit IKK. This suggests that the presence of κB-Ras either indirectly blocks phosphorylation of IκB in the ternary complex or affects ubiquitinylation/degradation of IκB. Further *in vitro* biochemical experiments revealed that κB-Ras masks the N-terminal IKK phosphorylation sites in IκBβ . While it is possible that this could make it a less efficient substrate for phosphorylation and inhibit activation of NF-κB/IκBβ complexes, it remains to be seen whether this observation bears any physiological significance.

6. POSSIBLE MECHANISM FOR REGULATION OF INACTIVE NF-κB/IκB COMPLEXES BY κB-RAS

It is known that IκBβ is not completely degraded in response to certain stimuli. One likely explanation for this phenomenon is that the IKK phosphorylation sites in some NF-κB/IκBβ complexes are concealed. It is also known that prolonged NF-κB activation requires degradation of IκBβ (Johnson et al., 1996; Thompson et al., 1995). As mentioned earlier, post-induction newly synthesized IκBα represses transcription by removing NF-κB dimers from DNA. Therefore, sustained NF-κB activity could be due to the inability of IκBα to remove certain NF-κB dimers from their DNA targets.

There exist in cells only five probable dimer combinations of NF-κB subunits that are inhibited by IκBα and IκBβ . These are p50/RelA, RelA/RelA, c-Rel/RelA, c-Rel/c-Rel and p50/c-Rel. Although IκBα binds to each of these dimers, it does so with significantly different affinities (Phelps et al., 2000). For example, IκBα binds p50/RelA 10-fold more tightly than RelA and c-Rel homodimers. Considering that p50/RelA heterodimer is more abundant than the other dimers, newly synthesized IκBα might be engaged in removing primarily the p50/RelA heterodimer. It is possible that IκBβ preferentially inhibits NF-κB RelA and c-Rel homodimers *in vivo* and that κB-Ras significantly enhances this interaction.

We have recently observed that κB-Ras interacts with IκBβ/c-Rel homodimer when the C-terminal PEST of IκBβ is phosphorylated by protein kinase CK2. Recent findings also reveal that IκBβ binds more specifically to RelA and c-Rel homodimers *in vivo*. Because p50/RelA heterodimers are most prevalent and transcriptionally active in cells, much less attention has been given to the biological role and regulation of NF-κB RelA and c-Rel homodimers. However, it has been shown that certain genes are regulated specifically by these homodimers. For example, c-Rel homodimer is key to the expression of the gene encoding the cytokine interleukin-2 (IL-2) (Kontgen et al., 1995). T-cell proliferation requires IL-2 expression and T cell receptor (TCR) co-stimulatory signal is essential to activate c-Rel. Therefore, it is possible that at least in T cells, the c-Rel homodimer remains complexed to IκBβ and κB-Ras as a ternary or multimeric complex. The presence of κB-Ras slows induction of NF-κB from the pool of transcription factor that is maintained strictly cytoplasmic through association with κB-Ras. Additional factors, such as phosphatases, may be involved in this process. A schematic representation of this model is depicted in Figure 3.

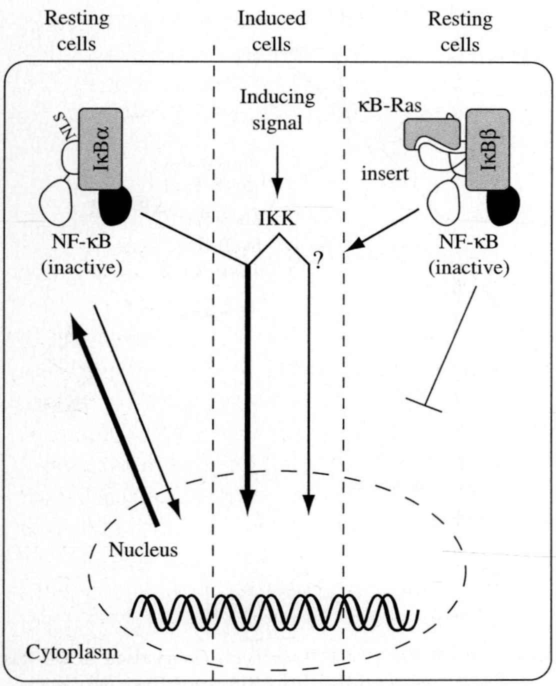

Figure 3. Model for the preferential regulation of NF-κB/IκBβ complexes by κB-Ras. In resting cells, inactive NF-κB/IκBα complexes shuttle between the nucleus and cytoplasm while NF-κB/IκBβ complexes are strictly cytoplasmic. Also, NF-κB/IκBβ complexes are known to react more slowly to inducing signals and mediate a prolonged NF-κB response. κB-Ras might contribute to these differences by binding to NF-κB/IκBβ complexes through the unique insert in the IκBβ protein. This could completely mask the inherent nuclear localization potential of NF-κB dimers. Additional signaling events, possibly involving κB-Ras and the phosphorylation state of the IκBβ inhibitor, are proposed to regulate the prolonged NF-κB response

7. κB-RAS AND NF-κB PROTEINS ARE EVOLUTIONARILY CORRELATED

The dimeric NF-κB transcription factors, their IκB inhibitors, and the IKK protein complex constitute the NF-κB signaling module (Hoffmann et al., 2002). This module is evolutionarily conserved in species as diverse as flies, fish, fowl, and mammals. For example, in *D. melanogaster* the well-characterized Dorsal/Cactus morphogen is highly similar to mammalian NF-κB/IκB. The *Drosophila* genome also contains the NF-κB orthologs Dorsal and Dif as well as a distinct IKK protein (Silverman and Maniatis, 2001). Interestingly, no homolog of any NF-κB signaling module proteins has been detected in nematodes or yeast. This suggests

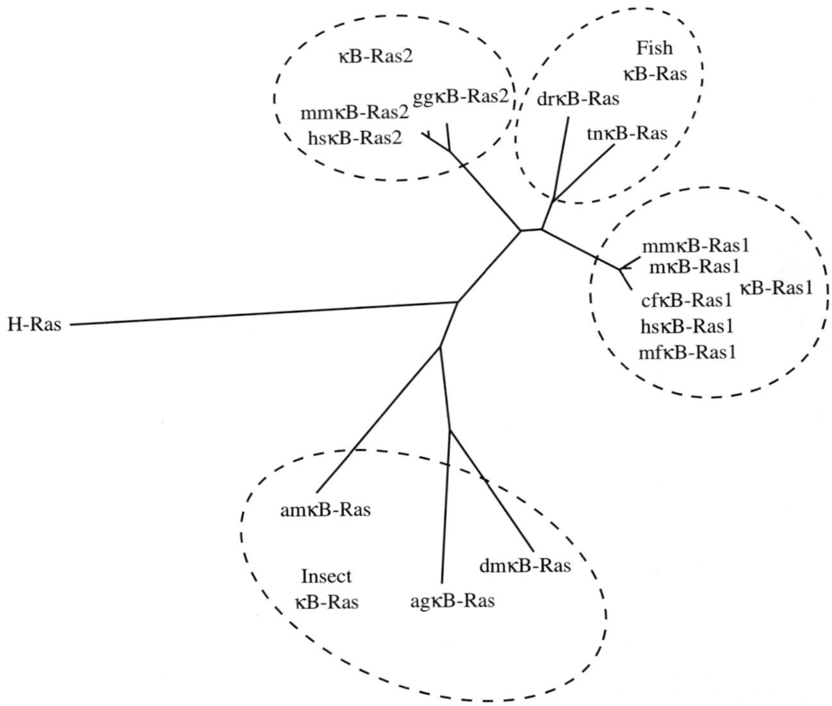

Figure 4. Unrooted phylogenetic tree of known κB-Ras protein sequences. Comparison of κB-Ras proteins from vaious species reveals four sub-classes: Insect κB-Ras, fish κB-Ras, κB-Ras1 and κB-Ras2. The insect κB-Ras (represented by fly-dm, mosquito-ag, and honeybee-am sequences) is more closely related to human H-Ras than are the other κB-Ras sub-classes. Fish κB-Ras (represented by zebrafish-dr and pufferfish-tn sequences) shows nearly equal homology to both κB-Ras1 (represented by human-hs, mouse-mm, rat-rn, dog-cf, and macaque-mf sequences) and κB-Ras2 (represented by human-hs, mouse-mm and chicken-gg sequences). This analysis suggests that κB-Ras1 and κB-Ras2 diverged relatively recently from a common progenitor of fish κB-Ras. Furthermore, these three sub-types and insect κB-Ras represent distant orthologs of the Ras GTPase sub-family

that evolution of the NF-κB signaling module is a relatively modern phenomenon and is most likely present and functional only in the so-called "higher eukaryotes".

In addition to being discovered by virtue of its ability to interact with NF-κB inhibitors, κB-Ras exhibits a striking evolutionary correlation with proteins of the NF-κB signaling module. Sequence database searches reveal that κB-Ras is present only in organisms that also contain some form of NF-κB, IκB, and IKK. κB-Ras1 and/or κB-Ras2 are present in mammals (human, mouse, rat, dog, macaque) and birds (chicken), while fish (zebrafish, pufferfish) appear to contain a single κB-Ras isoform with nearly equal homology to both κB-Ras1 and κB-Ras2. One κB-Ras ortholog is present in the *D. melanogaster* genome (dmκB-Ras). Close orthologs to dmκB-Ras can also be identified in mosquito and honeybee. Thus, the κB-Ras proteins can be divided into three distinct sub-groups: κB-Ras1, κB-Ras2 and insect κB-Ras (Figure 4). The single κB-Ras gene present in fish species is likely orthologous to an ancestral κB-Ras gene that through gene duplication has given rise to the κB-Ras1 and κB-Ras2 paralogs currently present in mammals and fowl. No recognizable κB-Ras orthologs have been identified in yeast or nematodes. Overall, the evolutionary correlation strongly suggests that κB-Ras and NF-κB are functionally linked.

8. DOES κB-RAS ACT AS A MOLECULAR SWITCH?

The differences between the GTP- and GDP-bound conformations of the small GTPases drastically influence their signaling properties. In their GTP-bound state, typical small GTPases interact with effector proteins and activate a multitude of signaling pathways (Bourne et al., 1990). In their GDP-bound states, on the other hand, the signal transduction pathways are turned off. By cycling between their GTP- and GDP-bound states, these small GTPases act as molecular switches (Vetter and Wittinghofer, 2001).

Although it appears that κB-Ras proteins are capable of binding both GTP and GDP, it is not clear whether they can hydrolyze GTP and cycle between GTP and GDP bound states. As mentioned earlier, residues at key positions, that clearly play significant roles in GTP hydrolysis, are different in κB-Ras (Li and Zhang, 2004). However, these residues (Gly12, Gly13 and Gln61), though highly conserved in small GTPases, are not invariant residues. Members of the GTPase superfamily with non-glycine residues at the two positions are still capable of functioning as GTPases. For example, it has been shown in the case of Rab5, which has an alanine at position 13, that any amino acid with the exception of proline can substitute at this position (Liang et al., 2000; Zhu et al., 2003). Similarly, Gln61 of Ras, which has been demonstrated to act as a catalytic base, is not stringently conserved. Rap1 does not have a glutamine at this position (Daumke et al., 2004). κB-Ras2 has a glutamine just one position later and the *Drosophila* κB-Ras also contains a glutamine at the same position (Figure 1).

Intrinsic GTP hydrolysis and GDP dissociation activities of small GTPases are usually very low and GTPase activating proteins (GAPs) facilitate GTP hydrolysis by proper positioning of the catalytic residues (Donovan et al., 2002; Scheffzek et al., 1998). Typically, the GAP provides key residues that are absent in GTPases to efficiently carry out the catalysis. Therefore although the GTPase activity of κB-Ras proteins are low, the presence of all other critical features suggests that these proteins may indeed function as GTPases in the presence of a GAP specific to κB-Ras. It is also important to note here that although the intrinsic GTPase activities are low, catalytic rates vary among the GTPases. Therefore, κB-Ras may represent a subclass where functional GTPase activity requires the absolute presence of a GAP.

Guanine nucleotide exchange factors (GEFs) are another class of proteins that typically facilitate GDP to GTP exchange activities in GTPases (Boriack-Sjodin et al., 1998; Chardin et al., 1993; Egan et al., 1993; Rossman et al., 2005). Although no GAP or GEF specific to κB-Ras has been found to date, the possibility of its presence cannot be excluded. Future investigation will determine if such GAP and GEF proteins exist and whether κB-Ras proteins function as cellular molecular switches.

9. CONCLUSION

By influencing membrane signaling, nuclear transport, small GTPases display broad functions affecting diverse biological activities. The κB-Ras proteins, on the other hand, appear to function in a much more specialized manner to influence NF-κB activation. The ability of κB-Ras proteins to interact with members of the IκB class proteins and their strict conservation within organisms that contain functional NF-κB signaling modules indicate an evolutionary correlation driven by function. Although our current understanding of their cellular role is extremely limited, current and future studies are certain to shed light on this interesting new member of the small GTPase superfamily.

REFERENCES

Baldwin, A.S., Jr. (1996) The NF-kappa B and I kappa B proteins: new discoveries and insights. Annu. Rev. Immunol. *14*, 649-683.

Barroga, C.F., Stevenson, J.K., Schwarz, E.M. and Verma, I.M. (1995) Constitutive phosphorylation of I kappa B alpha by casein kinase II. Proc. Natl. Acad. Sci. U.S.A. *92*, 7637-7641.

Boriack-Sjodin, P.A., Margarit, S.M., Bar-Sagi, D. and Kuriyan, J. (1998) The structural basis of the activation of Ras by Sos. Nature *394*, 337-343.

Bourne, H.R., Sanders, D.A. and McCormick, F. (1990) The GTPase superfamily: a conserved switch for diverse cell functions. Nature *348*, 125-132.

Bourne, H.R., Sanders, D.A. and McCormick, F. (1991) The GTPase superfamily: conserved structure and molecular mechanism. Nature *349*, 117-127.

Casey, P.J., Solski, P.A., Der, C.J. and Buss, J.E. (1989) p21ras is modified by a farnesyl isoprenoid. Proc. Natl. Acad. Sci. U.S.A. *86*, 8323-8327.

Chardin, P., Camonis, J.H., Gale, N.W., van Aelst, L., Schlessinger, J., Wigler, M.H. and Bar-Sagi, D. (1993) Human Sos1: a guanine nucleotide exchange factor for Ras that binds to GRB2. Science *260*, 1338-1343.

Chen, Y., Wu, J. and Ghosh, G. (2003) kappaB-Ras binds to the unique insert within the ankyrin repeat domain of IkappaBbeta and regulates cytoplasmic retention of IkappaBbeta/NF-kappaB complexes. J. Biol. Chem. *278*, 23101-23106.

Chen, Y., Vallee, S., Wu, J., Vu, D., Sondek, J. and Ghosh, G. (2004) Inhibition of NF-kappaB activity by IkappaBbeta in association with kappaB-Ras. Mol. Cell. Biol. *24*, 3048-3056.

Daumke, O., Weyand, M., Chakrabarti, P.P., Vetter, I.R. and Wittinghofer, A. (2004) The GTPase-activating protein Rap1GAP uses a catalytic asparagine. Nature *429*, 197-201.

Donovan, S., Shannon, K.M. and Bollag, G. (2002) GTPase activating proteins: critical regulators of intracellular signaling. Biochim. Biophys. Acta *1602*, 23-45.

Egan, S.E., Giddings, B.W., Brooks, M.W., Buday, L., Sizeland, A.M. and Weinberg, R.A. (1993) Association of Sos Ras exchange protein with Grb2 is implicated in tyrosine kinase signal transduction and transformation. Nature *363*, 45-51.

Fenwick, C., Na, S.Y., Voll, R.E., Zhong, H., Im, S.Y., Lee, J.W. and Ghosh, S. (2000) A subclass of Ras proteins that regulate the degradation of IkappaB. Science *287*, 869-873.

Frech, M., Darden, T.A., Pedersen, L.G., Foley, C.K., Charifson, P.S., Anderson, M.W. and Wittinghofer, A. (1994) Role of glutamine-61 in the hydrolysis of GTP by p21H-ras: an experimental and theoretical study. Biochemistry *33*, 3237-3244.

Fujiyama, A. and Tamanoi, F. (1990) RAS2 protein of Saccharomyces cerevisiae undergoes removal of methionine at N terminus and removal of three amino acids at C terminus. J. Biol. Chem. *265*, 3362-3368.

Ghosh, S., May, M.J. and Kopp, E.B. (1998) NF-kappaB and Rel proteins: evolutionarily conserved mediators of immune responses. Annu. Rev. Immunol. *16*, 225-260.

Gutierrez, L., Magee, A.I., Marshall, C.J. and Hancock, J.F. (1989) Post-translational processing of p21ras is two-step and involves carboxyl-methylation and carboxy-terminal proteolysis. EMBO J. *8*, 1093-1098.

Hancock, J.F., Magee, A.I., Childs, J.E. and Marshall, C.J. (1989) All ras proteins are polyisoprenylated but only some are palmitoylated. Cell *57*, 1167-1177.

Hoffmann, A., Levchenko, A., Scott, M.L. and Baltimore, D. (2002) The IkappaB-NF-kappaB signaling module: temporal control and selective gene activation. Science *298*, 1241-1245.

Huang, T.T., Kudo, N., Yoshida, M. and Miyamoto, S. (2000) A nuclear export signal in the N-terminal regulatory domain of IkappaBalpha controls cytoplasmic localization of inactive NF-kappaB/IkappaBalpha complexes. Proc. Natl. Acad. Sci. U.S.A. *97*, 1014-1019.

Huxford, T., Huang, D.B., Malek, S. and Ghosh, G. (1998) The crystal structure of the IkappaBalpha/NF-kappaB complex reveals mechanisms of NF-kappaB inactivation. Cell *95*, 759-770.

Jacobs, M.D. and Harrison, S.C. (1998) Structure of an IkappaBalpha/NF-kappaB complex. Cell *95*, 749-758.

Johnson, C., Van Antwerp, D. and Hope, T.J. (1999) An N-terminal nuclear export signal is required for the nucleocytoplasmic shuttling of IkappaBalpha. EMBO J. *18*, 6682-6693.

Johnson, D.R., Douglas, I., Jahnke, A., Ghosh, S. and Pober, J.S. (1996) A sustained reduction in IkappaB-beta may contribute to persistent NF-kappaB activation in human endothelial cells. J. Biol. Chem. *271*, 16317-16322.

Karin, M. and Ben-Neriah, Y. (2000) Phosphorylation meets ubiquitination: the control of NF-kappaB activity. Annu. Rev. Immunol. *18*, 621-663.

Kontgen, F., Grumont, R.J., Strasser, A., Metcalf, D., Li, R., Tarlinton, D. and Gerondakis, S. (1995) Mice lacking the c-rel proto-oncogene exhibit defects in lymphocyte proliferation, humoral immunity, and interleukin-2 expression. Genes Dev. *9*, 1965-1977.

Li, G. and Zhang, X.C. (2004) GTP hydrolysis mechanism of Ras-like GTPases. J. Mol. Biol. *340*, 921-932.

Liang, Z., Mather, T. and Li, G. (2000) GTPase mechanism and function: new insights from systematic mutational analysis of the phosphate-binding loop residue Ala30 of Rab5. Biochem. J.*346* Pt2, 501-508.

Malek, S., Huxford, T. and Ghosh, G. (1998) IkappaBalpha functions through direct contacts with the nuclear localization signals and the DNA binding sequences of NF-kappaB. J. Biol. Chem. *273*, 25427-25435.

Malek, S., Chen, Y., Huxford, T. and Ghosh, G. (2001) IkappaBbeta, but not IkappaBalpha, functions as a classical cytoplasmic inhibitor of NF-kappaB dimers by masking both NF-kappaB nuclear localization sequences in resting cells. J. Biol. Chem. *276*, 45225-45235.

Malek, S., Huang, D.B., Huxford, T., Ghosh, S. and Ghosh, G. (2003) X-ray crystal structure of an IkappaBbeta/NF-kappaB p65 homodimer complex. J. Biol. Chem. *278*, 23094-23100.

McElhinny, J.A., Trushin, S.A., Bren, G.D., Chester, N. and Paya, C.V. (1996) Casein kinase II phosphorylates IkappaBalpha at S-283, S-289, S-293, and T-291 and is required for its degradation. Mol. Cell. Biol. *16*, 899-906.

Milburn, M.V., Tong, L., deVos, A.M., Brunger, A., Yamaizumi, Z., Nishimura, S. and Kim, S.H. (1990) Molecular switch for signal transduction: structural differences between active and inactive forms of protooncogenic ras proteins. Science *247*, 939-945.

Pai, E.F., Krengel, U., Petsko, G.A., Goody, R.S., Kabsch, W. and Wittinghofer, A. (1990) Refined crystal structure of the triphosphate conformation of H-ras p21 at 1.35 Å resolution: implications for the mechanism of GTP hydrolysis. EMBO J. *9*, 2351-2359.

Phelps, C.B., Sengchanthalangsy, L.L., Huxford, T. and Ghosh, G. (2000) Mechanism of IkappaBalpha binding to NF-kappaB dimers. J. Biol. Chem. *275*, 29840-29846.

Rossman, K.L., Der, C.J. and Sondek, J. (2005) GEF means go: turning on RHO GTPases with guanine nucleotide-exchange factors. Nat. Rev. Mol. Cell Biol. *6*, 167-180.

Saraste, M., Sibbald, P.R. and Wittinghofer, A. (1990) The P-loop – a common motif in ATP- and GTP-binding proteins. Trends Biochem. Sci. *15*, 430-434.

Scheffzek, K., Ahmadian, M.R. and Wittinghofer, A. (1998) GTPase-activating proteins: helping hands to complement an active site. Trends Biochem. Sci. *23*, 257-262.

Silverman, N. and Maniatis, T. (2001) NF-kappaB signaling pathways in mammalian and insect innate immunity. Genes Dev. *15*, 2321-2342.

Tam, W.F. and Sen, R. (2001) IkappaB family members function by different mechanisms. J. Biol. Chem. *276*, 7701-7704.

Thompson, J.E., Phillips, R.J., Erdjument-Bromage, H., Tempst, P. and Ghosh, S. (1995) I kappa B-beta regulates the persistent response in a biphasic activation of NF-kappa B. Cell *80*, 573-582.

Vetter, I.R. and Wittinghofer, A. (2001) The guanine nucleotide-binding switch in three dimensions. Science *294*, 1299-1304.

Zhu, G., Liu, J., Terzyan, S., Zhai, P., Li, G. and Zhang, X.C. (2003) High resolution crystal structures of human Rab5a and five mutants with substitutions in the catalytically important phosphate-binding loop. J. Biol. Chem. *278*, 2452-2460.

CHAPTER 15

ANTI-RAS STRATEGIES FOR CANCER TREATMENT

JAMES J. FIORDALISI, CHANNING J. DER AND ADRIENNE D. COX
Departments of Radiation Oncology and Pharmacology, Lineberger Comprehensive Cancer Center, University of North Carolina at Chapel Hill Chapel Hill, NC 27599

Abstract: Ras is the most frequently mutated oncogene in human malignancies (30%) and is associated most strongly with the development of neoplasms in which therapeutic approaches have met with limited success (e.g., lung, colon, and pancreatic cancers). Extensive experimental analyses in cell culture and animal models have established strong and compelling validation of a causal role for aberrant Ras function in tumor progression and maintenance. Consequently, there has been considerable interest in and effort towards the development of strategies to block Ras function for cancer treatment. Despite intensive investigation, no strategies have yet been devised that can selectively block oncogenic Ras function in cells or animals. In this review, we summarize the approaches that have been considered, why the problem is so difficult, and the current status of these efforts

Keywords: target-based drug discovery, farnesyltransferase inhibitors, Raf, MEK, reovirus

1. INTRODUCTION

Mutational activation of Ras is associated with 30% of human tumors, with the prevalence as high as 90% in pancreatic cancers and 50% in colon cancers (Cox, A. D. and Der, C. J., 2002, Malumbres, M. and Barbacid, M., 2003). Even in cancers in which Ras is not mutated, Ras activity is often increased as a result of other genetic lesions, in particular, overexpression and/or mutational activation of receptor tyrosine kinases that function upstream of Ras. A vast of body of evidence has accumulated implicating Ras in virtually every aspect of malignant tumorigenesis, including increased cell proliferation, acquisition of anchorage-independence, survival, motility and invasion, and metastasis (Hanahan, D. and Weinberg, R. A., 2000, Campbell, P. M. and Der, C. J., 2004). Collectively these observations have made Ras an attractive target for anti-cancer drug development efforts (Cox, A. D. and Der, C. J., 2002, Downward, J., 2003). To block Ras

C. Der (ed.), RAS Family GTPases, 353–380.

function effectively, a number of approaches have been considered and developed. Of these efforts, inhibitors of Ras association with the plasma membrane or of Ras-mediated signaling have shown the greatest promise. In this chapter, we summarize the current status of these and other approaches to block the aberrant function of Ras in human cancers.

2. VALIDATION OF THE IMPORTANCE OF ABERRANT RAS ACTIVATION IN CANCER DEVELOPMENT AND GROWTH

There is now considerable experimental evidence that validates Ras as an excellent target for anti-cancer drug development (Malumbres, M. and Barbacid, M., 2003). First, cell culture studies demonstrate that mutated Ras can promote growth transformation of a wide variety of cell types. Of particular importance is the recent utilization of primary human cells (Zhao, J. J., et al., 2004). Ectopic expression of oncogenic Ras alone in primary human cells is not sufficient to cause tumorigenic transformation. However, prior immortalization by ectopic expression of the catalytic subunit of telomerase (hTERT) together with SV40 large T antigen (to inhibit Rb and p53 tumor suppressor function) and small t antigen (to inhibit protein phosphatase 2A function) renders a variety of primary human cell types sensitive to Ras-mediated growth transformation. This three-step genetic approach, and related experimental schemes, have shown that oncogenic Ras can cause growth transformation of human fibroblasts; mammary, kidney, lung, and ovarian epithelial cells; endothelial cells; and other cell types (Hahn, W. C., et al., 1999, Elenbaas, B., et al., 2001, Rich, J. N., et al., 2001, Lundberg, A. S., et al., 2002, Liu, J., et al., 2004). Finally, the use of anti-sense, interfering RNA and other genetic approaches to prevent mutant Ras function in Ras mutation-positive human tumor cell lines demonstrates that correction of the Ras mutation alone, in cells that harbor multiple genetic lesions, can impair the growth of tumor cells (Shirasawa, S., et al., 1993, Brummelkamp, T. R., et al., 2002).

Second, mouse models have provided further evidence for a causal role for Ras activation in cancer development. These models include those in which treatment with carcinogens causes mutational activation of Ras and tumor formation, as well as transgenic models in which mutant Ras expression is targeted to specific tissues (Mangues, R. and Pellicer, A., 1992, Rangarajan, A. and Weinberg, R. A., 2003). The recent development of knock-in models, in which activation of an endogenous mutant K-Ras allele is initiated in a controlled manner, support the ability of Ras to cause lung and pancreatic cancers (Johnson, L., et al., 2001, Guerra, C., et al., 2003, Hingorani, S. R., et al., 2003). Finally, the reversible activation of Ras in a melanoma mouse model demonstrated that the continued expression of activated Ras was essential for tumor maintenance (Chin, L., et al., 1999). In summary, the extensive and comprehensive body of cell culture and animal analyses provide compelling evidence for the causal role of mutated Ras in cancer development and a key role in tumor maintenance. Thus, the validation of Ras as an important target for therapeutic intervention is strong and is widely accepted.

3. REGULATION OF THE FUNCTION OF NORMAL RAS – PROVIDING CLUES FOR ANTI-RAS STRATEGIES

The three human *RAS* genes (H-, K- and N-Ras) encode four highly related 21 kDa proteins (K-Ras encodes the related 4A and 4B splice variants) (Barbacid, M., 1987). Ras proteins are GTPases that bind and hydrolyze GTP and cycle between a GTP-bound, active state and a GDP-bound, inactive state (Repasky, G. A., et al., 2004). In response to extracellular stimuli that regulate cell growth, guanine nucleotide exchange factors (GEFs) stimulate transient formation of Ras-GTP. Binding of GTP causes conformational changes in two regions of the Ras protein, designated switch 1 (residues 32-38) and switch 2 (residues 57-63), that increase the affinity of Ras for its downstream effector targets. GTPase activating proteins (GAPs) stimulate the intrinsic GTPase activity to hydrolyze GTP and cause formation of GDP-bound Ras. Thus, the coordinated action of GEFs and GAPs control the activation state of wild-type Ras. Naturally occurring mutations, typically found at positions 12, 13 or 61, render Ras proteins resistant to inactivation by GAPs. These observations provided one of the first clues used for the development of putative anti-Ras drugs, to develop an approach to render mutant Ras responsive to GAP (Figure 1). However, while there is some evidence for the feasibility of this approach (Fischbach, M. A. and Settleman, J., 2003), no drug has been developed successfully to target any Ras GAP.

Ras proteins are synthesized initially as cytosolic proteins that then undergo a rapid series of post-translational modifications that are vital to their normal and oncogenic functions (Figure 2) (Cox, A. D. and Der, C. J., 2002, Sebti, S. M. and Der, C. J., 2003, Winter-Vann, A. M. and Casey, P. J., 2005). First, farnesyltransferase (FTase) covalently attaches a farnesyl isoprenoid lipid to the cysteine located in the C-terminal CAAX motif (C = cysteine, A = aliphatic amino acid, X = terminal amino acid) that is found in all Ras proteins, in other members of the Ras superfamily, and in many other proteins. Farnesylation is the obligate first step in this process, as inhibition prevents all subsequent processing steps. However, farnesylation alone is not sufficient for full Ras function. Both proteolytic cleavage and removal of the AAX residues by Ras converting enzyme 1 (Rce) and carboxymethylation of the now-terminal farnesylated cysteine by isoprenylcysteine carboxymethyltransferase (Icmt) are necessary for proper Ras membrane localization and full activity. These modifications, together with a second membrane targeting signal positioned immediately upstream of the CAAX motif: either cysteine residues modified by the fatty acid palmitate (H-Ras, N-Ras and K-Ras4A) or a series of basic residues (K-Ras4B), increase the hydrophobic nature of Ras and are critical for targeting Ras to the inner face of the plasma membrane. The finding that structural mutants of Ras that fail to undergo the initial FTase modification (e.g., by mutation of the cysteine residue of the CAAX motif, generally to serine, hence "SAAX" mutants) are completely cytosolic and inactive, prompted an intensive effort to identify and characterize inhibitors of FTase as possible anti-Ras drugs.

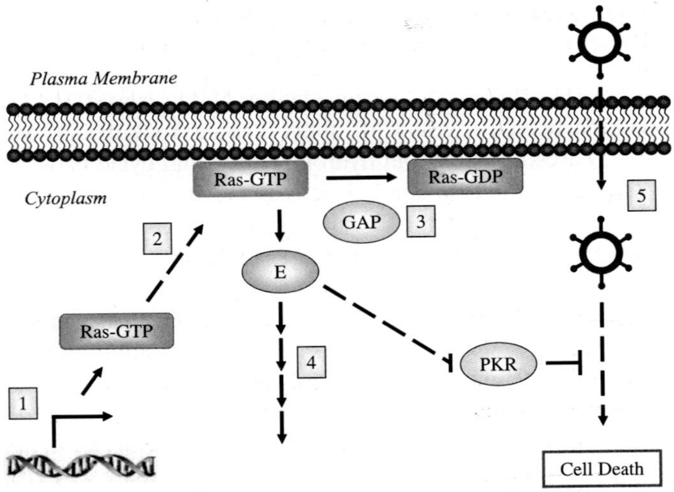

Figure 1. Inhibition of oncogenic Ras function. (1) Selective inhibition of mutant Ras gene expression. Use of anti-sense, ribozyme, and interfering RNA have been assessed as possible approaches to block the expression of mutant Ras. (2) Inhibition of posttranslational processing and membrance association. Inhibitors of the three enzymes that catalyze CAAX-signaled modification of Ras to promote association with the plasma membrane have been developed and analyzed in preclinical and clinical studies. (3) Promotion of GTP hydrolysis. Mutationally activated Ras (with mutations at residues G12, G13 or Q61) are GAP-insensitive and persistently GTP-bound. Therefore, the development of GTP analogues that can be hydrolyzed by oncogenic Ras proteins or a GAP that can stimulate the GTP hydrolysis activity of mutant Ras has been considered. (4) Inhibition of Ras-mediated cytoplasmic signaling. Inhibitors of one effector (E) signaling pathway, the Raf-MEK-ERK protein kinase cascade, have been developed and are under clinical evaluation. Since other effector pathways (e.g., P13K, RalGEF, and PLCε) also contribute to Ras oncogenesis, inhibitors of these pathways may also be effective anti-Ras agents. (5) Reovirus oncolysis of cells with activated Ras. In normal cells, the reoviral genome is transcribed into mRNA during viral replication, but viral protein synthesis is blocked by double-stranded RNA-activated protein kinase (PKR). Ras activation inhibits PKR function, allowing viral replication and induction of host cell death, followed by release of virus

4. TARGETING RAS PROCESSING

Inhibiting the enzymes that catalyze the post-translational modifications of Ras and its translocation to the plasma membrane constitutes an indirect but promising approach to blocking Ras function. While inhibitors of FTase have received the most attention to date, inhibitors of the modification steps catalyzed by Rce and Icmt have recently been considered as promising approaches for blocking Ras transformation (Winter-Vann and Casey, 2005).

4.1 Farnesyltransferase Inhibitors (FTIs)

FTIs are currently the most extensively developed approach to targeting Ras (Brunner, T. B., et al., 2003, Doll, R. J., et al., 2004, Sebti, S. M. and Adjei, A. A.,

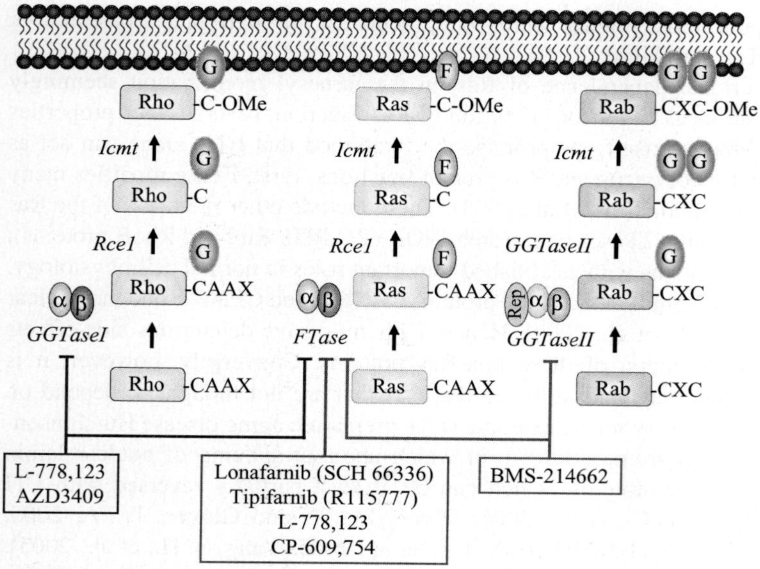

Figure 2. Inhibitors of protein prenylation. There are three human protein prenyltransferases. FTase and GGTase I catalyze the addition of farnesyl or geranylgeranyl isoprenoid groups, respectively, to proteins that terminate in CAAX tetrapeptide sequences. FTase preferentially recognizes CAAX motifs where X = M, S, A or Q, whereas GGTase I preferentially recongnizes CAAX motifs where X = L or I. Both FTase and GGTase I substrates undergo further modification by Rce1 and proteolytic cleavage of the AAX residues, and by Icmt and covalent addition of a methyl group to the now terminal prenylated cysteine residue. GGTase II (also called RabGGTase) recognizes C-terminal motifs typically with tandem cysteine residues: CXC, CC, CCXX, CCXXX but also CXXX. Both cysteine residues are modified by sequential addition of a geranylgeranyl isoprenoid. FTase and GGTase I are heterodimeric enzymes that share a common α subunit, but distinct β subunits. GGTase II is comprised of distinct α and β subunits, and require a third component (Rep) for recognition and prenylation of Rab proteins. The only known substrates are Rab small GTPases. Clinical evaluations of the FTIs L-778,123 and CP-609,754 have been discontinued due to toxicity. Some Rab proteins are also Icmt substrates. Several FTIs have been evaluated in clinical trials. Two FTIs currently under extended evaluation are lonafarnib and tipifarnib, which are selective inhibitors of FTase. L-778,123 and AZD3409 are dual-specificity inhibitors of both FTase and GGTase I, whereas BMS-214662 is a potent inhibitor of FTase and additionally of GGTase II. The GGTase II inhibitory activity contributes to the apoptotic activity of this FTI

2004). FTIs inhibit the enzyme FTase, which catalyzes the first step in Ras post-translational processing (Figure 1). Such compounds have been identified and developed by both rational drug design and by high throughput screens. Since the CAAX tetrapeptide sequence of Ras is necessary and sufficient to signal FTase-catalyzed prenylation, and since CAAX peptides are potent inhibitors of FTase activity *in vitro*, some FTIs have been rationally designed to mimic CAAX tetrapeptide sequences. Interestingly, however, the two FTIs for which there is the most clinical trial analyses and information (Sarasar®/lonafarnib/SCH66336 and

Zarnestra®/tipifarnib/R115777) were both identified through large-scalerandom screens for FTase inhibitors.

While the critical dependence of Ras on the farnesyl modification seemingly makes FTase an ideal target for inhibition of Ras function, several other properties of the FTase enzyme itself compromise the likelihood that FTIs alone can act as selective inhibitors of oncogenic Ras protein functions. First, FTase modifies many cellular proteins (Reid, T. S., et al., 2004). These include other members of the Ras superfamily of small GTPases (e.g., Rheb, NOEY2/ARHI, Rnd and RhoB proteins), as well as other proteins with established important roles in normal cell physiology, such as the mitotic spindle-associated proteins CENP-E and CENP-F and the nuclear lamins (Ashar, H. R., et al., 2000). Hence, FTIs may have deleterious side-effects by blocking the function of these non-Ras proteins. Conversely, however, it is also true that FTIs may find utility in diseases that are not thought to depend on oncogenic Ras. One very recent example is the premature aging disease Hutchinson-Gilford progeria syndrome (HGPS), in which aberrant splicing of nuclear lamin A leads to disease symptoms, which can be at least partially reversed with FTI treatment (Capell, B. C., et al., 2005, Glynn, M. W. and Glover, T. W., 2005, Mallampalli, M. P., et al., 2005, Toth, J. I., et al., 2005, Yang, S. H., et al., 2005). Second, normal Ras function is also dependent on farnesylation, and normal Ras function is expected to be critical for normal cell viability. Therefore, FTIs are not expected to selectively inhibit the function of mutated Ras. Finally, recent observations with FTase-deficient mice hint at further potential complications with the use of FTIs as anti-Ras inhibitors (Mijimolle, N., et al., 2005), as will be mentioned below.

A vast amount of preclinical data demonstrating the effectiveness of FTIs in a variety of experimental systems has been reviewed in detail elsewhere (Cox, A. D. and Der, C. J., 2002, Sebti, S. M. and Der, C. J., 2003, Doll, R. J., et al., 2004, Sebti, S. M. and Adjei, A. A., 2004). Briefly, FTIs inhibit the activity of purified FTase *in vitro*, block the prenylation of several farnesylated proteins in cell culture systems, and inhibit Ras signaling to transcription factors and the expression of Ras-dependent genes. FTIs also reverse many aspects of the Ras-transformed phenotype including increased proliferation, altered morphology, loss of actin stress fibers, and anchorage-independent cell growth. In nude mouse tumor xenograft models, FTIs inhibit the growth of several tumor types including colon, pancreas, prostate, lung, blood, bladder and brain, but produce little tumor regression. In transgenic mice whose activating mutations in N-Ras or K-Ras give rise to mammary, lymphoid and salivary tumors, FTIs also slow tumor growth and prevent the establishment of new tumors. Most impressively, FTIs cause regression in H-Ras transgenic tumor models, suggesting that, in certain circumstances, FTIs might kill tumors rather than simply slowing their growth. These observations also support the concept that transgenic tumor models more accurately reflect the conditions found in naturally-occurring tumors than do explant models.

FTIs whose clinical trial results have been reported include R115777 (tipifarnib/Zarnestra®; Johnson&Johnson), SCH66336 (lonafarnib/Sarasar®;

Schering-Plough), L778,123 (Merck), BMS-214662 (Bristol Myers-Squibb (BMS)), CP-609754 (Pfizer) and AZD3409 (AstraZeneca) (Brunner, T. B., et al., 2003, Doll, R. J., et al., 2004, Sebti, S. M. and Adjei, A. A., 2004). The Merck, BMS and Pfizer compounds are no longer being advanced for clinical use. As predicted by preclinical studies, FTIs have been particularly effective in combination with standard cancer therapies, especially with taxanes, which affect microtubule dynamics. Clinical trials using FTIs in combination rather than monotherapy are likely to be more successful. Many partial responses have been reported, particularly in breast cancer, multiple myeloma and related hematopoietic dyscrasias, and leukemias, and trials continue in these tumor types.

While these successes clearly demonstrate the potential utility of FTIs, the story of our understanding of their mechanism of action is convoluted and illustrates some of the potential pitfalls of targeted drug development. Much of the early work with FTIs focused on H-Ras-dependent phenotypes (Cox, A. D. and Der, C. J., 1997). FTIs were initially shown to effectively block the prenylation, membrane localization and function of oncogenic H-Ras *in vitro*, in cell culture systems, and in a variety of mouse tumor models. At that point, one of the strengths of FTIs was their apparent lack of general toxicity. This observation allayed concerns that inhibition of wild-type Ras might adversely affect normal cells, which also utilize Ras signaling pathways for their metabolic processes. One possible explanation for this lack of toxicity came with the observation that N-Ras and K-Ras were able to bypass functional FTI inhibition by becoming alternatively prenylated in the presence of FTI (Lerner, E. C., et al., 1997, Rowell, C. A., et al., 1997, Whyte, D. B., et al., 1997). The CAAX motifs of N-Ras (CVVM) and of K-Ras4B (CVIM) contain a C-terminal methionine that allows them to be utilized as substrates by geranylgeranyltransferase I (GGTase I)(James, G. L., et al., 1995), whereas the C-terminal serine of H-Ras (CVLS) does not. As a result, N-Ras and K-Ras are still prenylated in the presence of FTI, albeit by a non-native geranylgeranyl lipid that appears to be functionally equivalent to the normal farnesyl group (Cox, A. D., et al., 1992). It was speculated that endogenous N-Ras and K-Ras are able to substitute functionally for H-Ras when it is inhibited by FTI treatment of normal cells. While this might explain the relatively low toxicity of FTIs, the phenomenon of alternative prenylation also implied that tumors involving oncogenic N-Ras or K-Ras, which are much more common than those involving H-Ras, might not be affected by FTIs.

However, this suggestion was belied by evidence showing FTIs to be quite effective against cells (Sepp-Lorenzino, L., et al., 1995) and tumors expressing oncogenic N-Ras (Mangues, R., et al., 1998) or K-Ras (Sun, J., et al., 1995). How could FTIs, which fail to inhibit N-Ras and K-Ras activity, also block N-Ras or K-Ras-induced transformation? Inhibition of non-Ras farnesylated proteins by FTIs is the likely explanation for this previously unexpected finding. As mentioned earlier, it is important to remember that FTIs, while designed initially for the purpose of inhibiting Ras function, are not Ras-selective inhibitors. By blocking FTase, FTIs inhibit the prenylation and presumably the function of numerous other cellular proteins containing a CAAX motif that are known to be FTase substrates (Reid, T. S., et al., 2004). Thus, while Ras proteins were the original targets of FTIs, there are several other farnesylated proteins which are known or suspected to have roles in cancer development and whose inhibition may help to account for the effects of FTIs. These include several other members of the Ras superfamily (i.e RhoB, Rheb, Rnd3/RhoE, TC10), as well as proteins that are unrelated to Ras including the phosphatases of regenerating liver (PRL-1, -2 and -3), the centromere binding protein CENP-E, and several others. One of the ongoing mysteries surrounding FTIs is the identity of the farnesylated protein target(s) that accounts for the observed effects of FTIs on cellular transformation. Complicating the matter is the fact that it may be necessary for FTIs to block more than one farnesylated protein simultaneously, and that the relevant combination of target proteins may not be the same in every tumor type. This uncertainty has made the design of FTI clinical trials difficult and may explain the limited success of FTIs in several of those trials. Certainly it is clear that Ras mutation status is not predictive of FTI sensitivity (Sepp-Lorenzino, L., et al., 1995).

One controversial issue in the development of FTIs has been the extent to which they should exhibit specificity for FTase and limited activity for the related enzyme geranylgeranyltransferase I (GGTase I). Like FTase, GGTase I recognizes CAAX-terminating peptides, but those in which the X residue is leucine. In contrast, FTase prefers CAAX sequences in which X is serine, glutamine, methionine or alanine. A few CAAX sequences (e.g., in which X is phenylalanine or isoleucine) can be recognized by both enzymes. In addition to many Rho family GTPases, the Ras family members Ral, Rap and R-Ras terminate in leucine and are also modified by GGTase I. The dependence of Ras transformation on signaling from some of these GTPases as well as on other geranylgeranylated proteins has led to the suggestion that inhibiting both enzymes could improve the efficacy of antitumor treatment, whereas the great number of proteins in and out of the Ras superfamily that are modified by GGTase I has led to concerns that toxicity would be greatly increased by dual-specificity inhibitors. There is some evidence both for and against this theory (Lerner, E. C., et al., 1997, Sun, J., et al., 1998, Lobell, R. B., et al., 2001). Nevertheless, despite the similar nature of the substrate recognition sequences for FTase and GGTase I, it has been a relatively straightforward matter to design FTIs that are potent and selective for FTase, and most FTIs advancing to clinical practice are indeed FTase-selective. However, in light of the ability of K-Ras and N-Ras to undergo alternative prenylation, dual-specificity inhibitors of both FTase and GGTaseI have also been evaluated in phase I clinical trials. L-778, 123 is a potent inhibitor of FTase (*in vitro* IC50 = 2 nM) with some activity against GGTase-I (*in vitro* IC50 = 98 nM) (Buser, C. A., et al., 2001). Although L-778, 123 did show partial inhibition of GGTaseI activity in patients, this degree of inhibition was not sufficient to block K-Ras prenylation (Lobell, R. B., et al., 2002). AZD3409 is another dual-specificity inhibitor undergoing evaluation in phase I clinical trials. It remains to be determined whether such dual-specificity inhibitors, or the combination use of a FTI together with a GGTI (Sebti, S. M. and Hamilton, A. D., 2000), will be effective inhibitors of K-Ras prenylation and function.

The third human prenyltransferase, Rab GGTase or GGTase II, has a substrate recognition sequence very distinct from those of FTase and GGTase I, and with one exception, no FTI has exhibited any inhibitory activity for GGTase II. Interestingly enough, however, that FTI, BMS-214662, has a unique anti-apoptotic effect that is mediated in part via GGTase II (Rose, W. C., et al., 2001, Lackner, M. R., et al., 2005).

Clearly, Ras itself is not the only target of FTI action, but the most critical biologically relevant target(s) remain to be identified. This may allow us to characterize tumors by their dependence on individual farnesylated proteins (or groups of farnesylated proteins) and to distinguish FTI-sensitive tumors from FTI-insensitive tumors, improving the application of this still promising class of anti-cancer drugs. Meanwhile, the search is ongoing for other classes of anti-Ras inhibitors.

4.2 Inhibitors of Rce1 and Icmt

The inability of FTIs to effectively block K-Ras and N-Ras function has stimulated increasing interest in the validation and development of approaches to block other steps in CAAX-signaled processing (Winter-Vann, A. M. and Casey, P. J., 2005). Earlier studies suggested that, unlike FTase inhibition, inhibition of the AAX proteolytic or carboxylmethylation steps did not completely block Ras function and transforming activity (Hrycyna, C. A., et al., 1991, Kato, K., et al., 1992), and hence, these were not felt to be attractive targets for anti-Ras inhibitor development. However, recent studies with mouse embryonic fibroblasts (MEFs) derived from mice deficient for Rce1 or Icmt function suggests that oncogenic Ras-mediated transformation is impaired when these processing steps do not occur. Thus, while much additional target validation is still needed, we will review here the current status of the feasibility and development of inhibitors of Icmt and Rce1 inhibitors for inhibition of oncogenic Ras function.

To address the importance of the Rce1-mediated processiong step in Ras function, Young and colleagues established mice with a conditional Rce1 allele and generated MEFs for *in vitro* analyses (Bergo, M. O., et al., 2002). Cre recombinase-mediated disruption of Rce1 function in MEFs caused a significant (~50%) reduction in H-Ras and K-Ras membrane association. The growth of H-Ras and K-Ras-transformed MEFs was also reduced (65% and 30%, respectively) upon Rce1 deletion. By comparison, treatment with the FTI SCH66336 caused a complete loss of growth in soft agar. Thus, while Rce1-mediated processing is not as critical as the farnesylation step, these results nevertheless suggest that inhibitors of Rce1 may be effective as anti-Ras inhibitors.

Bergo et al. utilized a similar approach to study the consequences of conditional disruption of Icmt on Ras function in MEFs (Bergo, M. O., et al., 2004). Inactivation of Icmt partially reduced K-Ras association with membranes in MEFs as well as in ES cells (Bergo, M. O., et al., 2000). This corresponded with an inhibition in anchorage-independent growth and tumor formation in K-Ras-transformed MEFs

(Bergo, M. O., et al., 2004). Additional evidence that Icmt may be an important target for anti-cancer therapy was provided by the observation that the anti-folate methotrexate, a widely used and successful cancer chemotherapeutic, may inhibit cell proliferation, in part, by indirectly inhibiting Icmt function (Winter-Vann, A. M., et al., 2003). Various inhibitors of Icmt have been shown to impair Ras transformation; however, these compounds are not specific inhibitors of Icmt, and consequently have not provided reliable evidence regarding the specific consequences of Icmt inhibition. Recently, Casey and colleagues identified a small molecule inhibitor of Icmt, designated cysmethynil (Winter-Vann, A. M., et al., 2005). Cysmethynil treatment reduced K-Ras membrane association and caused Icmt-dependent impairment of the soft agar growth of K-Ras mutation-positive colorectal carcinoma cells.

Somewhat surprising have been the more drastic consequences of Icmt deficiency when compared to an Rce1 deficiency (Bergo, M. O., et al., 2001), which is expected to prevent both the AAX proteolysis step as well as the subsequent Icmt-catalyzed modification. An Icmt deficiency did impair the growth of MEFs and, surprisingly, inhibited B-Raf transformation (Bergo, M. O., et al., 2004). This may be due to impairment of the function of non-Ras proteins that cooperate with activated Raf, such as Rho GTPases.

While these preliminary studies support the potential value of targeting Rce1 and Icmt for inhibition of Ras-mediated oncogenesis, further studies with human tumor cells and mouse models are needed to strengthen the validation of these enzymes as targets. Will the limited reductions seen in Ras function be sufficient to significantly impact human tumor cell growth? Furthermore, to better appreciate the potential nonspecific effects of these agents, the consequences of Rce1 and Icmt inhibition on the function of other targets need to be addressed. Unlike FTIs or GGTIs that target only CAAX-terminating substrates of FTase or GGTase I, respectively, the function of proteins in both classes of prenylated proteins is expected to depend on Rce1 and Icmt function. Icmt is needed, additionally, for some of the Rab substrates of GGTaseII (Bergo, M. O., et al., 2001). The embryonic lethality seen in mice deficient in Rce1 and Icmt, (after embryonic day 15.5 and 11.5, respectively) supports this concern (Kim, E., et al., 1999, Bergo, M. O., et al., 2001). However, FTase is also essential for mouse embryogenesis (Mijimolle, N., et al., 2005), yet FTIs have exhibited relatively limited normal cell toxicity, so perhaps the need for each of these processing enzymes may be less critical for normal adult tissue homeostasis. Consistent with this possibility, a deficiency in Rce1 did not significantly impair the growth of untransformed MEFs. Also, while tissue-specific reduction of Rce1 resulted in a lethal cardiomyopathy (Bergo, M. O., et al., 2004), tissue-specific reduction in the liver showed no deleterious consequences, and normal hematopoietic cell function was not compromised by Rce1 deficiency (Aiyagari, A. L., et al., 2003). Hence, Rce1 and Icmt inhibitors may be better tolerated *in vivo* than previously expected.

As with FTIs, inhibitors of Rce1 and Icmt also should not be considered simply anti-Ras inhibitors, and other tissue type differences in sensitivity are anticipated (Figure 2). For example, the processing of two other prenylated proteins, the farnesylated Gγ1 subunit of transducin and geranylgeranylated Rap1B, was also blocked in Rce1-deficient MEFs (Kim, E., et al., 1999). Hence, the inhibition of Ras transformation seen in MEFs may be due, in part, to the impaired function of Rho (e.g., Rac, RhoA) and Ral GTPases that are known to contribute to Ras transformation. Consistent with this possibility, a lack of Icmt modification decreased the membrane association and function of the geranylgeranylated RhoA protein (Harrington, E. O., et al., 2004, Lu, Q., et al., 2004), due possibly to decreased protein stability (Bergo, M. O., et al., 2004). However, a recent study by Philips and colleagues found, surprisingly, that the membrane association and function of geranylgeranylated Rho GTPases may not be critically dependent on either Rce1 or Icmt function (Michaelson, D., et al., 2005). Furthermore, an experimentally-generated variant of Ras modified by geranylgeranylation was found to be independent of Rce1 or Icmt function. If this holds true for other geranylgeranylated proteins, then the critical targets of Rce1 may be restricted to farnesylated proteins, thereby reducing the number of targets and possibly the toxicity of Rce1 inhibitors. However, this also means that Rce1 or Icmt inhibitors may not be useful to overcome the alternative prenylation of K-Ras caused by FTI treatment. Clearly, the importance of Rce1 and Icmt-catalyzed modifications for the function of other prenylated proteins needs to be addressed in more detail to better assess the promise and potential of Rce1- and Icmt-based therapies.

5. TARGETING RAS EXPRESSION AND/OR STABILITY

Inhibiting the enzymes that carry out the post-translational processing of Ras prevents the maturation and function of Ras proteins. But functional inhibition of Ras can also be achieved by preventing the initial expression of Ras or by blocking its function after synthesis and maturation are complete.

5.1 Targeting Ras Membrane Association and Stability: Farnesylthiosalicylic Acid (FTS)

One such approach involves farnesylthiosalicylic acid (FTS), a molecule that mimics the structure of the fully-processed, farnesylated Ras C-terminus. Rather than preventing the processing of Ras proteins, FTS is thought to compete with fully processed Ras for membrane targeting sites. The resulting dislodgment of mature Ras from membranes is followed by its increased degradation (Haklai, R., et al., 1998). Thus, FTS reduces Ras signaling by at least two means: inhibition of localization and protein stability. Interestingly, although the structure of FTS is somewhat related to that of N-acetyl cysteine and therefore might be expected to act by competing with the carboxymethyltransferase enzyme Icmt, this has been shown not to be the case. FTS was also able to dislodge and increase degradation of a mutant form of oncogenic K-Ras (12V/187Y) that is not a substrate for Icmt (Elad, G., et al., 1999). FTS inhibits the growth of a multitude of tumor types in preclinical models *in vitro* and *in vivo* (Kloog, Y. and Cox, A. D., 2000), and can synergize with conventional cytotoxic agents (Gana-Weisz, M., et al., 2002). Preclinical studies suggest the potential of FTS to decrease not only tumor cell growth but also possibly invasiveness and even metastatic potential (Blum, R., et al., 2005). Clinical trials of FTS are expected to begin in the fall of 2006 (Y. Kloog, personal communication). However, as with the other inhibitors of Ras processing, FTS is not strictly an anti-Ras drug, and is also expected to compete with the membrane attachment of other farnesylated proteins. Truly specific anti-Ras drugs may come instead from inhibiting the expression of specific oncogenic alleles of Ras.

5.2 Targeting Oncogenic Ras Expression

5.2.1 RNAi

RNA interference (RNAi) is a process in which small interfering RNAs (siRNAs) are designed to complement and anneal to the mRNA for a particular protein, thereby preventing translation and creating a functional "knock-out" of that protein. RNAi has become widely popular in the functional evaluation of proteins across many fields of research, and may also find utility as a cancer treatment (reviewed in (Duursma, A. M. and Agami, R., 2003) and (Dykxhoorn, D. M. and Lieberman, J., 2005). Preclinically, small interfering RNA oligonucleotides are delivered directly to cells in culture or animal models. Increasingly, short hairpin RNAs (shRNAs) expressed from retrovirus and lentivirus plasmids are packaged for introduction into tumor cells where they are then processed into siRNAs by the RNAi machinery including DICER.

An early proof-of-principle study (Brummelkamp, T. R., et al., 2002) demonstrated that inhibition of Ras expression might have therapeutic value when it showed that an siRNA directed against a G12V mutant of K-Ras4B could inhibit the expression of the oncogenic K-Ras allele and reduce tumorigenicity in the human pancreatic cancer cell line Capan-1. This study also demonstrated clearly the dependence of this particular tumor cell line on the continued expression of oncogenic Ras for maintenance of its tumorigenic phenotype. The ability of siRNA to distinguish between wild-type K-Ras and oncogenic K-Ras, which differ in mRNA sequence by a single base, could have beneficial implications for cancer treatment, since it is obviously preferable to inhibit only the oncogenic Ras allele while leaving the wild-type allele to perform its normal cellular functions. Subsequent studies showed that ablation of H-Ras expression by RNAi also inhibited the growth of human ovarian cancer cell lines (Yang, G., et al., 2003) and a melanoma cell line (Yin, J. Q., et al., 2003), confirming both the utility of this technique against cancer cells generally and against Ras-associated cancer cells specifically.

As with many other gene therapy-style approaches, a limitation of the RNAi approach to targeting Ras function is the difficulty associated with delivering unstable siRNAs into the tumor cells. Engineered viruses are most often employed as a delivery system to overcome this problem in preclinical models. It remains to be seen how well siRNA delivery can be accomplished for clinical use.

5.2.2 *Anti-Ras ribozymes and anti-Ras antibodies*

Hammerhead ribozymes that specifically target the mutant K-Ras allele (K-Ras12V) have been shown *in vitro* and in nude mouse models to cause tumor growth inhibition or even regression of pancreatic, colon and lung tumor cell lines (Funato, T., et al., 2000, Zhang, Y. A., et al., 2000, Kijima, H., et al., 2004). Inhibition of oncogenic Ras function has also been achieved using Ras antibodies expressed intracellularly. Expression in human cancer cells of a single chain Fv fragment of the monoclonal Ras antibody Y13-259, which is known to be a neutralizing antibody, has been reported to inhibit Ras signaling, induce apoptosis in Ras-transformed, but not normal cells, cause tumor regression in nude mice, and cause radiosensitization (Cochet, O., et al., 1998, Cochet, O., et al., 1999, Russell, J. S., et al., 1999, Tanaka, T. and Rabbitts, T. H., 2003). This may be the result of either the physical interference of the antibody with Ras function or the accelerated degradation of antibody-associated Ras by the cell, or both. The coding sequences for the proteins of interest (antibody or ribozyme) were introduced using viruses. While interesting and worthy of mention for their technical elegance, these techniques do not as yet represent viable approaches to treating human disease. However, they may be used to study Ras function in the laboratory setting.

6. REOVIRUS THERAPY

In recent years the power of lytic viruses to efficiently infect and kill human cells has emerged as an intriguing and promising new approach to cancer treatment (reviewed in (Zwiebel, J. A., 2001) and (Norman, K. L. and Lee, P. W., 2005)). Oncolytic viruses currently under development include adenovirus, adeno-associated virus, Sindbis virus, herpes simplex virus, Epstein-Barr virus, Newcastle disease virus, and reovirus. An advantage of using viruses is their natural ability to survive *in vivo*, and to enter and kill cells, thus avoiding some of the major pharmacokinetic and pharmacodynamic problems associated with small molecule anti-cancer agents. Moreover, the use of viruses avoids the need for potentially toxic systemic compounds to kill cells because these viruses are selectively toxic to the cells they infect. But, as with any anti-cancer drug, if viruses are not made highly selective for either delivery to or killing of tumor cells vs. normal cells, then general toxicity will limit their usefulness.

Attempts to improve tumor cell selectivity have been addressed using several approaches, including engineering viral coat proteins to bind to cell surface molecules preferentially expressed on a desired cell type, placing viral proteins under the control of tissue-specific promoters and thus limiting the cell types in which they will replicate, or engineering viruses that will replicate only in tumor cells due to the tumor-specific activation or inactivation of particular signaling pathways. The latter approach is used by the adenovirus ONYX-015, which is engineered to replicate only in cells inactive for the tumor suppressor p53, thus targeting tumor cells with this common genetic lesion but leaving normal cells unharmed (McNeish, I. A., et al., 2004). Similarly, it is the ability of reovirus to replicate selectively in cells with high levels of active Ras signaling that may be exploited to treat Ras-dependent cancers.

Asymptomatic in adults, reovirus is a common respiratory virus that selectively infects and kills cells in which activated Ras signaling is elevated. This feature of reovirus is natural and thus requires no genetic manipulation of the virus. Moreover, the reovirus receptor protein, junctional adhesion molecule-1 (JAM-1), is similar to that of adenovirus and is ubiquitously expressed (Stehle, T. and Dermody, T. S., 2004), such that reovirus may potentially be effective against Ras-associated tumors.

Interest in reovirus as an anti-cancer agent grew from studies on the mechanism of reoviral infection. These early studies (Strong, J. E., et al., 1993, Strong, J. E. and Lee, P. W., 1996) suggested that signaling pathways activated downstream of the epidermal growth factor receptor (EGFR) permitted reoviral infection and/or replication and subsequent host cell lysis. Because Ras is a major pathway activated downstream of the EGFR, its role in reoviral infections was analyzed further.

Overexpression of either the Ras-GEF Sos or mutationally active Ras itself also rendered cells susceptible to reoviral infection, thereby confirming a central role for Ras activity in the infection process (Strong, J. E., et al., 1998) and suggesting the possibility of using reovirus for anti-Ras therapy (Coffey, M. C., et al., 1998). Further, the mechanism of Ras-dependent reoviral susceptibility was shown to involve double stranded RNA-activated protein kinase (PKR) (Strong, J. E., et al., 1998). Activated in normal cells by the reoviral double-stranded RNA genome, this intracellular host-cell resistance factor prevents translation of viral mRNA into viral proteins by phosphorylating and inactivating the translation initiation factor protein eIF2a. Ras activity inhibits PKR, thereby permitting viral protein translation and replication, and host cell lysis. Whether inhibition of PKR by Ras is necessary for cellular transformation is not known.

The first proof-of-principle studies for reovirus anti-tumor activity showed that the Ras-dependence of reoviral lysis could be exploited selectively to target tumor cells with high levels of Ras activity, while leaving normal cells unaffected (Coffey, M. C., et al., 1998). In this study, intratumoral injections of reovirus into mouse tumor xenografts derived from either viral ErbB-transformed NIH 3T3 cells, the human glioblastoma cell line U87, or Ras-transformed C3H10T1/2 mouse fibroblasts led to regressions in 65-80% of tumors. Subsequent studies in a variety of other tumor types *in vitro* and *in vivo*, including gliomas, breast, lymphoid, colon, pancreas, and medulloblastoma produced similarly impressive results (reviewed in Norman, K. L. and Lee, P. W., 2005). Reovirus caused extensive cell death in cultured tumor cell lines and in primary tumor-derived cells, and regression of tumor explants in mice with significant increases in long-term survival rates. Whereas activating Ras mutations are common in pancreas and colon tumors, they are rare in breast, lymphoid and medulloblastoma tumors. Nevertheless, reovirus infected and significantly inhibited these tumor types as well, presumably due to activation of wild-type Ras by upstream receptor tyrosine kinase activation. Moreover, susceptibility of host cells to reoviral infection appears to be independent of the activated Ras isoform. For example, the medulloblastoma cell line ONS76, which has high levels of active N-Ras, but low levels of H- or K-Ras, is as susceptible to reoviral infection as other medulloblastoma cell lines in which the activation levels of the different Ras isoforms is reversed (Yang, W. Q., et al., 2003). Thus the utility of reovirus may not be limited to tumors in which one Ras isoform is preferentially activated. Perhaps most importantly, reovirus was shown both to prevent and to cause regression of tumor metastases (Hirasawa, K., et al., 2003, Yang, W. Q., et al., 2003), which would be of tremendous benefit since the majority of cancer deaths are caused by the effects of metastases and not the primary tumors.

Different treatment regimens were also studied. In some case, mouse xenograft tumors responded even when reovirus was injected at sites distant from the primary tumor including by the intravenous route (Hirasawa, K., et al., 2002, Norman, K. L., et al., 2002) offering hope that reovirus might be used to treat tumors that are difficult to reach. Finally, the potency of reovirus as an anti-tumor agent was improved when given in combination with immunosuppressants to reduce the host immune response to the virus and to effectively increase viral titer (Hirasawa, K., et al., 2003).

Although results using reovirus to treat tumors in preclinical models have been impressive, there are still important gaps in our knowledge of the mechanism of Ras-mediated reovirus sensitivity. For example, while it is clear that host cell susceptibility to reoviral infection correlates with Ras activity levels when analyzed across several cell types (Yang, W. Q., et al., 2003), it is not clear if there is an absolute threshold level of Ras activity that determines whether a particular tumor would be susceptible or resistant to reoviral treatment. This problem is complicated by the fact that in these studies not all tumors derived from cells with high levels of active Ras were sensitive to reovirus. Not surprisingly, factors other than Ras activity may also influence the outcome of reoviral treatment. Furthermore, it is not clear what Ras signaling pathway(s) is required to promote reovirus sensitivity. ERK activation alone is not a reliable predictor of tumor cell sensitivity. There is evidence that the RalGEF-Ral effector pathway (discussed below; Figure 3) is sufficient and necessary for reovirus sensitivity in Ras-transformed NIH 3T3 cells (Norman, K. L., et al., 2004). Whether this pathway is also important for dictating reovirus sensitivity of human tumor cells is not known.

It should also be noted that although reovirus only replicates in and lyses host cells with overactive Ras, normal cells also become infected. In normal cells with low levels of active Ras, the viral genome is transcribed into mRNA, but viral protein synthesis is blocked by PKR. It is not clear what side effects, if any, non-productive viral infection of normal cells may have. In some of the mouse studies described earlier, side effects of reoviral treatment such as hydrocephalus and inflammation were reported (Wilcox, M. E., et al., 2001, Yang, W. Q., et al., 2003). However, it is not clear whether these effects were the result of a host immune response to the virus, viral damage to normal tissue resulting from non-productive infection, damage by debris from dead tumor cells, contaminants in the viral preparation used,

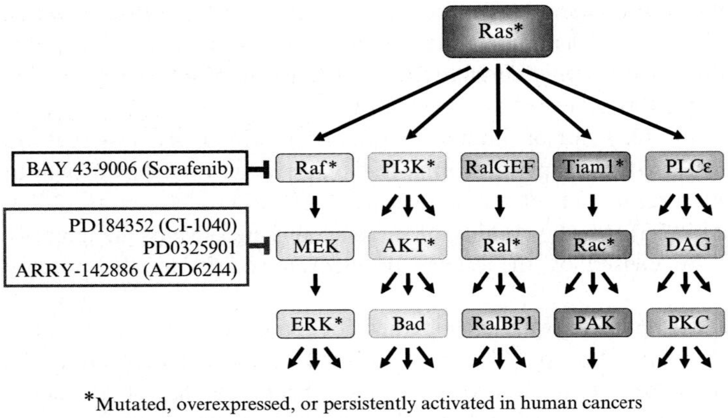

Figure 3. Ras effector pathways involved in oncogenesis. The best characterized effectors are the Raf serine/threonine kinases, whch activate the MEK>ERK MAPK cascade. The next best characterized effectors are the p110 catalytic subunits of PI3K. The main activity of PI3K involves conversion of phosphatidylinositol 3,5-bisphosphate (PIP2) to phosphatidylinositol 4,5-bisphosphate (PIP3). PIP3 promotes the activation of multiple, functionally diverse proteins, with the AKT serine/threonine kinases the best characterized. AKT phophorylates a spectrum of proteins, many of which are involved in regulation of cell survival (e.g., Bad). RalGEFs are activators of the RalA and RalB small GTPases. Activated Ral interacts with multiple, downstream effectors, including RalBP1, which in turn interacts with a variety of proteins that include Rho GTPases. Tiam1 is GEF for the Rac small GTPases (Rac1, Rac2 and Rac3), and activated Rac regulates multiple downstream effectors, including the PAK serine/threonine kinases. PLCe has two distinct catalytic functions: as a GEF for Ras family proteins, and also as a phospholipase to stimulate phosphotidylinositol-4,5 biphosphate hydrolysis and formation of the second messengers inositol-1,4,5 triphosphate (IP3) and diacylglycerol (DAG). IP3 activates calcium release, whereas the membrane-bound DAG regulates multiple proteins, in particular, memebers of the protein kinase C (PKC) family

or some other cause. In any case, potentially detrimental side effects will need to be assessed carefully.

Phase I and II clinical trials examining reovirus therapy (Reolysin) have been reported. A phase I evaluation showed no serious adverse toxicity, with only occasional transient flu-like symptoms and headache observed (Norman, K. L. and Lee, P. W., 2005). Consequently, no dose-limiting toxicity or maximum tolerated dose was found. A phase II trial of Reolysin for prostate cancer has also been completed, with the primary endpoints being to examine safety and the histopathology of prostate tissue post-therapy and post-prostactectomy. Patients were treated with a single reovirus injection directly into the prostate gland. Evidence of apoptotic tumor cell death was seen in four of six patients, and no safety concerns were reported. A phase I trial with malignant glioma patients is ongoing to further assess the safety of reovirus therapy. Whether activation of Ras or of a specific Ras signaling pathway will be useful to predict patient response to Reolysin has not been addressed in clinical trials to date.

7. TARGETING RAS SIGNALING

7.1 The Raf>MEK>ERK Mitogen-activated Protein Kinase Cascade

Yet another approach to blocking oncogenic Ras function involves inhibition of Ras-mediated signaling (Figures 1 and 3). The initial discovery of the Raf serine/threonine kinases (c-Raf-1, A-Raf, and B-Raf) as key effectors of Ras signaling and transformation identified this as a promising direction for the development of inhibitors of the Raf-MEK-ERK protein kinase cascade. Studies in rodent fibroblasts established that this signaling pathway is sufficient and necessary for Ras-mediated transformation (Repasky, G. A., et al., 2004). The recent identification of mutationally activated B-Raf in a non-overlapping pattern with that of mutationally activated Ras in melanoma and in colon and other human carcinomas provided further validation of the key role of this effector pathway in Ras-mediated oncogenesis (Garnett, M. J. and Marais, R., 2004). This pathway promotes cellular proliferation and survival, as well as tumor cell invasion and metastasis (Cox, A. D. and Der, C. J., 2003, Campbell, P. M. and Der, C. J., 2004). Thus, while Ras also utilizes a multitude of other effectors, the Raf>MEK>ERK pathway has attracted the greatest interest and has seen the most significant progress to date in terms of anti-cancer drug discovery (Sebolt-Leopold, J. S. and Herrera, R., 2004). Phase I-II clinical trials have been reported for kinase inhibitors of both Raf and MEK (Figure 3).

Currently, clinical trials of only one Raf inhibitor have been reported. The bi-aryl urea BAY43-9006 was identified in a screen for inhibitors of the serine/threonine kinase p38 MAPK, and was developed originally as an inhibitor of Raf-1 (Lyons, J. F., et al., 2001). It exhibited anti-tumor activity against various human tumor cell lines, including colon, lung, breast, ovarian and pancreatic carcinomas, and melanomas, in cell culture and in mouse models (Wilhelm, S. M., et al., 2004). Although tumor inhibition was associated with inhibition of ERK activity, BAY43-9006 also has significant activity against other protein kinases. In particular, BAY43-9006 potently inhibits several receptor tyrosine kinases involved in tumor angiogenesis, including the vascular endothelial growth factor receptors (VEGFR) 2 and 3, the platelet-derived growth factor receptor (PDGFR), Flt-3, and c-Kit. Thus, the anti-tumor activity of BAY43-9006 may be due to inhibition of Raf, of angiogenesis-related kinases, or of yet other non-Raf kinase targets.

Phase I clinical trials indicated that BAY43-9006 (sorafenib) is safe and well-tolerated, with the most common toxicities involving simply skin rash and diarrhea (Awada, A., et al., 2005, Moore, M., et al., 2005). Stable disease was seen in several patients, with one partial response in a patient with renal cell carcinoma (RCC). Hence, a major focus of subsequent phase II and III trials has been in RCC. One phase II trial showed a 40% response rate in the 41 RCC patients evaluable (Ahmad, T. and Eisen, T., 2004). Neither Ras nor B-Raf mutations are seen in RCC, but these tumors are heavily dependent on VEGF-mediated signaling as a consequence of frequent loss of the von Hippel-Lindau (VHL) tumor suppressor. These tumors also respond fairly well to more conventional anti-angiogenesis therapy such as Avastin, so it is speculated that the anti-tumor activity of BAY43-9006 in RCC may be due to inhibition of targets that regulate angiogenesis, and not due to inhibition of Raf. Based on these promising results, phase III trials in patients with advanced RCC are ongoing.

Melanomas harbor the highest frequency of B-Raf mutations (70%) of any tumor type. Hence, together with the 20% frequency of N-Ras mutations seen in melanomas, hyperactivation of the Raf->MEK>ERK pathway may be associated with up to 90% of this cancer. Since BAY43-9006 can inhibit mutant B-Raf proteins, and interfering RNA studies have demonstrated a critical role for mutant B-Raf function in melanoma growth *in vitro* (Hingorani, S. R., et al., 2003, Karasarides, M., et al., 2004, Sharma, A., et al., 2005), BAY43-9006 clinical trials have also focused on melanomas (Danson, S. and Lorigan, P., 2005). However, phase II studies found limited activity for BAY 43-9006 as a single agent (Ahmad et al., ASCO 2004) or in combination with carboplatin and paclitaxel in this disease (Flaherty et al., ASCO 2004). The development and application of more potent and selective Raf inhibitors will be needed before it can be determined whether Raf is a good therapeutic target for B-Raf- and Ras-mutation positive cancers.

To date, the only known catalytic substrates of Raf kinase activity are the dual-specificity kinases MEK1 and MEK2 (Chong, H., et al., 2003, Mercer, K. E. and Pritchard, C. A., 2003). Similarly, the only known substrates of MEK1 and MEK2 are the ERK1 and ERK2 kinases. Thus, in principle, inhibition of MEK should be equivalent to inhibition of Raf. However, there is some evidence that the Raf>MEK>ERK pathway is not simply linear, and that Raf possesses MEK-independent functions that contribute to oncogenesis, growth control and survival. Nevertheless, the usefulness of MEK inhibition to block Ras transformation has been demonstrated by both genetic and pharmacologic approaches (Shields, J. M., et al., 2000). In particular, the widely used, small molecule, non-ATP competitive inhibitors of MEK, PD98059 and U0126, have been shown to be very effective inhibitors of Ras transformation when assessed in a wide variety of cell types in cell culture. However, while these compounds are highly selective for MEK (Davies, S. P., et al., 2000), pharmacologic limitations have prevented the application of these particular inhibitors to in vivo models.

CI-1040 (formerly PD184352) is a potent and specific, orally available, ATP non-competitive inhibitor of MEK1/2 that has been shown to potently inhibit the growth of mouse and human colon carcinoma cells in nude mice (Sebolt-Leopold, J. S., et al., 1999). Similarly, CI-1040 inhibited human melanoma metastatic growth in immune compromised mice (Collisson, E. A., et al., 2003). A phase I clinical trial found that CI-1040 was well-tolerated. Even in the phase I study, one partial response was reported in a patient with pancreatic cancer, and 19 of 66 patients (28%) achieved stable disease lasting a median of 5.5 months. Importantly, partial inhibition of ERK activity was seen both in white blood cells and in tumor tissue (46-100%; median 73% inhibition) from selected patients.

A phase II study then assessed the antitumor activity and safety of CI-1040 in patients with several types of solid cancers in which mutational or upstream RTK activation of Ras is common: breast, colon, lung and pancreas. No patient achieved a CR or PR, so the anti-tumor activity was deemed insufficient to warrant further development. However, no post-treatment evaluation of changes in ERK activity was done, so it is unclear whether the lack of anti-tumor activity in this trial was due simply to failure to achieve sufficient target inhibition, or because MEK is not a therapeutically useful target for the treatment of these cancers. This important question is likely to be answered in the near future, however.

Two additional MEK inhibitors are currently under clinical evaluation (Sebolt-Leopold, J. S. and Herrera, R., 2004). PD0325901 is a second generation oral MEK inhibitor that has considerably improved pharmacologic and biopharmaceutical properties, in particular a 50-fold greater MEK inhibition potency than CI-1040, much improved oral bioavailability, and a longer duration of target suppression. It is currently under phase I and II clinical analysis. AZD6244 (ARRY-142886) is another potent oral ATP non-competitive inhibitor of MEK1 and MEK2 that has shown potent anti-tumor efficacy in xenograft tumor models of a variety of human tumor cell lines, and is currently in phase I clinical trials.

Another approach to blocking the Raf>MEK>ERK signaling cascade involves inhibition of KSR1. KSR1 is a scaffolding protein that interacts with all three components in this cascade, and that regulates the intensity and duration of Ras activation of ERK. MEFs deficient in KSR1 showed impaired sensitivity to Ras transformation (Kortum, R. L. and Lewis, R. E., 2004), and anti-sense suppression of KSR1 inhibited the growth of Ras mutation-positive human tumor xenografts (Xing, H. R., et al., 2003). Finally, in addition to KSR1, other modulators of this kinase cascade (e.g., RKIP, IMP) have been described. Whether these also represent useful targets for inhibition of this pathway will be interesting to determine.

7.2 Targeting Other Ras Effector Signaling Pathways

In addition to Raf, there is evidence for the role of other effectors in promoting Ras-mediated oncogenesis (Repasky, G. A., et al., 2004)(Figure 3). PI3K activation of AKT has been shown to be important for oncogenic Ras function, for example, in protecting against anoikis (apoptosis in response to deprivation of matrix attachment) (Khwaja, A., et al., 1997). The recent identification of mutationally activated p110 alpha catalytic subunits of PI3K in human tumors, as well as the long-appreciated and very common loss of the tumor suppressor PTEN (a negative regulator of PI3K) in many cancers (Steelman, L. S., et al., 2004), support the important contribution of aberrant PI3K activation in cancer development and growth (Samuels, Y. and Velculescu, V. E., 2004). Hence, inhibitors of the AKT serine/threonine kinase may also be useful for blocking this important survival pathway of Ras.

Members of another effector family, RalGEFs, which are guanine nucleotide exchange factors that activate the Ral small GTPases, have recently been implicated in Ras-mediated oncogenesis (Feig, L. A., 2003). RalGEF, rather than Raf or PI3K, activation was found to be sufficient to mimic Ras transformation of human cells (Hamad, N. M., et al., 2002). Ral GTPase function was found to contribute to tumor cell anchorage-independent growth and cell survival (Chien, Y. and White, M. A., 2003). Persistent activation of RalA was found in pancreatic cancers and was important for the anchorage-independent and tumorigenic growth of these and other human cancer cells (Lim, K. H., et al., 2005). Mice deficient in one RalGEF (RalGDS) are viable, but are refractory to Ras-induced skin tumor formation (Gonzalez-Garcia, A., et al., 2005). Since Ral GTPases are GGTase I substrates, one possible pharmacologic approach to blocking this pathway is the use of GGTIs.

Finally, two other effector pathways implicated in Ras-mediated oncogenesis include GEFs for Rac (Tiam1) and for Rap (phospholipase C epsilon:,PLCε). PLCε additionally functions as a lipase for the generation of the second messengers IP3 and DAG. Mice deficient in Tiam1 or PLCε are viable, but are impaired in Ras-induced tumor formation (Malliri, A., et al., 2002, Bai, Y., et al., 2004). A small molecule inhibitor of Tiam1 activation of Rac has recently been described and shown to impair Ras transformation, thus providing proof-of-principle that such inhibitors may become useful for anti-Ras treatment (Gao, Y., et al., 2004). Whether inhibitors of PLC or DAG-mediated activation of PKC would also be effective inhibitors of Ras transformation has not been determined.

8. CONCLUSIONS AND FUTURE PROSPECTS

Despite intensive basic research and pharmaceutical efforts, no anti-Ras therapeutic strategies have successfully transitioned to the cancer patient. Although FTIs still hold promise for cancer treatment and may soon be added to our repertoire of anti-neoplastic drugs, these inhibitors are not effective due to anti-Ras activity, and other approaches to selectively blocking oncogenic Ras must still be sought. Although promising observations have been made with other approaches to block Ras membrane association and transformation, these efforts are still very much in their infancy. In light of the successful clinical development of other kinase inhibitors for cancer treatment (eg., Gleevec, Iressa, Tarceva) (Druker, B. J., 2004, Baselga, J. and Arteaga, C. L., 2005), inhibitors of the Raf-MEK-ERK protein kinase cascade are perhaps the most promising candidates for the effective treatment of Ras mutation-positive cancers. However, since the anti-tumor activity of BAY 43-9006 may not be due to inhibition of Raf kinases, only when more potent and selective inhibitors of the Raf kinases enter clinical trials will we have a better assessment of Raf as a therapeutic target. Additionally, as other effector signaling pathways are also implicated in Ras-mediated oncogenesis, in particular the RalGEF-Ral and Tiam1-Rac pathways, this raises the issue of whether concurrent inhibition of these pathways will be needed together with inhibition of Raf or MEK. While mouse models support these Ras effector pathways as promising directions for anti-Ras strategies, further validation of these effectors as drug targets is still needed both preclinically in human tumor cells and in cancer patients. Reovirus therapy remains an attractive possibility, although further understanding of the mechanism by which Ras renders tumor cells sensitive to reovirus-induced cell killing is needed. Finally, as our knowledge of Ras function proceeds at a rapid pace, new clues and approaches to blocking Ras-mediated oncogenesis will most certainly arise. Anti-cancer drug discovery is a complex, time-consuming, and risky process, and the road to success is filled with unexpected twists and turns. Anti-Ras drug discovery has certainly demonstrated these attributes. Nevertheless, the strong promise of oncogenic Ras as a therapeutic target should continue to fuel comprehensive efforts to develop anti-Ras drugs, and optimism remains high that success will be achieved.

ACKNOWLEDGEMENTS

We thank Misha Rand for assistance in manuscript preparation. We apologize for not being able to cite original work of many colleagues due to space constraints. Our studies are supported by grants from the National Institutes of Health to C.J.D. (CA42978 and CA67771) and to A.D.C. (CA67771 and CA109550).

REFERENCES

Ahmad, T. and Eisen, T. (2004). Kinase inhibition with BAY 43-9006 in renal cell carcinoma. Clin Cancer Res. *10*, 6388S-6392S.

Ahmad, T., Marais, R., Pyle, L., James, M., Schwartz, B., Gore, M., Eisen, T. (2004). Journal of Clinical Oncology, ASCO Annual Meeting Proceedings (Post-Meeting Edition). Vol 22, No 14S (July 15 Supplement), 2004:7506.

Aiyagari, A.L., Taylor, B.R., Aurora, V., Young, S.G. and Shannon, K.M. (2003). Hematologic effects of inactivating the Ras processing enzyme Rce1. Blood. *101*, 2250-2252.

Ashar, H.R., James, L., Gray, K., Carr, D., Black, S., Armstrong, L., Bishop, W.R. and Kirschmeier, P. (2000). Farnesyl transferase inhibitors block the farnesylation of CENP-E and CENP-F and alter the association of CENP-E with the microtubules. J Biol Chem. *275*, 30451-30457.

Awada, A., Hendlisz, A., Gil, T., Bartholomeus, S., Mano, M., de Valeriola, D., Strumberg, D., Brendel, E., Haase, C.G., Schwartz, B. and Piccart, M. (2005). Phase I safety and pharmacokinetics of BAY 43-9006 administered for 21 days on/7 days off in patients with advanced, refractory solid tumours. Br J Cancer. *92*, 1855-1861.

Bai, Y., Edamatsu, H., Maeda, S., Saito, H., Suzuki, N., Satoh, T. and Kataoka, T. (2004). Crucial role of phospholipase Cepsilon in chemical carcinogen-induced skin tumor development. Cancer Res. *64*, 8808-8810.

Barbacid, M. (1987). Ras genes. Annu Rev Biochem. *56*, 779-827.

Baselga, J. and Arteaga, C.L. (2005). Critical update and emerging trends in epidermal growth factor receptor targeting in cancer. J Clin Oncol. *23*, 2445-2459.

Bergo, M.O., Ambroziak, P., Gregory, C., George, A., Otto, J.C., Kim, E., Nagase, H., Casey, P.J., Balmain, A. and Young, S.G. (2002). Absence of the CAAX endoprotease Rce1: effects on cell growth and transformation. Mol Cell Biol. *22*, 171-181.

Bergo, M.O., Gavino, B.J., Hong, C., Beigneux, A.P., McMahon, M., Casey, P.J. and Young, S.G. (2004). Inactivation of Icmt inhibits transformation by oncogenic K-Ras and B-Raf. J Clin Invest. *113*, 539-550.

Bergo, M.O., Leung, G.K., Ambroziak, P., Otto, J.C., Casey, P.J., Gomes, A.Q., Seabra, M.C. and Young, S.G. (2001). Isoprenylcysteine carboxyl methyltransferase deficiency in mice. J Biol Chem. *276*, 5841-5845.

Bergo, M.O., Leung, G.K., Ambroziak, P., Otto, J.C., Casey, P.J. and Young, S.G. (2000). Targeted inactivation of the isoprenylcysteine carboxyl methyltransferase gene causes mislocalization of K-Ras in mammalian cells. J Biol Chem. *275*, 17605-17610.

Bergo, M.O., Lieu, H.D., Gavino, B.J., Ambroziak, P., Otto, J.C., Casey, P.J., Walker, Q.M. and Young, S.G. (2004). On the physiological importance of endoproteolysis of CAAX proteins: heart-specific RCE1 knockout mice develop a lethal cardiomyopathy. J Biol Chem. *279*, 4729-4736.

Blum, R., Jacob-Hirsch, J., Amariglio, N., Rechavi, G. and Kloog, Y. (2005). Ras inhibition in glioblastoma down-regulates hypoxia-inducible factor-1alpha, causing glycolysis shutdown and cell death. Cancer Res. *65*, 999-1006.

Brummelkamp, T.R., Bernards, R. and Agami, R. (2002). Stable suppression of tumorigenicity by virus-mediated RNA interference. Cancer Cell. *2*, 243-247.

Brunner, T.B., Hahn, S.M., Gupta, A.K., Muschel, R.J., McKenna, W.G. and Bernhard, E.J. (2003). Farnesyltransferase inhibitors: an overview of the results of preclinical and clinical investigations. Cancer Res. *63*, 5656-5668.

Buser, C.A., Dinsmore, C.J., Fernandes, C., Greenberg, I., Hamilton, K., Mosser, S.D., Walsh, E.S., Williams, T.M. and Koblan, K.S. (2001). High-performance liquid chromatography/mass spectrometry characterization of Ki4B-Ras in PSN-1 cells treated with the prenyltransferase inhibitor L-778,123. Anal Biochem. *290*, 126-137.

Campbell, P.M. and Der, C.J. (2004). Oncogenic Ras and its role in tumor cell invasion and metastasis. Semin Cancer Biol. *14*, 105-114.

Capell, B.C., Erdos, M.R., Madigan, J.P., Fiordalisi, J.J., Varga, R., Conneely, K.N., Gordon, L.B., Der, C.J., Cox, A.D. and Collins, F.S. (2005). Inhibiting farnesylation of progerin prevents the characteristic nuclear blebbing of Hutchinson-Gilford progeria syndrome. Proc Natl Acad Sci U S A. *102*, 12879-12884.

Chien, Y. and White, M.A. (2003). RAL GTPases are linchpin modulators of human tumour-cell proliferation and survival. EMBO Rep. *4*, 800-806.

Chin, L., Tam, A., Pomerantz, J., Wong, M., Holash, J., Bardeesy, N., Shen, Q., O'Hagan, R., Pantginis, J., Zhou, H., Horner, J.W., 2nd, Cordon-Cardo, C., Yancopoulos, G.D. and DePinho, R.A. (1999). Essential role for oncogenic Ras in tumour maintenance. Nature. *400*, 468-472.

Chong, H., Vikis, H.G. and Guan, K.L. (2003). Mechanisms of regulating the Raf kinase family. Cell Signal. *15*, 463-469.

Cochet, O., Gruel, N., Fridman, W.H. and Teillaud, J.L. (1999). Ras and p53 intracellular targeting with recombinant single-chain Fv (scFv) fragments: a novel approach for cancer therapy? Cancer Detect Prev. *23*, 506-510.

Cochet, O., Kenigsberg, M., Delumeau, I., Duchesne, M., Schweighoffer, F., Tocque, B. and Teillaud, J.L. (1998). Intracellular expression and functional properties of an anti-p21Ras scFv derived from a rat hybridoma containing specific lambda and irrelevant kappa light chains. Mol Immunol. *35*, 1097-1110.

Coffey, M.C., Strong, J.E., Forsyth, P.A. and Lee, P.W. (1998). Reovirus therapy of tumors with activated Ras pathway. Science. *282*, 1332-1334.

Collisson, E.A., De, A., Suzuki, H., Gambhir, S.S. and Kolodney, M.S. (2003). Treatment of metastatic melanoma with an orally available inhibitor of the Ras-Raf-MAPK cascade. Cancer Res. *63*, 5669-5673.

Cox, A.D. and Der, C.J. (2002). Ras family signaling: therapeutic targeting. Cancer Biol Ther. *1*, 599-606.

Cox, A.D. and Der, C.J. (2002). Farnesyltransferase inhibitors: promises and realities. Curr Opin Pharmacol. *2*, 388-393.

Cox, A.D. and Der, C.J. (1997). Farnesyltransferase inhibitors and cancer treatment: targeting simply Ras? Biochim Biophys Acta. *1333*, F51-71.

Cox, A.D. and Der, C.J. (2003). The dark side of Ras: regulation of apoptosis. Oncogene. *22*, 8999-9006.

Cox, A.D., Hisaka, M.M., Buss, J.E. and Der, C.J. (1992). Specific isoprenoid modification is required for function of normal, but not oncogenic, Ras protein. Mol Cell Biol. *12*, 2606-2615.

Danson, S. and Lorigan, P. (2005). Improving outcomes in advanced malignant melanoma: update on systemic therapy. Drugs. *65*, 733-743.

Davies, S.P., Reddy, H., Caivano, M. and Cohen, P. (2000). Specificity and mechanism of action of some commonly used protein kinase inhibitors. Biochem J. *351*, 95-105.

Doll, R.J., Kirschmeier, P. and Bishop, W.R. (2004). Farnesyltransferase inhibitors as anticancer agents: critical crossroads. Curr Opin Drug Discov Devel. *7*, 478-486.

Downward, J. (2003). Targeting RAS signalling pathways in cancer therapy. Nat Rev Cancer. *3*, 11-22.

Druker, B.J. (2004). Imatinib as a paradigm of targeted therapies. Adv Cancer Res. *91*, 1-30.

Duursma, A.M. and Agami, R. (2003). Ras interference as cancer therapy. Semin Cancer Biol. *13*, 267-273.

Dykxhoorn, D.M. and Lieberman, J. (2005). The silent revolution: RNA interference as basic biology, research tool, and therapeutic. Annu Rev Med. *56*, 401-423.

Elad, G., Paz, A., Haklai, R., Marciano, D., Cox, A. and Kloog, Y. (1999). Targeting of K-Ras 4B by S-trans,trans-farnesyl thiosalicylic acid. Biochim Biophys Acta. *1452*, 228-242.

Elenbaas, B., Spirio, L., Koerner, F., Fleming, M.D., Zimonjic, D.B., Donaher, J.L., Popescu, N.C., Hahn, W.C. and Weinberg, R.A. (2001). Human breast cancer cells generated by oncogenic transformation of primary mammary epithelial cells. Genes Dev. *15*, 50-65.

Feig, L.A. (2003). Ral-GTPases: approaching their 15 minutes of fame. Trends Cell Biol. *13*, 419-425.

Fischbach, M.A. and Settleman, J. (2003). Specific biochemical inactivation of oncogenic Ras proteins by nucleoside diphosphate kinase. Cancer Res. *63*, 4089-4094.

Flaherty, K.T., Brose, M., Schuchter, L., Tuveson, D., Lee, R., Schwartz, B., Lathia, C., Weber, B., O'Dwyer, P. (2004) Journal of Clinical Oncology, ASCO Annual Meeting Proceedings (Post-Meeting Edition). Vol 22, No 14S (July 15 Supplement), 2004:7507.

Funato, T., Ishii, T., Kambe, M., Scanlon, K.J. and Sasaki, T. (2000). Anti-K-ras ribozyme induces growth inhibition and increased chemosensitivity in human colon cancer cells. Cancer Gene Ther. *7*, 495-500.

Gana-Weisz, M., Halaschek-Wiener, J., Jansen, B., Elad, G., Haklai, R. and Kloog, Y. (2002). The Ras inhibitor S-trans,trans-farnesylthiosalicylic acid chemosensitizes human tumor cells without causing resistance. Clin Cancer Res. *8*, 555-565.

Gao, Y., Dickerson, J.B., Guo, F., Zheng, J. and Zheng, Y. (2004). Rational design and characterization of a Rac GTPase-specific small molecule inhibitor. Proc Natl Acad Sci U S A. *101*, 7618-7623.

Garnett, M.J. and Marais, R. (2004). Guilty as charged: B-RAF is a human oncogene. Cancer Cell. *6*, 313-319.

Glynn, M.W. and Glover, T.W. (2005). Incomplete processing of mutant lamin A in Hutchinson-Gilford progeria leads to nuclear abnormalities, which are reversed by farnesyltransferase inhibition. Hum Mol Genet. *14*, 2959-2969.

Gonzalez-Garcia, A., Pritchard, C.A., Paterson, H.F., Mavria, G., Stamp, G. and Marshall, C.J. (2005). RalGDS is required for tumor formation in a model of skin carcinogenesis. Cancer Cell. *7*, 219-226.

Guerra, C., Mijimolle, N., Dhawahir, A., Dubus, P., Barradas, M., Serrano, M., Campuzano, V. and Barbacid, M. (2003). Tumor induction by an endogenous K-ras oncogene is highly dependent on cellular context. Cancer Cell. *4*, 111-120.

Hahn, W.C., Counter, C.M., Lundberg, A.S., Beijersbergen, R.L., Brooks, M. W. and Weinberg, R.A. (1999). Creation of human tumour cells with defined genetic elements. Nature. *400*, 464-468.

Haklai, R., Weisz, M.G., Elad, G., Paz, A., Marciano, D., Egozi, Y., Ben-Baruch, G. and Kloog, Y. (1998). Dislodgment and accelerated degradation of Ras. Biochemistry. *37*, 1306-1314.

Hamad, N.M., Elconin, J.H., Karnoub, A.E., Bai, W., Rich, J.N., Abraham, R.T., Der, C.J. and Counter, C.M. (2002). Distinct requirements for Ras oncogenesis in human versus mouse cells. Genes Dev. *16*, 2045-2057.

Hanahan, D. and Weinberg, R.A. (2000). The hallmarks of cancer. Cell. *100*, 57-70.

Harrington, E.O., Newton, J., Morin, N. and Rounds, S. (2004). Barrier dysfunction and RhoA activation are blunted by homocysteine and adenosine in pulmonary endothelium. Am J Physiol Lung Cell Mol Physiol. *287*, L1091-1097.

Hingorani, S.R., Jacobetz, M.A., Robertson, G.P., Herlyn, M. and Tuveson, D.A. (2003). Suppression of BRAF(V599E) in human melanoma abrogates transformation. Cancer Res. *63*, 5198-5202.

Hingorani, S.R., Petricoin, E.F., Maitra, A., Rajapakse, V., King, C., Jacobetz, M.A., Ross, S., Conrads, T.P., Veenstra, T.D., Hitt, B.A., Kawaguchi, Y., Johann, D., Liotta, L.A., Crawford, H.C., Putt, M.E., Jacks, T., Wright, C.V., Hruban, R. H., Lowy, A.M. and Tuveson, D.A. (2003). Preinvasive and invasive ductal pancreatic cancer and its early detection in the mouse. Cancer Cell. *4*, 437-450.

Hirasawa, K., Nishikawa, S.G., Norman, K.L., Alain, T., Kossakowska, A. and Lee, P.W. (2002). Oncolytic reovirus against ovarian and colon cancer. Cancer Res. *62*, 1696-1701.

Hirasawa, K., Nishikawa, S.G., Norman, K.L., Coffey, M.C., Thompson, B.G., Yoon, C.S., Waisman, D.M. and Lee, P.W. (2003). Systemic reovirus therapy of metastatic cancer in immune-competent mice. Cancer Res. *63*, 348-353.

Hrycyna, C.A., Sapperstein, S.K., Clarke, S. and Michaelis, S. (1991). The Saccharomyces cerevisiae STE14 gene encodes a methyltransferase that mediates C-terminal methylation of a-factor and RAS proteins. Embo J. *10*, 1699-1709.

James, G.L., Goldstein, J.L. and Brown, M.S. (1995). Polylysine and CVIM sequences of K-RasB dictate specificity of prenylation and confer resistance to benzodiazepine peptidomimetic in vitro. J Biol Chem. *270*, 6221-6226.

Johnson, L., Mercer, K., Greenbaum, D., Bronson, R.T., Crowley, D., Tuveson, D.A. and Jacks, T. (2001). Somatic activation of the K-ras oncogene causes early onset lung cancer in mice. Nature. *410*, 1111-1116.

Karasarides, M., Chiloeches, A., Hayward, R., Niculescu-Duvaz, D., Scanlon, I., Friedlos, F., Ogilvie, L., Hedley, D., Martin, J., Marshall, C.J., Springer, C.J. and Marais, R. (2004). B-RAF is a therapeutic target in melanoma. Oncogene. *23*, 6292-6298.

Kato, K., Cox, A.D., Hisaka, M.M., Graham, S.M., Buss, J.E. and Der, C.J. (1992). Isoprenoid addition to Ras protein is the critical modification for its membrane association and transforming activity. Proc Natl Acad Sci U S A. *89*, 6403-6407.

Khwaja, A., Rodriguez-Viciana, P., Wennstrom, S., Warne, P.H. and Downward, J. (1997). Matrix adhesion and Ras transformation both activate a phosphoinositide 3-OH kinase and protein kinase B/Akt cellular survival pathway. Embo J. *16*, 2783-2793.

Kijima, H., Yamazaki, H., Nakamura, M., Scanlon, K.J., Osamura, R.Y. and Ueyama, Y. (2004). Ribozyme against mutant K-ras mRNA suppresses tumor growth of pancreatic cancer. Int J Oncol. *24*, 559-564.

Kim, E., Ambroziak, P., Otto, J.C., Taylor, B., Ashby, M., Shannon, K., Casey, P.J. and Young, S.G. (1999). Disruption of the mouse Rce1 gene results in defective Ras processing and mislocalization of Ras within cells. J Biol Chem. *274*, 8383-8390.

Kloog, Y. and Cox, A.D. (2000). RAS inhibitors: potential for cancer therapeutics. Mol Med Today. *6*, 398-402.

Kortum, R.L. and Lewis, R.E. (2004). The molecular scaffold KSR1 regulates the proliferative and oncogenic potential of cells. Mol Cell Biol. *24*, 4407-4416.

Lackner, M.R., Kindt, R.M., Carroll, P.M., Brown, K., Cancilla, M.R., Chen, C., de Silva, H., Franke, Y., Guan, B., Heuer, T., Hung, T., Keegan, K., Lee, J.M., Manne, V., O'Brien, C., Parry, D., Perez-Villar, J.J., Reddy, R.K., Xiao, H., Zhan, H., Cockett, M., Plowman, G., Fitzgerald, K., Costa, M. and Ross-Macdonald, P. (2005). Chemical genetics identifies Rab geranylgeranyl transferase as an apoptotic target of farnesyl transferase inhibitors. Cancer Cell. *7*, 325-336.

Lerner, E.C., Zhang, T.T., Knowles, D.B., Qian, Y., Hamilton, A.D. and Sebti, S.M. (1997). Inhibition of the prenylation of K-Ras, but not H- or N-Ras, is highly resistant to CAAX peptidomimetics and requires both a farnesyltransferase and a geranylgeranyltransferase I inhibitor in human tumor cell lines. Oncogene. *15*, 1283-1288.

Lim, K.H., Baines, A.T., Fiordalisi, J.J., Shipitsin, M., Feig, L.A., Cox, A.D., Der, C.J. and Counter, C.M. (2005). Activation of RalA is critical for Ras-induced tumorigenesis of human cells. Cancer Cell. *7*, 533-545.

Liu, J., Yang, G., Thompson-Lanza, J.A., Glassman, A., Hayes, K., Patterson, A., Marquez, R.T., Auersperg, N., Yu, Y., Hahn, W.C., Mills, G.B. and Bast, R.C., Jr. (2004). A genetically defined model for human ovarian cancer. Cancer Res. *64*, 1655-1663.

Lobell, R.B., Liu, D., Buser, C.A., Davide, J.P., DePuy, E., Hamilton, K., Koblan, K.S., Lee, Y., Mosser, S., Motzel, S.L., Abbruzzese, J.L., Fuchs, C.S., Rowinsky, E.K., Rubin, E.H., Sharma, S., Deutsch, P.J., Mazina, K.E., Morrison, B.W., Wildonger, L., Yao, S.L. and Kohl, N.E. (2002). Preclinical and clinical pharmacodynamic assessment of L-778, 123, a dual inhibitor of farnesyl:protein transferase and geranylgeranyl:protein transferase type-I. Mol Cancer Ther. *1*, 747-758.

Lobell, R.B., Omer, C.A., Abrams, M.T., Bhimnathwala, H.G., Brucker, M.J., Buser, C.A., Davide, J.P., deSolms, S.J., Dinsmore, C.J., Ellis-Hutchings, M.S., Kral, A.M., Liu, D., Lumma, W.C., Machotka, S.V., Rands, E., Williams, T.M., Graham, S.L., Hartman, G.D., Oliff, A.I., Heimbrook, D.C. and Kohl, N.E. (2001). Evaluation of farnesyl:protein transferase and geranylgeranyl:protein transferase inhibitor combinations in preclinical models. Cancer Res. *61*, 8758-8768.

Lu, Q., Harrington, E.O., Hai, C.M., Newton, J., Garber, M., Hirase, T. and Rounds, S. (2004). Isoprenylcysteine carboxyl methyltransferase modulates endothelial monolayer permeability: involvement of RhoA carboxyl methylation. Circ Res. *94*, 306-315.

Lundberg, A.S., Randell, S.H., Stewart, S.A., Elenbaas, B., Hartwell, K.A., Brooks, M.W., Fleming, M.D., Olsen, J.C., Miller, S.W., Weinberg, R.A. and Hahn, W.C. (2002). Immortalization and transformation of primary human airway epithelial cells by gene transfer. Oncogene. *21*, 4577-4586.

Lyons, J.F., Wilhelm, S., Hibner, B. and Bollag, G. (2001). Discovery of a novel Raf kinase inhibitor. Endocr Relat Cancer. *8*, 219-225.

Mallampalli, M.P., Huyer, G., Bendale, P., Gelb, M.H. and Michaelis, S. (2005). Inhibiting farnesylation reverses the nuclear morphology defect in a HeLa cell model for Hutchinson-Gilford progeria syndrome. Proc Natl Acad Sci U S A. *102*, 14416-14421.

Malliri, A., van der Kammen, R.A., Clark, K., van der Valk, M., Michiels, F. and Collard, J.G. (2002). Mice deficient in the Rac activator Tiam1 are resistant to Ras-induced skin tumours. Nature. *417*, 867-871.

Malumbres, M. and Barbacid, M. (2003). RAS oncogenes: the first 30 years. Nat Rev Cancer. *3*, 459-465.

Mangues, R., Corral, T., Kohl, N.E., Symmans, W.F., Lu, S., Malumbres, M., Gibbs, J.B., Oliff, A. and Pellicer, A. (1998). Antitumor effect of a farnesyl protein transferase inhibitor in mammary and lymphoid tumors overexpressing N-ras in transgenic mice. Cancer Res. *58*, 1253-1259.

Mangues, R. and Pellicer, A. (1992). Ras activation in experimental carcinogenesis. Semin Cancer Biol. *3*, 229-239.

McNeish, I.A., Bell, S.J. and Lemoine, N.R. (2004). Gene therapy progress and prospects: cancer gene therapy using tumour suppressor genes. Gene Ther. *11*, 497-503.

Mercer, K.E. and Pritchard, C.A. (2003). Raf proteins and cancer: B-Raf is identified as a mutational target. Biochim Biophys Acta. *1653*, 25-40.

Michaelson, D., Ali, W., Chiu, V.K., Bergo, M., Silletti, J., Wright, L., Young, S.G. and Philips, M. (2005). Postprenylation CAAX processing is required for proper localization of Ras but not Rho GTPases. Mol Biol Cell. *16*, 1606-1616.

Mijimolle, N., Velasco, J., Dubus, P., Guerra, C., Weinbaum, C.A., Casey, P.J., Campuzano, V. and Barbacid, M. (2005). Protein farnesyltransferase in embryogenesis, adult homeostasis, and tumor development. Cancer Cell. *7*, 313-324.

Moore, M., Hirte, H.W., Siu, L., Oza, A., Hotte, S.J., Petrenciuc, O., Cihon, F., Lathia, C. and Schwartz, B. (2005). Phase I study to determine the safety and pharmacokinetics of the novel Raf kinase and VEGFR inhibitor BAY 43-9006, administered for 28 days on/7 days off in patients with advanced, refractory solid tumors. Ann Oncol. *16*, 1688-94.

Norman, K.L., Coffey, M.C., Hirasawa, K., Demetrick, D.J., Nishikawa, S.G., DiFrancesco, L.M., Strong, J.E. and Lee, P.W. (2002). Reovirus oncolysis of human breast cancer. Hum Gene Ther. *13*, 641-652.

Norman, K.L., Hirasawa, K., Yang, A.D., Shields, M.A. and Lee, P.W. (2004). Reovirus oncolysis: the Ras/RalGEF/p38 pathway dictates host cell permissiveness to reovirus infection. Proc Natl Acad Sci U S A. *101*, 11099-11104.

Norman, K.L. and Lee, P.W. (2005). Not all viruses are bad guys: the case for reovirus in cancer therapy. Drug Discov Today. *10*, 847-855.

Rangarajan, A. and Weinberg, R.A. (2003). Opinion: Comparative biology of mouse versus human cells: modelling human cancer in mice. Nat Rev Cancer. *3*, 952-959.

Reid, T.S., Terry, K.L., Casey, P.J. and Beese, L.S. (2004). Crystallographic analysis of CaaX prenyltransferases complexed with substrates defines rules of protein substrate selectivity. J Mol Biol. *343*, 417-433.

Repasky, G.A., Chenette, E.J. and Der, C.J. (2004). Renewing the conspiracy theory debate: does Raf function alone to mediate Ras oncogenesis? Trends Cell Biol. *14*, 639-647.

Rich, J.N., Guo, C., McLendon, R.E., Bigner, D.D., Wang, X.F. and Counter, C.M. (2001). A genetically tractable model of human glioma formation. Cancer Res. *61*, 3556-3560.

Rose, W.C., Lee, F.Y., Fairchild, C.R., Lynch, M., Monticello, T., Kramer, R.A. and Manne, V. (2001). Preclinical antitumor activity of BMS-214662, a highly apoptotic and novel farnesyltransferase inhibitor. Cancer Res. *61*, 7507-7517.

Rowell, C.A., Kowalczyk, J.J., Lewis, M.D. and Garcia, A.M. (1997). Direct demonstration of geranylgeranylation and farnesylation of Ki-Ras in vivo. J Biol Chem. *272*, 14093-14097.

Russell, J.S., Lang, F.F., Huet, T., Janicot, M., Chada, S., Wilson, D.R. and Tofilon, P.J. (1999). Radiosensitization of human tumor cell lines induced by the adenovirus-mediated expression of an anti-Ras single-chain antibody fragment. Cancer Res. *59*, 5239-5244.

Samuels, Y. and Velculescu, V.E. (2004). Oncogenic mutations of PIK3CA in human cancers. Cell Cycle. *3*, 1221-1224.

Sebolt-Leopold, J.S., Dudley, D.T., Herrera, R., Van Becelaere, K., Wiland, A., Gowan, R.C., Tecle, H., Barrett, S.D., Bridges, A., Przybranowski, S., Leopold, W.R. and Saltiel, A.R. (1999). Blockade of the MAP kinase pathway suppresses growth of colon tumors in vivo. Nat Med. *5*, 810-816.

Sebolt-Leopold, J.S. and Herrera, R. (2004). Targeting the mitogen-activated protein kinase cascade to treat cancer. Nat Rev Cancer. *4*, 937-947.

Sebti, S.M. and Adjei, A.A. (2004). Farnesyltransferase inhibitors. Semin Oncol. *31*, 28-39.

Sebti, S.M. and Der, C.J. (2003). Opinion: Searching for the elusive targets of farnesyltransferase inhibitors. Nat Rev Cancer. *3*, 945-951.

Sebti, S.M. and Hamilton, A.D. (2000). Farnesyltransferase and geranylgeranyltransferase I inhibitors and cancer therapy: lessons from mechanism and bench-to-bedside translational studies. Oncogene. *19*, 6584-6593.

Sepp-Lorenzino, L., Ma, Z., Rands, E., Kohl, N.E., Gibbs, J.B., Oliff, A. and Rosen, N. (1995). A peptidomimetic inhibitor of farnesyl:protein transferase blocks the anchorage-dependent and independent growth of human tumor cell lines. Cancer Res. *55*, 5302-5309.

Sharma, A., Trivedi, N.R., Zimmerman, M.A., Tuveson, D.A., Smith, C.D. and Robertson, G.P. (2005). Mutant V599EB-Raf regulates growth and vascular development of malignant melanoma tumors. Cancer Res. *65*, 2412-2421.

Shields, J.M., Pruitt, K., McFall, A., Shaub, A. and Der, C.J. (2000). Understanding Ras: 'it ain't over 'til it's over'. Trends Cell Biol. *10*, 147-154.

Shirasawa, S., Furuse, M., Yokoyama, N. and Sasazuki, T. (1993). Altered growth of human colon cancer cell lines disrupted at activated Ki-ras. Science. *260*, 85-88.

Steelman, L.S., Bertrand, F.E. and McCubrey, J.A. (2004). The complexity of PTEN: mutation, marker and potential target for therapeutic intervention. Expert Opin Ther Targets. *8*, 537-550.

Stehle, T. and Dermody, T.S. (2004). Structural similarities in the cellular receptors used by adenovirus and reovirus. Viral Immunol. *17*, 129-143.

Strong, J.E., Coffey, M.C., Tang, D., Sabinin, P. and Lee, P.W. (1998). The molecular basis of viral oncolysis: usurpation of the Ras signaling pathway by reovirus. Embo J. *17*, 3351-3362.

Strong, J.E. and Lee, P.W. (1996). The v-erbB oncogene confers enhanced cellular susceptibility to reovirus infection. J Virol. *70*, 612-616.

Strong, J.E., Tang, D. and Lee, P.W. (1993). Evidence that the epidermal growth factor receptor on host cells confers reovirus infection efficiency. Virology. *197*, 405-411.

Sun, J., Qian, Y., Hamilton, A.D. and Sebti, S.M. (1998). Both farnesyltransferase and geranylgeranyltransferase I inhibitors are required for inhibition of oncogenic K-Ras prenylation but each alone is sufficient to suppress human tumor growth in nude mouse xenografts. Oncogene. *16*, 1467-1473.

Sun, J., Qian, Y., Hamilton, A.D. and Sebti, S.M. (1995). Ras CAAX peptidomimetic FTI 276 selectively blocks tumor growth in nude mice of a human lung carcinoma with K-Ras mutation and p53 deletion. Cancer Res. *55*, 4243-4247.

Tanaka, T. and Rabbitts, T.H. (2003). Intrabodies based on intracellular capture frameworks that bind the RAS protein with high affinity and impair oncogenic transformation. Embo J. *22*, 1025-1035.

Toth, J.I., Yang, S.H., Qiao, X., Beigneux, A.P., Gelb, M.H., Moulson, C.L., Miner, J.H., Young, S.G. and Fong, L.G. (2005). Blocking protein farnesyltransferase improves nuclear shape in fibroblasts from humans with progeroid syndromes. Proc Natl Acad Sci U S A. *102*, 12873-12878.

Whyte, D.B., Kirschmeier, P., Hockenberry, T.N., Nunez-Oliva, I., James, L., Catino, J.J., Bishop, W.R. and Pai, J.K. (1997). K- and N-Ras are geranylgeranylated in cells treated with farnesyl protein transferase inhibitors. J Biol Chem. *272*, 14459-14464.

Wilcox, M.E., Yang, W., Senger, D., Rewcastle, N.B., Morris, D.G., Brasher, P.M., Shi, Z.Q., Johnston, R.N., Nishikawa, S., Lee, P.W. and Forsyth, P.A. (2001). Reovirus as an oncolytic agent against experimental human malignant gliomas. J Natl Cancer Inst. *93*, 903-912.

Wilhelm, S.M., Carter, C., Tang, L., Wilkie, D., McNabola, A., Rong, H., Chen, C., Zhang, X., Vincent, P., McHugh, M., Cao, Y., Shujath, J., Gawlak, S., Eveleigh, D., Rowley, B., Liu, L., Adnane, L., Lynch, M., Auclair, D., Taylor, I., Gedrich, R., Voznesensky, A., Riedl, B., Post, L.E., Bollag, G. and Trail, P.A. (2004). BAY 43-9006 exhibits broad spectrum oral antitumor activity and targets the RAF/MEK/ERK pathway and receptor tyrosine kinases involved in tumor progression and angiogenesis. Cancer Res. *64*, 7099-7109.

Winter-Vann, A.M., Baron, R.A., Wong, W., dela Cruz, J., York, J.D., Gooden, D.M., Bergo, M.O., Young, S.G., Toone, E.J. and Casey, P.J. (2005). A small-molecule inhibitor of isoprenylcysteine carboxyl methyltransferase with antitumor activity in cancer cells. Proc Natl Acad Sci U S A. *102*, 4336-4341.

Winter-Vann, A.M. and Casey, P.J. (2005). Post-prenylation-processing enzymes as new targets in oncogenesis. Nat Rev Cancer. *5*, 405-412.

Winter-Vann, A.M., Kamen, B.A., Bergo, M.O., Young, S.G., Melnyk, S., James, S.J. and Casey, P.J. (2003). Targeting Ras signaling through inhibition of carboxyl methylation: an unexpected property of methotrexate. Proc Natl Acad Sci U S A. *100*, 6529-6534.

Xing, H.R., Cordon-Cardo, C., Deng, X., Tong, W., Campodonico, L., Fuks, Z. and Kolesnick, R. (2003). Pharmacologic inactivation of kinase suppressor of ras-1 abrogates Ras-mediated pancreatic cancer. Nat Med. *9*, 1266-1268.

Yang, G., Thompson, J.A., Fang, B. and Liu, J. (2003). Silencing of H-ras gene expression by retrovirus-mediated siRNA decreases transformation efficiency and tumorgrowth in a model of human ovarian cancer. Oncogene. *22*, 5694-5701.

Yang, S.H., Bergo, M.O., Toth, J.I., Qiao, X., Hu, Y., Sandoval, S., Meta, M., Bendale, P., Gelb, M.H., Young, S.G. and Fong, L.G. (2005). Blocking protein farnesyltransferase improves nuclear blebbing in mouse fibroblasts with a targeted Hutchinson-Gilford progeria syndrome mutation. Proc Natl Acad Sci U S A. *102*, 10291-10296.

Yang, W.Q., Senger, D., Muzik, H., Shi, Z.Q., Johnson, D., Brasher, P.M., Rewcastle, N.B., Hamilton, M., Rutka, J., Wolff, J., Wetmore, C., Curran, T., Lee, P.W. and Forsyth, P.A. (2003). Reovirus prolongs survival and reduces the frequency of spinal and leptomeningeal metastases from medulloblastoma. Cancer Res. *63*, 3162-3172.

Yin, J.Q., Gao, J., Shao, R., Tian, W.N., Wang, J. and Wan, Y. (2003). siRNA agents inhibit oncogene expression and attenuate human tumor cell growth. J Exp Ther Oncol. *3*, 194-204.

Zhang, Y.A., Nemunaitis, J., Scanlon, K.J. and Tong, A.W. (2000). Anti-tumorigenic effect of a K-ras ribozyme against human lung cancer cell line heterotransplants in nude mice. Gene Ther. *7*, 2041-2050.

Zhao, J.J., Roberts, T.M. and Hahn, W.C. (2004). Functional genetics and experimental models of human cancer. Trends Mol Med. *10*, 344-350.

Zwiebel, J.A. (2001). Cancer gene and oncolytic virus therapy. Semin Oncol. *28*, 336-343.

INDEX

Proteins and Cell Regulation

1. R.A. Kahn (ed.): *ARF Family GTPases*. 2004 ISBN 1-4020-1719-7
2. J. Frampton (ed.): *Myb Transcription Factors: Their Role in Growth, Differentiation and Disease*. 2004. ISBN 1-4020-2779-6
3. E. Manser (ed.): *RHO Family GTPases*. 2005 ISBN 1-4020-3461-X
4. C. Der (ed.): *RAS Family GTPases*. 2006 ISBN 1-4020-4328-7